W9-CFI-144

Failure of Materials
in Mechanical Design

Failure of Materials in Mechanical Design

Analysis
Prediction
Prevention

J. A. Collins
The Ohio State University

A WILEY-INTERSCIENCE PUBLICATION

JOHN WILEY & SONS
New York • Chichester • Brisbane • Toronto • Singapore

Copyright © 1981 by John Wiley & Sons, Inc.

All rights reserved. Published simultaneously in Canada.

Reproduction or translation of any part of this work
beyond that permitted by Sections 107 or 108 of the
1976 United States Copyright Act without the permission
of the copyright owner is unlawful. Requests for
permission or further information should be addressed to
the Permissions Department, John Wiley & Sons, Inc.

Library of Congress Cataloging in Publication Data:

Collins, Jack A

  Failure of materials in mechanical design.

  "A Wiley-Interscience publication."
  Includes index.
  1. Metals—Fracture. 2. Machine parts—Materials
—Fracture. 3. Machinery—Design. 4. Strains and
stresses. I. Title.

TA460.C63   620.1'12   80-20674
ISBN 0-471-05024-5

Printed in the United States of America

10 9

To my wife, Jo Ann, my children, Mike, Jennifer, Joan, and Greg, and my parents

Preface

The threefold purpose to which this book is dedicated includes service as a textbook for advanced undergraduate and beginning graduate-level engineering students, use as the basis for a continuing education course for graduate engineers, and use as a reference book for mechanical designers engaged in engineering practice. The reader is presumed to be acquainted with the elements of solid mechanics, strength of materials, and machine design traditionally included in undergraduate engineering curricula. Building on these elements, the book leads the reader to appreciate the potential for mechanical failure brought about by the stresses, strains, and energy transfers in machine parts that result from the forces, deflections, and energy inputs applied.

Recognition of the potential for failure and identification of the modes of mechanical failure that persist in the real engineering world are absolutely essential to prediction and prevention of mechanical failure, the cornerstone objectives of every mechanical designer. Thus the book identifies the modes of mechanical failure early in the presentation and later devotes full-chapter discussions to the more important failure modes. Because of its extreme practical importance throughout engineering practice, the topic of fatigue failure is treated extensively in this text, with attention devoted to both the high-cycle and low-cycle range of fatigue failure. Brittle fracture, creep, stress rupture, fretting fatigue, fretting wear, impact, buckling, and several other failure modes are also discussed at some length, and a chapter is devoted to the influence of stress concentration effects on failure response. The basic concepts of fracture mechanics are integrated with the discussions of brittle fracture and fatigue failure.

This book has developed over more than 15 years through the teaching of undergraduate and graduate-level courses in mechanical engineering, first at Arizona State University and more recently at The Ohio State University, through the teaching of annual short courses in fatigue and failure for engineering designers in industry, primarily at the University of Wisconsin—Madison, and through a steady dialogue with engineers in industry relative to mechanical failure during consulting activity and research in this area.

The philosophy at both Arizona State University and Ohio State University developed in much the same way, asserting that the mechanical engineering student would be given the opportunity to capitalize on fundamental course work in solid mechanics and machine design by providing a broad perspective on mechanical failures in the real world, while at the same time extending the depth of his analytical capabilities to include the three-dimensional viewpoint in states of stress and strain, application of failure theories, and failure prediction and prevention. Students who have seen a concentration of engineering science in their undergraduate curriculum are thereby afforded a somewhat more practical tie with the real world of engineering. That this philosophy has met with success is attested to by the many favorable comments from engineering managers of former students participating in the courses using this manuscript in various phases of its development.

The more practical aspects of this book, distilled from a variety of contacts with industrial problems encountered in consulting, research, and short course interaction with engineers in industry, are intended to make the presentation useful as a professional reference.

I have found that as the years pass it becomes impossible to distinguish my own original thoughts from the thoughts gathered through reading and discussing the works of others. For the many contributions of others who find their essence in these pages without specific reference, I wish to express my appreciation. I wish to express particular appreciation to two of my own professors during my years as a student, Professor W. L. Starkey and the late Professor S. M. Marco: No doubt much of their philosophy has been adopted as my own. Finally, the many helpful suggestions of former graduate students and from colleagues in industry are gratefully acknowledged.

JACK A. COLLINS

Columbus, Ohio
December 1980

Contents

**Chapter 1 The Role of Failure Prevention Analysis in Mechanical
Design**   1

    1.1   Introduction, 1
    1.2   A Definition of Design, 2
    1.3   A Challenge, 2
    1.4   Some Design Objectives, 3
    1.5   Closure, 4

Chapter 2 Modes of Mechanical Failure   6

    2.1   Definition of Failure Mode, 6
    2.2   Failure Modes Observed in Practice, 7
    2.3   A Glossary of Mechanical Failure Modes, 9

Chapter 3 Strength and Deformation of Engineering Metals   15

    3.1   Introduction, 15
    3.2   Deformation Response to Applied Shear, 20
    3.3   Elastic Deformation, 23
    3.4   Plastic Deformation, 23

        Slip, 23
        Critical Resolved Shearing Stress for Slip, 25
        Twinning, 30
        Grain Boundary Sliding and Diffusional
        Creep, 31
        Effects of Grain Boundaries in Polycrystals, 32
        Strain Rate Effects, 32

    3.5   Fracture, 33
    3.6   An Introduction to Dislocation Theory, 37

        Dislocation Geometry, 37
        Dislocation Motion, 42
        Dislocation Pinning, Generation, and
        Interaction, 45

3.7   An Introduction to Linear Elastic
       Fracture Mechanics, 49
3.8   Use of Fracture Mechanics in Design, 60
3.9   Elastic-Plastic Fracture Mechanics, 67
3.10   Using the Ideas, 68
3.11   Closure, 71

Chapter 4   State of Stress   **76**

4.1   Introduction, 76
4.2   State of Stress at a Point, 76
4.3   Principal Normal Stresses, 81
4.4   Principal Shearing Stresses, 86
4.5   Using the Ideas, 91

Chapter 5   Relationships Between Stress and Strain   **98**

5.1   Introduction, 98
5.2   Concepts of Engineering Stress-Strain
       and True Stress-Strain, 98
5.3   Elastic Stress-Strain Relationships, 104
5.4   Plastic Stress-Strain Relationships, 111
5.5   Using the Ideas, 116

**Chapter 6   Combined Stress Theories of Failure and Their Use
in Design**   **126**

6.1   Introduction, 126
6.2   Maximum Normal Stress Theory
       (Rankine's Theory), 128
6.3   Maximum Shearing Stress Theory
       (Tresca-Guest Theory), 130
6.4   Maximum Normal Strain Theory
       (St. Venant's Theory), 132
6.5   Total Strain Energy Theory
       (Beltrami Theory), 134
6.6   Distortion Energy Theory
       (Huber-Von Mises-Hencky Theory), 137
6.7   Failure Theory Comparison in Biaxial
       State of Stress, 142
6.8   Mohr's Failure Theory, 143
6.9   Summary of Failure Theory Evaluation, 149
6.10   Combined Stress Failure Theories as
       Design Tools, 149
6.11   Using the Ideas, 151

Chapter 7   High-Cycle Fatigue   **164**

    7.1   Introduction,   164
    7.2   Historical Remarks,   165
    7.3   The Nature of Fatigue,   167
    7.4   Fatigue Loading,   170
    7.5   Laboratory Fatigue Testing,   174
    7.6   The S-N-P Curves—A Basic Design Tool,   180
    7.7   Factors That Affect S-N-P Curves,   184

         Material Composition,   185
         Grain Size and Grain Direction,   185
         Heat Treatment,   188
         Welding,   188
         Geometrical Discontinuities,   193
         Surface Conditions,   193
         Size Effect,   199
         Residual Surface Stresses,   199
         Operating Temperature,   203
         Corrosion,   206
         Fretting,   207
         Operating Speed,   207
         Configuration of Stress-Time Pattern,   209
         Nonzero Mean Stress,   212
         Damage Accumulation,   213

    7.8   Using the Factors in Design,   214
    7.9   The Influence of Nonzero Mean Stress,   215
    7.10   Using the Ideas,   222
    7.11   Multiaxial Fatigue Stresses,   224

         Maximum Normal Stress Multiaxial
         Fatigue Failure Theory,   225
         Maximum Shearing Stress Multiaxial
         Fatigue Failure Theory,   226
         Distortion Energy Multiaxial
         Fatigue Failure Theory,   227

    7.12   Use of Multiaxial Fatigue Failure
           Theories,   230
    7.13   Using the Ideas,   230

**Chapter 8   Concepts of Cumulative Damage, Life Prediction,
and Fracture Control**   **240**

    8.1   Introduction,   240
    8.2   The Linear Damage Theory,   241

8.3   Cumulative Damage Theories,   243

        Marco-Starkey Cumulative Damage Theory,   243
        Henry Cumulative Damage Theory,   246
        Gatts Cumulative Damage Theory,   249
        Corten-Dolan Cumulative Damage Theory,   254
        Marin Cumulative Damage Theory,   263
        Manson Double Linear Damage Rule,   266

8.4   Using the Ideas,   268
8.5   Life Prediction Based on Local Stress-Strain
        and Fracture Mechanics Concepts,   275

        Local Stress-Strain Approach to Crack
        Initiation,   275

8.6   Fracture Mechanics Approach to Crack
        Propagation,   288
8.7   Service Loading Simulation and Full-Scale
        Fatigue Testing,   295
8.8   Damage Tolerance and Fracture Control,   297
8.9   Using the Ideas,   302

Chapter 9   Use of Statistics in Fatigue Analysis   319

9.1   Introduction,   319
9.2   Definitions,   319
9.3   Population Distributions,   322
9.4   Sampling Distributions,   327
9.5   Statistical Hypotheses,   336
9.6   Confidence Limits,   340
9.7   Properties of Good Estimators,   342
9.8   Sample Size for Desired Confidence,   342
9.9   Probability Paper,   344
9.10   Comparison of Means and Variances,   351
9.11   In Summary,   357

Chapter 10   Fatigue Testing Procedures and Statistical Interpretations
             of Data   360

10.1   Introduction,   360
10.2   Standard Method,   360
10.3   Constant Stress Level Testing,   361
10.4   Response or Survival Method
        (Probit Method),   363
10.5   Step-Test Method,   365
10.6   Prot Method,   367
10.7   Staircase or Up-and-Down Method,   369

10.8   Extreme Value Method,   374
10.9   In Summary,   376

Chapter 11   Low-Cycle Fatigue   **379**

11.1   Introduction,   379
11.2   The Strain Cycling Concept,   380
11.3   The Strain-Life Curve and Low-Cycle Fatigue
        Relationships,   384
11.4   The Influence of Nonzero Mean Strain
        and Nonzero Mean Stress,   387
11.5   Cumulative Damage in Low-Cycle Fatigue,   390
11.6   Influence of Multiaxial States of Stress,   391
11.7   Relationship of Thermal Fatigue to Low-Cycle
        Fatigue,   391
11.8   In Summary,   393
11.9   Using the Ideas,   393

Chapter 12   Stress Concentration   **400**

12.1   Introduction,   400
12.2   Stress Concentration Effects,   402
12.3   Stress Concentration Factors for the
        Elastic Range,   404
12.4   Stress Concentration Factors and Strain
        Concentration Factors for the Plastic
        Range,   413
12.5   Stress Concentration Factors for Multiple
        Notches,   415
12.6   Fatigue Stress Concentration Factors
        and Notch Sensitivity Index,   416
12.7   Using the Ideas,   422

Chapter 13   Creep, Stress Rupture, and Fatigue   **435**

13.1   Introduction,   435
13.2   Prediction of Long-Term Creep Behavior,   437

        Abridged Method,   438
        Mechanical Acceleration Method,   439
        Thermal Acceleration Method,   439

13.3   Theories for Predicting Creep Behavior,   440

        Larson-Miller Parameter,   440
        Manson-Haferd Parameter,   441

13.4   Creep Under Uniaxial State of Stress,   442

13.5   Creep Under Multiaxial State of Stress, 447
13.6   Cumulative Creep Concepts, 449
13.7   Combined Creep and Fatigue, 451

Chapter 14   Fretting, Fretting Fatigue, and Fretting Wear   **479**

14.1   Introduction, 479
14.2   Variables of Importance in the Fretting
        Process, 480
14.3   Fretting Fatigue, 481
14.4   Fretting Wear, 491
14.5   Fretting Corrosion, 494
14.6   Minimizing or Preventing Fretting
        Damage, 496

Chapter 15   Shock and Impact   **500**

15.1   Introduction, 500
15.2   Energy Method of Approximating Stress
        and Deflection Under Impact Loading
        Conditions, 501
15.3   Stress Wave Propagation Under Impact
        Loading Conditions, 507
15.4   Particle Velocity and Wave Propagation
        Velocity, 511
15.5   Stress Wave Behavior at Free and
        Fixed Ends, 516
15.6   Stress Wave Propagation in a Bar Under
        Suddenly Applied Axial Force, 518
15.7   Stress Wave Attenuation Due to
        Hysteretic Damping, 520
15.8   Stress Wave Propagation in a Bar Struck
        on the End by a Moving Mass, 524
15.9   Maximum Stress in a Bar Struck on the End
        by a Moving Mass, 531
15.10   Stress Wave Propagation When the Yield
        Point Is Exceeded, 535
15.11   Changes in Material Properties Under Impact
        Loading, 536
15.12   Spalling or Scabbing Under Impact
        Loading, 543
15.13   The Effects of Stress and Strain Concentrations
        Under Impact Loading Conditions, 546
15.14   Using the Ideas, 549
15.15   In Summary, 552

Chapter 16   Buckling and Instability   **557**

    16.1   Introduction,   557
    16.2   Buckling of a Simple Pin-Jointed
            Mechanism,   557
    16.3   Buckling of a Pinned-End Column,   559
    16.4   The Influence of End Support on Column
            Buckling,   563
    16.5   Inelastic Behavior in Column Buckling,   565
    16.6   Using the Ideas,   568
    16.7   Lateral Buckling of Deep, Narrow Beams
            Subjected to Bending,   572
    16.8   Lateral Buckling of Thin, Circular Shaft Subjected
            to Torsion,   576
    16.9   Other Buckling Phenomena,   579

Chapter 17   Wear, Corrosion, and Other Important Failure Modes   **583**

    17.1   Introduction,   583
    17.2   Wear,   583

            Adhesive Wear,   584
            Abrasive Wear,   590
            Corrosion Wear,   594
            Surface Fatigue Wear,   594
            Deformation Wear, Fretting Wear,
            and Impact Wear,   595

    17.3   Empirical Model for Zero Wear,   596
    17.4   Using the Ideas,   601
    17.5   Corrosion,   603

            Direct Chemical Attack,   604
            Galvanic Corrosion,   606
            Crevice Corrosion,   608
            Pitting Corrosion,   608
            Intergranular Corrosion,   609
            Selective Leaching,   610
            Erosion Corrosion,   610
            Cavitation Corrosion,   611
            Hydrogen Damage,   611
            Biological Corrosion,   612

    17.6   Stress Corrosion Cracking,   612
    17.7   Closure,   613

Index   **619**

Failure of Materials
in Mechanical Design

CHAPTER 1

The Role of Failure Prevention Analysis in Mechanical Design

Have you heard of the wonderful one-hoss shay
That was built in such a logical way
It ran a hundred years to a day————?

It went to pieces all at once—
All at once, and nothing first,
Just as bubbles do when they burst————

—Oliver Wendell Holmes, Jr.,
"Deacon's Masterpiece"

1.1 INTRODUCTION

Mechanical failure may be defined as any change in the size, shape, or material properties of a structure, machine, or machine part that renders it incapable of satisfactorily performing its intended function. It is the primary responsibility of any mechanical designer to ensure that his design functions as intended for the prescribed design lifetime and, at the same time, that it is competitive in the marketplace. Success in designing competitive products while averting premature mechanical failures can be consistently achieved only by recognizing and evaluating all potential modes of failure that might govern the design. If the designer is to recognize potential failure modes, he must at least be acquainted with the array of failure modes observed in the field and with the conditions leading to these failures. If the designer is to be effective in averting failure, he must have a good working knowledge of analytical and/or empirical techniques of predicting failure so that he can design to prevent failure during the prescribed design life. Thus it is clear that

1

failure analysis, prediction, and prevention are of critical importance to any designer if he is to achieve success.

1.2 A DEFINITION OF DESIGN

Engineering design is an *iterative decision-making* process that has as its objective the *creation* and *optimization* of a *new* or *improved* engineering system or device for the *fulfillment of a human need or desire*, with due regard for *conservation of resources* and *environmental impact*. The definition just given includes several key ideas that characterize all design activity and that exert a substantial influence on the way the material is presented in this text. Any engineering design project has as its first objective the fulfillment of a human need or desire—otherwise, as engineers, we would be wasting our time. Whether a designer is creating a new device or improving an existing design, he must strive to provide the "best," or optimum, design consistent with the constraints of time and money placed on him by the marketplace. Unfortunately, an absolute optimum design may be impossible to define, much less produce, in a complex engineering system. Even if an optimum design can be defined, it is often very expensive to do so. Yet competition often demands that performance be improved, life be extended, weight be reduced, or cost be lowered. That is, competition often demands that a better engineering job be done of optimizing the design with respect to the criteria of performance, life, weight, cost, or all of these, while fulfilling our responsibilities to conserve resources and preserve the earth's environment.

1.3 A CHALLENGE

Technological advances have swept our society so regularly for the past three or four decades that future advances of major significance are routinely expected. Designers will be challenged as never before if they are to meet the ever-increasing demands of society. The introduction of new materials and the need for higher operating speeds, higher temperatures, lighter weight, smaller volume, longer life, lower cost, and improved ecological compatibility all serve to demand better design techniques.

For example, rotational shaft speeds of 30,000 rpm and higher are becoming commonplace. Operating temperatures of 2000°F and higher are increasingly more common. Supersonic flight regimes and space environments confront many designers. Nuclear environments, coupled with high operating temperatures and long-term dynamic loads and motions, are today's problems for many designers. Equally demanding are the problems of providing subminiature equipment or replacement prostheses in the human cardiovascular system or body organs.

These severe service conditions have forced designers to study the behavior of materials more carefully, to better assess the nature of actual service conditions, and to better understand the many modes of mechanical failure. Designers have been forced to develop a better understanding of stresses and strains produced by dynamic loading in adverse environments and the effects of residual stress fields produced by manufacturing processes. The recognition of preexisting cracklike flaws in all real materials and structures has forced the development of new design tools to deal with crack propagation under conditions of both monotonic and fluctuating loads. Inspectability and maintainability have joined reliability and availability as important design criteria.

Opposing demands of greater capacity and smaller size can be met either (1) by developing stronger and more rigid materials or (2) by more efficiently utilizing the strength and rigidity of available materials. The first alternative lies in the realm of the materials scientist and will not be pursued here. The second alternative is a challenge to the designer and, indeed, is the motivation for this text. In more efficiently utilizing the strength and rigidity of existing materials to meet the ever-increasing demands of tomorrow's technology, the designer will be asked to use his analytical tools to the fullest measure, to draw upon empiricisms of the highest available accuracy, and to exercise ingenuity, creativity, and engineering judgment if he is to be successful. This is the challenge.

1.4 SOME DESIGN OBJECTIVES

In one sense, a "perfect" design would be one in which the entire machine would fail completely at a given preselected life. That is, every part of every member would be so designed that it would disintegrate into dust at a precisely predictable time. There are many practical reasons why such a design is not possible and perhaps not even desirable, even though such a design would fully utilize the material.

If it were possible to produce such a "perfect" design, it would surely require a highly refined analysis, extensive experimental development, exceedingly consistent material properties, precisely defined operational conditions, and a meticulous designer to coordinate the effort. Since a highly refined design effort is costly, both in time and money, the successful engineering designer must balance the cost of analysis and design effort against the need for it in each particular case. Obviously, a so-called "perfect" design that markets at 10 times the cost of the competitor's "not-quite-so-perfect" but acceptable design is not the hallmark of a successful designer. When to stop calculating and start building is an issue of major concern in the work of every designer and engineering manager. In assessing when a design has achieved its major objectives, the designer usually considers the following factors:

1. All machine parts must be capable of transmitting the necessary forces and performing the necessary motions efficiently and economically.
2. Failure must not occur in any part before a predetermined span of operating life has elapsed.
3. Each machine part must perform its function without interfering with any other part of the machine.
4. It must be possible to manufacture the part and assemble it in the machine.
5. The cost of the finished part must be consistent with the application.
6. The weight and space occupied must be consistent with the application.
7. It must be possible to service and maintain all parts requiring service during the design life.
8. The machine must not only function satisfactorily for its design lifetime, but must also be competitive in the marketplace as well as profitable to the manufacturer.

1.5 CLOSURE

Although the following chapters do not deal specifically with design philosophy, all of the analyses and procedures presented are to help the designer meet his objectives. Analysis is presented where possible, but when design empiricism is necessary to perform an acceptable failure analysis or prediction, it will be found. It must be recognized also that techniques of failure prevention analysis are constantly being improved and changed. This text is therefore not the "last word" in any sense, but it aspires to be a useful tool, both academically and also in the real world of engineering design, which may be improved by the user as he grows in understanding and experience.

QUESTIONS

1. In your opinion, what does Oliver Wendell Holmes, Jr.'s "Deacon's Masterpiece," quoted in part as the prologue to Chapter 1, have to do with the philosophy of engineering design?
2. Elaborate on the responsibilities of a mechanical designer who is involved in the design of high-performance machines.
3. Define *engineering design* and elaborate on each important concept in the definition.

4. List several factors that might be used to judge how well a design meets its objectives.

5. List all the considerations you can think of that would be involved in establishing the selling price of an engineering product, for example, an aircraft gas turbine. Of these considerations, which ones would be directly influenced by the quality of engineering performed by the mechanical designer?

6. Write a review list of strength of materials formulas useful to a stress analyst. Include direct axial loading, bending, bearing, direct shear, torsional shear, and transverse shear loading in your list.

7. When to stop calculating and start building is an engineering judgment of critical importance. Write about 300 words discussing your views on what factors influence such a judgment.

8. Inspectability, maintainability, reliability, and availability are claimed to be important design criteria. Define each of them and briefly discuss why they are important considerations to the designer.

9. Under what circumstances might a designer utilize "empiricisms" rather than highly refined analytical techniques?

CHAPTER 2

Modes of Mechanical Failure

2.1 DEFINITION OF FAILURE MODE

In the first chapter it was suggested that mechanical failure might be defined as any change in the size, shape, or material properties of a structure, machine, or machine part that renders it incapable of satisfactorily performing its intended function. With this definition in mind, one might define *failure mode* as the physical process or processes that take place or combine their effects to produce failure.

It has been suggested* that a systematic classification might be devised by which all possible failure modes could be predicted. Such a classification is based on defining three categories: (1) manifestations of failure, (2) failure-inducing agents, and (3) locations of failure. These categories are specifically defined in the text that follows. Each specific failure mode is then identified as a combination of one or more manifestations of failure together with one or more failure-inducing agents and a failure location. Literally hundreds of combinations can be systematically listed. To explain the system in more detail, we may develop the three categories in more detail, as follows.

The four *manifestations of failure*, some with subcategories, are:

1. Elastic deformation
2. Plastic deformation
3. Rupture or fracture
4. Material change
    A. Metallurgical
    B. Chemical
    C. Nuclear

*Classification suggested by Prof. W. L. Starkey at The Ohio State University.

The four *failure-inducing agents*, each with subcategories, are:

1. Force
    A. Steady
    B. Transient
    C. Cyclic
    D. Random
2. Time
    A. Very short
    B. Short
    C. Long

3. Temperature
    A. Low
    B. Room
    C. Elevated
    D. Steady
    E. Transient
    F. Cyclic
    G. Random
4. Reactive environment
    A. Chemical
    B. Nuclear

The two *failure locations* are:

1. Body type
2. Surface type

To be precise in describing a specific mode of failure, it is necessary to select appropriate categories from those just listed without omitting any of the three major categories. For example, one might select *plastic deformation* from the first category, *steady force* and *room temperature* from the second category, and *body type* from the third category. Thus, the failure mode selected could be properly described as body-type plastic deformation under steady force at room temperature. This failure mode is commonly called *yielding*. Note, however, that the term *yielding* does not imply all of these restrictions; it is more general than that.

Many other failure modes of special interest have been defined that refer to general patterns of the three categories listed. To be useful, these terms require additional description and elaboration, but the terms are commonly used and very useful because of the importance of the failure phenomena that they represent. Twenty-three such specific failure modes are listed in Section 2.2. Later in the text, entire chapters are devoted to some of the more important failure modes.

2.2 FAILURE MODES OBSERVED IN PRACTICE

The following list of failure modes includes those most commonly observed in practice. In reviewing the list, it may be noted that certain failure modes are unilateral phenomena, whereas others are combined phenomena. For example, corrosion is listed as a failure mode, fatigue is listed as a failure mode, and corrosion-fatigue is listed as still another failure mode. Such combinations are included because they are commonly observed, important, and usually *synergistic*. That is, in the case of corrosion-fatigue, for example, the presence of active corrosion aggravates the fatigue process, and at the same time the presence of fluctuating fatigue loads aggravates the corrosion process. The following list is not presented in any special order, but it includes all commonly observed modes of mechanical failure.

1. Force and/or temperature-induced elastic deformation
2. Yielding
3. Brinnelling
4. Ductile rupture
5. Brittle fracture
6. Fatigue
   A. High-cycle fatigue
   B. Low-cycle fatigue
   C. Thermal fatigue
   D. Surface fatigue
   E. Impact fatigue
   F. Corrosion fatigue
   G. Fretting fatigue
7. Corrosion
   A. Direct chemical attack
   B. Galvanic corrosion
   C. Crevice corrosion
   D. Pitting corrosion
   E. Intergranular corrosion
   F. Selective leaching
   G. Erosion corrosion
   H. Cavitation corrosion
   I. Hydrogen damage
   J. Biological corrosion
   K. Stress corrosion
8. Wear
   A. Adhesive wear
   B. Abrasive wear
   C. Corrosive wear
   D. Surface fatigue wear
   E. Deformation wear
   F. Impact wear
   G. Fretting wear

9. Impact
   A. Impact fracture
   B. Impact deformation
   C. Impact wear
   D. Impact fretting
   E. Impact fatigue
10. Fretting
    A. Fretting fatigue
    B. Fretting wear
    C. Fretting corrosion
11. Creep
12. Thermal relaxation
13. Stress rupture
14. Thermal shock
15. Galling and seizure
16. Spalling
17. Radiation damage
18. Buckling
19. Creep buckling
20. Stress corrosion
21. Corrosion wear
22. Corrosion fatigue
23. Combined creep and fatigue

A brief glossary describing these terms is presented in Section 2.3. Much more complete descriptions of many of the more important failure modes are included in later chapters.

2.3   A GLOSSARY OF MECHANICAL FAILURE MODES

As they are used in this text and commonly in engineering practice, the failure modes just listed may be defined and described briefly as follows.

Force and/or temperature-induced elastic deformation failure occurs whenever the elastic (recoverable) deformation in a machine member, brought about by the imposed operational loads or temperatures, becomes great enough to interfere with the ability of the machine to satisfactorily perform its intended function.

Yielding failure occurs when the plastic (unrecoverable) deformation in a ductile machine member, brought about by the imposed operational loads or motions, becomes great enough to interfere with the ability of the machine to satisfactorily perform its intended function.

Brinnelling failure occurs when the static forces between two curved surfaces in contact result in local yielding of one or both mating members to produce a permanent surface discontinuity of significant size. For example, if a ball bearing is statically loaded so that a ball is forced to permanently indent the race through local plastic flow, the race is brinnelled. Subsequent operation of the bearing might result in intolerably increased vibration, noise, and heating; and, therefore, failure would have occurred.

Ductile rupture failure occurs when the plastic deformation, in a machine part that exhibits ductile behavior, is carried to the extreme so that the member separates into two pieces. Initiation and coalescence of internal voids slowly propagate to failure, leaving a dull, fibrous rupture surface.

Brittle fracture failure occurs when the elastic deformation, in a machine part which exhibits brittle behavior, is carried to the extreme so that the primary interatomic bonds are broken and the member separates into two or more pieces. Preexisting flaws or growing cracks form initiation sites for very rapid crack propagation to catastrophic failure, leaving a granular, multi-faceted fracture surface.

Fatigue failure is a general term given to the sudden and catastrophic separation of a machine part into two or more pieces as a result of the application of fluctuating loads or deformations over a period of time. Failure takes place by the initiation and propagation of a crack until it becomes unstable and propagates suddenly to failure. The loads and deformations that typically cause failure by fatigue are far below the static failure levels. When loads or deformations are of such magnitude that more than about 10,000 cycles are required to produce failure, the phenomenon is usually termed *high-cycle fatigue*. When loads or deformations are of such magnitude that less than about 10,000 cycles are required to produce failure, the phenomenon is usually termed *low-cycle fatigue*. When load or strain cycling is produced by a fluctuating temperature field in the machine part, the process is usually termed *thermal fatigue*. *Surface fatigue* failure, usually associated with rolling surfaces in contact, manifests itself as pitting, cracking, and spalling of the contacting surfaces as a result of the cyclic Hertz contact stresses that result in maximum values of cyclic shear stresses slightly below the surface. The cyclic subsurface shear stresses generate cracks that propagate to the contacting surface, dislodging particles in the process to produce surface pitting. This phenomenon is often viewed as a type of wear. Impact fatigue, corrosion fatigue, and fretting fatigue are described later.

Corrosion failure, a very broad term, implies that a machine part is rendered incapable of performing its intended function because of the undesired deterioration of the material as a result of chemical or electrochemical interaction with the environment. Corrosion often interacts with other failure modes such as wear or fatigue. The many forms of corrosion include the following. *Direct chemical attack*, perhaps the most common type of corrosion, involves corrosive attack of the surface of the machine part exposed to the corrosive media, more or less uniformly over the entire exposed surface. *Galvanic corrosion* is an accelerated electrochemical corrosion that occurs when two dissimilar metals in electrical contact are made part of a circuit completed by a connecting pool or film of electrolyte or corrosive medium, leading to current flow and ensuing corrosion. *Crevice corrosion* is the accelerated corrosion process highly localized within crevices, cracks, or joints where small volume regions of stagnant solution are trapped in contact with the corroding metal. *Pitting corrosion* is a very localized attack that leads to the development of an array of holes or pits that penetrate the metal. *Intergranular corrosion* is the localized attack occurring at grain boundaries of certain copper, chromium, nickel, aluminum, magnesium, and zinc alloys when they are improperly heat treated or welded. Formation of local galvanic cells that precipitate corrosion products at the grain boundaries seriously degrade the material strength because of the intergranular corrosive process. *Selective leaching* is a corrosion process in which one element of a solid alloy is removed, such as in dezincification of brass alloys or graphitization of gray cast irons. *Erosion corrosion* is the accelerated chemical attack that results when abrasive or viscid material flows past a containing surface, continuously baring fresh, unprotected material to the corrosive medium. *Cavitation corrosion* is the accelerated chemical corrosion that results when, because of differences in vapor pressure, certain bubbles and cavities within a fluid collapse adjacent to the pressure vessel walls, causing particles of the surface to be expelled, baring fresh, unprotected surface to the corrosive medium. *Hydrogen damage*, while not considered to be a form of direct corrosion, is induced by corrosion. Hydrogen damage includes hydrogen blistering, hydrogen embrittlement, hydrogen attack, and decarburization. *Biological corrosion* is a corrosion process that results from the activity of living organisms, usually by virtue of their processes of food ingestion and waste elimination, in which the waste products are corrosive acids or hydroxides. *Stress corrosion*, an extremely important type of corrosion, is described separately later.

Wear is the undesired cumulative change in dimensions brought about by the gradual removal of discrete particles from contacting surfaces in motion, usually sliding, predominantly as a result of mechanical action. Wear is not a single process, but a number of different processes that can take place independently or in combination, resulting in material removal from contacting surfaces through a complex combination of local shearing, plowing, gouging, welding, tearing, and others. *Adhesive wear* takes place because of high local pressure and welding at asperity contact sites, followed by motion-induced plastic deformation and rupture of asperity junctions, with resulting metal removal or transfer. *Abrasive wear* takes place when the wear particles are removed from the surface by plowing, gouging and cutting action of the asperities of a harder mating surface or by hard particles entrapped between the mating surfaces. When the conditions for either adhesive wear or abrasive wear coexist with conditions that lead to corrosion, the processes interact synergistically to produce *corrosive wear*. As described earlier, *surface fatigue wear* is a wear phenomenon associated with curved surfaces in rolling or sliding contact, in which subsurface cyclic shear stresses initiate microcracks that propagate to the surface to spall out macroscopic particles and form wear pits. *Deformation wear* arises as a result of repeated *plastic* deformation at the wearing surfaces, producing a matrix of cracks that grow and coalesce to form wear particles. Deformation wear is often caused by severe impact loading. *Impact wear* is impact-induced repeated *elastic* deformation at the wearing surface that produces a matrix of cracks that grow in accordance with the surface fatigue description just given. Fretting wear is described later.

Impact failure results when a machine member is subjected to nonstatic loads that produce in the part stresses or deformations of such magnitude that the member no longer is capable of performing its function. The failure is brought about by the interaction of stress or strain waves generated by dynamic or suddenly applied loads, which may induce local stresses and strains many times greater than would be induced by static application of the same loads. If the magnitudes of the stresses and strains are sufficiently high to cause separation into two or more parts, the failure is called *impact fracture*. If the impact produces intolerable elastic or plastic deformation, the resulting failure is called *impact deformation*. If repeated impacts induce cyclic elastic strains that lead to initiation of a matrix of fatigue cracks, which grow to failure by the surface fatigue phenomenon described earlier, the process is called *impact wear*. If fretting action, as described in the next paragraph, is induced by the small lateral relative displacements between two surfaces as they impact together, where the small displacements are caused by Poisson strains or small tangential "glancing" velocity components, the phenomenon is called *impact fretting*. *Impact fatigue* failure occurs when impact loading is repetitively applied to a machine member until failure occurs by the nucleation and propagation of a fatigue crack.

Fretting action may occur at the interface between any two solid bodies whenever they are pressed together by a normal force and subjected to small-amplitude cyclic relative motion with respect to each other. Fretting usually takes place in joints that are not intended to move but, because of vibrational loads or deformations, experience minute cyclic relative motions. Typically, debris produced by fretting action is trapped between the surfaces because of the small motions involved. *Fretting fatigue* failure is the premature fatigue fracture of a machine member subjected to fluctuating loads or strains together with conditions that simultaneously produce fretting action. The surface discontinuities and microcracks generated by the fretting action act as fatigue crack nuclei that propagate to failure under conditions of fatigue loading that would otherwise be acceptable. Fretting fatigue failure is an insidious failure mode because the fretting action is usually hidden within a joint where it cannot be seen and leads to premature, or even unexpected, fatigue failure of a sudden and catastrophic nature. *Fretting wear* failure results when the changes in dimensions of the mating parts, because of the presence of fretting action, become large enough to interfere with proper design function or large enough to produce geometrical stress concentration of such magnitude that failure ensues as a result of excessive local stress levels. *Fretting corrosion* failure occurs when a machine part is rendered incapable of performing its intended function because of the surface degradation of the material from which the part is made, as a result of fretting action.

Creep failure results whenever the plastic deformation in a machine member accrues over a period of time under the influence of stress and temperature until the accumulated dimensional changes interfere with the ability of the machine part to satisfactorily perform its intended function. Three stages of creep are often observed: (1) transient or primary creep during which time the rate of strain decreases, (2) steady state or secondary creep during which time the rate of strain is virtually constant, and (3) tertiary creep during which time the creep strain rate increases, often rapidly, until rupture occurs. This terminal rupture is often called creep rupture and may or may not occur, depending on the stress-time-temperature conditions.

Thermal relaxation failure occurs when the dimensional changes due to the creep process result in the relaxation of a prestrained or prestressed member until it no longer is able to perform its intended function. For example, if the prestressed flange bolts of a high-temperature pressure vessel relax over a period of time because of creep in the bolts, so that finally the peak pressure surges exceed the bolt preload to violate the flange seal, the bolts will have failed because of thermal relaxation.

Stress rupture failure is intimately related to the creep process except that the combination of stress, time, and temperature is such that rupture into two parts is assured. In stress rupture failures the combination of stress and temperature is often such that the period of steady-state creep is short or nonexistent.

Thermal shock failure occurs when the thermal gradients generated in a machine part are so pronounced that differential thermal strains exceed the ability of the material to sustain them without yielding or fracture.

Galling failure occurs when two sliding surfaces are subjected to such a combination of loads, sliding velocities, temperatures, environments, and lubricants that massive surface destruction is caused by welding and tearing, plowing, gouging, significant plastic deformation of surface asperities, and metal transfer between the two surfaces. Galling may be thought of as a severe extension of the adhesive wear process. When such action results in significant impairment to intended surface sliding, or in seizure, the joint is said to have failed by galling. *Seizure* is an extension of the galling process to such severity that the two parts are virtually welded together so that relative motion is no longer possible.

Spalling failure occurs whenever a particle is spontaneously dislodged from the surface of a machine part so as to prevent the proper function of the member. Armor plate fails by spalling, for example, when a striking missile on the exposed side of an armor shield generates a stress wave that propagates across the plate in such a way as to dislodge or spall a secondary missile of lethal potential on the protected side. Another example of spalling failure is manifested in rolling contact bearings and gear teeth because of the action of surface fatigue as described earlier.

Radiation damage failure occurs when the changes in material properties induced by exposure to a nuclear radiation field are of such a type and magnitude that the machine part is no longer able to perform its intended function, usually as a result of the triggering of some other failure mode, and often related to loss in ductility associated with radiation exposure. Elastomers and polymers are typically more susceptible to radiation damage than are metals whose strength properties are sometimes enhanced rather than damaged by exposure to a radiation field, though ductility is usually decreased.

Buckling failure occurs when, because of a critical combination of magnitude and/or point of load application, together with the geometrical configuration of a machine member, the deflection of the member suddenly increases greatly with only a slight change in load. This nonlinear response results in a buckling failure if the buckled member is no longer capable of performing its design function.

Creep buckling failure occurs when, after a period of time, the creep process results in an unstable combination of the loading and geometry of a machine part so that the critical buckling limit is exceeded and failure ensues.

Stress corrosion failure occurs when the applied stresses on a machine part in a corrosive environment generate a field of localized surface cracks, usually along grain boundaries, that render the part incapable of performing its function, often because of triggering some other failure mode. Stress corrosion is a very important type of corrosion failure mode because so many different metals are susceptible to it. For example, a variety of iron, steel, stainless steel, copper, and aluminum alloys are subject to stress corrosion cracking if placed in certain adverse corrosive media.

Corrosion wear failure is a combination failure mode in which corrosion and wear combine their deleterious effects to incapacitate a machine part. The corrosion process often produces a hard, abrasive corrosion product that accelerates the wear, while the wear process constantly removes the protective corrosion layer from the surface, baring fresh metal to the corrosive medium and thus accelerating the corrosion. The two modes combine to make the result more serious than either of the modes would have been otherwise.

Corrosion fatigue is a combination failure mode in which corrosion and fatigue combine their deleterious effects to cause failure of a machine part. The corrosion process often forms pits and surface discontinuities that act as stress raisers that in turn accelerate fatigue failure. Further, cracks in the usually brittle corrosion layer also act as fatigue crack nuclei that propagate into the base material. On the other hand, the cyclic loads or strains cause cracking and flaking of the corrosion layer, which bares fresh metal to the corrosive medium. Thus, each process accelerates the other, often making the result disproportionately serious.

Combined creep and fatigue failure is a combination failure mode in which all of the conditions for both creep failure and fatigue failure exist simultaneously, each process influencing the other to produce failure. The interaction of creep and fatigue is probably synergistic but is not well understood.

QUESTIONS

1.   Define the terms *mechanical failure* and *failure mode*.

2.   List the four *manifestations of failure* and give an example of each one.

3.   List the four *failure-inducing agents* and give an example of each one.

4.   List the two *failure locations* and give an example of each one.

5.   Through library research and/or personal experience, determine whether there are additional modes of failure that have not been listed in Section 2.2.

6.   Taking a passenger automobile as an example of an engineering system, list all failure modes that you think might be significant and indicate where in the auto you think each failure mode might be active.

7.   Select three of the failure modes of greatest interest to you and, through library research, write a discussion of approximately 300 words for each one.

8.   Describe what is meant by a synergistic failure mode, give three examples, and for each example describe how the synergistic interaction proceeds.

9.   Select five of the failure modes in Section 2.2 and classify them according to the system suggested in Section 2.1.

10.   For each of the following applications, list three of the more likely failure modes, describing why each might be expected: (a) high-performance automotive racing engine (b) pressure vessel for commercial power plant (c) domestic washing machine.

CHAPTER 3

Strength and Deformation of Engineering Metals

3.1 INTRODUCTION

No attempt will be made in this text to explore thoroughly the solid state physics of material behavior, but some attention to a simple atomic model of metallic behavior should provide insight important to the evaluation of various failure modes. It may be somewhat surprising to the reader to find that the nature of the metallic bond is still virtually unknown in a quantitative theoretical sense. Many definitions have been given for the metallic bond in terms of its chemistry, properties, and differences from other types of atomic bonds, but a simple exact definition of the bonding forces cannot yet be given because of the complexity of the metallic state.

It is widely agreed that all metals, indeed, all materials, are composed of atoms. Atoms are, to a large extent, electrical structures whose diameters are of the order of one Angstrom unit (1 A = 10^{-8} cm). An atom is composed of a nucleus, in which the mass is concentrated, and a complement of electrons in a cloud around the nucleus, which determines the chemical nature of the atom. Atoms are held in the solid state by various interatomic forces that are, in general, functions of temperature and pressure. Interatomic forces can be either attractive or repulsive, and a solid has a well defined equilibrium spacing of its atoms at any given temperature and pressure, due to a balance between the repulsive and attractive interatomic forces.

If a potential energy datum is established for a pair of atoms at infinite separation (for practical purposes a few hundred Angstroms) and the potential energy arbitrarily defined to be zero at the datum, then as the two atoms are brought closer together, the resultant *attractive* interatomic forces become operative and the potential energy of the two-atom system becomes negative with respect to the datum since work is being done by the atoms. After reaching a certain critical separation distance (the equilibrium position), the resultant *repulsive* interatomic forces become positive with respect to the datum since work must be done on the atoms to bring them closer together.

Investigations have shown that the potential energy of a two-atom system may be expressed as a function of the distance by which they are separated in the following way (4):*

$$V = -\frac{A}{r^n} + \frac{B}{r^m}$$ (3-1)

V = potential energy
r = distance between atoms
A = proportionality constant for attraction
B = proportionality constant for repulsion
n = attraction exponent
m = repulsion exponent

From this expression for potential energy, the resultant force F between the atoms may be obtained by differentiating the energy expression with respect to separation distance, in the following form:

$$F = -\frac{\partial V}{\partial r} = \frac{nA}{r^{n+1}} + \frac{mB}{r^{m+1}}$$ (3-2)

By establishing the arbitrary definitions

$$nA = a$$
$$nB = b$$
$$n + 1 = N$$
$$m + 1 = M$$

the expression (3-2) may be rewritten as:

$$F = -\frac{a}{r^N} + \frac{b}{r^M}$$ (3-3)

Equations (3-1) and (3-3) are clearly of the same form and may be plotted as shown in Figure 3.1. These curves are known as the *Condon-Morse curves*. The value of separation distance r that corresponds to the minimum potential energy is the equilibrium spacing r_0 for the two atoms. The net force is zero at the spacing r_0, and any attempt toward a displacement in either direction from the r_0 position will call restoring forces into play.

Although these curves are developed for an isolated pair of atoms, the same type of behavior is exhibited when a free atom is brought into the vicinity of an existing crystal lattice. That is, at first a net attractive force builds up as the atom is brought closer to the crystal lattice, with an attendant decrease in the potential energy of the system. The attractive force then levels out and decreases to zero at the time the atom reaches its equilibrium separation r_0. At this time the potential energy of the system also reaches a minimum value. If the interatomic spacing is further decreased, a net repulsive force builds up

*Numbers in parentheses refer to References at the end of the chapter.

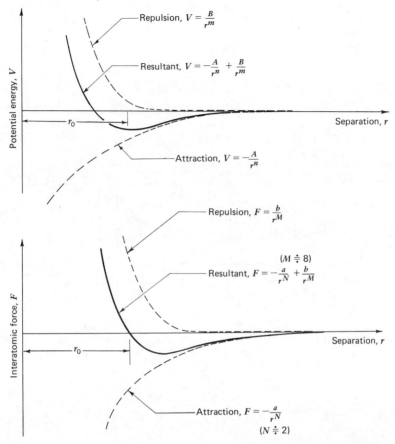

FIGURE 3.1. Condon-Morse curves showing the variation in potential energy and interatomic force as a function of separation distance for a two-atom system. (After ref. 4; Reprinted with permission from John Wiley & Sons, Inc.)

and the tendency is to return the atom to its equilibrium spacing. For this reason it is observed that the atoms in any crystal structure tend to be arrayed in a definite pattern with respect to their neighboring atoms. The various arrangements of atoms are referred to as *space lattices*. At first thought it might seem that there would be a large number of space lattices that could be formed, but it was shown in 1848 by Bravais that only 14 different networks of lattice geometry are possible. These are illustrated in Figure 3.2.

Small changes in the interatomic spacing of a material are manifested macroscopically as elastic strain. The engineering definition of strain is

$$\varepsilon = \frac{l - l_0}{l_0} \tag{3-4}$$

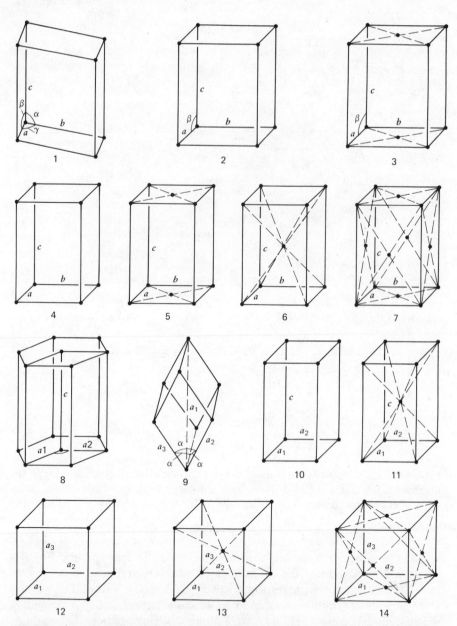

FIGURE 3.2. The 14 space lattices illustrated by a unit cell of each; (1) triclinic, simple; (2) monoclinic, simple; (3) monoclinic, base centered; (4) orthorhombic, simple; (5) orthorhombic, base centered; (6) orthorhombic, body centered; (7) orthorhombic, face centered; (8) hexagonal; (9) rhombohedral; (10) tetragonal, simple; (11) tetragonal, body centered; (12) cubic, simple; (13) cubic body centered; (14) cubic, face centered. (From ref. 1; Reprinted with permission of McGraw Hill Book Company.)

where ε is the macroscopic engineering strain, l_0 is the original unstrained length and l is the strained length of the member. This macroscopic strain is equal to the average fractional change in interatomic spacing in the same direction, or

$$\eta = \frac{r - r_0}{r_0} \tag{3-5}$$

where η is the fractional change in interatomic spacing, r_0 is the equilibrium spacing, and r is the strained spacing. It may, therefore, be deduced that Young's modulus of elasticity should be proportional to the slope of the Condon-Morse curve in the vicinity of r_0. The normal range of elastic strain observed in crystalline materials rarely exceeds 0.5 percent. As illustrated in Figure 3.3, the tangent to the Condon-Morse curve very nearly coincides with the force curve in this small range of strain, and for all practical purposes force is a linear function of strain as would be expected from observations in theory of elasticity.

Although the maximum elastic strain in crystalline solids, including the engineering metals, is typically very small, the force required to produce the small strain is usually large; hence the stress is large. This ratio of stress to strain is high because the applied force works in direct opposition to primary interatomic bonds. Certain noncrystalline materials, such as glass and cross-linked polymers, may also exhibit linear elasticity because their structure is such that distortion is opposed from the beginning by primary bonds. On the

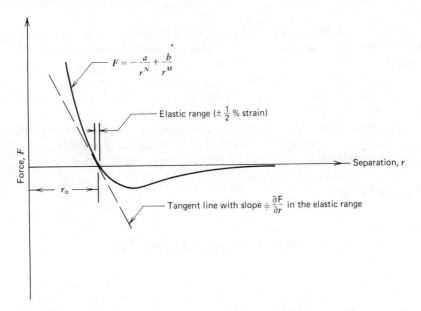

FIGURE 3.3.   The linear elastic range on the Condon-Morse force curve. (After ref. 4; Reprinted with permission from John Wiley & Sons, Inc.)

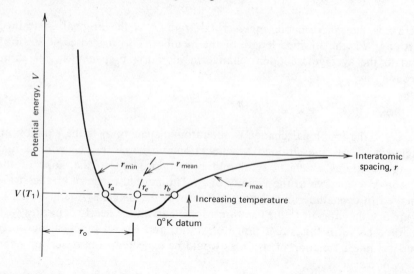

FIGURE 3.4. Condon-Morse potential energy curve showing influence of temperature on mean interatomic spacing. (After ref. 4; Reprinted with permission from John Wiley & Sons, Inc.)

other hand, certain other noncrystalline materials, such as rubber, are composed of intertangled long-chain molecules that may exhibit recoverable (but not necessarily linear) strains of several hundred percent. Such materials are usually called elastomers.

In the Condon-Morse potential energy curve shown in Figure 3.4, it should be noted that the interatomic spacing r_0 is the equilibrium distance between atoms at a temperature of absolute zero. As thermal energy is added to the system of two atoms, they begin to oscillate about their equilibrium position r_e. The minimum and maximum interatomic spacing at temperature T_1, for example, is at r_a and r_b in Figure 3.4. The mean interatomic spacing at that temperature is shown as r_e. Because of the asymmetry of the potential energy curve, the mean spacing increases with temperature (except for certain allotropic transformations). This change in the mean interatomic spacing with temperature is observed on the macroscopic level as a change in dimension due to thermal expansion.

3.2 DEFORMATION RESPONSE TO APPLIED SHEAR

Deformation behavior in a more complex atomic arrangement than the two-atom model is illustrated in Figure 3.5. The application of shearing stresses τ on planes within the crystal causes a displacement δ of atoms from their original positions. If the displacement is small, the strain is *elastic* and

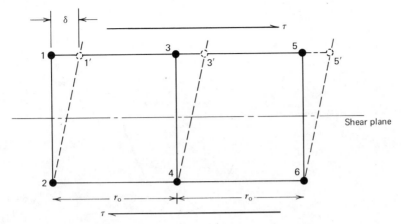

FIGURE 3.5. Deformation response to applied shearing stress in a simple cubic array of atoms. (After ref. 4; Reprinted with permission from John Wiley & Sons, Inc.)

recoverable. That is, the removal of the applied shearing stress results in the return of the atoms to their original positions. However, if the magnitude of the shearing stress is large enough to move atom 1 to a position midway between 2 and 4, atom 1 is in a state of metastable equilibrium with respect to atoms 2 and 4 and could as well take up a new equilibrium position directly over atom 4 as to return to its original position above atom 2.

Qualitatively, the metastable equilibrium position of atom 1 when it lies midway between atoms 2 and 4 is illustrated in Figure 3.6. It may be observed that the potential energy of the system may be lowered by moving either direction from the metastable position, and virtually zero shearing stress is needed to cause displacement in either direction. If atom 1 does take up a new position over atom 4 in Figure 3.5, the symmetry of the lattice is restored, but atoms on either side of the shear plane will have new nearest neighbors. The crystal is then said to have *slipped*, or to have undergone a *plastic* deformation of one interatomic distance.

If two complete planes of perfectly arrayed atoms are to be sheared over each other, the applied shearing stress must be large enough to overcome the forces between each atom in one plane and its neighbors in the adjacent plane. It has been calculated in various ways that the theoretical shearing stress necessary to accomplish such a sliding of one atomic plane over another (plastic deformation) should theoretically be on the order of 1–2 million psi for typical engineering metals. Experimentally measured values of only 10,000–50,000 psi are commonly observed. The question that immediately arises is "*Why is there such a big discrepancy between theoretical and measured values of critical shearing stress required to initiate plastic deformation?*"

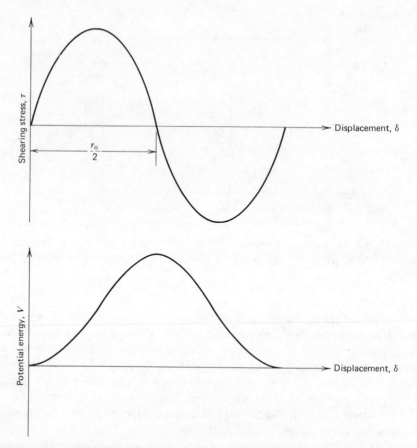

FIGURE 3.6. Qualitative illustration of the metastable equilibrium of an atom in a cubic array when displaced to a position midway between two neighboring atoms. (After ref. 4; Reprinted with permission from John Wiley & Sons, Inc.)

A suitable explanation for the discrepancy was not found until the concept of dislocations was postulated in the early 1930s by G. I. Taylor, E. Orowan, and M. Polanyi. Since the postulate was first presented, extensive research effort in the area of dislocation behavior has resulted in the experimental observation of dislocations and their movement, as well as an extensive literature on the mathematical description and prediction of dislocation interaction. Among other things, the experimentally observed magnitudes of shearing stress required to initiate plastic deformation can now be suitably estimated through application of the principles of dislocation theory. Some of the elementary ideas of dislocation theory will be discussed later in this chapter.

3.3 ELASTIC DEFORMATION

A brief discussion of failure by reason of elastic deformation was presented in Section 2.3. It may now be observed that such failures, a result of excessive elastic deformation, are brought about by the accumulated effect of slightly displacing the atoms from their equilibrium positions. So long as the forces and resulting atomic displacements are small, the atoms will return to their original equilibrium positions, and on a macroscopic scale, the machine part will return to its original dimensions. The majority of engineering design problems still lie within the elastic range. For this reason several of the following chapters will dwell on stress-strain relationships and material behavior in the elastic range.

3.4 PLASTIC DEFORMATION

Plastic deformation in crystalline materials generally occurs by one or more of four processes. These include (1) slip, (2) twinning, (3) grain boundary sliding, and (4) diffusional creep.

The principal mode of plastic deformation is *slip*. If the slip process is constrained or inhibited, *twinning* may contribute significantly to plastic deformation. At high temperatures and low rates of strain, polycrystalline materials may also deform plastically by *grain boundary sliding* and *diffusional creep*.

Slip

The most common mechanism of plastic deformation is the gliding of one plane of atoms over another, commonly called slip. Certain crystallographic planes and directions within a given crystal lattice are more susceptible to slip than others, resulting in the appearance of bands of fine parallel slip lines at a crystal surface when plastic straining is induced. The slip planes are usually the most densely packed atomic planes, and the slip directions are the most densely packed atomic directions in the lattice. The combination of a slip direction and the plane containing it is called a *slip system*. The appearance of these slip lines and slip bands is illustrated schematically in Figure 3.7. At higher magnification these slip lines may be seen to result from the relative parallel displacement of crystal planes that are on the order of about 100 atomic diameters apart. Displacement distances associated with the appearance of slip lines are commonly on the order of 1000 atomic diameters, as illustrated in Figure 3.7.

At lower temperatures the continued application of external shearing stresses to cause more and more plastic deformation results in the creation of

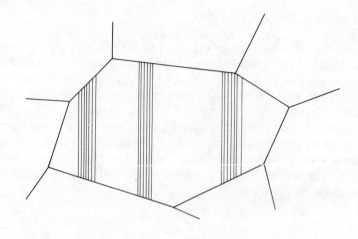

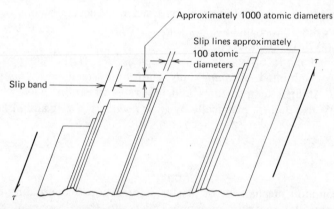

FIGURE 3.7. Schematic illustration of slip lines and slip bands on a crystal surface under the influence of an applied shearing stress.

many new slip lines, rather then to extend those that were originally formed. This indicates that slip planes are made more resistant to the shearing process as a result of the slip process itself. At higher temperatures, however, slip lines tend to cluster together, forming coarse slip bands with little slip occurring between these bands. Under these conditions the slip is confined to extension of lines within each band to produce relatively large local steps. It has been observed that the slip plane may be influenced by temperature, chemical composition, and amount of prior plastic deformation. The slip direction is not a function of these factors. Slip is generally an abrupt movement that can sometimes be heard as an audible "cry" or "tick" as it takes place.

A schematic representation of the lattice array before and after slip is shown in Figure 3.8. It may be noted that the slip process takes place by

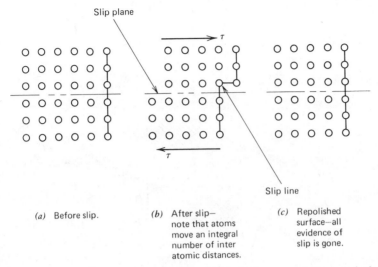

Slip plane

τ

τ

Slip line

(a) Before slip.

(b) After slip—
note that atoms
move an integral
number of inter
atomic distances.

(c) Repolished
surface—all
evidence of
slip is gone.

FIGURE 3.8.   Schematic representation of a cubic lattice array before and after slip.

atoms moving an integral number of interatomic distances. Thus, after the slip has occurred, the general symmetry of the lattice is restored but a slip step is evident at the free surface. If a polishing procedure were employed to smooth the free surface, all evidence of slip would be destroyed and the crystal lattice would be indistinguishable from its configuration prior to the slip process.

Critical Resolved Shearing Stress for Slip

If a selection of test specimens were cut from a single crystal at random with respect to orientation within the crystal, it would be found that physical properties such as proportional limit, yield point, tensile strength, and ductility would all vary over a range of values. A careful correlation of these physical properties with orientation in the crystal would indicate a strong dependence on specimen orientation. Of special interest in the consideration of slip is the establishment of a criterion for predicting the onset of plastic deformation in an externally loaded specimen cut from a single crystal. The establishment of this criterion may be accomplished in view of the observation just cited that physical properties vary with the orientation of specimens taken from a single crystal ingot.

A single crystal of simple geometrical shape is shown in Figure 3.9. The cylindrical specimen of normal cross-sectional area A is subjected to an axial tensile force, F. The active slip plane in the crystal is defined by its normal, which intersects the specimen axis of symmetry at angle ψ. The active slip direction in the slip plane is defined by the angle λ between the specimen axis

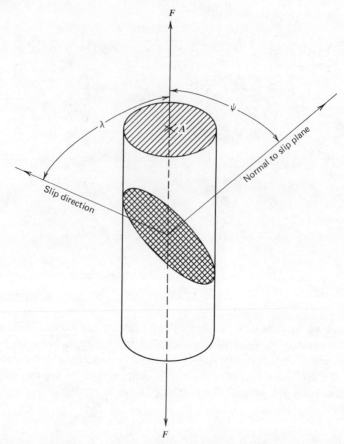

FIGURE 3.9. Schematic illustration of slip system in a single crystal specimen of simple geometry.

of symmetry and direction of slip. With these definitions the area of the slip plane A_{sp} may be expressed as

$$A_{sp} = \frac{A}{\cos \psi} \qquad (3\text{-}6)$$

and the component F_{sd} of the tensile force F resolved in the slip direction is

$$F_{sd} = F \cos \lambda \qquad (3\text{-}7)$$

The shearing stress τ_r resolved on the slip plane and in the slip direction may therefore be expressed as

$$\tau_r = \frac{F_{sd}}{A_{sp}} = \frac{F}{A} \cos \lambda \cos \psi \qquad (3\text{-}8)$$

It has been hypothesized and experimentally well verified that the magnitude of the resolved shearing stress τ_r required to initiate slip in a pure and perfect single crystal of a given material is a constant at any given temperature. This critical magnitude at which slip is initiated is called the *critical resolved shearing stress*, and the rule expressed in (3-8) is known as *Schmid's law*. Schmid's law has been verified for a large number of different metallic single crystals.

It may have occurred to the reader to question the influence and effect of the *normal stress* component on the magnitude of critical resolved shearing stress. An expression may be written for the normal stress σ_n on the slip plane by considering the normal component F_n of the applied force F as shown in Figure 3.9. This expression becomes

$$\sigma_n = \frac{F_n}{A_{sp}} = \frac{F\cos\psi}{A/\cos\psi} = \frac{F}{A}\cos^2\psi \tag{3-9}$$

Experimental investigations have shown the effect of normal stress on the magnitude of critical resolved shearing stress to be insignificant. However, the

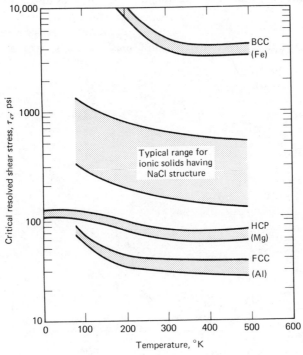

FIGURE 3.10. Temperature variation of critical resolved shearing stress for several classes of crystalline materials. (From ref. 4; Reprinted with permission from John Wiley & Sons, Inc.)

magnitude of the normal stress σ_n will be of interest in any discussion of fracture strength.

In Figure 3.10 the magnitude of critical resolved shearing stress is plotted as a function of temperature for several classes of crystalline materials. As the temperature approaches the melting temperature of the material, the critical resolved shearing stress drops rapidly toward zero.

The phenomena of *strain hardening* and *recovery* also exert a strong influence on the critical resolved shearing stress. *Strain hardening* is the process whereby the stress required to produce further plastic deformation is *increased* because of prior plastic strain. Thus the material becomes harder and stronger in a sense and is said to have been work hardened or strain hardened. *Recovery* is the process by which the stress level required to produce plastic flow is *reduced*. As will be seen, both strain hardening and recovery are explainable by the application of dislocation theory.

The strain hardening process and the recovery process interact and combine their effects in widely different ways, depending on the material, the amount of prior strain, and the temperature. Figure 3.11 illustrates three widely differing strain hardening responses to prior strain as shown by the resolved shear-stress–shear-strain curves. The magnesium is a *linear strain hardening* material, which exhibits a constant hardening rate of low magnitude all the way to fracture. The iron exhibits a decreasing rate of strain hardening over its entire range of deformation and is typical of *parabolic*

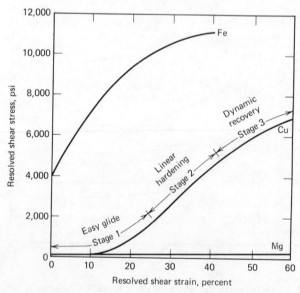

FIGURE 3.11. Three different strain-hardening characteristics exhibited by single crystals of iron, copper, and magnesium. (Adapted from ref. 4; with permission from John Wiley & Sons, Inc.)

strain hardening materials. The copper displays three stages of strain hardening, usually referred to as easy glide (stage 1), linear hardening (stage 2), and
dynamic recovery (stage 3). The characteristics of each of these stages is
demonstrated in Figure 3.11.

Strain hardening variation as a function of temperature is illustrated for
single crystal aluminum in Figure 3.12. The change in balance between the
strain hardening and recovery processes is clearly evident. At lower temperatures the strain hardening process predominates, whereas at higher temperatures the recovery process virtually renders strain hardening ineffective.

It is also of importance to note that recovery is a *time dependent* process.
Figure 3.13 illustrates the difference in recovery of single crystals of zinc
when the load was applied and completely removed for one-half minute
before reloading, as compared with similar tests, except that the load was
applied and removed for a 24-hour period before reloading. The time dependence of recovery is illustrated in the return of the stress-strain behavior to
virtually the same as before testing when a 24-hour rest period was allowed.
When only one-half minute of rest was allowed, the strain hardening effect
and diminished recovery influence are observed as a shift of the stress-strain
curve to a higher position.

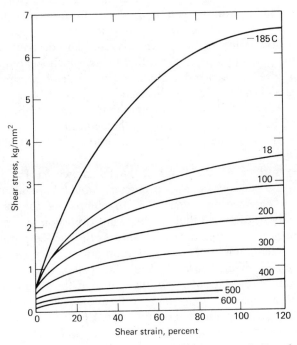

FIGURE 3.12. Changes in the strain-hardening characteristics of single crystal
aluminum as a function of temperature. (From ref. 28)

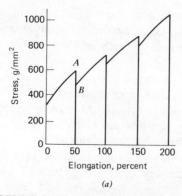

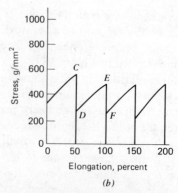

FIGURE 3.13. Stress-strain behavior of single crystal zinc repetitively loaded with intermediate test periods of different duration. (*a*) Load completely removed at point *A* for one-half minute. On subsequent reloading, yielding initiated at point *B*. (*b*) Load completely removed at *C* for 24 hours. On subsequent reloading, yielding initiated at point *D*. (From ref. 29)

Twinning

Plastic deformation by twinning is substantially different from deformation due to slip. Twinning results when, upon the application of a shearing stress, one portion of a crystal lattice takes up an orientation that is a mirror image of the lattice orientation in the parent crystal. (The description here is for mechanical twins as opposed to annealing twins, which sometime form upon the annealing of a cold-worked material.) The result of twinning due to the application of a shearing stress is illustrated in Figure 3.14. The lower sketch depicts a twin band and defines the twinning faces and twinning direction on a gross basis. The upper sketch illustrates the details of the shift in position of the atoms in the twin band to form the mirror image structure of twin deformation.

If the free surface of the twinned crystal shown in Figure 3.14 were polished flat, a noticeable atomic disregistry would be evident, unlike the slipped crystal depicted in Figure 3.8. In comparing slip with twinning, therefore, it may be observed that (1) the twinned portion of a crystal grain is the mirror image of the original lattice, whereas the slipped portion of a grain has the same orientation as the original grain, and (2) slip consists of a step shear displacement of an entire block of the crystal, whereas twinning is a distributed uniform displacement. By the nature of the geometry of twinning it is limited to a maximum deformation of only a few percent, whereas slip may result in a deformation of several hundred percent. It is suspected, but not well proved, that a critical twinning stress exists, just as a critical resolved shearing stress for slip does. The existence of a critical twinning stress is

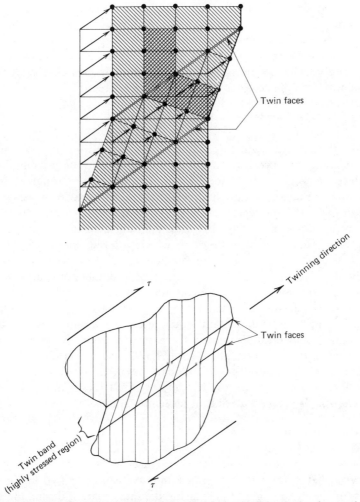

FIGURE 3.14. Sketches illustrating the process of plastic deformation by mechanical twinning. (Adapted from ref. 2; with permission from John Wiley & Sons, Inc.)

difficult to verify since in most cases slip deformation overshadows the twinning process.

Grain Boundary Sliding and Diffusional Creep

At high temperatures and low strain rates, sliding occurs along the grain boundaries of polycrystalline materials. In creep tests on pure metals at low stresses, as much as 30 percent of the total strain may be due to grain boundary sliding.

At temperatures close to the melting point where there is a high equilibrium concentration of lattice vacancies, and self diffusion is rapid, polycrystalline materials may deform by diffusional creep. In this deformation mode atoms diffuse to the boundaries normal to the force and vacancies migrate to boundaries parallel to the force axis. The result is crystal elongation in the direction of the applied force due to diffusional creep.

Effects of Grain Boundaries in Polycrystals

Most of the discussion in this chapter has centered on the deformation of single crystals. The presence of grain boundaries in polycrystalline materials introduces new constraints to the deformations that strongly affect their stress-strain behavior, but qualitatively the general behavior is very similar to single crystals. If the integrity of the grain boundaries is to be preserved, each grain must deform in a way that is compatible with its neighboring grains, and a complex readjustment takes place over a large number of grains.

At low temperatures the grain boundaries in polycrystalline metals are usually stronger than the grains themselves and therefore, in most metals, fracture at low and normal temperatures is *transcrystalline*. That is, the fracture passes across the grain rather than along the grain boundaries. At elevated temperatures the grain boundaries are usually weaker than the grains; high temperature fractures, therefore, are typically *intergranular* fractures. That is, they tend to follow along the grain boundaries. Polycrystalline nonmetals usually exhibit grain boundary strengths lower than the strength within their grains even at low temperature, and intergranular fractures are characteristic for these materials at all temperatures.

Strain Rate Effects

The resistance to plastic deformation increases with an increase in the rate of straining. That is, the tensile stress-strain curve can be shifted to higher stress values for the same strain by increasing the rate of strain during the test. The effect is slight, however, for ordinary rates. For example, Nadai* found that for a 0.35 carbon steel a 10 thousandfold increase in strain rate only doubled the flow resistance. Nevertheless, the effect is of sufficient importance that standard testing speeds have been established for obtaining materials property data so that results from various testing laboratories are directly comparable. Some materials are far more sensitive to strain rate influence than others. The ultimate strength may be significantly increased or remain unaffected. The yield point may be raised significantly, or very little. The ductility and energy-absorbing capacity may respond in widely different ways for different materials and rates of strain—increasing, decreasing, or remaining

*See p. 535 of ref. 5.

unchanged. These responses to strain rate will be examined more thoroughly in a later chapter.

3.5 FRACTURE

Fracture phenomena in crystalline solids are very complex. In many instances fracture takes place only after a complex history of prior plastic deformation. It is clear from the discussion of plastic deformation earlier in this chapter that the deformation of a simple single crystal is difficult to describe, and polycrystalline materials are even more complicated. It should not be surprising to find that the terminal conditions of a complex phenomenon are equally difficult to characterize. If conditions and materials are such that fracture takes place without prior plastic deformation, the analysis is made somewhat easier.

In the discussion of plastic deformation by slip, the concept of a critical resolved shearing stress for the onset of plastic flow was developed. As an adjunct to that analysis it was suggested that an expression for normal stress on the slip plane could be written as

$$\sigma_n = \frac{F}{A}\cos^2\psi \tag{3-9}$$

If a material is so constituted that the normal stress σ_n reaches a magnitude large enough to cause cleavage of the crystal before the resolved shearing stress reaches its critical value, then the crystal will be cleaved into two parts. The magnitude of σ_n that results in cleavage fracture is called the *critical normal fracture stress* and has been shown to be a material constant similar to the critical resolved shearing stress. The rule which states that cleavage results when the expression for σ_n in (3-9) reaches the critical normal fracture stress is known as *Sohncke's law*. As in the case of deformation processes, it is found that the most common cleavage planes are those of high atomic density.

Fracture is generally characterized as being either *brittle* or *ductile*. Brittle fracture manifests itself as the very rapid propagation of a crack after little or no plastic deformation. The speed at which cracks propagate in brittle behavior after initialtion rises rapidly from zero to a limiting velocity of about one-third the speed of sound in the material. In polycrystalline materials the fracture proceeds along cleavage planes within each crystal, giving the fracture surface a granular appearance because of the change in orientation of the crystals and their cleavage planes within the matrix. Brittle fracture sometimes proceeds primarily along grain boundaries and is then called *intergranular fracture*.

Ductile rupture is fracture that takes place after extensive plastic deformation. Ductile rupture proceeds by slow crack propagation resulting from the formation and coalescence of voids. The fracture surface appearance associated with ductile rupture is characteristically dull and fibrous. Three distinct stages are observed in the ductile rupture of most polycrystalline metals. First, the sample begins to "neck down" locally, and small discrete cavities form in the necked region. Next, the cavities coalesce into a crack in the center of the cross section with the direction of the crack generally perpendicular to the direction of applied stress. Finally, the crack spreads to the surface of the sample along shear planes oriented at approximately 45 degrees to the direction of the tensile axis. The result of this sequence of events is often the well known *cup-and-cone* failure surface.

In some materials, notably the body-centered cubic transition metals, use at low temperatures, high rates of strain, or with the presence of certain notches may result in a transition from ductile behavior to brittle behavior. In using such materials it is important to avoid situations in which brittle behavior is induced. A classic example of transition from ductile to brittle behavior is afforded by certain welded ships and tankers of the World War II era that underwent such a transition because of the low temperatures of the North Atlantic and were literally torn in half by a rapidly propagating brittle crack induced by a slight impact loading together with the built-in residual stresses of welding. Other similar examples have been observed in bridges, pressure vessels, smoke stacks, penstocks, and gas transmission pipe lines.

In Section 3.2 it was noted that the estimation of shearing strength of the crystalline metals by theoretical consideration of atomic bonding forces leads to strength estimates of several million psi. For example, in 1929 Frenkel utilized a simple atomic model to estimate the theoretical yield strength to be about one-tenth of Young's modulus of elasticity. This would indicate theoretical yield points around 3×10^6 psi for steel alloys. Observed yield strength values of one or two orders of magnitude less than this are commonly observed in the laboratory. Other materials exhibit even greater discrepancies, ranging to five orders of magnitude in some cases.

Other questions also required answers. For example observed elastic deformations determined experimentally were much greater for a given applied load than would be predicted on a theoretical basis. Crystals exhibited greater strength after deformation than before. Mechanical properties varied with changes in temperature. All of these questions stimulated much investigation in an attempt to explain the behavior.

The discrepancy between theoretical strength and actual strength seemed to imply that something within the matrix of a real material must locally concentrate the stress to an extent that the theoretical failure strength is locally exceeded to initiate failure. A. A. Griffith in 1920 postulated that brittle materials contain many submicroscopic cracks that are caused to grow to a macroscopic size upon the application of a sufficiently high stress, finally causing brittle fracture. The Griffith theory, and other similar theories, assumed that these microcracks or other lattice defects resulted in local stress concentrations. By utilizing the theory of elasticity estimates for stress distribution near the end of the major axis of a flat elliptical hole, Griffith determined the strain energy released by the growth in length of an existing crack. He then utilized the concept that the process of fracturing produces two new surfaces, each having an energy per unit area of, say, W_α. He observed that the energy available for propagating a crack should be the difference between the energy required to form new fracture surface and the strain energy released by the increase in crack length. The Griffith expression for the energy W required to propagate a crack of unit width and length $2c$ was then derived as

$$W = 4cW_\alpha - \frac{\pi(1 - \mu^2)\sigma^2c^2}{E} \tag{3-10}$$

and is plotted as a dashed line in Figure 3.15. It may be noted that μ is Poisson's ratio and E is Young's modulus of elasticity.

When the energy required for crack propagation does not increase with increasing crack length, that is, when $dW/dc = 0$, the crack will become spontaneously self-propagating. This is indicated in Figure 3.15 by noting that crack propagation releases energy in the self-propagating range. From (3-10) it may be determined that the onset of spontaneous crack propagation is predicted when

$$\sigma_p = \sqrt{\frac{2EW_\alpha}{\pi(1 - \mu^2)c}} \tag{3-11}$$

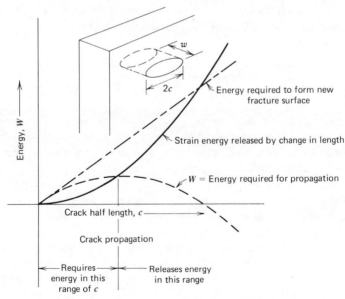

FIGURE 3.15. Sketch showing a Griffith crack and the energy W required to propagate a crack. (Adapted from ref. 17)

where σ_p = tensile stress to propagate a crack
          E = Young's modulus of elasticity
          μ = Poisson's ratio
          W_α = energy per unit area of newly formed crack
         $2c$ = crack length

Griffith checked his theory by producing artificial cracks in sheets of silicate glass and found $\sigma^2 c$ to be a constant, in agreement with (3-11).

Determination of theoretical fracture strengths by employing energy balance equations between strain energy released by cracking on one hand and energy required to form new surface on the other has been widely applied. It was determined independently by Irwin and Orowan in 1948 that when dealing with engineering metals a modification of the Griffith theory is required to account for the inherent ductility. Even when fracture is classified as brittle, they asserted that there is always some plastic flow in the region bordering the fracture surface. It was proposed then that energy irreversibly dissipated as plastic flow per unit area, say W_p, should be added to the surface energy term W_α. Thus, it was proposed that the expression for spontaneous crack propagation should be

$$\sigma_p = \sqrt{\frac{2E(W_\alpha + W_p)}{\pi(1 - u^2)c}} \tag{3-12}$$

However, it was further shown that W_p is orders of magnitude larger than W_α so that (3-12) may be written as the approximate expression

$$\sigma_p \doteq \sqrt{\frac{EW_p}{c}} \tag{3-13}$$

This expression is referred to as the Griffith-Irwin-Orowan criterion for spontaneous crack propagation. This theory was tested by artificially introducing cracks into mild steel specimens, and it was found that the fracture stress σ_p was proportional to $c^{-1/2}$ as predicted.

These developments of Griffith, Irwin, and Orowan did much to improve the agreement between theory and experiment and aid in understanding the behavior of engineering metals under applied loads. Ultimately, two major engineering tools developed from this work, namely, dislocation theory and fracture mechanics. Other theories postulating solid state defects, lattice vacancies, and mosaic block defects were advanced. Yet none of the proposed theories fully explained the discrepancies between theoretical and experimental values of strength without making many questionable assumptions, until dislocation theory was developed.

3.6   AN INTRODUCTION TO DISLOCATION THEORY

In 1934, G. I. Taylor, E. Orowan, and M. Polanyi (26), three investigators working independently, all proposed virtually the same postulate to explain the troublesome discrepancies. They postulated that it is possible for crystal lattice imperfections to exist within the crystal structure that can be moved by the application of suprisingly low stress levels to cause plastic deformation. These lattice imperfections were called *dislocations*, and the concept of *mobility* of the dislocation was the missing ingredient that has ultimately made this theory of dislocations so superior to all preceding theories. Following the postulation of mobile dislocations, more than a quarter century of intensive investigation has led to the firm establishment of the existence of dislocations in all engineering materials. Significant contributions in defining types of dislocations, how they interact, and how they may be generated were made by Frank, Read, Burgers, and Shockley. Dislocations were first observed in the early 1950's by Hedges and Mitchell, who utilized a decorating technique to make them visible in silver halide crystals. Dislocations are now commonly observed by use of the electron microscope, using a transmission technique developed in 1956 by Hirsch, Horne, and Whelan and independently by Bellman. Important contributions continue to be made.

Dislocation Geometry

Metallic crystal lattice defects may be classified into four broad categories including point defects, line defects, surface defects, and volume defects.

A *point defect* is a highly local lattice defect whose influence extends only one or a few atomic diameters from its center. Point defects include vacancies (missing atoms), interstitial atoms, solute atoms, and misplaced atoms in an otherwise ordered super lattice structure. A *line defect* is a *dislocation*. This type of defect will be considered in detail in following paragraphs. A *surface defect* is a plane or a curved surface of defects arrayed within a crystal. Surface defects include grain boundaries, subgrain boundaries, twin boundaries, and stacking faults in the atomic planes within a crystal. A *volume defect* is a three-dimensional defect such as a void, a bubble, a particle oriented differently from the surrounding matrix, or a cluster of point defects in an otherwise perfect matrix.

Of the four types of defects identified, the only one of immediate interest in this discussion is the line defect or *dislocation*. From the simplest geometrical considerations, dislocations are of three types: (1) *edge dislocations*, sometimes called Taylor dislocations; (2) *screw dislocations*, sometimes called Burgers dislocations; and (3) hybrid or *mixed dislocations*. All three types of dislocations are forms of atomic disregistry along a line within the crystal lattice. The mixed dislocation is simply a mixture of the edge and screw types.

To visulaize the *edge dislocation*, consider the rubberlike rectangular parallelepiped of cubic lattice structure shown in Figure 3.16. Imagine the block to be cut half-way through, spread apart at the slit, and an extra half-plane of atoms inserted in the slit. The parallelepiped is then released and allowed to mold itself as best it can back around the added half-plane of atoms. This results in an atomic disregistry near the edge of the extra half-plane of atoms, as shown in Figure 3.16, which is called the *edge dislocation line*. Edge dislocations are defined to be positive if there are $n + 1$ atoms above the line and n atoms below the line. If there are n atoms above the line and $n + 1$ atoms below the line, the edge dislocation is defined to be negative. The symbol $\perp$ is used to indicate a positive edge dislocation and $\top$ is used to denote a negative edge dislocation.

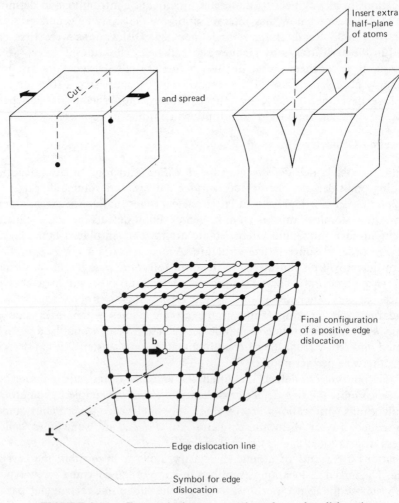

FIGURE 3.16. Geometrical representation of an edge dislocation.

Screw dislocation geometry may also be visualized by considering a rubber-like model as shown in Figure 3.17. Again imagine the rectangular block to be cut half-way through as before. However, rather than spreading the block at the cut surface it is forced to slide upon itself parallel to the direction of the slit and pinned back together with a built-in disregistry of one interatomic distance. This results in an atomic disregistry near the edge of the slit that is called the *screw dislocation line*. The definition of the sense of a screw dislocation is arbitrary, but one convention is to term the screw dislocation right-handed if a clockwise circuit around the dislocation line results in movement away from the observer and left-handed if the circuit results in movement toward the observer.

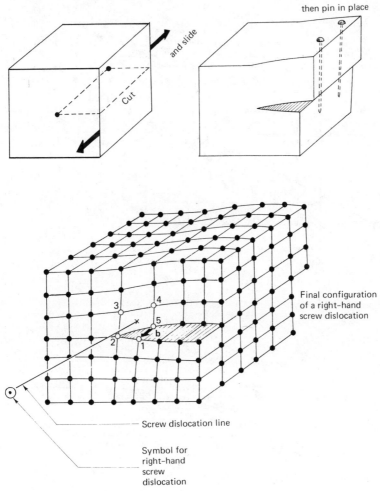

FIGURE 3.17.   Geometrical representation of a screw dislocation.

A hybrid or mixed dislocation is a dislocation having elements of both an edge dislocation and a screw dislocation at various positions along the dislocation line. As illustrated in Figure 3.18, a dislocation line may be pure edge at one place, pure screw at another place, and mixed at still other places. In all cases the dislocation line is the boundary between slipped and unslipped crystal. It is therefore true that a dislocation line cannot end within a region of otherwise perfect crystal. It must terminate at a free surface, a grain boundary, another dislocation line, or some other defect, or it must close back upon itself to form a dislocation *loop*.

When a dislocation passes any given point, the magnitude and direction of the shear displacement that occurs is described by a vector called the *Burgers vector*, **b**. The direction of the Burgers vector **b** with respect to the direction of

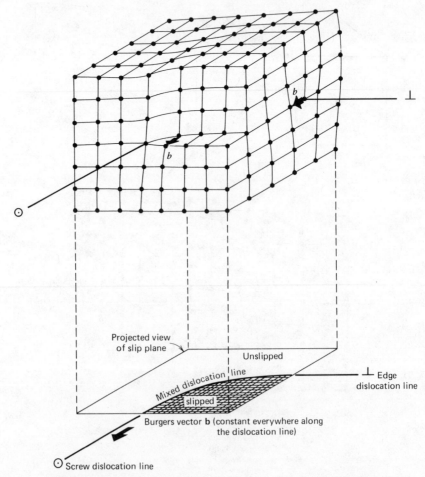

FIGURE 3.18.   Geometrical representation of a mixed dislocation.

the dislocation line characterizes the type of dislocation, and its magnitude defines the magnitude of shear displacement. If the Burgers vector is perpendicular to the dislocation line, the dislocation is of the edge type. If the Burgers vector is parallel to the dislocation line, the dislocation is of the screw type. If the Burgers vector is neither parallel nor perpendicular to the dislocation line, the dislocation is of the mixed type, having components of both edge and screw. The Burgers vector **b** is *invariant* for any given dislocation line, as shown for example in Figure 3.18.

Dislocations may, therefore, be classified in a more formal way, as follows: If a unit vector ξ is used to describe the direction of a dislocation line, then the dislocation is of the edge type if

$$\mathbf{b} \cdot \xi = 0 \qquad\qquad (3\text{-}14)$$

the dislocation is of the screw type if

$$\mathbf{b} \times \xi = 0 \qquad\qquad (3\text{-}15)$$

and the dislocation is of the mixed type if

$$\mathbf{b} \cdot \xi \neq 0 \neq \mathbf{b} \times \xi \qquad\qquad (3\text{-}16)$$

The Burgers vector may be determined for any given dislocation by

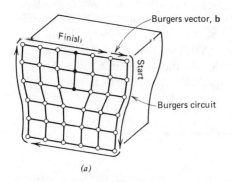

(a)

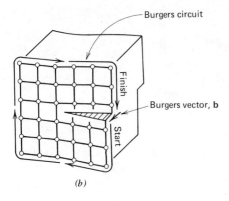

(b)

FIGURE 3.19. Burgers circuits for edge and screw dislocation. (From ref. 4; Reprinted with permission from John Wiley & Sons, Inc.)

constructing the *Burgers circuit*. The Burgers circuit is constructed by moving in a path, atom-by-atom, in a plane normal to the dislocation line all the way around the dislocation and then comparing with a similar atom-by-atom path in a perfect reference crystal, as shown in Figure 3.19. The amount by which the Burgers circuit fails to close as compared with the perfect crystal is the Burgers vector.

Dislocation Motion

Mobility has been cited as the key idea that led to the success of the dislocation model where earlier concepts of deficient crystal strength proved to be inadequate. To visualize the idea of dislocation mobility, consider the simple edge dislocation depicted in Figure 3.20. On application of a shearing stress τ, this positive edge dislocation is moved from left to right across the crystal along the slip plane. Note that the shearing stress must do work in pulling atom 1 farther away from atom 2, but simultaneously atom 3 moves closer to its equilibrium distance from atom 4. In so doing, the pair 3-4 give up an amount of stored elastic strain energy approximately equal to the energy newly stored in the pair 1-2. If the dotted position in Figure 3.20 represents a prior equilibrium position before application of shearing stress τ, the dislocation movement may be visualized. The prior dislocation position was centered at atom 3, whereas atoms 4 and 5 were neighbors. On application of a sufficiently large shearing stress, the relative positions of atoms 3, 4, and 5 will produce a force field of such character that the 4-5 bond will be broken and a new 3-4 bond will be made. At the same time, other energy adjustments will be made in the surrounding atoms that will approximately

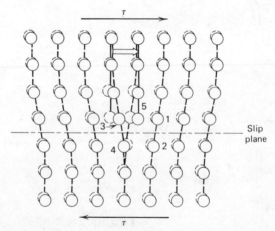

FIGURE 3.20. Sketch showing the rearrangement of atoms in the neighborhood of an edge dislocation as it moves under the influence of an applied shearing stress. (From ref. 4; Reprinted with permission from John Wiley & Sons, Inc.)

balance in terms of energy released versus energy stored. The net effect is that the dislocation moves to the right with a total external energy input much less than the energy that would be required to break the bonds associated with a block movement of all the atoms above the slip plane at one time. The concept is roughly analogous to the relative difficulty of sliding a heavy rug across the floor by application of brute force versus moving the rug by first pulling up a "wrinkle" and then moving the wrinkle easily along the rug until it pops out the other side. The net result in both cases is slip of the rug just as in the case of propagating a dislocation along the slip plane. The creation of a slip step by propagation of edge and screw dislocations is illustrated in Figure 3.21.

Referring again to Figure 3.21, we may observe that the edge dislocation moves parallel to its Burgers vector, whereas the screw dislocation moves perpendicular to its Burgers vector. In the case of the moving edge dislocation, the plane over which slip takes place, often called the *glide plane*, is uniquely determined. The glide plane is defined by its normal, $\mathbf{b} \times \boldsymbol{\xi}$, where $\mathbf{b}$ is the Burgers vector and $\boldsymbol{\xi}$ is the unit vector defining the sense of the dislocation line. Similarly, in the case of a mixed dislocation, the glide plane $\mathbf{b} \times \boldsymbol{\xi}$ is uniquely defined. However, for a screw dislocation $\mathbf{b}$ is parallel to $\boldsymbol{\xi}$ and $\mathbf{b} \times \boldsymbol{\xi}$ is zero; therefore the glide plane is indeterminate. In fact, any

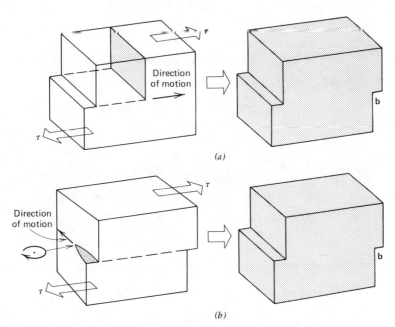

FIGURE 3.21. Creation of a slip step by (*a*) a moving edge dislocation and (*b*) a moving screw dislocation under the influence of applied shearing stress τ. (From ref. 4; Reprinted with permission from John Wiley & Sons, Inc.)

plane for which **b** is a zone axis is a potential glide plane for a screw dislocation. (A zone axis is the line of intersection of a set of planes that meet along a line or parallel lines.) Thus, if a screw dislocation were moving along a given glide plane and encountered an obstacle to further motion, it might avoid the obstacle by changing to another glide plane. Such a change to a new slip plane by a moving screw dislocation is called *cross slip* and is illustrated in Figure 3.22. This type of behavior is not observed in edge or mixed dislocations because their glide planes are uniquely determined. However, if by some means the bottom row of the extra half-plane of atoms of an edge dislocation is removed, or an extra row added to it, the half-plane then terminates on a new parallel slip plane. This process, shown in Figure 3.22, is called *dislocation climb*. Thus, if a moving edge dislocation were to encounter an obstacle on its slip plane, it might be able to *climb* to a new parallel slip plane and continue to move. Dislocation climb takes place by the diffusion of vacancies to the dislocation, as shown in Figure 3.22, or the diffusion of interstitials to the dislocation site. The phenomenon of dislocation climb is diffusion controlled and, therefore, temperature sensitive since the equilibrium concentration of vacancies increases with temperature increase.

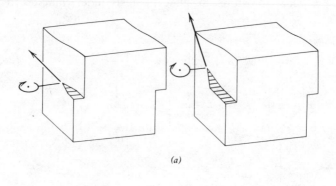

(a)

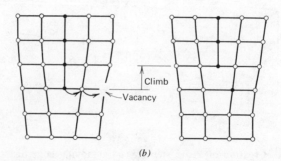

(b)

FIGURE 3.22. (*a*) Cross slip of a screw dislocation and (*b*) climb of an edge dislocation. (From ref. 4; Reprinted with permission from John Wiley & Sons, Inc.)

Dislocation Pinning, Generation, and Interaction

The strain energy of an edge dislocation has been estimated* to be approximately

$$E \doteq \frac{lGb^2}{1 - \nu} \tag{3-17}$$

where E = strain energy of edge dislocation
l = length of dislocation line
G = elastic shear modulus
$b = |\mathbf{b}|$, magnitude of Burgers vector
ν = Poisson's ratio

Further, it has been found that the strain energy of a screw dislocation in most engineering metals is about two-thirds that for an edge dislocation of the same length. From (3-17) the following two observations may be made: (1) The strain energy is proportional to b^2; hence the most stable dislocations (those of lowest energy) are those with minimum Burgers vectors. Since minimum Burgers vectors appear in planes of most closely packed atoms, one would usually expect to find a concentration of dislocations along the close-packed directions. (2) The energy of a dislocation is proportional to its length.

Associated with the concept of strain energy of a dislocation line is the concept of line tension T, which is a vector acting along the dislocation line, whose magnitude is

$$\mathbf{T} = \frac{\partial E}{\partial l} \doteq Gb^2 \tag{3-18}$$

Suppose that a moving edge dislocation is being moved across its slip plane by an applied shearing stress τ when it encounters a pair of obstacles, say two precipitate particles, as shown in Figure 3.23. The virtual normal force on the line segment between the obstacles B and C can be shown[†] to be equal to τbl. This force, which tends to bow the dislocation line out between the two pinning points B and C, must be in equilibrium with the parallel components of the dislocation line tension T so that

$$\tau bl = 2T \sin \theta \tag{3-19}$$

as illustrated in Figure 3.23 where θ is the angle between the line tension vector and the line connecting the pinning points. Substituting for T in (3-19) its expression from (3-18), the shearing stress may be obtained as

$$\tau = \frac{2Gb}{l} \sin \theta \tag{3-20}$$

---

*See pp. 66ff. of ref. 4.
†See pp. 68ff. of ref. 4.

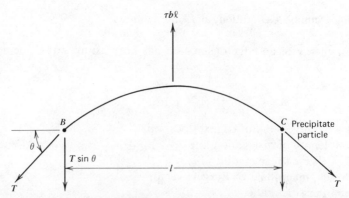

FIGURE 3.23. A schematic representation of a pinned dislocation line under the influence of an applied shearing stress.

From (3-20) it may be observed that to cause greater bowing (that is, to make θ larger) a larger and larger shearing stress is required. Thus, if the dislocation is pinned in this way, one would expect to observe an increase in the applied shearing stress required to cause further slip. This argument holds until θ reaches 90 degrees, at which time the shearing stress reaches its maximum value.

Furthermore, it may be observed from (3-20) that if a moving dislocation does not meet any obstacles or pinning points, the expression for τ is zero since $\sin\theta$ is zero. Thus the stress required to move the dislocation is essentially zero if no pinning points are encountered. If obstacles are encountered in pairs, the required shearing stress to move the dislocation will be greater for closer spacing than for large spacing between pinning points, since τ is inversely proportional to l in (3-20).

It was observed in the preceding paragraph that the shearing stress required to move a single dislocation through a crystal with no built-in obstacles is a very low stress. Also, the plastic strain produced when a single dislocation runs out of the crystal at a free surface is very small, a slip step of magnitude $|b|$ being produced. At first thought it would seem that the application of a relatively small shearing stress should push all of the dislocations out of the crystal, leaving it dislocation free. Early investigators in the field were, therefore, puzzled to find that instead of producing dislocation free crystals by the steady application of a shearing stress, the dislocation density actually seemed to be increased by deformation under the applied shearing stress. For example, in the case of very pure, well annealed single crystals, the dislocation density has been observed to be as low as $10^2 - 10^3$ lines per square centimeter. For typical annealed polycrystalline metals the dislocation density is about $10^7 - 10^8$ lines per square centimeter. By comparison the same polycrystalline metals after severe plastic deformation contain $10^{11} - 10^{12}$

dislocation lines per square centimeter, an increase of about four orders of magnitude.

Dislocation density is expressed as the length of dislocation line per unit volume, centimeters of line per cubic centimeter. Approximately, but not precisely, the same thing is expressed as the number of lines piercing a unit area of a random cross section, lines per square centimeter. The units in both cases are lines per square centimeter, and the two concepts of dislocation density are regarded as essentially equivalent.

The fact that dislocation density is substantially increased by plastic deformation was not successfully explained until 1950 when Frank and Read (27) proposed a mechanism whereby a succession of dislocation lines and loops could be generated and propagated across the slip plane. Figure 3.24

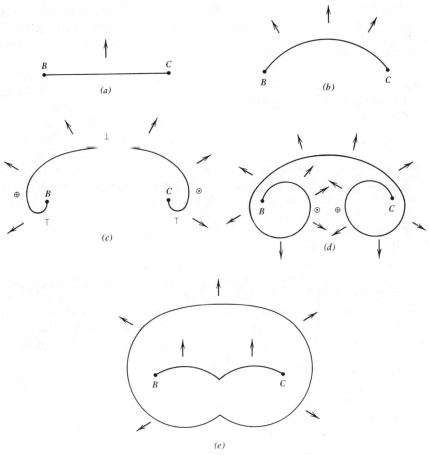

FIGURE 3.24. Generation of dislocation loops by a Frank-Read source.

illustrates the operation of a Frank-Read source. Consider the dislocation line to be pinned at points B and C. The line is forced to bow out as shown in Figure 3.24b by action of the applied shearing force. If the magnitude of the shearing force is large enough, the dislocation line will continue to expand and pass around B and C, as shown in Figure 3.24c. At this stage screw dislocation segments of opposite sense will exist, as shown. As the dislocation line continues to move, the screw dislocations of opposite sense at 1 and 2 attract and annihilate each other and leave a region of perfect lattice, as shown in Figure 3.24d. The remaining parts of the line, being edge dislocations, can lower their energy by joining together as shown in Figure 3.24e. At this stage the original dislocation has formed a complete loop, and in so doing has also reformed the line segment B-C as at the onset. The loop, which is a boundary between slipped and unslipped crystal, continues to grow and finally intersects the surface, where it produces a single unit of slip. The reformed line goes on to generate a succession of new loops and lines that in their turn expand to the free surface to produce additional slip steps. When many such loops add their effects at a free surface, they appear as slip lines on the crystal face.

Without discussing the geometrical details, it may be visualized that dislocation lines and loops moving and being generated on a variety of slip planes, some parallel and some that intersect, must interact with one another. Further, the processes of cross slip and dislocation climb complicate the interaction, producing jogs, helices, and complex three-dimensional interactions, sometimes referred to as a *tangled forest* of dislocations. The more dense the forest, the more energy is required to force a dislocation to pass through the forest, and the line may be immobilized or pinned by its interactions with other dislocations. Thus it may be observed that the increased density of dislocation lines and their interaction to immobilize dislocation motion manifests itself macroscopically as strain hardening. At the same time, the increased stress levels may activate the cross slip mechanism and release dislocation pileups. Also, dislocation climb may be activated, especially as temperatures are increased, to produce the macroscopic phenomenon called recovery. The yield point phenomenon can also be explained as the pileup of dislocations in certain regions until the applied stress level reaches a critical value when the dislocations are dislodged and suddenly move to give macroscopic plastic flow and the attendant sudden drop in load-carrying ability of some machine parts or specimens observed when the yield point is reached. Correlations have been observed between the distribution of dislocations in deformed alloys and their susceptibility to stress corrosion cracking. Crack propagation under conditions of fatigue loading can be qualitatively explained on the basis of dislocation motion and interaction. Creep phenomena are in some measure explainable by dislocation mobility and interaction. Much work remains, however, in providing a quantitative tie between dislocation interaction and macroscopic phenomena.

It must also be observed that all macroscopic behavior has not yet been satisfactorily explained by the dislocation model even in a qualitative sense, though strides are made almost daily in discovering new information of this kind.

3.7   AN INTRODUCTION TO LINEAR ELASTIC FRACTURE MECHANICS

The development of dislocation theory has made great strides in explaining the mechanisms of deformation and fracture of engineering materials at the atomic level, yet it has not provided engineers with the quantitative tools necessary to estimate potentially critical combinations of loading, geometry, and material properties. Thus, while one sector of engineering research concentrated on the microscopic pursuit of dislocation theory, another group worked at the macroscopic level to develop predictive models of fracture in engineering structures and machines, now well known as the area of fracture mechanics. Inspired by the work of Griffith, Orowan, and Irwin, work in the field of fracture mechanics was given impetus by the some 1289 failures (of which 233 were serious) out of 4694 Liberty Ships constructed during the 1940s, the Comet failures of the early 1950s, missile tank failures and large steam turbine failures of the mid-1950s, the Bomark helium tank failure in 1960, the large solid rocket motor case failures and Apollo tank failures of the mid-1960s, and others, though fewer in number, as recently as the mid-1970s.

The reasons behind these failures involved both the development of new materials and a thrust toward higher efficiency in design. The introduction of higher strength structural alloys, the use of welding, the use of thicker sections in some cases, and the use of limit design procedures combined their influence to reduce toward a critical level the capacity of structural members to accommodate local plastic strain without fracture. At the same time, fabrication by welding, residual stresses due to machining, and high-production assembly mismatch increased the need for accommodating local plastic strain to prevent failure. Fluctuating service loads of greater severity and more aggressive environments also contributed to unexpected fractures. Thus, the basic concepts of *fracture control* began to evolve. Fracture control consists, simply, of controlling the combination of nominal stress and existing crack size so that they are always below a critical level for the material being used in a given machine part or structure.

An important observation in studying fracture behavior is that the magnitude of the nominal applied stress which causes fracture is related to the size of the crack or cracklike flaw within the structure (12). For example, observations of the behavior of central through-the-thickness cracks, oriented normal to the applied tensile stress, in steel and aluminum plates yielded the results shown in Figure 3.25 and 3.26. In these tests, as the tensile loading on the

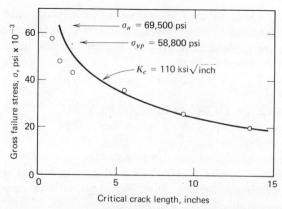

FIGURE 3.25. Influence of crack length on gross failure stress for center cracked aluminum plate, 24 in. wide, 0.1 in. thick, room temperature, 2219-T87 aluminum alloy, longitudinal direction. (After ref. 12, copyright ASTM; adapted with permission.)

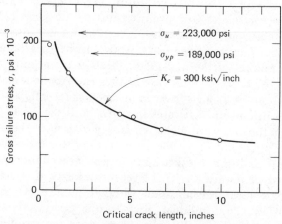

FIGURE 3.26. Influence of crack length on gross failure stress for center cracked steel plate, 36 in. wide, 0.14 in. thick, room temperature, 4330 M steel, longitudinal direction. (After ref. 12, copyright ASTM; adapted with permission.)

precracked plates was slowly increased, the crack extension slowly increased for a time and then abruptly extended to failure by rapid crack propagation. Slow stable crack growth was characterized by speeds of the order of fractions of an inch per minute. Rapid crack propagation was characterized by speeds of the order of hundreds of feet per second. The data of Figures 3.25 and 3.26 indicate that for longer initial crack lengths the fracture stress, that is, the stress corresponding to the onset of rapid crack extension, was lower. For the aluminum alloy the fracture stress was less than the yield strength for cracks longer than about three-quarter inch. For the steel alloy of Figure 3.26, the fracture stress was less than the yield strength for cracks longer than about one-half inch. In both cases, for shorter cracks the fracture stress approaches the ultimate strength of the material determined from a conventional uniaxial tension test.

Experience has shown that the abrupt change from slow crack propagation to rapid crack propagation establishes an important material property that is termed *fracture toughness*. The fracture toughness may be used as a design criterion in fracture prevention, just as the yield strength is used as a design criterion in prevention of yielding of a ductile material under static loading.

In many cases slow crack propagation is also of interest, especially under conditions of fluctuating loads and/or aggressive environments. Later, in Chapter 8, in discussions related to fatigue failure phenomena, characterization of the rate of slow crack extension and the initial flaw size, together with critical crack size, are used to determine the useful life of any component or structure subjected to fluctuating loads.

The most successful approach to prediction and prevention of fracture has been to model the behavior at the crack tip as simply as possible, yet include all *significant* measurable or calculable variables, such as crack length, state of stress, and fracture toughness. The potential influence of temperature, environment, loading rate, and fluctuating loads also must be ultimately recognized in terms of their influence on the variables and the response of the basic fracture relationships.

The simplest useful model for stress at the tip of a crack is based on the assumptions of linear elastic material behavior and a two-dimensional analysis; thus the procedure is often referred to as *linear elastic fracture mechanics*. Although the validity of the linear elastic assumption may be questioned in view of plastic zone formation at the tip of a crack in any real engineering material, as long as "small scale yielding" occurs, that is, as long as the plastic zone size remains small compared to the dimensions of the crack, the linear elastic model gives good engineering results, especially if a small correction factor is employed to account for crack-tip plasticity. Thus, the small scale yielding concept implies that the small plastic zone is confined within a linear elastic field surrounding the crack tip. If the material properties, section size, loading conditions, and environment combine in such a way that "large scale" plastic zones are formed, the basic assumptions of linear elastic fracture mechanics are violated, and elastic-plastic fracture mechanics methods must be employed. Although several promising elastic-plastic fracture mechanics analyses are available,* they are all relatively new and require further development before proven design tools emerge.

In developing stress field expressions for the regions around crack tips, it was recognized that the free crack surfaces have a primary influence on the stress *distribution* around a crack tip, whereas other remote boundaries and

---

*See chap. 6 of ref. 19.

applied forces affect only the *intensity* of the local stress field near the crack tip.

Three basic types of stress fields can be defined for crack-tip stress analysis, each one associated with a distinct mode of crack deformation, as illustrated in Figure 3.27. The crack opening mode, Mode I, is associated with local displacement in which the crack surfaces move directly apart, as shown in Figure 3.27a. The edge sliding or forward sliding mode, Mode II, is developed when crack surfaces slide over each other in a direction perpendic-

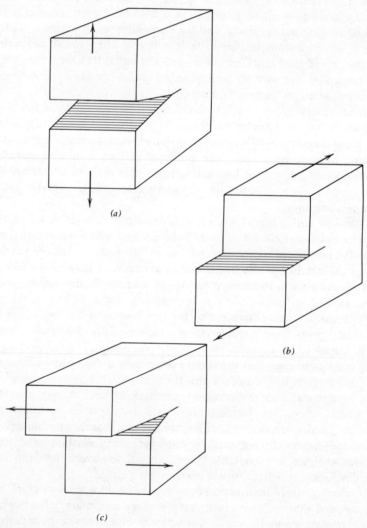

(a)

(b)

(c)

FIGURE 3.27. Basic modes of crack displacement. (*a*) Mode I. (*b*) Mode II. (*c*) Mode III.

ular to the leading edge of the crack, as shown in Figure 3.27b. The side sliding, or tearing, or parallel shear mode, Mode III, is characterized by crack surfaces sliding with respect to each other in a direction parallel to the leading edge of the crack, as shown in Figure 3.27c. Superposition of these three modes will fully describe the most general three-dimensional case of local crack tip deformation and stress field.

Based on the methods developed by Westergaard (13), Irwin (14) developed the two-dimensional stress field and displacement field equations for each of the three modes depicted in Figure 3.27, expressing them in terms of the coordinates shown in Figure 3.28. The stress notation is defined in Section 4.2.

For Mode I, crack opening, the stress components in the crack-tip stress field are

$$\sigma_x = \frac{K_I}{\sqrt{2\pi r}} \cos\frac{\theta}{2}\left[1 - \sin\frac{\theta}{2}\sin\frac{3\theta}{2}\right] + \sigma_{x0} + [O]r^{1/2} \qquad (3\text{-}21)$$

$$\sigma_y = \frac{K_I}{\sqrt{2\pi r}} \cos\frac{\theta}{2}\left[1 + \sin\frac{\theta}{2}\sin\frac{3\theta}{2}\right] + [O]r^{1/2} \qquad (3\text{-}22)$$

$$\tau_{xy} = \frac{K_I}{\sqrt{2\pi r}} \sin\frac{\theta}{2}\cos\frac{\theta}{2}\cos\frac{3\theta}{2} + [O]r^{1/2} \qquad (3\text{-}23)$$

For conditions of plane strain, where displacements in the z direction are constrained to be zero (thick members), the remaining three stress components are

$$\sigma_z = \nu(\sigma_x + \sigma_y) \qquad (3\text{-}24)$$

$$\tau_{xz} = 0 \qquad (3\text{-}25)$$

$$\tau_{yz} = 0 \qquad (3\text{-}26)$$

For Mode II, forward sliding, the stress components in the crack-tip stress field are

$$\sigma_x = \frac{-K_{II}}{\sqrt{2\pi r}} \sin\frac{\theta}{2}\left[2 + \cos\frac{\theta}{2}\cos\frac{3\theta}{2}\right] + \sigma_{x0} + [O]r^{1/2} \qquad (3\text{-}27)$$

$$\sigma_y = \frac{K_{II}}{\sqrt{2\pi r}} \sin\frac{\theta}{2}\cos\frac{\theta}{2}\cos\frac{3\theta}{2} + [O]r^{1/2} \qquad (3\text{-}28)$$

$$\tau_{xy} = \frac{K_{II}}{\sqrt{2\pi r}} \cos\frac{\theta}{2}\left[1 - \sin\frac{\theta}{2}\sin\frac{3\theta}{2}\right] + [O]r^{1/2} \qquad (3\text{-}29)$$

and, for plane strain conditions,

$$\sigma_z = \nu(\sigma_x + \sigma_y) \qquad (3\text{-}30)$$

$$\tau_{xy} = 0 \qquad (3\text{-}31)$$

$$\tau_{yz} = 0 \qquad (3\text{-}32)$$

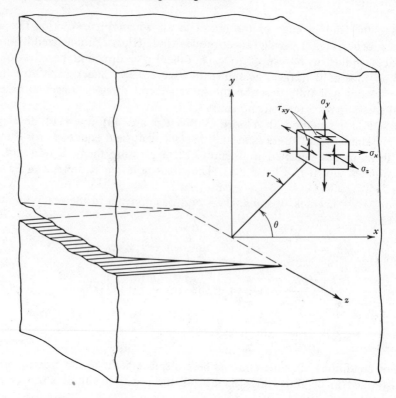

FIGURE 3.28.   Coordinates measured from leading edge of a crack.

For Mode III, side sliding, the stress components in the crack-tip stress field are

$$\tau_{xz} = \frac{K_{\mathrm{III}}}{\sqrt{2\pi r}} \sin\frac{\theta}{2} + \tau_{xz0} + [O]r^{1/2} \qquad (3\text{-}33)$$

$$\tau_{yz} = \frac{K_{\mathrm{III}}}{\sqrt{2\pi r}} \cos\frac{\theta}{2} + [O]r^{1/2} \qquad (3\text{-}34)$$

$$\sigma_x = \sigma_y = \sigma_z = \tau_{xy} = 0 \qquad (3\text{-}35)$$

In expressions (3-21) through (3-35), higher order terms such as the uniform stresses parallel to cracks, σ_{x0} and τ_{xz0}, and terms of the order of $r^{1/2}$, that is, $[O]r^{1/2}$, are indicated. These terms are usually neglected as higher order compared to the leading $1/\sqrt{r}$ term. The parameters K_{I}, K_{II}, and K_{III} are called crack-tip stress field intensity factors, or simply *stress intensity factors*. They represent the *strength* of the stress field surrounding the tip of the crack. Physically, K_{I}, K_{II}, and K_{III} may be interpreted as the intensity of load transmittal through the crack-tip region, induced by the introduction of a crack into a flaw-free body. Since fracture is induced by the crack-tip stress field, the stress intensity factors are primary correlation parameters in current practice.

An important feature of these results is that, in a plane member of any shape made of homogeneous linear elastic material, the stress distributions close to the tip (i.e., for r small compared with a) of a flat planar sharp crack are given by (3-21) through (3-25). Thus, a crack generates its own stress field, which differs from another crack-tip stress field by only a scaling factor, K. This scaling factor K, called stress intensity factor, involves the external loading and the geometry of the plate, including crack size. In Figure 3.29, the external loading is the stress σ applied at infinity, and the only geometric factor is the crack length, $2a$. Purely from dimensional considerations in (3-21) through (3-23), the stress intensity factor K must, in all cases, be of the form

$$K = C_1 \sigma \sqrt{a} \qquad (3\text{-}36)$$

For the infinite plane shown, the theory of elasticity solution gives $C_1 = \sqrt{\pi}$ and (3-36) becomes

$$K = \sigma \sqrt{\pi a} \qquad (3\text{-}37)$$

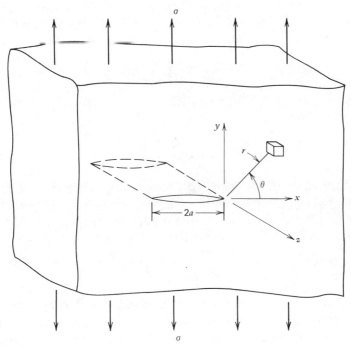

FIGURE 3.29.   Coordinate system for infinite plate containing a through-the-thickness crack of length $2a$.

For remote loadings in general, the expressions for stress intensity factor are of the form

$$K = C\sigma\sqrt{\pi a} \tag{3-38}$$

where C is dependent upon the type of loading and the geometry away from the crack. Much work has been completed in determining values of C for a wide variety of conditions.*

For a given cracked plate, for example Figure 3.29, the factor K increases proportionally with gross nominal stress and also is a function of the instantaneous crack length. Thus, K is a single parameter measure of the stress field around the crack tip. The value of K associated with the onset of rapid crack extension has been designated the *critical stress intensity*, K_c. As noted earlier in Figures 3.25 and 3.26, the onset of rapid crack propagation for specimens with different initial crack lengths occurs at different values of gross-section stress, but at a constant value of K_c. Thus, K_c provides a single parameter fracture criterion that allows the prediction of fracture based on (3-38). That is, *fracture is predicted to occur if*

$$C\sigma\sqrt{\pi a} \geq K_c \tag{3-39}$$

Often, a simple adjustment factor is applied to (3-38) to account for small scale plasticity at the tip of the crack. The simplest approach is to estimate the extent of yielding by treating the problem as one of plane stress and equating the y component of stress in (3-22) to the yield strength of the material, σ_{yp}. For $\theta = 0$, (3-22) gives

$$\sigma_{yp} = \frac{K}{\sqrt{2\pi r}} \tag{3-40}$$

Then, solving for r at yielding and defining it to be the plane stress plastic zone adjustment factor $r_{Y\sigma}$,

$$r_{Y\sigma} = \frac{1}{2\pi}\left(\frac{K}{\sigma_{yp}}\right)^2 \tag{3-41}$$

An estimate of plastic zone size under plane strain conditions may be made by considering the effective increase in yield strength by a factor of approximately $\sqrt{3}$ due to the plane strain elastic constraint (20). Thus, the plane strain plastic zone adjustment factor $r_{Y\epsilon}$ becomes

$$r_{Y\epsilon} = \frac{1}{6\pi}\left(\frac{K}{\sigma_{yp}}\right)^2 \tag{3-42}$$

Stress redistribution accompanies plastic yielding and causes the plastic zone to extend approximately $2r_Y$ ahead of the real crack tip to satisfy equilibrium conditions. Because of the crack-tip plastic deformation, the

*See, for example, ref. 15.

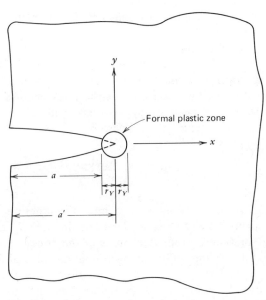

FIGURE 3.30. Definition of formal plastic zone at crack tip.

crack is "blunted," as indicated in Figure 3.30, and the stresses in the surrounding elastic medium "focus" on a virtual crack tip within the plastic zone as the "effective" elastic crack tip. To account for this effect, then, an effective crack length a' is inserted into (3-38) where

$$a' = a + r_Y \tag{3-43}$$

to yield

$$K = C\sigma\sqrt{\pi(a + r_Y)} \tag{3-44}$$

It should be noted that (3-44) requires an iterative solution since K depends on the magnitude of r_Y and r_Y depends on the magnitude of K. For stress levels and crack sizes well below critical values, the magnitude of r_Y is typically very small and often neglected in practice. For cases in which the stress intensity factor K approaches its critical magnitude K_c, the plastic zone correction factor becomes significant and should be included. However, it should be pointed out that if the plastic zone becomes "large" relative to the crack size, that is if r_Y is more than a few percent of a, the accuracy of linear elastic fracture mechanics techniques becomes questionable, and elastic-plastic fracture mechanics precedures should be considered.

In studying material behavior, one finds that for a given material, depending upon the state of stress at the crack tip, the critical stress intensity K_c decreases to a lower limiting value as the state of strain approaches the condition of plane strain. This lower limiting value defines a basic material property K_{Ic}, the *plane strain fracture toughness* for the material. Standard test methods have been established for the determination of K_{Ic} values (16). A few data are shown in Table 3.1.

For the plane strain fracture toughness K_{Ic} to be a valid failure prediction criterion for a specimen or a machine part, plane strain conditions must exist at the crack tip; that is, the material must be *thick* enough to ensure plane strain conditions. It has been estimated empirically that for plane strain conditions the minimum material thickness B must be

$$B \geq 2.5 \left(\frac{K_{Ic}}{\sigma_{yp}} \right)^2 \tag{3-45}$$

If the material is not thick enough to meet the criterion of (3-45), plane stress is a more likely state of stress at the crack tip; and K_c, the critical stress intensity factor for failure prediction under plane stress conditions, may be from two to ten times higher than the lower limiting value of K_{Ic}. Under such conditions, although it would be on the safe side from the failure prediction standpoint to use the value K_{Ic} as a failure criterion, it would be more efficient, and a better design approach, to try to employ elastic-plastic fracture mechanics methods.

It should be pointed out that the concepts of linear elastic fracture mechanics may be developed independently either in terms of stress intensity factor K, as has been done in the preceding presentation, or in terms of "crack extension force" or "strain energy release rate" G, the strain energy released by an incremental increase in crack length, shown as the final term of (3-10). Although the stress intensity factor approach is preferred for the purpose of this text, there are cases in which the strain energy release rate approach may be more useful, as, for example, in cases where different crack displacement modes are simultaneously induced, in compliance testing techniques, or in certain elastic-plastic fracture mechanics methods. The concept of a critical value of strain energy release rate, G_c, at which a crack becomes unstable or self-propagating, is discussed in the literature* and may be related directly to the concept of critical stress intensity factor K_c. In any case, the stress intensity factor K and strain energy release rate G are directly related as follows:

$$G = \frac{K^2}{E} \quad \text{(for plane stress)} \tag{3-46}$$

and

$$G = \frac{K^2(1 - \nu^2)}{E} \quad \text{(for plane strain)} \tag{3-47}$$

*See, for example ref. 18 or 19.

Table 3.1. Yield Strength and Plane Strain Fracture Toughness Data for Selected Engineering Alloys (18, 21)

Alloy	Form	Test Temperature		σ_{yp}		K_{Ic}	
		°F	°C	ksi	MPa	ksi$\sqrt{\text{in}}$	MPa$\sqrt{\text{m}}$
4340 (500°F temper) steel	Plate	70	21	217–238	1495–1640	45–57	50–63
4340 (800°F temper) steel	Forged	70	21	197–211	1360–1455	72–83	79–91
D6AC (1000°F temper) steel	Plate	70	21	217	1495	93	102
D6AC (1000°F temper) steel	Plate	−65	−54	228	1570	56	62
A 538 steel				250	1722	100	111
2014-T6 aluminum	Forged	75	24	64	440	28	31
2024-T351 aluminum	Plate	80	27	54–56	370–385	28–40	31–44
7075-T6 aluminum				85	585	30	33
7075-T651 aluminum	Plate	70	21	75–81	515–560	25–28	27–31
7075-T7351 aluminum	Plate	70	21	58–66	400–455	28–32	31–35
Ti-6Al-4V titanium	Plate	74	23	119	820	96	106

59

3.8   USE OF FRACTURE MECHANICS IN DESIGN

In predicting failure or designing a part so that failure will not occur, a designer must, at an early stage, identify the probable mode of failure, employ a suitable "modulus" by which severity of loading and environment may be represented analytically, select a material and geometry for the proposed part, and obtain pertinent critical material strength properties related to the probable failure mode. He must next calculate the magnitude of the selected "modulus" under applicable loading and environmental conditions and compare the calculated magnitude of the modulus with the proper critical material strength property. Failure is predicted to occur if the magnitude of the selected modulus equals or exceeds the critical material strength parameter.

For example, if a designer determines yielding to be a potential failure mode for his part, he would probably select stress (σ) as his "modulus" and the uniaxial yield point strength (σ_{yp}) as the critical material strength parameter. He would then assess the quality of his design by asserting that *failure is predicted to occur if*

$$\sigma \geq \sigma_{yp} \tag{3-48}$$

As discussed more fully in Chapter 6, he might choose strain, strain energy per unit volume, or something else as the "modulus" by which he judges loading severity on his part in terms of its critical failure strength.

The fracture mechanics approach is useful to the designer in precisely the same way when brittle fracture is a possible failure mode. He would select stress intensity factor K as his "modulus" and fracture toughness K_c as the appropriate critical strength parameter and assert that *failure is predicted to occur if*

$$K \geq K_c \tag{3-49}$$

Although the details of calculating K and determining K_c for some cases may be difficult, the basic concept of predicting failure by brittle fracture is no more complicated than this. It is worth noting that in most cases a designer would be well advised to consider *both* the possibility of failure by brittle fracture and also the possibility of failure by yielding as a routine procedure.

To utilize (3-49) as a design or failure prediction tool, the stress intensity factor must be determined for the particular loading and geometry of the part or structure under investigation. To illustrate the procedure, several configurations are mentioned here, with many more solutions available in the literature.*

For *central through-the-thickness cracks, double edge through-the-thickness cracks*, and *single edge through-the-thickness cracks* under *direct tension loading*

---

*See, for example, refs. 15, 21, and 22.

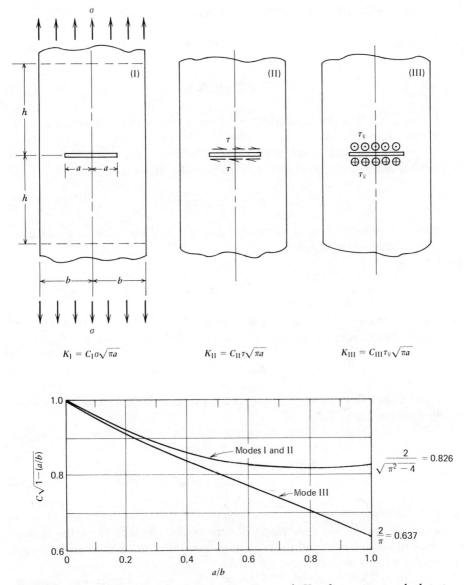

$$K_{\mathrm{I}} = C_{\mathrm{I}}\sigma\sqrt{\pi a} \qquad K_{\mathrm{II}} = C_{\mathrm{II}}\tau\sqrt{\pi a} \qquad K_{\mathrm{III}} = C_{\mathrm{III}}\tau_\varrho\sqrt{\pi a}$$

FIGURE 3.31. Stress intensity factors K_{I}, K_{II}, and K_{III} for center-cracked test specimen. (From ref. 15, copyright Del Research Corp.; adapted with permission.)

or *shear loading*, the form of the stress intensity factor is

$$K = C\sigma_t\sqrt{\pi a} \tag{3-50}$$

or

$$K = C\tau\sqrt{\pi a} \tag{3-51}$$

where C is a function of geometry and crack displacement mode, as given in Figures 3.31, 3.32, and 3.33.

For a *beam* with a *single through-the-thickness edge crack* under a *pure bending moment*, the form of the stress intensity factor is

$$K_I = C_1\sigma_b\sqrt{\pi a} \tag{3-52}$$

where C_1 is a function of geometry, as given in Figure 3.34, and the gross section bending stress σ_b is

$$\sigma_b = \frac{6M}{tb^2} \tag{3-53}$$

For a *through-the-thickness crack emanating from a circular hole* in an infinite plate under *biaxial tension*, the form of the stress intensity factor is

$$K_I = C_I\sigma\sqrt{\pi a} \tag{3-54}$$

where C_I is a function of geometry and the ratio of biaxial stress components, as shown in Figure 3.35.

For a *part-through thumbnail surface crack* in a plate subjected to *uniform tension loading*, the form of the stress intensity factor is

$$K_I = \frac{1.12}{\sqrt{Q}}\sigma_t\sqrt{\pi a} \tag{3-55}$$

where Q is a surface flaw shape parameter that depends upon the ratio of crack depth to length and the ratio of nominal applied stress to yield strength of the material, as shown in Figure 3.36.

Utilizing (3-50) through (3-55) and other similar expressions from the literature, together with fracture toughness properties for the material of interest, a designer may utilize (3-49) to predict failure or, more importantly, to design his part so that failure will not occur under service loading. It should be reiterated here that fracture toughness is not only a function of metallurgical factors such as alloy composition and heat treatment, but a function of service temperature, loading rate, and state of stress in the vicinity of the crack tip as well.

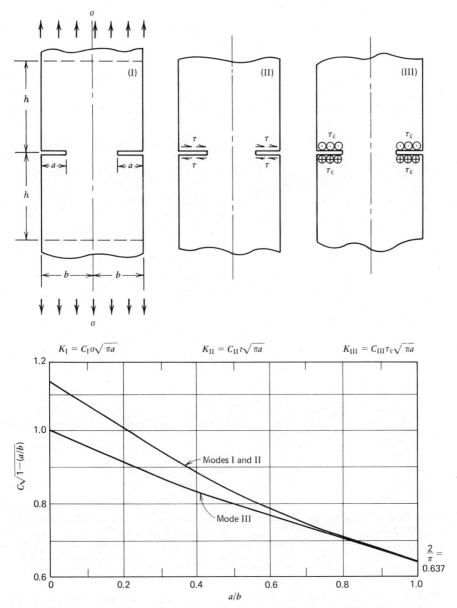

FIGURE 3.32. Stress intensity factors K_I, K_{II}, and K_{III} for double edge notch test specimen. (From ref. 15, copyright Del Research Corp.; adapted with permission.)

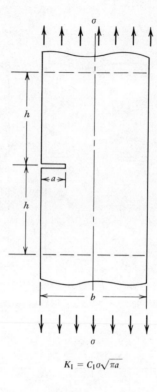

$$K_I = C_I \sigma \sqrt{\pi a}$$

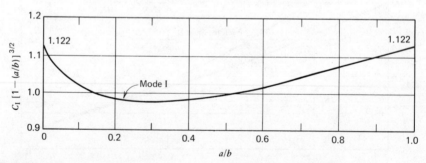

FIGURE 3.33. Stress intensity factor K_I for single edge notch test specimen. (From ref. 15, copyright Del Research Corp.; reprinted with permission.)

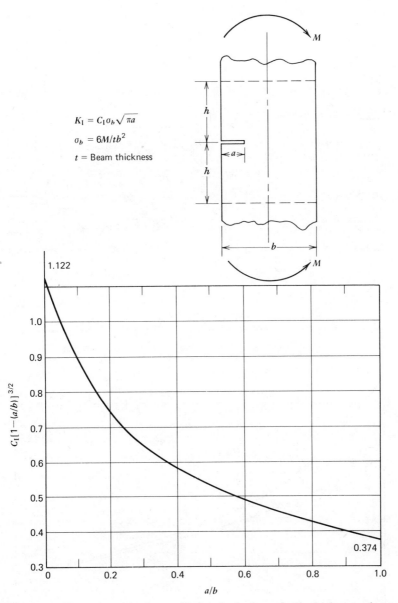

FIGURE 3.34. Stress intensity factor K_I for single through-the-thickness edge crack under pure bending moment. (From ref. 15, copyright Del Research Corp.; adapted with permission.)

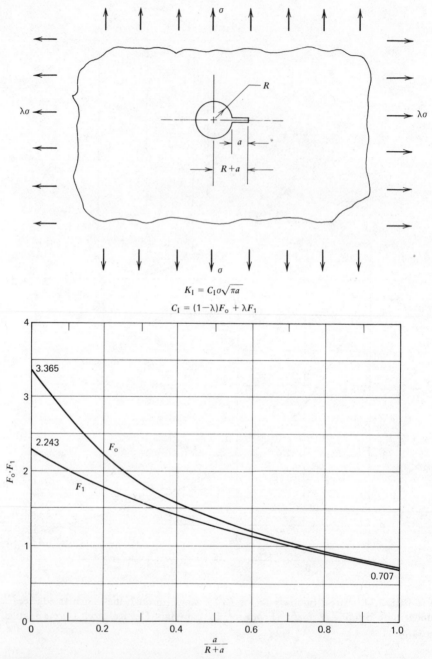

$$K_I = C_I \sigma \sqrt{\pi a}$$

$$C_I = (1-\lambda)F_o + \lambda F_1$$

FIGURE 3.35. Stress intensity factor K_1 for a through-the-thickness crack emanating from a circular hole in an infinite plate under biaxial tension. (From ref. 15, copyright Del Research Corp.; adapted with permission.)

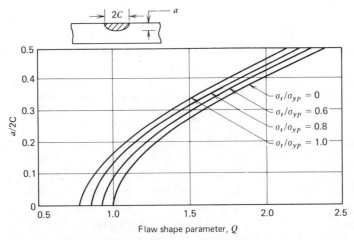

FIGURE 3.36. Surface flaw shape parameter. (From ref. 19; Stanley T. Rolfe, John M. Barsom, *Fracture and Fatigue Control In Structures*, © 1977, p. 41. Adapted by permission of Prentice-Hall, Inc., Englewood Cliffs, New Jersey.)

3.9 ELASTIC-PLASTIC FRACTURE MECHANICS

In many practical applications the plastic zone size ahead of the crack tip becomes so large that the assumption of small scale yielding is no longer valid, and the linear elastic analyses just described are no longer appropriate. A large percentage of low strength and medium strength steels used in the fabrication of modern ships, bridges, pressure vessels, structures, and machine parts are typically used in section sizes that are too thin to maintain plane strain conditions at the crack tip. The use of linear elastic fracture mechanics under such circumstances and the use of K_{Ic} for the failure criterion are, in these cases, not valid. Efforts to extend the techniques of fracture mechanics into this elastic-plastic regime have led to several promising procedures,* including (1) crack opening displacement (COD) methods, (2) R-curve methods, and (3) J-integral methods. Although it is beyond the scope of this discussion to develop these methods in detail, a brief conceptual description may be useful.

The basic concept of the COD method is that fracture behavior in the region of a sharp crack may be described in terms of the opening displacement of the crack faces, that is, in terms of crack opening displacement (23). For plane strain conditions a relationship between COD measurements and plane strain fracture toughness K_{Ic} may then be established. Because COD measurements can also be made for conditions where plane strain no longer

*See, for example, chap. 4 of ref. 19.

holds, that is, for large scale yielding, it has been proposed that quantitative methods similar to those of linear elastic fracture mechanics might be developed to predict critical stress levels or critical crack sizes under such conditions. Although much progress has been made, there seems to be a wide variation in the ability of COD measurements to accurately predict failure stresses in such structures as flawed pressure vessels (24). Many factors seem to be involved in the prediction inaccuracies, but the major factor is apparently related to the occurrence of slow stable crack extension prior to instability. The development of the R-curve concept to take slow stable crack extension into consideration holds promise for improving failure prediction accuracy under these conditions of large scale yielding around the crack tip.

Resistance curves, or R curves, are a means of characterizing the resistance of materials to failure during the period of slow stable crack extension under the influence of increasing applied loads. Under plane strain conditions the fracture toughness K_{Ic} of a material depends on only two variables: temperature and strain rate. In contrast, under plane stress conditions the fracture toughness K_c depends on not only temperature and strain rate but section thickness and crack size as well. An R curve describes the complete variation of K_c with change in crack size from its initial value. Thus, an R curve is a plot of crack growth resistance K_R as a function of crack extension for specified values of temperature, rate of loading, and section thickness. The state-of-the-art in R-curve testing is described in a recent ASTM special technical publication (25).

The J-integral technique proposes to characterize the stress and strain fields at the tip of a crack by a path-independent line integral close to the crack tip, which is evaluated by substituting an integration path that lies a distance away from the crack-tip yield zone. Behavior in the crack-tip region is then inferred by analyzing the region away from the crack tip. For linear elastic behavior, the J integral is identical to G, the energy release rate per unit crack extension, and $J_{Ic} = G_{Ic} = (1 - \nu^2)K_{Ic}^2/E$ are equivalent failure criteria under conditions of plane strain. Techniques for utilizing the J integral as a tool for extending the methods of linear elastic fracture mechanics to the regime of elastic-plastic behavior are being investigated and appear to hold promise as a future design tool.

3.10   USING THE IDEAS

To illustrate the approach to a relatively simple problem in which fracture mechanics methods would provide useful insight, consider the following example: A 7075-T6 aluminum member of rectangular cross section, which is 0.38 inch thick by 4 inches wide, is 20 inches long between pinned ends. The member is to be subjected to static tensile loading during operation. If inspection of this member has shown a 0.15 inch single edge through-the-thickness crack, what is the maximum allowable tensile load that may be supported by the member if a safety factor of 1.5 is required?

As a first step it should be recognized that at least two failure modes should be investigated namely, failure by yielding and failure by brittle fracture due to rapid crack extension.

To consider yielding first, the maximum allowable load $P_{y\text{-all}}$ may be calculated as

$$P_{y\text{-all}} = \sigma_{\text{all}} A_n = \left(\frac{\sigma_{yp}}{n}\right) A_n \qquad (3\text{-}56)$$

Using $\sigma_{yp} = 85{,}000$ psi from Table 3.1, $n = 1.5$ as specified, and the net cross-sectional area,

$$P_{y=\text{all}} = \left(\frac{85{,}000}{1.5}\right)(0.38)(4 - 0.15) = 82{,}900 \ 1b \qquad (3\text{-}57)$$

Next, considering brittle fracture by rapid crack extension, $P_{bf\text{-all}}$ may be calculated as

$$P_{bf\text{-all}} = \sigma_{bf\text{-all}} A_g = \left(\frac{\sigma_{bf}}{n}\right) A_g \qquad (3\text{-}58)$$

Using the same safety factor of $n = 1.5$ and the gross cross-sectional area, it remains to determine the magnitude of σ_{bf} by using (3-49) with proper values of K and K_c. For the single edge through-the-thickness crack under tensile loading, (3-50) gives the proper expression for K. From (3-45) it may be noted that plane strain conditions will be met if

$$B \geq 2.5 \left(\frac{K_{Ic}}{\sigma_{yp}}\right)^2 = 2.5 \left(\frac{30}{85}\right)^2 = 0.31 \text{ inch} \qquad (3\text{-}59)$$

using materials property data from Table 3.1. Since the 0.38 inch actual thickness "B" of the member does exceed 0.31 inch, plane strain conditions hold and K_{Ic} is the correct value to use for K_c in (3-49). Dropping the inequality sign in (3-49) and substituting (3-50) for K then gives the equation for incipient failure by brittle fracture as

$$C\sigma_{bf}\sqrt{\pi a} = K_{Ic} \qquad (3\text{-}60)$$

or, solving for σ_{bf},

$$\sigma_{bf} = \frac{K_{Ic}}{C\sqrt{\pi a}} \qquad (3\text{-}61)$$

From Figure 3.33 it may be determined that for

$$\frac{a}{b} = \frac{0.15}{4} = 0.0375 \qquad (3\text{-}62)$$

$$(1 - 0.0375)^{3/2} C = 1.07 \qquad (3\text{-}63)$$

or

$$C = 1.13 \tag{3-64}$$

Thus from (3-61)

$$\sigma_{bf} = \frac{30,000}{1.13\sqrt{\pi(0.15)}} = 38,700 \text{ psi} \tag{3-65}$$

and from (3-58), therefore,

$$P_{bf\text{-all}} = \left(\frac{38,700}{1.5}\right)(0.38)(4.0) = 39,200 \text{ lb} \tag{3-66}$$

Comparing the results of (3-57) with those of (3-66), it is clear that the controlling failure mode is brittle fracture and the design allowable load with the cracked member is only 39,200 lb. A small error has been made by ignoring notch-tip plasticity effects, and the reader should verify that the error is relatively small by utilizing (3-42).

Suppose that the 7075-T6 aluminum member were scheduled for service again with the external load limited to 39,200 lb tension, but a further inspection indicated a thumbnail-shaped surface flaw of depth 0.050 inch and surface length 0.125 inch. Would the discovery of this surface flaw change the recommended operating load of 39,200 lb?

To investigate the seriousness of the newly discovered surface flaw, the brittle fracture failure stress calculated in (3-65) must be reevaluated for a thumbnail surface flaw by using (3-55) and the data of Figure 3.36. To determine the flaw shape parameter Q, two ratios must be computed; they are

$$\frac{a}{2C} = \frac{0.050}{0.125} = 0.4 \tag{3-67}$$

and

$$\frac{\sigma_{\text{nom}}}{\sigma_{yp}} = \frac{P/A_g}{\sigma_{yp}} = \frac{39,200/(0.38)(4.0)}{85,000} = 0.3 \tag{3-68}$$

Utilizing the results of (3-67) and (3-68), we find the flaw shape parameter Q may be read from Figure 3.36 as $Q = 1.95$ and from (3-55); then

$$\sigma'_{bf} = \frac{30,000}{\frac{1.12}{\sqrt{1.95}}\sqrt{\pi(0.050)}} = 94,380 \text{ psi} \tag{3-69}$$

and the allowable load on the basis of this surface flaw is

$$P'_{bf\text{-all}} = \frac{94,380}{1.5}(0.38)(4.0) = 143,450 \text{ lb} \tag{3-70}$$

Thus, the surface flaw causes no additional concern, and the part may be returned to service as planned with the operational load restriction of

39,200 lb placed by the originally discovered single edge through-the-thickness crack. As a matter of fact, by checking back to (3-57), it may be noted that the surface flaw would be less serious than yielding.

Although the field of fracture mechanics is more extensive than indicated here, further considerations in this area will be deferred until discussions of fatigue are presented later in Chapter 8.

3.11 CLOSURE

Mechanical engineering designers are necessarily intensely interested in modes of mechanical failure that may govern their design. Although a microscopic study of material behavior is not directly useful in designing a specific piece of hardware, the qualitative understanding of how failure might manifest itself at the atomic level through the interactions and motions of dislocations is useful. The fracture mechanics model provides an important quantitative tool for prediction of failure by rapid crack extension. Other failure prediction models are developed in the following chapters.

QUESTIONS

1.   Through library research, list all the types of bonding found in solid materials and describe in detail the character of each type of bond.
2.   List the processes by which plastic deformation may proceed in engineering metals and describe the essence of each process.
3.   Conceive and describe in detail an experimental program from which one could prove the existence of a critical resolved shearing stress for pure single crystal aluminum.
4.   Compare and contrast ductile rupture and brittle fracture.
5.   Describe what is meant by strain hardening and recovery at the macroscopic level. How are these processes postulated to occur according to simplified dislocation theory?
6.   Describe the concepts of edge, screw, and mixed dislocations and tell what is meant by cross slip and climb. Use sketches to make each description clear.
7.   How does the dislocation model account for the apparent large discrepancy between theoretical shearing strength and experimentally observed shearing strength in engineering metals?
8.   Of what use is the dislocation concept to mechanical engineering designers?
9.   Using the Griffith expression for energy required to propagate a crack (3-10), derive an expression for the stress level at which the onset of spontaneous crack propagation would be predicted.

10. Construct Burgers circuits for the dislocations indicated in the front face and side face of the crystal lattice shown in Figure 3.18. What is the magnitude of the Burgers vector for each of these dislocations?

11. Measurements of energy stored in a 10 cm³ sample of strain-hardened copper indicate that 50 calories have been stored. Assuming that edge and screw dislocations are present in equal numbers, and all stored energy is due to the dislocation strain fields, estimate the dislocation density in lines per square centimeter. Note: $G = 6 \times 10^6$ psi, $b = 10^{-8}$/inch, 100 inch-pounds = 2.7 calories.

12. Explain the design implications of using yield strength as the governing material strength parameter for the 4330 M steel material of Figure 3.26, if center cracks longer than 0.5 inch are present in the 36-inch-wide steel plates.

13. Compare and contrast the basic philosophy of failure prediction for yielding type failure with failure by rapid crack extension. As a part of your discussion carefully define the terms *stress intensity factor*, *critical stress intensity*, and *fracture toughness*.

14. Describe the three basic crack displacement modes, using appropriate sketches.

15. Interpret the following equation and carefully define each symbol used:

Failure is predicted to occur if

$$C\sigma\sqrt{\pi a} \geq K_{Ic}$$

16. A very wide sheet of 7075-T651 aluminum plate, 5/16 inch thick, is found to have a single edge through-the-thickness crack 1.0 inch long. The loading produces a gross nominal tensile stress of 10,000 psi perpendicular to the plane of the crack.
(a) Calculate the stress intensity factor at the crack tip.
(b) Estimate the plastic zone size at the crack tip.
(c) Comment on the validity of plastic zone correction factor for this configuration.
(d) Determine critical stress intensity factor.
(e) Estimate the safety factor that exists in this case, based on rapid crack propagation (brittle fracture) as a failure mode.

17. Discuss all parts of problem 16 under conditions that are identical to those stated except that the sheet thickness is 1/8 inch.

18. A cylindrical thin-walled pressure vessel, such as shown in Figure 5.6, is closed at the ends and internally pressurized at $p = 5000$ psi. One of the $\frac{1}{4}$-inch diameter rivet holes in the cylindrical shell is found to have a crack of $\frac{1}{16}$ inch length emanating from it in the longitudinal direction. The crack is all the way through the thickness of the wall. The pressure vessel is fabricated of $\frac{1}{4}$-inch thick D6AC steel plate (1000°F temper) and is 10 inches in diameter by 20 inches long.

(a) Determine the stress intensity factor at the crack tip.

(b) Estimate the plastic zone size at the crack tip.

(b) Determine the critical stress intensity factor.

(d) Estimate the safety factor for this case, based on rapid crack extension (brittle fracture) and also based on yielding.

(e) Comment on your results.

19.   Repeat problem 18 with all conditions and dimensions the same except that the 1/16-inch crack emanates from the rivet hole in the circumferential direction. Compare results with those of problem 18.

20.   A steam generator in a remote power station is supported by two tension straps, each one 3 inches wide by 0.44 inch thick by 26 inches long. The straps are made of A538 steel. When in operation the fully loaded steam generator weighs 300,000 lb, equally distributed to the two support straps. The load may be regarded as static. Ultrasonic inspection has detected a through-the-thickness center crack 0.50 inch long oriented perpendicular to the 26-inch dimension (i.e., perpendicular to the tensile load). Would you allow the plant to be put back into operation? Support your answer with clear, complete engineering calculations.

21.   A cantilever beam of Ti-6A-4V titanium alloy has been designed for use in an aggresive chemical environment for which only titanium has good resistance to corrosion problems. The cantilever beam is 26 inches in length and has a *rectangular cross section* 10 inches deep by 2 inches thick (wide). A large fillet at the fixed end will allow you to neglect stress concentration there. The beam is to carry a static end load of 125,000 lb, which is applied by a large fragile glass vat of reactive chemical hanging on the beam for short-term storage. After the beam installation a 0.3-inch-long through-the-thickness edge crack has been discovered on the top (tension side) of the beam, at a location 2 inches from the fixed end of the beam. To replace the beam is very expensive. Would you insist on its replacement or allow it to be used? Support your answer with solid engineering calculations. Do not concern yourself with buckling. ($E = 16 \times 10^6$ psi; $\nu = 0.28$).

22.   A pinned end structural member in a high-performance tanker is made of a 0.375 -inch-thick by 5-inches-wide rectangular cross section titanium 6A1-4V bar 48 inches long. The member is normally subjected to a pure static tensile load of 154,400 lb. Inspection of the member has indicated a central through-the-thickness crack of 0.50 inch length, oriented perpendicular to the applied load. If a safety factor of $n = 1.7$ is required, what "reduced" load limit on the member would your recommend for safe operation (i.e., to give $n = 1.7$)? (Neglect plastic zone correction factor effects.)

23.   Cracks have been discovered in several locations along the edge of a reinforcing rib of a 6A1-4V titanium housing for the hydraulic power drive unit for the wing flap actuators of a commercial aircraft. The rib is 0.25 inch thick and 2.0 inch wide, as shown in the sketch. The cracks go all the way

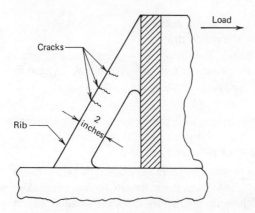

Load

Cracks

Rib

2 inches

FIGURE Q3.23.

through the thickness and are about 0.10 inch deep. A new part can not be delivered for at least six months.

(a) If the part is subjected to static loading only (no fatigue) and the maximum estimated loading produces a tensile load along the rib of 50,000 lb, would you recommend grounding the airplane?

(b) How confident are you in making the recommendation?

REFERENCES

1. Barrett, C. S., *The Structure of Metals*, McGraw-Hill, New York, 1966.

2. Sinnott, M. S., *The Solid State for Engineers*, John Wiley & Sons, New York, 1958.

3. Marin, J., *Mechanical Behavior of Engineering Materials*, Prentice-Hall, Englewood Cliffs, N. J., 1962.

4. Hayden, H. W., Moffatt, W. G., and Wulff, J., *The Structure and Properties of Materials*, Vol. III, Mechanical Behavior, John Wiley & Sons, New York, 1965.

5. Nadai, A., *Theory of Flow and Fracture of Solids, Vol. II*, McGraw-Hill, New York, 1963.

6. Averbach, B. L., Felbeck, D. K., Hahn, G. J., and Thomas, D. A., *Fracture*, M.I.T. Press, Cambridge, 1959.

7. Juvinall, R. C., *Engineering Considerations of Stress, Strain, and Strength*, McGraw-Hill, New York, 1967.

8. Nabarro, F. R. N., *Theory of Crystal Dislocations*, Oxford University Press, London, 1967.

9. Fridel, L., *Dislocations*, Addison-Wesley, Reading, Mass., 1964.

10. Hirth, J. P., and Lothe, J., *Theory of Dislocations*, McGraw-Hill, New York, 1968.

11. Kennedy, A. J., *Processes of Creep and Fatigue in Metals*, John Wiley & Sons, New York, 1963.

12. "Progress in Measuring Fracture Toughness and Using Fracture Mechanics," *Materials Research and Standards*, ASTM (March 1964): 103–119.

13. Westergaard, H. M., "Bearing Pressures and Cracks," *Journal of Applied Mechanics*, ASME Transactions, Series A, **66**, (1939): 49.

14. Irwin, G. R., "Analysis of Stresses and Strains Near the End of a Crack Traversing a Plate," *Journal of Applied Mechanics*, ASME Transactions, **24**, (1957): 361.

15. Tada, H., Paris, P. C. and Irwin, G. E., *The Stress Analysis of Cracks Handbook*, Del Research Corporation, Hellertown, Pa., 1973.

16. "Standard Method of Test for Plane-Strain Fracture Toughness of Metallic Materials," Designation: E399-72, Annual Book of ASTM Standards, Part 31, The American Society for Testing and Materials, Philadelphia, 1972.

17. Polakowski, N. H., and Ripling, E. J., *Strength and Structure of Engineering Materials*, Prentice-Hall, Englewood Cliffs, N. J., 1966.

18. Hertzberg, R. W., *Deformation and Fracture Mechanics of Engineering Materials*, John Wiley & Sons, New York, 1976.

19. Rolfe, S. T., and Barsom, J. M., *Fracture and Fatigue Control in Structures*, Prentice-Hall, Englewood Cliffs, N. J., 1977.

20. Irwin, G. R., "Linear Fracture Mechanics, Fracture Transition, and Fracture Control," *Engineering Fracture Mechanics*, No. 2, (August 1968): 1.

21. Cambell, J. E., Berry, W. E., and Fedderson C. E., *Damage Tolerant Design Handbook*, MCIC-HB-01, September 1973.

22. Sih, G. C., *Handbook of Stress Intensity Factors for Researchers and Engineers*, Institute of Fracture and Solid Mechanics, Lehigh University, Bethlehem, Pa., 1973.

23. Wells, A. A., "Unstable Crack Propagation in Metals-Cleavage and Fast Fracture," *Cranfield Crack Propagation Symposium*, (September 1, 1961): 210.

24. Hord, J. E., "Fracture Initiation in Tough Materials," Conference of Metallurgists, CIMM, Montreal, P. Q., August 31, 1971.

25. *Fracture Toughness Evaluation by R-Curve Methods*, STP-527, American Society for Testing and Materials, Philadelphia, 1973.

26. Taylor, G. I., *Proceedings, Royal Society* (London), A145, (1934): 362; Orowan, E., *Zeitschrift für Physik*, 89, (1934): 634; Polanyi, M., *Zeitschrift für Physik*, 89, (1934): 660.

27. Frank, F. C., and Read, W. T., *Physics Review*, 79, (1950): 722.

28. Boas, W., and Schmid, E., *Zeitschrift für Physik*, 71 (1931): 712.

29. Haase, O., and Schmid, E., *Zeitschrift für Physik*, 33 (1925): 413.

CHAPTER 4

State of Stress

4.1 INTRODUCTION

In designing a machine part, it is necessary to conceive a combination of material and geometry that will fulfill the design objective without failure. To do this a designer must have at his disposal a calculable mechanical modulus that is physically related to the governing failure mode in such a way that failure is accurately predictable when the mechanical modulus reaches a determinable critical value. The three most useful mechanical moduli for this purpose are stress, strain, and energy. For most failure modes it is found that stress may be utilized as a modulus by which failure can be predicted and, therefore, averted by a proper design configuration. The concept of state of stress at a point and consideration of the general triaxial state of stress therefore become topics of great importance to the designer.

4.2 STATE OF STRESS AT A POINT

A solid body of arbitrary size, shape, and material acted upon by an equilibrium system of body forces and/or surface forces will respond so as to develop a system of internal forces also in equilibrium at any point within the body. If the body were cut by an arbitrary plane, these internal forces would in general be distributed continuously over the cut surface and would vary across the surface in both direction and intensity. Furthermore, the internal force distribution would also be a function of the orientation of the plane chosen for investigation. *Stress* is the term used to define the intensity and direction of the internal forces acting at a given point on a particular plane. Stress is thus a second-order tensor quantity because not only the magnitude and direction of the stress is involved in its description but also the orientation of the plane on which it acts. A complete description of the magnitudes and directions of stresses on all possible planes through a point constitute the *state of stress* at the point. Although stress can be conveniently expressed and manipulated as a tensor, the more conventional strength of materials approach will be followed here.

Complete definition of the state of stress at a point may be effected by considering all of the components of stress that can occur on the faces of a infinitesimal cube of material placed on an arbitrary right-handed cartesian coordinate system. All components of stress may be expressed as either *normal stresses*, normal to the faces of the cube, or *shear stresses*, parallel to the faces of the cube. Since the purpose here is to investigate the stresses acting on planes that go precisely through a given point, the limiting values of stresses when the infinitesimal element dimensions approach zero are the only values of interest. Thus, stress gradients across the cube faces are here neglected as quantities of higher order. Also, since body forces decrease as the cube of linear dimensions and surface forces decrease only as the square of linear dimensions, the body forces are also neglected as higher order terms for this limiting case of state of stress at a point. Thus, the completely defined state of stress on an elemental volume of dimensions dx, dy, and dz is shown in Figure 4.1.

Normal stresses will be designated by the symbol σ, and shearing stresses by the symbol τ. The conventions used for subscript notation are the usual

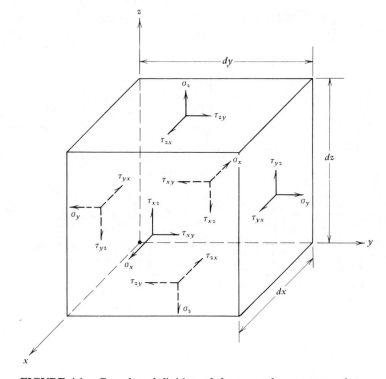

FIGURE 4.1. Complete definition of the state of stress at a point.

ones, defined as follows:

1. For normal stresses a single subscript will be used to correspond to the direction of the outward drawn normal to the plane on which it acts.
2. For shearing stresses two subscripts will be used, the first of which indicates the direction of the plane on which it acts and the second subscript to indicate the direction of the shear stress in the plane.
3. Normal stresses will be called positive ($+$) when they produce tension and negative ($-$) when they produce compression.
4. Shear stresses will be called positive ($+$) if they are in the direction of an axis whose sign is the same as the sign of the axis in the direction of the outward drawn normal to the plane on which the stresses act.

Surveying the sketch of Figure 4.1, we find that in general three normal stresses are active, namely, σ_x, σ_y, and σ_z. It may be further noted that six shear stresses are active, namely, τ_{xy}, τ_{yx}, τ_{yz}, τ_{zy}, τ_{zx}, and τ_{xz}. Thus a total of nine components of stress are apparently necessary to completely define the triaxial state of stress at a point: three normal stresses and six shear stresses. However, it can be shown on the basis of moment equilibrium that τ_{xy} and τ_{yx} are identical, τ_{yz} and τ_{zy} are identical, and τ_{zx} and τ_{xz} are identical. This may be demonstrated by referring to the view of the y-z plane shown in Figure 4.2.

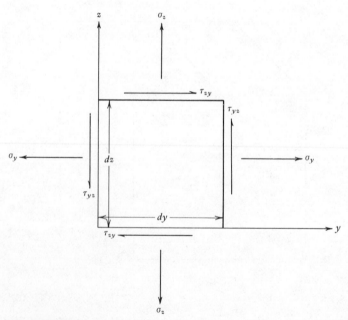

FIGURE 4.2. Sketch showing moment equilibrium about the x axis for purpose of demonstrating equivalence of τ_{yz} and τ_{zy}.

Writing the moment equilibrium equation about the origin, and recognizing that the equal and opposite normal force components nullify, one obtains

$$\tau_{yz}(dz\,dx)\,dy - \tau_{zy}(dy\,dx)\,dz = 0 \tag{4-1}$$

whence

$$\tau_{yz} = \tau_{zy} \tag{4-2}$$

Similar expressions are obtained by writing moment equilibrium equations for the remaining two coordinate planes to obtain

$$\tau_{xy} = \tau_{yx} \tag{4-3}$$

and

$$\tau_{xz} = \tau_{zx} \tag{4-4}$$

Thus the six apparent shear stress components required in a complete description of the state of stress at a point are reduced to three distinct shearing stresses. Consequently, to completely define the most general state of stress at a point requires the specification of six components of stress, three normal stresses, σ_x, σ_y, and σ_z and three shearing stresses, τ_{xy}, τ_{yz}, and τ_{zx}. If one knows the six components of stress at a point, it is possible to compute the stresses on *any* plane passing through the point from simple equilibrium concepts, as demonstrated in the following discussion:

Consider the sketch of Figure 4.3 in which an arbitrary plane *BCD* is passed through the differential element of volume shown in Figure 4.1, at a small distance N from the origin, which plane in the limit then will pass through the origin. To determine the magnitude of the normal stress acting on plane *BCD* in terms of the six components of stress and the direction cosines of the plane *BCD*, the following procedure may be followed:

Referring to Figure 4.3, we find the direction cosines for plane *BCD* (defined by its normal) are defined to be

$$\cos\alpha = l$$
$$\cos\beta = m \tag{4-5}$$
$$\cos\gamma = n$$

From purely geometrical considerations, if the area of plane *BCD* is A, the areas of the triangular faces of the elemental volume perpendicular to the x, y, and z axes are, respectively

$$A_x = Al$$
$$A_y = Am \tag{4-6}$$
$$A_z = An$$

The components of force acting on plane *BCD* in the three coordinate directions will be defined as shown in Figure 4.3 as F_x, F_y, and F_z.

Knowing the area of plane *BCD* to be A and knowing the three components of force acting on the plane, we can write expressions for the stress

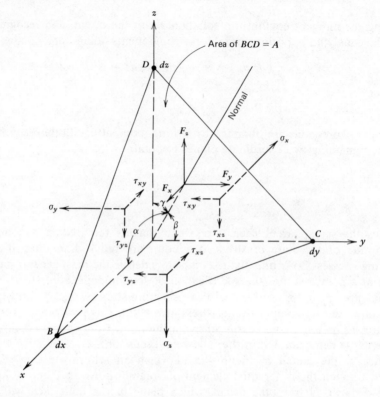

FIGURE 4.3. Sketch showing equilibrium forces on a differential volume element cut by arbitrary plane *BCD*.

components in the x, y, and z directions as

$$S_x = \frac{F_x}{A}$$

$$S_y = \frac{F_y}{A} \qquad (4\text{-}7)$$

$$S_z = \frac{F_z}{A}$$

Further, observing that the forces on the three remaining faces of the tetrahedron in Figure 4.3 may be expressed in each case by an area multiplied by the appropriate stress, we can write three equations of force equilibrium, one corresponding to each of the three coordinate directions. These equations are

$$F_x = A_x\sigma_x + A_y\tau_{xy} + A_z\tau_{xz}$$

$$F_y = A_x\tau_{xy} + A_y\sigma_y + A_z\tau_{yz} \qquad (4\text{-}8)$$

$$F_z = A_x\tau_{xz} + A_y\tau_{yz} + A_z\sigma_z$$

Substituting from (4-6) and (4-7) into (4-8) and dividing out the common term A, we write the expressions for the three components of stress acting on plane *BCD* as

$$S_x = l\sigma_x + m\tau_{xy} + n\tau_{xz}$$
$$S_y = l\tau_{xy} + m\sigma_y + n\tau_{yz} \qquad (4\text{-}9)$$
$$S_z = l\tau_{xz} + m\tau_{yz} + n\sigma_z$$

Referring again to Figure 4.3, we write expressions for the normal components S_{xn}, S_{yn}, and S_{zn} of each of the stresses S_x, S_y, and S_z as

$$S_{xn} = lS_x$$
$$S_{yn} = mS_y \qquad (4\text{-}10)$$
$$S_{zn} = nS_z$$

Whereupon, the expressions of (4-9) may be substituted into (4-10) to yield

$$S_{xn} = l^2\sigma_x + lm\tau_{xy} + ln\tau_{xz}$$
$$S_{yn} = lm\tau_{xy} + m^2\sigma_y + mn\tau_{yz} \qquad (4\text{-}11)$$
$$S_{zn} = ln\tau_{xz} + mn\tau_{yz} + n^2\sigma_z$$

The resultant normal stress σ_n on plane *BCD* then is the sum of these three components, whence

$$\sigma_n = S_{xn} + S_{yn} + S_{zn} \qquad (4\text{-}12)$$

or, substituting the expressions of (4-11) into (4-12),

$$\sigma_n = l^2\sigma_x + m^2\sigma_y + n^2\sigma_z + 2(lm\tau_{xy} + ln\tau_{xz} + mn\tau_{yz}) \qquad (4\text{-}13)$$

This expression yields the value of the normal stress σ_n on any specified plane through the point in terms of the six distinct components of stress and the direction of the plane. One could develop another expression for a resultant shearing stress on the plane by a similar technique, but it will be more convenient to develop the shearing stress expression later in a slightly different way.

4.3 PRINCIPAL NORMAL STRESSES

Principal normal stresses, sometimes called simply *principal stresses*, are the normal stresses that occur on planes where the shearing stresses are zero. The planes on which the principal normal stresses act are called *principal planes*. The principal normal stresses also are local extremes of stress that include the maximum value of normal stress that can occur on any plane through the point. Since failure moduli are often related to the principal stresses or to the maximum principal stress, it will be of interest to develop a means of determining the principal normal stresses.

To determine the principal normal stresses in terms of the six components of stress, it will be convenient to simply postulate that plane *BCD* in Figure 4.3 is a principal plane and determine the direction cosines to make the postulate true, then express the principal normal stress in terms of the six components of the general triaxial state of stress. Thus the plane *BCD* is postulated to be a principal plane; hence the shearing stress components are zero. The normal stress on the plane is therefore the resultant and may be defined as

$$\sigma = \text{resultant normal stress on principal plane } BCD \qquad (4\text{-}14)$$

Since the area of principal plane *BCD* is known to be *A*, the resultant normal force F_n on the plane is

$$F_n = \sigma A \qquad (4\text{-}15)$$

and, utilizing (4-7), the components of the normal force in the *x*, *y*, and *z* directions may be expressed as

$$F_{nx} = AS_x$$
$$F_{ny} = AS_y \qquad (4\text{-}16)$$
$$F_{nz} = AS_z$$

Also from purely geometrical considerations

$$F_{nx} = lF_n = l\sigma A$$
$$F_{ny} = mF_n = m\sigma A \qquad (4\text{-}17)$$
$$F_{nz} = nF_n = n\sigma A$$

combining (4-16) and (4-17), then,

$$S_x = l\sigma$$
$$S_y = m\sigma \qquad (4\text{-}18)$$
$$S_z = n\sigma$$

and substituting (4-9) into (4-18) yields

$$l\sigma = l\sigma_x + m\tau_{xy} + n\tau_{xz}$$
$$m\sigma = l\tau_{xy} + m\sigma_y + n\tau_{yz} \qquad (4\text{-}19)$$
$$n\sigma = l\tau_{xz} + m\tau_{yz} + n\sigma_z$$

which may be more conveniently expressed as

$$l(\sigma - \sigma_x) - m\tau_{xy} - n\tau_{xz} = 0$$
$$-l\tau_{xy} + m(\sigma - \sigma_y) - n\tau_{yz} = 0 \qquad (4\text{-}20)$$
$$-l\tau_{xz} - m\tau_{yz} + n(\sigma - \sigma_z) = 0$$

From the postulate leading to the development of this set of three equations, then, it may be asserted that if σ is to be a principal stress, it must satisfy these equations. One must further recognize that these three equations are not independent since l, m, and n are geometrically related by

$$l^2 + m^2 + n^2 = 1 \qquad (4\text{-}21)$$

This geometrical compatibility equation simply states that the three direction cosines can not all be zero simultaneously, and the definition of any two uniquely determines the third.

Since the equations of (4-10) are not independent, the determinant of their coefficients must vanish; and hence

$$\begin{vmatrix} (\sigma - \sigma_x) & -\tau_{xy} & -\tau_{xz} \\ -\tau_{xy} & (\sigma - \sigma_y) & -\tau_{yz} \\ -\tau_{xz} & -\tau_{yz} & (\sigma - \sigma_z) \end{vmatrix} = 0 \qquad (4\text{-}22)$$

Finally, expanding the determinant, one obtains the cubic equation

$$\sigma^3 - \sigma^2(\sigma_x + \sigma_y + \sigma_z) + \sigma(\sigma_x\sigma_y + \sigma_y\sigma_z + \sigma_x\sigma_z - \tau_{xy}^2 - \tau_{xz}^2 - \tau_{yz}^2)$$
$$- (\sigma_x\sigma_y\sigma_z + 2\tau_{xy}\tau_{xz}\tau_{yz} - \sigma_x\tau_{yz}^2 - \sigma_y\tau_{xz}^2 - \sigma_z\tau_{xy}^2) = 0 \qquad (4\text{-}23)$$

This expression is called the *general stress cubic equation* for the triaxial state of stress. From the conditions leading to its development, namely, that all normal and shearing stress components are real numbers, one may be assured mathematically that at least one of the three solutions for σ from this equation will be real; and, based upon physical realities, all three solutions will be real.

It should be recognized that for any given equilibrium system of forces applied to a body, the principal stresses and principal planes are uniquely determined and must be independent of the orientation of the cartesian coordinate axes. For this reason the coefficients in (4-23) are constant or *invariant*, no matter what choice is made for the orientation of the coordinate axes. These coefficients have been named the first, second, and third stress invariants*, specifically,

First stress invariant: $(\sigma_x + \sigma_y + \sigma_z) = \text{const.}$ \qquad (4-24)

Second stress invariant: $(\sigma_x\sigma_y + \sigma_y\sigma_z + \sigma_x\sigma_z - \tau_{xy}^2 - \tau_{xz}^2 - \tau_{yz}^2) = \text{const.}$

$$(4\text{-}25)$$

Third stress invariant: $(\sigma_x\sigma_y\sigma_z + 2\tau_{xy}\tau_{xz}\tau_{yz} - \sigma_x\tau_{yz}^2 - \sigma_y\tau_{xz}^2 - \sigma_z\tau_{xy}^2) = \text{const.}$

$$(4\text{-}26)$$

Summarizing the results of this discussion, then, we may observe that the three solutions of the stress cubic equation (4-23) give the three principal normal stresses. Substitution of these three solutions into (4-20), together with

*See p. 217 of ref. 1.

the geometrical compatibility equation (4-21), allows the determination of the direction cosines, defining explicitly the principal planes.

Further, it may be deduced that the three principal planes are mutually perpendicular. To demonstrate this, one may transform the x-y-z coordinate axes into a new set of coordinate axes x'-y'-z', so orienting the x' axis that it is perpendicular to the principal plane BCD. Since BCD is by hypothesis a principal plane, the shearing stresses on this plane must be zero so that

$$\tau_{x'y'} = \tau_{x'z'} = 0 \tag{4-27}$$

In view of (4-27) and the principle of stress invariance defined in (4-24), (4-25), and (4-26), the stress cubic equation (4-23) may be written as

$$\sigma^3 - \sigma^2(\sigma_{x'} + \sigma_{y'} + \sigma_{z'}) + \sigma(\sigma_{x'}\sigma_{y'} + \sigma_{y'}\sigma_{z'} + \sigma_{x'}\sigma_{z'} - \tau_{y'z'}^2)$$
$$- (\sigma_{x'}\sigma_{y'}\sigma_{z'} - \sigma_{x'}\tau_{y'z'}^2) = 0 \tag{4-28}$$

Expanding and factoring this equation then yields

$$[\sigma - \sigma_{x'}][\{(\sigma - \sigma_{y'})(\sigma - \sigma_{z'})\} - \{\tau_{y'z'}^2\}] = 0 \tag{4-29}$$

The solution of (4-29) leads to the three roots

$$\sigma_1 = \sigma_{x'} \tag{4-30}$$

$$\sigma_2 = \frac{\sigma_{y'} + \sigma_{z'}}{2} + \sqrt{\left(\frac{\sigma_{y'} - \sigma_{z'}}{2}\right)^2 + \tau_{y'z'}^2} \tag{4-31}$$

$$\sigma_3 = \frac{\sigma_{y'} + \sigma_{z'}}{2} - \sqrt{\left(\frac{\sigma_{y'} - \sigma_{z'}}{2}\right)^2 + \tau_{y'z'}^2} \tag{4-32}$$

It may be observed that the right-hand sides of all three of these equations are real, and therefore the three principal stresses are all real and will always be real.

The principal planes may now be defined with respect to the primed axes by utilizing (4-20) and (4-21). Recall that plane BCD is by hypothesis a principal plane and normal to the x' axis, which yields the result of (4-27). In view of this, the equations of (4-20) for this condition become

$$l(\sigma - \sigma_{x'}) = 0 \tag{4-33}$$

$$m(\sigma - \sigma_{y'}) - n\tau_{y'z'} = 0 \tag{4-34}$$

$$- m\tau_{y'z'} + n(\sigma - \sigma_{z'}) = 0 \tag{4-35}$$

and, of course, the geometrical compatibility condition of (4-21) must still be invoked.

Postulating BCD as a principal plane and specifying the x' axis to be normal to the principal plane was at the outset the equivalent of specifying the direction cosines $l_1 = 1$, $m_1 = 0$, and $n_1 = 0$, where the subscript corresponds to the definition of the plane upon which the principal stress σ_1 acts. Note that this is confirmed by (4-30), where σ_1 is equal to $\sigma_{x'}$.

Now, if σ is either of the other two principal stresses, σ_2, or σ_3, it is clear that σ in (4-33) can not be σ_1, which is equal to $\sigma_{x'}$. Thus, the only way in which (4-33) can be satisfied is for l_2 and l_3 to both be zero. From this it must be concluded that the principal planes on which σ_2 and σ_3 act are perpendicular to the plane on which σ_1 acts. There are an infinite number of planes perpendicular to the σ_1 plane, but the only ones of interest are those upon which the shearing stresses are zero.

The y' axis may now be specified to coincide with the direction of σ_2, and the following arguments made based on (4-21), (4-33), (4-34), and (4-35): Since σ_2 is not equal, in general, to σ_1 (4-33) can be satisfied only if l_2 is zero. Further, since the y' direction has been chosen to coincide with σ_2, σ_2 is not, in general, equal to $\sigma_{z'}$. Also, $\tau_{y'z'}$ is independent of $(\sigma - \sigma_{z'})$ so that (4-35) can be satisfied only if n_2 is zero. If l_2 and n_2 are both zero, (4-21) must be invoked to properly determine m_2. From (4-21), then, m_2 is found to be ± 1. Since m_2 is not zero, from (4-35) it must be concluded that $\tau_{y'z'}$ is zero; and from (4-31), then, it is found that principal stress σ_2 is equal to $\sigma_{y'}$, verifying that the y' axis has been properly selected as a principal axis. With these restrictions imposed, (4-33) (4-34), and (4-35) may then be further specialized to yield

$$l(\sigma - \sigma_{x'}) = 0 \qquad (4\text{-}36)$$

$$m(\sigma - \sigma_{y'}) = 0 \qquad (4\text{-}37)$$

$$n(\sigma - \sigma_{z'}) = 0 \qquad (4\text{-}38)$$

and, as before, the geometrical compatibility equation (4-21) must be satisfied at the same time.

Recognizing that in general σ_3 will not be equal to either σ_1 or σ_2, and hence not equal to either $\sigma_{x'}$ or $\sigma_{y'}$, it may be concluded from (4-36) and (4-37) that both l_3 and m_3 must be zero. Finally, invoking (4-21) again, if l_3 and m_3 are zero, n_3 must be ± 1 and σ_3 must be $\sigma_{z'}$.

Summarizing the direction cosines associated with the three principal stresses σ_1, σ_2, and σ_3, we get

$$
\begin{array}{lll}
l_1 = 1 & l_2 = 0 & l_3 = 0 \\
m_1 = 0 & m_2 = 1 & m_3 = 0 \\
n_1 = 0 & n_2 = 0 & n_3 = 1
\end{array}
\qquad (4\text{-}39)
$$

From (4-39) it may be readily deduced that the three principal planes are, indeed, mutually perpendicular.

It is of great importance to the mechanical designer to recognize that *the largest of the three principal stresses is the maximum normal stress that can occur on any plane passing through the point*. It is also important to recognize that that the minimum principal stress is the minimum normal stress that can occur on any plane passing through the point.

4.4 PRINCIPAL SHEARING STRESSES

Principal planes were defined in Section 4.3, as planes upon which the shearing stresses are zero. Thus, the resultant stress on the principal planes is entirely accounted for by the normal stress component. On any other plane through the point the resultant stress will in general have both a normal stress component and a shearing stress component. It is clear that on at least one of these other planes the shearing stress will attain a maximum value. These extreme values of shearing stresses are called *principal shearing stresses* and the planes upon which they act are called *planes of principal shear*.

To determine the magnitudes of the principal shearing stresses and the directions of the planes on which they act, it will be convenient to consider a differential element oriented with its faces parallel to the principal planes, as shown in Figure 4.4, cut by an arbitrary plane *EFG*, as shown. In this orientation all shearing stress components on the coordinate planes are zero and only the principal normal stresses remain to be considered. Defining the plane *EFG* by its direction cosines, the components of stress acting on plane

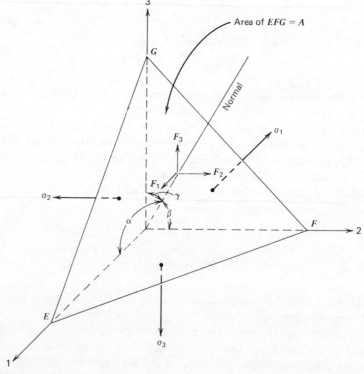

FIGURE 4.4. Sketch showing equilibrium forces on a differential volume element aligned on the principal axes and cut by arbitrary plane *EFG*.

EFG in the direction of the 1, 2, and 3 axes are readily obtained from the general expressions of (4-9) as

$$S_1 = l\sigma_1$$
$$S_2 = m\sigma_2 \tag{4-40}$$
$$S_3 = n\sigma_3$$

Also, for this principal element the expression of (4-13) reduces to

$$\sigma_n = l^2\sigma_1 + m^2\sigma_2 + n^2\sigma_3 \tag{4-41}$$

where σ_n is the normal stress acting on plane *EFG*.

The resultant force F_r on plane *EFG* may be expressed as the vector sum of its components, F_1, F_2, and F_3, in the three coordinate directions as

$$F_r^2 = F_1^2 + F_2^2 + F_3^2 \tag{4-42}$$

Dividing both sides of this expression by the square of the area A of plane *EFG* then yields

$$\sigma_r^2 = S_1^2 + S_2^2 + S_3^2 \tag{4-43}$$

Substituting the stresses from the equations of (4-40) into (4-43),

$$\sigma_r^2 = l^2\sigma_1^2 + m^2\sigma_2^2 + n^2\sigma_3^2 \tag{4-44}$$

It may also be recognized that the resultant stress σ_r may be expressed as the vector sum of its normal stress component σ_n and its shearing stress component τ, thus

$$\sigma_r^2 = \sigma_n^2 + \tau^2 \tag{4-45}$$

Substituting into this relationship for σ_n from (4-41) and solving for τ^2 finally yields

$$\tau^2 = l^2\sigma_1^2 + m^2\sigma_2^2 + n^2\sigma_3^2 - \left(l^2\sigma_1 + m^2\sigma_2 + n^2\sigma_3\right)^2 \tag{4-46}$$

This is the expression for the shearing stress on any plane passing through the point in terms of the principal normal stresses and the direction cosines for the plane on which τ acts, referenced to the principal axes.

To obtain the maximum value of the resultant shearing stress τ, (4-46) will be differentiated partially with respect to the direction cosines l, m, and n. The derivatives will be equated to zero and solutions for l, m, and n will be obtained for the extreme values of shearing stress τ. It must again be recognized that the direction cosines are related by (4-21), and only two of the three may be treated as independent variables for any given differentiation. To proceed, (4-21) is first solved for $n^2 = 1 - l^2 - m^2$, and this expression is substituted for n^2 in (4-46) to yield

$$\tau^2 = l^2\left(\sigma_1^2 - \sigma_3^2\right) + m^2\left(\sigma_2^2 - \sigma_3^2\right) + \sigma_3^2$$
$$- \left[l^2(\sigma_1 - \sigma_3) + m^2(\sigma_2 - \sigma_3) + \sigma_3\right]^2 \tag{4-47}$$

Partially differentiating (4-47) with respect to l gives

$$2\tau\frac{\partial\tau}{\partial l} = 2l(\sigma_1^2 - \sigma_3^2) - \left[2\{l^2(\sigma_1 - \sigma_3) + m^2(\sigma_2 - \sigma_3) + \sigma_3\}2l(\sigma_1 - \sigma_3)\right]$$

(4-48)

which may be combined, rearranged, and set equal to zero to give

$$\frac{\partial\tau}{\partial l} = -\frac{2l(\sigma_1 - \sigma_3)}{\tau}\left[l^2(\sigma_1 - \sigma_3) + m^2(\sigma_2 - \sigma_3) - \frac{1}{2}(\sigma_1 - \sigma_3)\right] = 0$$

(4-49)

In a similar way, differentiation of (4-47) with respect to m yields, finally,

$$\frac{\partial\tau}{\partial m} = -\frac{2m(\sigma_2 - \sigma_3)}{\tau}\left[l^2(\sigma_1 - \sigma_3) + m^2(\sigma_2 - \sigma_3) - \frac{1}{2}(\sigma_2 - \sigma_3)\right] = 0$$

(4-50)

A second pair of similar equations may be obtained by solving (4-21) for $m^2 = 1 - l^2 - n^2$, substituting this expression for m^2 into (4-46) and partially differentiating with respect to l and n to obtain

$$\frac{\partial\tau}{\partial l} = -\frac{2l(\sigma_1 - \sigma_2)}{\tau}\left[l^2(\sigma_1 - \sigma_2) + n^2(\sigma_3 - \sigma_2) - \frac{1}{2}(\sigma_1 - \sigma_2)\right] = 0$$

(4-51)

and

$$\frac{\partial\tau}{\partial n} = -\frac{2n(\sigma_3 - \sigma_2)}{\tau}\left[l^2(\sigma_1 - \sigma_2) + n^2(\sigma_3 - \sigma_2) - \frac{1}{2}(\sigma_3 - \sigma_2)\right] = 0$$

(4-52)

A third and final pair of similar equations may be obtained by solving (4-21) for $l^2 = 1 - m^2 - n^2$, substituting this expression for l^2 into (4-46) and differentiating partially with respect to m and n to yield

$$\frac{\partial\tau}{\partial m} = -\frac{2m(\sigma_2 - \sigma_1)}{\tau}\left[m^2(\sigma_2 - \sigma_1) + n^2(\sigma_3 - \sigma_1) - \frac{1}{2}(\sigma_2 - \sigma_1)\right] = 0$$

(4-53)

and

$$\frac{\partial\tau}{\partial n} = -\frac{2n(\sigma_3 - \sigma_1)}{\tau}\left[m^2(\sigma_2 - \sigma_1) + n^2(\sigma_3 - \sigma_1) - \frac{1}{2}(\sigma_3 - \sigma_1)\right] = 0$$

(4-54)

Solving (4-49) and (4-50) together, assuming that the three principal stresses are distinct and different, and invoking the geometrical compatibility prescribed by (4-21), the following sets of direction cosines are found to satisfy this pair of equations:

$$l = 0 \qquad l = 0 \qquad l = \pm\sqrt{1/2}$$
$$m = 0 \qquad m = \pm\sqrt{1/2} \qquad m = 0$$
$$n = 1 \qquad n = \pm\sqrt{1/2} \qquad n = \pm\sqrt{1/2}$$

Each of these three sets of direction cosines, therefore, defines the direction of a plane on which the shearing stress τ will have an extreme value. In a similar way, the equation pair (4-51) and (4-52) may be solved together and equation pair (4-53) and (4-54) may be solved together to yield additional planes on which the shearing stresses reach extreme values. Considering all such solutions and eliminating duplicate results, we find the results shown in Table 4.1. The last three columns of this table correspond to the principal normal planes, and the value of shearing stress is known to be zero on these planes. The first three columns define planes upon which the shearing stresses reach maximum or relative maximum values. If the direction cosines from column 1 are substituted into (4-46), the corresponding value of τ_1 becomes

$$\tau_1 = \pm 1/2\,(\sigma_2 - \sigma_3) \tag{4-55}$$

and utilizing columns 2 and 3 in (4-46) we determine the corresponding values for τ_2 and τ_3 as

$$\tau_2 = \pm 1/2\,(\sigma_1 - \sigma_3) \tag{4-56}$$

and

$$\tau_3 = \pm 1/2\,(\sigma_1 - \sigma_2) \tag{4-57}$$

From these last three expressions it may be deduced that the maximum shearing stress is in magnitude equal to half the difference between the maximum and minimum principal normal stress, and from the table of corresponding direction cosines, it may be further deduced that the plane on which the maximum shearing stress acts is the plane that bisects the angle between the maximum and minimum normal stress vectors. The three shearing stresses, τ_1, τ_2, and τ_3, as just defined, are called the *principal shearing stresses*; and the planes on which they act are called the *planes of principal*

Table 4.1. Planes on Which Shearing Stresses Reach Extreme Values

Direction	Plane					
Cosine	1	2	3	4	5	6
l	0	$\pm\sqrt{1/2}$	$\pm\sqrt{1/2}$	± 1	0	0
m	$\pm\sqrt{1/2}$	0	$\pm\sqrt{1/2}$	0	± 1	0
n	$\pm\sqrt{1/2}$	$\pm\sqrt{1/2}$	0	0	0	± 1

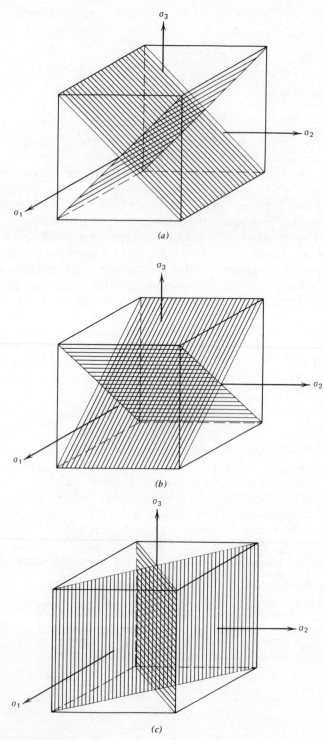

FIGURE 4.5. Sketches showing the planes of principal shearing stress. (*a*) τ_1 planes. (*b*) τ_2 planes. (*c*) τ_3 planes.

shear. The planes of principal shear are illustrated in Figure 4.5 for all three of the principal shearing stresses.

It is of great significance to the mechanical designer to recognize that *the largest of the three principal shearing stresses is the maximum value of shearing stress that can occur on any plane passing through the point.*

4.5 USING THE IDEAS

Consider a solid cylindrical bar fixed at one end and subjected to a pure torsional moment M_t at the free end, as shown in Figure 4.6. Find for this loading the principal normal stresses and the corresponding principal planes at an arbitrarily specified point P within the bar.

To solve the problem a suitable coordinate system must be selected. For example, select a cartesian coordinate system with origin at point P, x axis parallel to the longitudinal axis of the bar and the z axis on a radial line through the center of the bar cross section, as shown in Figure 4.6. An elemental cube at point P may be extracted as a free body and, in view of the applied loading, the only nonzero stress component on the element is τ_{xy}, as illustrated in Figure 4.6. From elementary considerations the shearing stress τ_{xy} may be written in terms of the applied torsional moment M_t, the outer radius of the bar a, and the radial distance r from the center of the bar to point P, as

$$\tau_{xy} = \frac{M_t r}{(\pi a^4/2)} = \frac{2 M_t r}{\pi a^4} \tag{4-58}$$

For this problem the general stress cubic equation (4-23) reduces to

$$\sigma^3 + \sigma\left(-\tau_{xy}^2\right) = 0 \tag{4-59}$$

which may be solved to obtain three roots as

$$\sigma_1 = \tau_{xy} = \frac{2 M_t r}{\pi a^4} \tag{4-60}$$

$$\sigma_2 = 0 \tag{4-61}$$

$$\sigma_3 = -\tau_{xy} = -\frac{2 M_t r}{\pi a^4} \tag{4-62}$$

Further, (4-20) and (4-21) for determining the direction cosines of the principal planes reduce for this case to

$$l\sigma - m\tau_{xy} = 0 \tag{4-63}$$

$$-l\tau_{xy} + m\sigma = 0 \tag{4-64}$$

$$n\sigma = 0 \tag{4-65}$$

and

$$l^2 + m^2 + n^2 = 1 \tag{4-66}$$

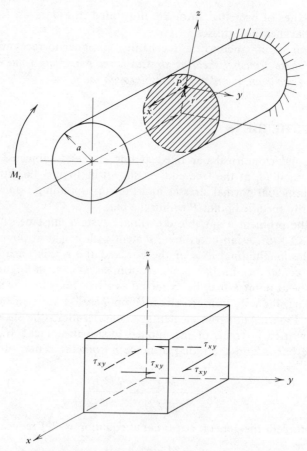

FIGURE 4.6. A solid circular bar subjected to a pure torsional moment, and an elemental free body taken at an arbitrary point P in the bar.

From (4-60) it is clear that $\sigma = \tau_{xy}$ is not in general zero, so from (4-65) n_1 must be zero. Also, since $\sigma_1 = \tau_{xy}$, from (4-63) or (4-64) $l_1 - m_1 = 0$ or $l_1 = m_1$. Then from (4-66) $l_1 = m_1 = \pm \sqrt{1/2}$.

From (4-61) $\sigma_2 = 0$, and since τ_{xy} is not in general zero, from (4-63) m_2 must be zero and from (4-64) l_2 must be zero. Then from (4-66), with m_2 and l_2 equal to zero, n_2 must be ± 1.

Finally from (4-62) $\sigma_3 = -\tau_{xy}$ and since $\sigma_3 = -\tau_{xy}$ is not in general zero, from (4-65) n_3 is zero. Also, since $\sigma_3 = -\tau_{xy}$, from (4-63) or (4-64) $l_3 + m_3 = 0$ or $l_3 = -m_3$. Then from (4-66) $l_3 = \pm \sqrt{1/2}$ and $m_3 = \mp \sqrt{1/2}$. Note that it is important to properly carry the sign order. That is, $+l_3$ and $-m_3$ are an acceptable combination, as are $-l_3$ and $+m_3$, but $+l_3$ and $+m_3$ or $-l_3$ and $-m_3$ are not. It should also be noted that angles are measured from the positive (+) axes and are always less than 180 degrees.

The results of this analysis may be summarized as

$$\sigma_1 = \frac{2M_t r}{\pi a^4} \qquad\qquad \sigma_2 = 0 \qquad\qquad \sigma_3 = \frac{-2M_t r}{\pi a^4}$$

$$
\begin{aligned}
l_1 &= \pm\sqrt{1/2} & l_2 &= 0 & l_3 &= \pm\sqrt{1/2} \\
m_1 &= \pm\sqrt{1/2} & m_2 &= 0 & m_3 &= \mp\sqrt{1/2} \\
n_1 &= 0 & n_2 &= \pm 1 & n_3 &= 0
\end{aligned}
\tag{4-67}
$$

Consider another example: Suppose that a proper stress analysis of a certain machine part has indicated that at a particular point of interest $\sigma_x = 6000$ psi, $\tau_{xy} = 4000$ psi, and all other components of stress at the point are zero. It is desired to determine the principal stresses and the principal planes, as well as the maximum principal shearing stress and the plane on which it acts.

Going back again to the general stress cubic equation (4-23), it becomes for this case

$$\sigma^3 - \sigma^2 \sigma_x + \sigma\left(-\tau_{xy}^2\right) = 0 \tag{4-68}$$

Solutions to this equation are

$$\sigma_1 = \frac{\sigma_x}{2} + \sqrt{\left(\frac{\sigma_x}{2}\right)^2 + \tau_{xy}^2} \tag{4-69}$$

$$\sigma_2 = 0 \tag{4-70}$$

$$\sigma_3 = \frac{\sigma_x}{2} - \sqrt{\left(\frac{\sigma_x}{2}\right)^2 + \tau_{xy}^2} \tag{4-71}$$

Substituting the numerical values of $\sigma_x = 6000$ and $\tau_{xy} = 4000$ we find that these equations yield

$$\sigma_1 = \frac{6000}{2} + \sqrt{\left(\frac{6000}{2}\right)^2 + 4000^2} = 8000 \text{ psi} \tag{4-72}$$

$$\sigma_2 = 0 \tag{4-73}$$

$$\sigma_3 = \frac{6000}{2} - \sqrt{\left(\frac{6000}{2}\right)^2 + 4000^2} = -2000 \text{ psi} \tag{4-74}$$

To determine the direction cosines, reference is again made to (4-20) and (4-21), which for this case become

$$l(\sigma - \sigma_x) - m\tau_{xy} = 0 \tag{4-75}$$

$$-l\tau_{xy} + m\sigma = 0 \tag{4-76}$$

$$n\sigma = 0 \tag{4-77}$$

and

$$l^2 + m^2 + n^2 = 1 \tag{4-78}$$

Considering first $\sigma_1 = 8000$ psi, we may rewrite the expressions of (4-75), (4-76), and (4-77) as

$$l_1(8000 - 6000) - m_1(4000) = 0 \qquad (4\text{-}79)$$

$$- l_1(4000) + m_1(8000) = 0 \qquad (4\text{-}80)$$

$$n_1(8000) = 0 \qquad (4\text{-}81)$$

From (4-81) n_1 must be zero, and from (4-79) $2l_1 - 4m_1 = 0$, or $l_1 = 2m_1$. Then from (4-78) $m_1 = \pm 1/\sqrt{5}$ and $l_1 = \pm 2/\sqrt{5}$.

In a similar way, considering $\sigma_2 = 0$, the expressions of (4-75), (4-76), and (4-77) may be written as

$$l_2(-6000) - m_2(4000) = 0 \qquad (4\text{-}82)$$

$$- l_2(4000) = 0 \qquad (4\text{-}83)$$

$$n_2(0) = 0 \qquad (4\text{-}84)$$

From (4-83) l_2 must be zero, whence from (4-82) m_2 must be zero, and therefore from (4-78) n_2 is ± 1.

Finally, for $\sigma_3 = -2000$ psi, from (4-75), (4-76), and (4-77) one obtains

$$l_3(-2000 - 6000) - m_3(4000) = 0 \qquad (4\text{-}85)$$

$$- l_3(4000) + m_3(-2000) = 0 \qquad (4\text{-}86)$$

$$n_3(-2000) = 0 \qquad (4\text{-}87)$$

From (4-87) n_3 must be zero. From (4-85) $m_3 = -2l_3$, and therefore from (4-78) l_3 is $\pm 1/\sqrt{5}$ amd m_3 is $\mp 2/\sqrt{5}$.

The principal normal stresses and the planes on which they act may therefore be summarized as

$$
\begin{array}{lll}
\sigma_1 = 8000 \text{ psi} & \sigma_2 = 0 & \sigma_3 = -2000 \text{ psi} \\[2mm]
l_1 = \pm \dfrac{2}{\sqrt{5}} & l_2 = 0 & l_3 = \pm \dfrac{1}{\sqrt{5}} \\[4mm]
m_1 = \pm \dfrac{1}{\sqrt{5}} & m_2 = 0 & m_3 = \mp \dfrac{2}{\sqrt{5}} \\[4mm]
n_1 = 0 & n_2 = \pm 1 & n_3 = 0
\end{array}
\qquad (4\text{-}88)
$$

To find the maximum principal shearing stress, reference is made to (4-56), noting that σ_1 is the maximum principal normal stress for the case at hand and σ_3 is the minimum principal normal stress. Thus the maximum principal shearing stress is τ_2, whence

$$\tau_2 = \frac{\sigma_1 - \sigma_3}{2} = \frac{8000 - (-2000)}{2} = 5000 \text{ psi} \qquad (4\text{-}89)$$

and because the planes of maximum shearing stress bisect the angle between σ_1 and σ_3, for this case the planes of maximum shear are parallel to the z axis and make angles of about 71.5 degrees and 161.5 degrees with respect to the x axis.

QUESTIONS

1. Using appropriate sketches and words, define the most general state of stress at a point.

2. Define the terms *principal normal stress* and *principal shearing stress*.

3. Write the complete set of equations required to determine the principal normal stresses at any arbitrary point.

4. Explain the importance of principal normal stresses and principal shearing stresses to a mechanical designer.

5. Make two neat, clear sketches illustrating two ways of completely defining the state of stress at a point. Define all symbols used.

6. How does one determine the maximum shearing stress at a point and the plane on which it acts, if the principal normal stresses are known?

7. Starting with a biaxial state of stress at a point, including σ_x, σ_y, and τ_{xy}, derive from basic principles an expression for the maximum *normal* stress at the point and define the direction of the plane on which it acts.

8. Starting with a biaxial state of stress at a point, including σ_x, σ_y, and τ_{xy}, derive from basic principles an expression for the maximum *shearing* stress at the point and define the direction of the plane on which it acts.

9. Compare the results of problems 7 and 8 with the results obtained using the methods of Sections 4.3 and 4.4 degenerated to a biaxial state of stress.

10. Consider a solid cylindrical bar fixed at one end and subjected to both a pure torsional moment M_t and a pure bending moment M_b at the free end. For this loading find the principal normal stresses and corresponding principal planes at an arbitrary point P within the bar.

11. A solid cylindrical member is fixed at one end and subjected to a pure torsional moment M_t and an axial force F at the free end. For this loading determine
(a) the principal normal stresses and corresponding principal planes
(b) the principal shearing stresses, the maximum shearing stress, and the plane of maximum shear.

12. From the stress analysis of a machine part at a specified critical point, it has been found that $\sigma_z = 9000$ psi, $\tau_{xy} = 3000$ psi, and $\tau_{yz} = 7000$ psi. For this state of stress, determine the principal stresses, principal planes, maximum shearing stress, and plane of maximum shear.

13. A hollow cylindrical 4340 steel member of 1 inch o.d. and 30 inches long with a one-quarter-inch wall is simply supported at the ends and symmetrically loaded at the one-third points by 1000 lb loads. The bar is simulta-

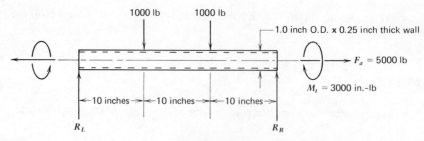

FIGURE Q4.13.

neously subjected to an axial force of 5000 lb and a torsional moment of 3000 in-lb. For the critical point at the center of the bar, determine the principal stresses, principal planes, principal shearing stresses, and planes of principal shear.

14. A solid cylindrical bar of 1020 steel is 30 inches long and 1.5 inches in diameter. The bar is simultaneously subjected to a pure bending moment of 10,600 in-lb and a pure torsional moment of 9900 in-lb.
(a) Establish the location(s) of the critical point(s).
(b) Using a coordinate system of your choice, and the usual elemental volume concept, draw a sketch to completely define the state of stress at each critical point.
(c) Determine all the principal stresses.
(d) Determine the direction cosines for the principal planes upon which the *largest* principal stress acts.
(e) Calculate the magnitude of the *maximum* principal shearing stress.

15. The square cantilever beam shown in Figure Q4.15 is subjected to pure bending moments M_y and M_z as shown. Stress concentration effects are negligible.

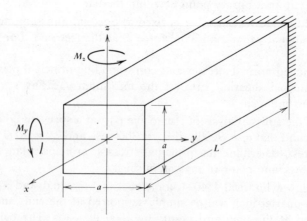

FIGURE Q4.15.

(a) For the critical point make a complete sketch depicting the state of stress.
(b) Determine the magnitude of the principal stresses and the directions of the principal planes.

REFERENCES

1. Timoshenko, S., and Goodier, J. N., *Theory of Elasticity*, McGraw-Hill, New York, 1951.
2. Juvinall, R. C., *Stress, Strain, and Strength*, McGraw-Hill, New York, 1967.

CHAPTER **5**

Relationships Between Stress and Strain

5.1 INTRODUCTION

Just as stress is an important mechanical modulus in assessing potential failure in a machine part, strain is also very important. *Strain* is the term used to define the intensity and direction of the deformation at a given point with respect to a specified plane within the solid body. Thus strain, like stress, is a second-order tensor quantity. Analogous to state of stress, state of strain at a point constitutes a complete description of the magnitudes and directions of strain on all possible planes passing through the point. All the concepts of principal stresses and planes of principal stress have direct analogs in terms of principal strains and planes of principal strain.

Relationships between stress and strain, both in the elastic and plastic ranges of behavior of engineering materials, are important and useful tools of the engineering designer. Such relationships also form the basis for many of the theories of failure discussed in Chapter 6.

The distinction between engineering stress and strain compared with true stress and strain is important to recognize, especially when machine parts operate in the plastic range. This distinction is identified in the following section, together with the related concepts of resilience, toughness, and instability.

5.2 CONCEPTS OF ENGINEERING STRESS-STRAIN AND TRUE STRESS-STRAIN

Usually, when nominal material properties are evaluated or specified, the concepts of engineering stress and engineering strain are employed. *Engineering stress*, or nominal stress, is defined to be the force per unit of *original* cross-sectional area, or

$$S = \frac{P}{A_o} \tag{5-1}$$

where S is engineering stress, P is the applied load, and A_o is the original cross-sectional area normal to P in a simple tensile test.

Similarly, *engineering strain*, or nominal strain, is defined as the elongation per unit of original length, or

$$\varepsilon = \frac{\Delta l}{l_o} \tag{5-2}$$

where ε is engineering strain, Δl is the change in gage length under load, and l_o is the original gage length prior to application of the load.

If a simple tension test is performed to measure engineering stress and engineering strain over a range of loading, an engineering stress-strain diagram may be plotted from the data, as shown, for example, in Figure 5.1.

The concept of resilience may be interpreted in terms of the stress-strain diagram of Figure 5.1. *Resilience* is defined as the ability of a material to absorb energy in the elastic range. The *modulus of resilience* is defined as the strain energy per unit volume stored in the specimen when it reaches incipient yielding. The modulus of resilience R may be expressed as

$$R = \frac{S_{yp}^2}{2E} \tag{5-3}$$

where S_{yp} is the yield point of the material and E is Young's modulus of elasticity. The resilience may be represented in Figure 5.1 as the area under the stress-strain curve up to the yield point.

Toughness is defined as the ability of a material to absorb energy prior to fracture. The *modulus of toughness* is defined as the strain energy per unit volume stored in the specimen when it reaches incipient fracture. The modulus of toughness T may be expressed as

$$T = \int_0^{\varepsilon_r} S d\varepsilon \tag{5-4}$$

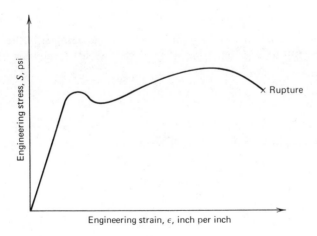

FIGURE 5.1.   An engineering stress-strain diagram.

where S is stress, ε is strain, and ε_r is the strain associated with incipient fracture. For certain ductile materials this integral is approximately equal to the ultimate strength multiplied by the strain associated with incipient fracture. This value is often called the *toughness merit number*. Referring again to Figure 5.1, we may state that the toughness is represented by the area under the stress-strain curve all the way out to fracture.

In the definition of engineering stress and engineering strain, the *original* cross-sectional area and the *original* gage length were used. Since the dimensions do, in fact, change as the load is applied, these calculations of engineering stress and strain are subject to errors. For ductile materials in the plastic range, and for certain brittle materials, these errors in stress and strain calculations based on A_o and l_o often become intolerable. For ductile materials in the elastic range, the errors are generally so small as to be negligible.

To provide a more accurate measure of stress and strain, the quantities true stress and true strain have been defined in the following way: *True stress S'* is the actual stress based on the *actual area A* corresponding at every instant to the current value of load P. Thus, true stress is given by

$$S' = \frac{P}{A} \tag{5-5}$$

Similarly, the *true strain* is associated not with the original gage length but with an *instantaneous value of gage length*, which changes with increase in the applied load. Consider, for example, that an increment of load is applied to change the load from zero to P_i, thereby changing the gage length from l_o to l_i. Suppose then that the load is again increased by a differential amount dP_i, which produces a change in gage length of dl_i. The differential strain for this general ith interval is dl_i/l_i. As the load is changed over its full range from zero to the final value, then the true strain δ is expressed as

$$\delta = \int_{l_o}^{l_f} \frac{dl_i}{l_i} \tag{5-6}$$

where l_o is the original gage length under no load and l_f is the final gage length under full load. Integration of (5-6) then yields true strain as

$$\delta = \ln\left(\frac{l_f}{l_o}\right) \tag{5-7}$$

True strain may be related to engineering strain by writing engineering strain from (5-2) as

$$\varepsilon = \frac{\Delta l}{l_o} = \frac{l_f - l_o}{l_o} = \frac{l_f}{l_o} - 1 \tag{5-8}$$

whereupon

$$\frac{l_f}{l_o} = 1 + \varepsilon \tag{5-9}$$

Taking the natural logarithm of both sides of (5-9) and introducing the result of (5-7), one obtains

$$\delta = \ln(1 + \varepsilon) \tag{5-10}$$

which can also be written as

$$\varepsilon = e^{\delta} - 1 \tag{5-11}$$

where e is the base of the Napierian logarithms.

Also, the true stress may be expressed in terms of engineering stress and engineering strain. To do so, the experimentally well verified assumption is made that in the plastic range the volume of the material remains constant throughout the loading, thus

$$A_o l_o = A l \tag{5-12}$$

or

$$\frac{l}{l_o} = \frac{A_o}{A} \tag{5-13}$$

Substituting from (5-13), then, (5-7) may be written as

$$\delta = \ln\left(\frac{A_o}{A}\right) \tag{5-14}$$

which may be combined with (5-10) to yield

$$\frac{A_o}{A} = 1 + \varepsilon \tag{5-15}$$

or

$$A = \frac{A_o}{1 + \varepsilon} \tag{5-16}$$

Recognizing from (5-1) that engineering stress S is P/A_o, we may use the expression of (5-16) in (5-5) to yield the true stress S' as

$$S' = S(1 + \varepsilon) \tag{5-17}$$

Experimental investigations have shown that many engineering alloys exhibit an approximately linear relationship between the logarithm of true stress and the logarithm of true strain. For these alloys the relationship between S' and δ may be expressed as

$$S' = k \delta^n \tag{5-18}$$

In this empirical equation, k and n are materials constants. The constant k is called the *strength coefficient*, and the constant n is called the *strain-hardening exponent*. A table of values for k and n for several engineering alloys is shown in Figure 5.2, together with a log-log plot of the relationship between true stress and true strain for these alloys. Not all materials follow the empirical relationship of (5-18), and other empirical equations have been devised to better characterize certain other materials. For example, the behavior of some

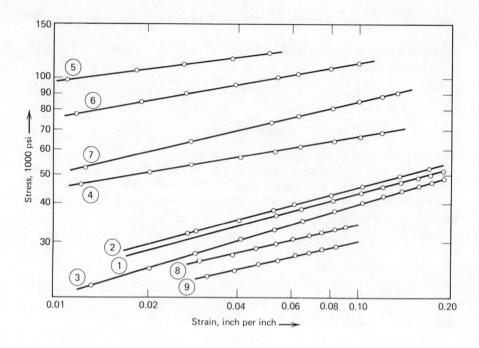

Material Constants *n* and *k* for Different Sheet Materials

	Material	Treatment	*n*	*k* (psi)	Thickness (inch)
1.	0.05 per cent Carbon rimmed steel	Annealed	0.261	77,100	0.037
2.	0.05 per cent Carbon killed steel	Annealed and temper-rolled	.234	73,100	.037
3.	Same as #2, completely decarburized	Annealed in wet hydrogen	.284	75,500	.037
4.	0.05/0.07 per cent Phosphorus low-carbon steel	Annealed	.156	93,330	.037
5.	SAE 4130 steel	Annealed	.118	169,400	.037
6.	SAE 4130 steel	Normalized and temper-rolled	.156	154,500	.037
7.	Type 430 stainless steel (17 per cent Cr)	Annealed	.229	143,000	.050
8.	Alcoa 24-S aluminum	Annealed	.211	55,900	.040
9.	Reynolds R-301 aluminum	Annealed	.211	48,450	.040

Tests by J. R. Low and F. Garafalo.

FIGURE 5.2. Relationship between true stress and true strain for several engineering alloys. Numbers in the table correspond to numbers on the curves. (From ref. 1; Joseph Marin, *Mechanical Behavior of Engineering Materials,* © 1962, pp. 39 and 40. Reprinted by permission of Prentice-Hall, Inc., Englewood Cliffs, New Jersey.)

plastic materials is better described by a relationship of the type $\delta = k_1 + k_2 \log S'$.

When many materials are loaded, for example in a simple tension test, it is noted that as the load is increased a point is reached where localized plastic deformation begins. At this point the load required to produce deformation reaches a maximum value and then drops off until rupture takes place. This point of maximum load and the onset of localized plastic deformation is often called the *point of instability*. Designers would never be interested in designing at stress levels that exceed the point of instability because rupture is spontaneous. The point of instability may be determined experimentally, or it may be predicted in the following way: From (5-14) one may write

$$A = \frac{A_o}{e^\delta} \qquad (5\text{-}19)$$

which may be combined with (5-5) to yield

$$P = \frac{S'A_o}{e^\delta} \qquad (5\text{-}20)$$

Recognizing that the point of instability occurs when the change in load to produce increased strain becomes zero, that is, when $dP/d\delta$ becomes zero, (5-20) may be differentiated with respect to δ and set equal to zero to give

$$\frac{dP}{d\delta} = \frac{A_o}{e^\delta}\left[\frac{dS'}{d\delta} - S'\right] = 0 \qquad (5\text{-}21)$$

The point of instability is, therefore, the point at which

$$\frac{dS'_u}{d\delta} = S'_u \qquad (5\text{-}22)$$

where the subscript u has been added to identify the particular point at which the instability appears; it is in keeping with the concept of an experimentally determined ultimate strength. Hence, from (5-22) it would be predicted that the point of instability occurs when the slope of the true stress-strain curve equals the true stress at the same point. A graphical technique* for locating the instability point on a true stress-strain diagram is illustrated in Figure 5.3, where the maximum load corresponds to point A where $\overline{CB}$ is 1.0 inch per inch. The justification for this construction is made on the basis that

$$\frac{dS'_u}{d\delta_u} = \tan \underline{/ABC} = \frac{\overline{AC}}{\overline{CB}} = \overline{AC} = S'_u \qquad (5\text{-}23)$$

While the instability point can be determined graphically, as shown in Figure 5.3, it is more convenient for materials that are adequately characterized by (5-18) to determine the point analytically. From (5-18) applied to

*Developed by Marin. See pp. 42–43 of ref. 1.

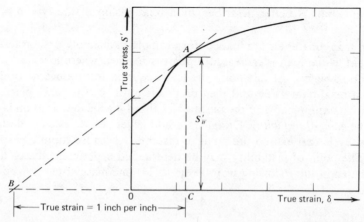

FIGURE 5.3. A geometrical construction for determining the point of load instability on a true stress true strain diagram. (From ref. 4; Joseph Marin, *Mechanical Behavior of Engineering Materials*, © 1962, p. 43. Reprinted by permission of Prentice-Hall, Inc., Englewood Cliffs, New Jersey.)

the instability point,

$$S'_u = k \, \delta_u^n \qquad (5\text{-}24)$$

Substituting this expression back into (5-22) and performing the differentiation yields

$$nk \, \delta_u^{n-1} = k \, \delta_u^n \qquad (5\text{-}25)$$

from which

$$\delta_u = n \qquad (5\text{-}26)$$

That is, the point of load instability occurs at a value of true strain equal to the strain-hardening exponent. Many of these concepts will be useful in later developments.

5.3   ELASTIC STRESS-STRAIN RELATIONSHIPS

The linear relationship between stress and strain in the elastic range for homogeneous and isotropic engineering materials has been well established experimentally. The relationship was first presented in the seventeenth century by Robert Hooke, an Englishman, who described the linear relationship in connection with his invention of applying a hairspring to a clock; hence, the linear relationships between stress and strain in the elastic range have come to be known as the Hooke's law relationships.

Consider an elemental parallelepiped with faces parallel to the coordinate planes of an x-y-z coordinate system, as shown in Figure 5.4. Experiments

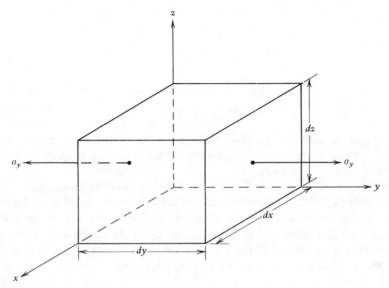

FIGURE 5.4. An elemental parallelepiped subjected to a pure tensile force in the y-direction.

have shown that if such an element is subjected to a uniform normal stress σ_y as shown, *no angular distortions are produced*. Further, it may be observed experimentally that for the case of a uniform normal tensile stress σ_y applied as shown in Figure 5.4, the element will increase in length in the y direction and decrease in its dimensions in both the x direction and the z direction. Similarly, if the sense of σ_y were compressive, the y direction dimension would shorten, whereas the x and z dimensions would increase. This experimentally observed change of length $\Delta(dy)$ in the y direction may be expressed as a function of applied stress σ_y as

$$\Delta(dy) = \frac{\sigma_y \, dy}{E} \tag{5-27}$$

where dy is the unstressed dimension in the y direction and E is an experimentally determined constant of proportionality called Young's modulus.

The unit elongation, or elastic strain, in the y direction caused by application of the stress σ_y may be expressed as a change in length $\Delta(dy)$ divided by the initial length dy, whence, utilizing (5-27),

$$\varepsilon_{yy} = \frac{\Delta dy}{dy} = \frac{\sigma_y}{E} \tag{5-28}$$

where ε_{yy} is the elastic strain in the y direction (first subscript) caused by a stress in the y direction (second subscript).

The lateral changes in dimensions in the x and z directions as a result of the y-direction stress σ_y may, on the basis of experimental evidence, be expressed as

$$\varepsilon_{xy} = -\nu\varepsilon_{yy} = -\nu\frac{\sigma_y}{E} \tag{5-29}$$

and

$$\varepsilon_{zy} = -\nu\varepsilon_{yy} = -\frac{\nu\sigma_y}{E} \tag{5-30}$$

where ε_{ij} is the unit change in dimension, or strain, in the i direction caused by the application of a stress σ_j in the j direction, and ν is another material constant called Poisson's ratio.

Referring again to Figure 5.4, we see that if σ_y were removed and replaced with a stress σ_x in the x direction, an entirely analogous pattern of deformations and strains would develop. A similar analogous pattern would also be observed for the case of a σ_z stress acting alone. If the results of all these experiments were tabulated, the strains produced in the x, y, and z directions by each of the stresses, σ_x, σ_y, and σ_z, applied one at a time, would be as shown in Table 5.1.

The small strains associated with the elastic range are linear functions of applied stress and, therefore, the principle of superposition holds. That is, the total strain in any given direction due to the simultaneous application of σ_x, σ_y, and σ_z is the same as the sum of the strain in that direction produced by σ_x acting alone plus the strain in that direction produced by σ_y acting alone plus the strain in that direction due to σ_z acting alone. This observation may be more precisely stated as

$$\varepsilon_x = \varepsilon_{xx} + \varepsilon_{xy} + \varepsilon_{xz}$$
$$\varepsilon_y = \varepsilon_{yx} + \varepsilon_{yy} + \varepsilon_{yz} \tag{5-31}$$
$$\varepsilon_z = \varepsilon_{zx} + \varepsilon_{zy} + \varepsilon_{zz}$$

where ε_x, ε_y, and ε_z are, respectively, the total strains in the x, y, and z directions due to the combined effect of all three normal stresses. If appropriate substitutions are made in (5-31) from Table 5.1, the following

Table 5.1. Components of Strain Produced by Normal Stresses in the x, y, and z Directions

Direction of Strain	Stress Causing the Strain		
	σ_x	σ_y	σ_z
x	$\varepsilon_{xx} = \dfrac{\sigma_x}{E}$	$\varepsilon_{xy} = -\dfrac{\nu\sigma_y}{E}$	$\varepsilon_{xz} = -\dfrac{\nu\sigma_z}{E}$
y	$\varepsilon_{yx} = -\dfrac{\nu\sigma_x}{E}$	$\varepsilon_{yy} = \dfrac{\sigma_y}{E}$	$\varepsilon_{yz} = -\dfrac{\nu\sigma_z}{E}$
z	$\varepsilon_{zx} = -\dfrac{\nu\sigma_x}{E}$	$\varepsilon_{zy} = -\dfrac{\nu\sigma_y}{E}$	$\varepsilon_{zz} = \dfrac{\sigma_z}{E}$

expressions result:

$$\varepsilon_x = \frac{1}{E}\left[\sigma_x - \nu(\sigma_y + \sigma_z)\right]$$

$$\varepsilon_y = \frac{1}{E}\left[\sigma_y - \nu(\sigma_x + \sigma_z)\right] \qquad (5\text{-}32)$$

$$\varepsilon_z = \frac{1}{E}\left[\sigma_z - \nu(\sigma_x + \sigma_y)\right]$$

These three equations are the Hooke's law equations. They relate the normal or elongational strains to the applied normal stresses through two constants, E and ν. One may deduce directly from (5-32) that the relationships for the principal strains ε_1, ε_2, and ε_3 in terms of the principal stresses are

$$\varepsilon_1 = \frac{1}{E}\left[\sigma_1 - \nu(\sigma_2 + \sigma_3)\right]$$

$$\varepsilon_2 = \frac{1}{E}\left[\sigma_2 - \nu(\sigma_1 + \sigma_3)\right] \qquad (5\text{-}33)$$

$$\varepsilon_3 = \frac{1}{E}\left[\sigma_3 - \nu(\sigma_1 + \sigma_2)\right]$$

Just as elastic extensional strains are linearly related to the normal stresses through two materials constants, linear relationships may also be established between shearing strains and shearing stresses using the same two materials constants. It may be recalled from elementary definitions of shearing strain that, for example, the shearing strain γ_{xy} is the change in the angle between two lines in the x-y plane that were perpendicular to each other in the unstrained state. Such a shearing strain would be produced by the shearing stress τ_{xy} acting on an elastic element.

To develop the elastic shearing stress-strain relationships, the case of pure shear in the x-y plane will be discussed and then extended by analogy to the x-z and y-z planes. Considering the case of pure shear γ_{xy}, we find that the stress cubic equation (4-23), with all stress components zero except τ_{xy}, yields

$$\sigma^3 - \sigma\tau_{xy}^2 = 0 \qquad (5\text{-}34)$$

which, when solved, results in the three roots

$$\sigma_1 = \tau_{xy} \qquad (5\text{-}35)$$

$$\sigma_2 = 0 \qquad (5\text{-}36)$$

$$\sigma_3 = -\tau_{xy} \qquad (5\text{-}37)$$

Utilizing (4-20) and (4-21) to determine the direction cosines associated with these roots, one finds that the principal planes make angles of 45 degrees with respect to the x-y axes. To state it another way, if an element is subjected to a normal stress field in which $\sigma = \sigma_1 = -\sigma_3$ and $\sigma_2 = 0$, it is found that on planes at 45 degrees to the 1-3 axes there will be induced a pure shear equal in magnitude to principal stress σ.

Referring to Figure 5.5, we note that the angles between the τ_{xy} planes and the σ_1 or σ_3 planes are 45 degrees prior to the application of the stresses. When the stresses are applied, the points a, b, c, and d move to position a', b', c', and d' respectively. For purposes of the following geometrical argument, the angle abc has been shown in an enlarged view at the bottom of Figure 5.5. Comparison of angles has been made more convenient by drawing the apexes coincident in that sketch. The shearing strain γ_{xy} in the x-y plane is measured by the change in angle abc upon the application of the stress field; thus the shearing strain is the difference between angle abc and angle $a'b'c'$. The angular change is brought about upon the application of σ_1 and σ_3 because diagonal $\overline{bd}$ elongates a small amount and diagonal $\overline{ac}$ shortens a small

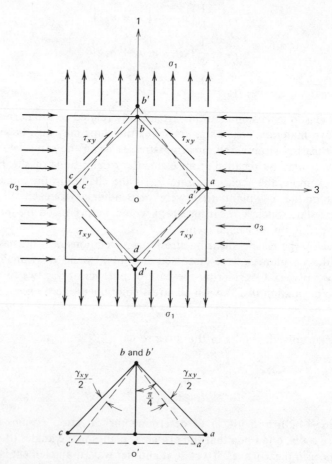

FIGURE 5.5. Illustration of planes of pure shear produced by the application of equal and opposite normal stresses on a biaxial element.

amount. From the lower sketch of Figure 5.5 it may be deduced that

$$\tan\left(\frac{\pi}{4} - \frac{\gamma_{xy}}{2}\right) = \frac{\overline{o'a'}}{\overline{o'b'}} \tag{5-38}$$

However, by utilizing the definition of strain, the right-hand side of (5-38) may be written as

$$\frac{\overline{o'a'}}{\overline{o'b'}} = \frac{oa(1 + \varepsilon_3)}{ob(1 + \varepsilon_1)} \tag{5-39}$$

But from the upper sketch of Figure 5.5, $\overline{oa}$ and $\overline{ob}$ are equal, so that combining (5-38) and (5-39) yields

$$\tan\left(\frac{\pi}{4} - \frac{\gamma_{xy}}{2}\right) = \frac{1 + \varepsilon_3}{1 + \varepsilon_1} \tag{5-40}$$

Now, utilizing the trigonometric identity

$$\tan(\alpha - \beta) = \frac{\tan\alpha - \tan\beta}{1 + \tan\alpha\tan\beta} \tag{5-41}$$

the left side of (5-40) may be expressed as

$$\tan\left(\frac{\pi}{4} - \frac{\gamma_{xy}}{2}\right) = \frac{\tan\dfrac{\pi}{4} - \tan\dfrac{\gamma_{xy}}{2}}{1 + \tan\dfrac{\pi}{4}\tan\dfrac{\gamma_{xy}}{2}} \tag{5-42}$$

Now, since γ_{xy} is a very small angle, it is accurate to set $\tan(\gamma_{xy}/2)$ equal to $\gamma_{xy}/2$, whereupon (5-42) becomes

$$\tan\left(\frac{\pi}{4} - \frac{\gamma_{xy}}{2}\right) = \frac{1 - \dfrac{\gamma_{xy}}{2}}{1 + \dfrac{\gamma_{xy}}{2}} \tag{5-43}$$

Equating (5-43) to (5-40) then gives

$$\frac{1 - \dfrac{\gamma_{xy}}{2}}{1 + \dfrac{\gamma_{xy}}{2}} = \frac{1 + \varepsilon_3}{1 + \varepsilon_1} \tag{5-44}$$

Since $\sigma_1 = -\sigma_3$, it may be determined from Hookes' law that $\varepsilon_3 = -\varepsilon_1$ and therefore (5-44) may be written as

$$\frac{1 - \dfrac{\gamma_{xy}}{2}}{1 + \dfrac{\gamma_{xy}}{2}} = \frac{1 - \varepsilon_1}{1 + \varepsilon_1} \tag{5-45}$$

which, when solved for γ_{xy}, yields

$$\gamma_{xy} = 2\varepsilon_1 \tag{5-46}$$

Substituting the first equation of (5-33) into (5-46) for ε_1, the shearing strain γ_{xy} becomes

$$\gamma_{xy} = \frac{2}{E}\left[\sigma_1 - \nu(\sigma_2 + \sigma_3)\right] \tag{5-47}$$

Recalling from (5-35), (5-36), and (5-37) that $\sigma_2 = 0$ and $\sigma_1 = -\sigma_3$, the expression of (5-47) becomes

$$\gamma_{xy} = \frac{2(1 + \nu)}{E}\sigma_1 \tag{5-48}$$

or, since $\sigma_1 = \tau_{xy}$,

$$\gamma_{xy} = \frac{2(1 + \nu)}{E}\tau_{xy} \tag{5-49}$$

It is customary to define a term G, called the shear modulus of elasticity, as

$$G = \frac{E}{2(1 + \nu)} \tag{5-50}$$

whereupon (5-49) may be expressed as

$$\gamma_{xy} = \frac{\tau_{xy}}{G} \tag{5-51}$$

Similar reasoning leads to similar expressions for the shearing strains γ_{xz} and γ_{yz}. To summarize these results, it is possible to express the shearing strains γ_{ij} in terms of the corresponding shearing stresses τ_{ij} and a material constant G. Thus,

$$\gamma_{xy} = \frac{\tau_{xy}}{G}$$

$$\gamma_{yz} = \frac{\tau_{yz}}{G} \tag{5-52}$$

$$\gamma_{xz} = \frac{\tau_{xz}}{G}$$

Further, if one defines the principal shearing strain γ_3 as the shearing strain caused by principal shearing stress $\tau_3 = \pm 1/2(\sigma_1 - \sigma_2)$, the following expressions may be written:

$$\gamma_3 = \pm\frac{\tau_3}{G} = \pm\frac{\sigma_1 - \sigma_2}{2G}$$

$$\gamma_1 = \pm\frac{\tau_1}{G} = \pm\frac{\sigma_2 - \sigma_3}{2G} \tag{5-53}$$

$$\gamma_2 = \pm\frac{\tau_2}{G} = \pm\frac{\sigma_1 - \sigma_3}{2G}$$

Finally, the principal shearing strains may be written in terms of the principal normal strains by combining (5-33) with (5-53) to yield

$$\gamma_3 = \varepsilon_1 - \varepsilon_2$$

$$\gamma_1 = \varepsilon_2 - \varepsilon_3 \qquad\qquad (5\text{-}54)$$

$$\gamma_2 = \varepsilon_1 - \varepsilon_3$$

It is important and useful to recognize that the elongational or normal strains of (5-32) and the distortional or shear strains of (5-52) are independent of each other. Thus, in the general case of a triaxial state of strain in the linear elastic region, the resultant state of strain can be obtained by superposition of the shearing strains of (5-52) upon the normal strains of (5-32).

Finally, it is important to recall that in Section 5.1 it was stated that the state of strain at a point is directly analogous to state of stress at a point. Thus it is that the state of strain at a point is completely defined by six components of strain, three normal strain components and three shearing strain components. The normal strain components are ε_x, ε_y, and ε_z as previously defined in (5-31). The shearing strain components are γ_{xy}, γ_{yz}, and γ_{xz} as shown in (5-52). Just as the state of stress at a point may be completely defined in terms of three principal stresses and their directions, the state of strain at a point may be completely defined in terms of three principal strains and their direction. These principal strains may be determined from a general strain cubic equation corresponding to the general stress cubic equation (4-23). The strain cubic equation is

$$\varepsilon^3 - \varepsilon^2 \big[\, \varepsilon_x + \varepsilon_y + \varepsilon_z \,\big] + \varepsilon \big[\, \varepsilon_x \varepsilon_y + \varepsilon_y \varepsilon_z + \varepsilon_x \varepsilon_z - 1/4 \big(\gamma_{xy}^2 + \gamma_{yz}^2 + \gamma_{xz}^2 \big) \big]$$

$$- \big[\, \varepsilon_x \varepsilon_y \varepsilon_z + 1/4 \big(\gamma_{xy} \gamma_{yz} \gamma_{xz} - \varepsilon_x \gamma_{yz}^2 - \varepsilon_y \gamma_{xz}^2 - \varepsilon_z \gamma_{xy}^2 \big) \big] = 0$$

$$(5\text{-}55)$$

This equation is of the same form as the stress cubic equation of (4-23) except that the normal strains and one-half the shear strains replace the normal stresses and shear stresses. The three solutions of (5-55) will provide the three principal strains ε_1, ε_2, and ε_3. The directions of the planes of principal strain may be determined by following the technique leading to (4-20) and (4-21) for stresses.

5.4 PLASTIC STRESS-STRAIN RELATIONSHIPS

Since machine parts and structures are sometimes operated beyond the yield point, it is important to investigate the stress-strain relationships in the plastic range where the linear elastic relationships no longer are applicable. The stress-strain relationships in the plastic region are not generally independent of time. Any exact theory of plastic deformation should take into account the entire history of deformation from the time the plastic flow was initiated. Such relationships would be complex, involving the stress and the time rate of strain. The equations would be analogous to the equations for flow of a viscous fluid, and the strain at any given time would be determined through a step-by-step integration over the entire strain history. This process results in a cumbersome analysis, even for the simplest problems of plastic deformation. Because of the complexity of such an analysis, it is usual to make certain simplifying assumptions that permit a relatively simple analysis of plastic deformation that is reasonably accurate so long as the temperature is below the creep temperature and the rate of strain is "normal."

Two simplified theories have been proposed for use in the plastic range. These are (1) the *proportional deformation* theory and (2) the *incremental strain* theory. The proportional deformation theory is actually a simplified case of the incremental strain theory in which the ratio of principal shearing strains to corresponding shearing stresses are assumed equal at any time during the deformation. As long as temperatures are below the creep range and strain rates are reasonably low, the proportional deformation theory yields relatively accurate results.

In developing the expressions for the plastic stress-strain relationships using the proportional deformation theory, we make the following five assumptions:

1. It is assumed that the change in volume as a result of plastic deformation is zero; that is,

$$\Delta V = 0 \tag{5-56}$$

This assumption has been experimentally well verified, with observed volume changes of less than 0.25 percent for up to tenfold plastic extensions.

2. It is assumed that the true principal strains remain parallel to the true principal stresses throughout the deformation;* that is,

$$\delta_1 \| \sigma_1'$$
$$\delta_2 \| \sigma_2' \tag{5-57}$$
$$\delta_3 \| \sigma_3'$$

3. It is assumed that the ratios of the principal shearing strains to the corresponding principal shearing stresses are equal at any stage of plastic deformation, or

$$\frac{\delta_1 - \delta_2}{\sigma_1' - \sigma_2'} = \frac{\delta_2 - \delta_3}{\sigma_2' - \sigma_3'} = \frac{\delta_3 - \delta_1}{\sigma_3' - \sigma_1'} = C_i \tag{5-58}$$

This assumption is true for a great many cases, but caution is urged in making sure that this assumption is a fair estimate of the actual material behavior. For moderate temperatures and rates of plastic strain (the "usual" case), this assumption is acceptable.

---

*The case of severe plastic torsion is one in which the directions of stress and strain differ. See Chap. 14 of ref. 3.

4. It is assumed that the elastic components of strain are negligible compared to the plastic components of strain.

5. It is assumed that the plastic flow assumptions and equations apply only for the case of increasing loads. For decreasing loads the behavior is assumed to be elastic; hence, the Hooke's law relationships hold.

From the first assumption, that plastic flow results in a volume change of zero, it may be shown that the sum of the three principal true strains is zero. To show this, consider a small parallelepiped of dimensions $a \times b \times c$. The volume V_o of the parallelepiped in its original unstrained condition is

$$V_o = abc \tag{5-59}$$

Next, let the parallelepiped be strained by an amount ε_1 in the a direction, ε_2 in the b direction, and ε_3 in the c direction. The dimensions of the cube in the strained conditions would then be $a(1 + \varepsilon_1) \times b(1 + \varepsilon_2) \times c(1 + \varepsilon_3)$, and the volume V_s in the strained condition would be

$$V_s = abc(1 + \varepsilon_1)(1 + \varepsilon_2)(1 + \varepsilon_3) \tag{5-60}$$

However, by the assumption of no volume change in (5-56)

$$V_o = V_s \tag{5-61}$$

Utilizing (5-59) and (5-60), we see that the expression of (5-61) becomes

$$(1 + \varepsilon_1)(1 + \varepsilon_2)(1 + \varepsilon_3) = 1 \tag{5-62}$$

Taking the natural logarithm of both sides and recalling from (5-10) that $\delta = \ln(1 + \varepsilon)$,

$$\delta_1 + \delta_2 + \delta_3 = 0 \tag{5-63}$$

Also, two independent expressions from (5-58) may be written as

$$\delta_1 - \delta_2 = C_i(\sigma_1' - \sigma_2') \tag{5-64}$$

$$\delta_1 - \delta_3 = C_i(\sigma_1' - \sigma_3') \tag{5-65}$$

Solving equations (5-63), (5-64), and (5-65) simultaneously yields

$$\delta_1 = \frac{2C_i}{3}\left[\sigma_1' - \frac{1}{2}(\sigma_2' + \sigma_3')\right] \tag{5-66}$$

$$\delta_2 = \frac{2C_i}{3}\left[\sigma_2' - \frac{1}{2}(\sigma_1' + \sigma_3')\right] \tag{5-67}$$

$$\delta_3 = \frac{2C_i}{3}\left[\sigma_3' - \frac{1}{2}(\sigma_1' + \sigma_2')\right] \tag{5-68}$$

This set of equations would be formally the same as the Hooke's law equations of (5-33) if one defined an instantaneous modulus of plasticity $D = 3/2C_i$ and Poisson's ratio of $v = 1/2$. It must be recognized that the modulus of plasticity D is not a constant but is a variable function of the amount of prior plastic strain. The limiting value of $\frac{1}{2}$ for Poisson's ratio is widely confirmed for plastic behavior. Thus the set of equations (5-66), (5-67), and (5-68) yields the principal true strains in terms of the principal true stresses and a quantity C_i yet to be determined. If the quantity C_i were known, the principal true strains would be completely defined in terms of the principal true stresses.

In (5-18) an expression was written for true stress as a function of true strain in a simple tension test. If one knew how the multiaxial states of plastic stress and strain were related to the simple tensile relationship of (5-18), the expressions of (5-66), (5-67), and (5-68) could be evaluated in a more useful form. To relate the behavior of a material in a multiaxial state of stress or strain to simple uniaxial behavior requires the postulation of a combined stress theory. Combined stress theories are discussed in detail in Chapter 6, where they are used to develop theories of failure for the general multiaxial state of stress. It will be observed in Chapter 6 that the best combined stress theory for ductile behavior is the distortion energy theory, or octahedral shear stress theory. Accepting for now that the octahedral shear stress theory is the most appplicable to the region of plastic deformation, the expressions for octahedral shearing stress τ_o and octahedral shearing strain γ_o have been derived by Nadai* as

$$\tau_o = \frac{1}{3}\sqrt{(\sigma_1' - \sigma_2')^2 + (\sigma_2' - \sigma_3')^2 + (\sigma_3' - \sigma_1')^2} \qquad (5\text{-}69)$$

and

$$\gamma_o = \frac{2}{3}\sqrt{(\delta_1 - \delta_2)^2 + (\delta_2 - \delta_3)^2 + (\delta_3 - \delta_1)^2} \qquad (5\text{-}70)$$

For the case of simple uniaxial tension, the only nonzero stress would be σ_1', and the octahedral shearing stress would reduce to

$$\tau_o = \frac{1}{3}\sqrt{(\sigma_1')^2 + (\sigma_1')^2} = \frac{\sqrt{2}}{3}\sigma_1' \qquad (5\text{-}71)$$

Likewise, for the case of simple uniaxial tension, with Poisson's ratio of one-half, the octahedral shearing strain would reduce to

$$\gamma_o = \left(\frac{2}{3}\right)\sqrt{\left[\delta_1 - \left(-\frac{\delta_1}{2}\right)\right]^2 + \left[\left(-\frac{\delta_1}{2}\right) - \left(-\frac{\delta_1}{2}\right)\right]^2 + \left[-\frac{\delta_1}{2} - \delta_1\right]^2} = \sqrt{2}\,\delta_1$$

$$(5\text{-}72)$$

Also, for the uniaxial case, (5-66) becomes

$$\delta_1 = \frac{2C_i}{3}\sigma_1' \qquad (5\text{-}73)$$

*See pp. 103 and 115 of ref. 2.

whence

$$C_i = \frac{3}{2} \frac{\delta_1}{\sigma_1'} \tag{5-74}$$

Substituting in this expression for δ_1 from (5-72) and for σ_1' from (5-71) gives

$$C_i = \frac{3}{2} \frac{\dfrac{\gamma_o}{\sqrt{2}}}{\dfrac{3\tau_o}{\sqrt{2}}} = \frac{\gamma_o}{2\tau_o} \tag{5-75}$$

If the results of (5-71) and (5-72) are put into (5-18), it becomes

$$\frac{3}{\sqrt{2}} \tau_o = k \left(\frac{\gamma_o}{\sqrt{2}} \right)^n \tag{5-76}$$

which may be rewritten as

$$\gamma_o = \sqrt{2} \left(\frac{3\tau_o}{\sqrt{2}\,k} \right)^{1/n} \tag{5-77}$$

Utilizing this expression for octahedral shear strain in terms of octahedral shear stress, we can write expression (5-75) as

$$C_i = \frac{\sqrt{2}}{2\tau_o} \left(\frac{3\tau_o}{\sqrt{2}\,k} \right)^{1/n} \tag{5-78}$$

Finally, making the assumption that the expression for C_i will remain the same for the multiaxial state of stress, the multiaxial expression for τ_o in (5-69) is substituted into (5-78) to obtain

$$C_i = 3(2)^{(-1-n)/2n} \left(\frac{1}{k} \right)^{1/n} \left[(\sigma_1' - \sigma_2')^2 + (\sigma_2' - \sigma_3')^2 + (\sigma_3' - \sigma_1')^2 \right]^{(1-n)/2n} \tag{5-79}$$

To simplify the following plastic stress-strain expressions the following ratios are defined:

$$\alpha = \frac{\sigma_2'}{\sigma_1'} \tag{5-80}$$

$$\beta = \frac{\sigma_3'}{\sigma_1'} \tag{5-81}$$

where

$$\sigma_1' \geq \sigma_2' \geq \sigma_3' \tag{5-82}$$

With these definitions the equations of (5-66), (5-67), and (5-68) may be written as

$$\delta_1 = \left[\frac{\sigma_1'}{k}\right]^{1/n}\left[\alpha^2 + \beta^2 + 1 - \alpha\beta - \alpha - \beta\right]^{(1-n)/2n}\left[1 - \frac{\alpha}{2} - \frac{\beta}{2}\right] \quad (5\text{-}83)$$

$$\delta_2 = \left[\frac{\sigma_1'}{k}\right]^{1/n}\left[\alpha^2 + \beta^2 + 1 - \alpha\beta - \alpha - \beta\right]^{(1-n)/2n}\left[\alpha - \frac{\beta}{2} - \frac{1}{2}\right] \quad (5\text{-}84)$$

$$\delta_3 = \left[\frac{\sigma_1'}{k}\right]^{1/n}\left[\alpha^2 + \beta^2 + 1 - \alpha\beta - \alpha - \beta\right]^{(1-n)/2n}\left[\beta - \frac{\alpha}{2} - \frac{1}{2}\right] \quad (5\text{-}85)$$

For more general conditions of plastic flow, for example when the ratios of principal shearing strains to principal shearing stresses are not constant, the incremental strain theory should be utilized. In this case, the expressions would be written in terms of true strain increments, and it would be necessary to sum the increments over the strain history to obtain results comparable to (5-83), (5-84), and (5-85). The reader is referred to references at the end of this chapter for a more complete treatment of the incremental strain relationships in plastic analysis.

5.5   USING THE IDEAS

Consider the following problem: A thin-walled pressure vessel is to be constructed from SAE 4130 steel sheet, normalized and temper-rolled. The vessel is to be 18 inches in outside diameter with a 1/8-inch-thick wall, as illustrated in Figure 5.6. Utilizing the proportional deformation theory of plastic flow, and employing the octahedral shear stress theory for combined stress, determine the pressure for which plastic instability occurs. Compare this pressure with the pressure at which yielding is initiated. Consider only the cylindrical wall and neglect end effects.

Referring to the sketch of Figure 5.6, we observe that the radial stress σ_3' will be small compared to σ_1' and σ_2' and will be neglected in the following simple analysis. The validity of this assumption may be verified at the end of this calculation by comparing numerical values of $\sigma_3' \doteq p$ with σ_1' and σ_2'. Thus, by assumption,

$$\sigma_3' = 0 \quad (5\text{-}86)$$

whence, from (5-81),

$$\beta = \frac{\sigma_3'}{\sigma_1'} = 0 \quad (5\text{-}87)$$

From elementary considerations of static equilibrium, the hoop stress σ_1' in the wall of the vessel of Figure 5.6 is obtained from

$$2(\sigma_1')tl = pD_i l \quad (5\text{-}88)$$

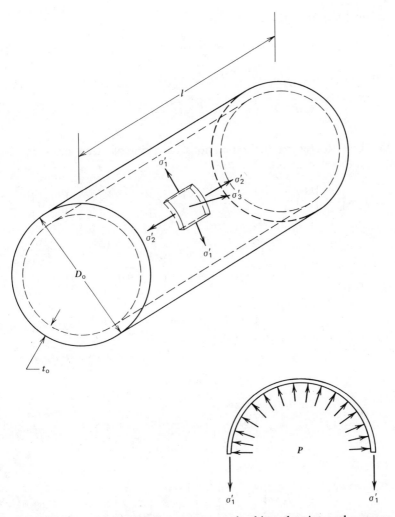

FIGURE 5.6.   A thin-walled pressure vessel subjected to internal pressure.

or

$$\sigma_1' = \frac{pD_i}{2t} \tag{5-89}$$

Again, on the basis of static equilibrium, the longitudinal stress is obtained from

$$\sigma_2'(\pi D_m t) = p\frac{\left(\pi D_i^2\right)}{4} \tag{5-90}$$

For a thin-walled vessel the mean diameter, D_m, and inside diameter, D_i, are for practical purposes the same, whence

$$\sigma_2' = \frac{pD_i}{4t}$$ (5-91)

Thus, from (5-80), (5-89), and (5-91) the expression for α becomes

$$\alpha = \frac{\sigma_2'}{\sigma_1'} = \frac{pD_i/4t}{pD_i/2t} = \frac{1}{2}$$ (5-92)

For this case, therefore, the expressions for principal true strain in (5-83), (5-84), and (5-85) become

$$\delta_1 = \left[\frac{\sigma_1'}{k}\right]^{1/n}\left[\frac{1}{4} + 0 + 1 - 0 - \frac{1}{2} - 0\right]^{(1-n)/2n}\left[1 - \frac{1}{4} - 0\right]$$ (5-93)

or

$$\delta_1 = \left[\frac{\sigma_1'}{k}\right]^{1/n}\left[\frac{3}{4}\right]^{(1-n)/2n}\left[\frac{3}{4}\right]$$ (5-94)

and

$$\delta_2 = \left[\frac{\sigma_1'}{k}\right]^{1/n}\left[\frac{3}{4}\right]^{(1-n)/2n}\left[\frac{1}{2} - 0 - \frac{1}{2}\right]$$ (5-95)

or

$$\delta_2 = 0$$ (5-96)

and

$$\delta_3 = \left[\frac{\sigma_1'}{k}\right]^{1/n}\left[\frac{3}{4}\right]^{(1-n)/2n}\left[0 - \frac{1}{4} - \frac{1}{2}\right]$$ (5-97)

or

$$\delta_3 = \left[\frac{\sigma_1'}{k}\right]^{1/n}\left[\frac{3}{4}\right]^{(1-n)/2n}\left[-\frac{3}{4}\right]$$ (5-98)

Thus the true hoop strain δ_1 is equal and opposite to the true radial strain, whereas the longitudinal strain is zero. This result has been experimentally verified.

Substituting for σ_1' its expression from (5-89) into the true strain equations yields

$$\delta_1 = \left[\frac{3}{4}\right]^{(1+n)/2n}\left[\frac{pD_i}{2tk}\right]^{1/n}$$ (5-99)

$$\delta_2 = 0$$ (5-100)

$$\delta_3 = -\delta_1$$ (5-101)

Now, by the definition of true strain, the circumferential (hoop) true strain may be written as

$$\delta_1 = \ln\left(\frac{l}{l_o}\right) = \ln\left(\frac{\pi D}{\pi D_o}\right) \tag{5-102}$$

where D_o is the original mean diameter of the vessel before pressurizing and D is the strained diameter corresponding to any other pressure p. From (5-102), then,

$$D = D_o e^{\delta_1} \tag{5-103}$$

By a similar argument the radial true strain may be written as

$$\delta_3 = \ln\left(\frac{t}{t_o}\right) \tag{5-104}$$

where t_o is the original wall thickness before pressurizing and t is the wall thickness corresponding to any other pressure p. Solving (5-104) for t and in view of (5-101),

$$t = t_o e^{\delta_3} = t_o e^{-\delta_1} \tag{5-105}$$

Substituting the results of (5-103) and (5-105) back into (5-99) gives

$$\delta_1 = \left[\frac{3}{4}\right]^{(1+n)/2n}\left[\frac{pD_o e^{\delta_1}}{2kt_o e^{-\delta_1}}\right]^{1/n} \tag{5-106}$$

which may be solved for p as

$$p = \left[\frac{2kt_o}{D_o}\right]e^{-2\delta_1}\left[\frac{\delta_1}{\left(\dfrac{3}{4}\right)^{(1+n)/2n}}\right]^n \tag{5-107}$$

From (5-107) it may be reasoned that the true strain δ_1 increases with increasing pressure until the point of instability is reached, where no additional pressure increase is required to produce an additional increase in plastic strain. That is, the instability condition is reached when $dp/d\delta_1$ becomes zero. To determine this point of instability, the expression of (5-107) is first differentiated with respect to δ_1, giving

$$\frac{dp}{d\delta_1} = \left[\frac{2kt_o}{D_o}\right]\left[\frac{1}{\left(\dfrac{3}{4}\right)^{(1+n)/2n}}\right]^n\left[-2e^{-2\delta_1}\delta_1^n + n\delta_1^{n-1}e^{-2\delta_1}\right] \tag{5-108}$$

The derivative (5-108) is then set equal to zero and solved for the instability value of true strain δ_{1_u} as

$$\delta_{1_u} = \frac{n}{2} \tag{5-109}$$

Thus the point of instability is reached when the true hoop strain δ_1 reaches a value of one-half the strain-hardening exponent for the material. To obtain the pressure p_u required to produce instability, the result of (5-109) is substituted back into (5-107) to give

$$p_u = \left[\frac{2kt_o}{D_o}\right] e^{-n} \left[\frac{n}{2\left(\frac{3}{4}\right)^{(1+n)/2n}}\right]^n \qquad (5\text{-}110)$$

The values of strength coefficient k and strain-hardening exponent n for the SAE 4130 steel material may be read from the table in Figure 5.2 as $k = 154,500$ psi and $n = 0.156$. The original thickness $t_o = 0.125$ inch and the original diameter $D_o = 18$ inches were specified. Substituting these numerical values into (5-110) gives

$$p_u = \frac{2(154,500)(0.125)}{18.0} e^{-0.156} \left[\frac{0.156}{2\left(\frac{3}{4}\right)^{1.156/0.312}}\right]^{0.156} \qquad (5\text{-}111)$$

or

$$p_u = 1455 \text{ psi} \qquad (5\text{-}112)$$

The conclusion, then, is that the pressure vessel can sustain pressures up to 1455 psi, at which time failure is imminent. Any slight increase in pressure above this value would trigger a spontaneous catastrophic failure by plastic flow, probably characterized by a local bulge and blowout.

For purposes of comparison, the pressure to initiate first yielding may be estimated by again utilizing the octahedral shearing stress theory. As will be shown in detail in Chapter 6, the octahedral shearing stress theory predicts that yielding will occur in the multiaxial state of stress when the octahedral shearing stress τ_o of expression (5-69) reaches the critical value of τ_{o_y} in a simple tensile test when yielding begins. The critical value τ_{o_y} at which yielding begins in a simple tensile test may be readily evaluated by recognizing that only one nonzero prinicpal stress exists, say σ'_1, which at the time of incipient yielding must equal the yield strength σ_{yp} for the material. Thus one may evaluate τ_{o_y} by setting $\sigma'_2 = \sigma'_3 = 0$ in (5-69) and $\sigma'_1 = \sigma_{yp}$, giving

$$\tau_{o_y} = \frac{1}{3}\sqrt{\sigma_{yp}^2 + \sigma_{yp}^2} = \frac{1}{3}\sqrt{2\sigma_{yp}^2} \qquad (5\text{-}113)$$

The octahedral shearing stress theory predicts that yielding will initiate when

$$\tau_o = \tau_{o_y} \qquad (5\text{-}114)$$

Substituting values for τ_o from (5-69) and τ_{o_y} from (5-113) into the expression of (5-114), yielding is predicted to initiate when

$$(\sigma'_1 - \sigma'_2)^2 + (\sigma'_2 - \sigma'_3)^2 + (\sigma'_3 - \sigma'_1)^2 = 2\sigma_{yp}^2 \qquad (5\text{-}115)$$

From (5-89) and (5-90) it may be observed that $\sigma_2' = \sigma_1'/2$ and from (5-86) that $\sigma_3' = 0$. With these values for principal stresses, (5-115) predicts yielding when

$$\left(\frac{\sigma_1'}{2}\right)^2 + \left(\frac{\sigma_1'}{2}\right)^2 + (-\sigma_1')^2 = 2\sigma_{yp}^2 \tag{5-116}$$

or when

$$\sigma_1' = \frac{2}{\sqrt{3}}\sigma_{yp} \tag{5-117}$$

Noting that the changes in vessel diameter D_o and wall thickness t_o are negligibly small in the elastic range, we can substitute (5-89) into (5-117) for σ_1' to give

$$\frac{P_{yp}D_o}{2t_o} = \frac{2}{\sqrt{3}}\sigma_{yp} \tag{5-118}$$

when P_{yp} is the pressure in the vessel required to initiate yielding. Solving equation (5-118) for P_{yp} gives

$$P_{yp} = \frac{4t_o}{\sqrt{3}\,D_o}\sigma_{yp} \tag{5-119}$$

For this material the ultimate tensile strength is found to be about 98,000 psi and the yield strength is about 89,000 psi. The pressure P_{yp} to initiate yielding, then, from (5-119) is

$$P_{yp} = \frac{4(0.125)}{\sqrt{3}\,(18.0)}(89,000) \tag{5-120}$$

or

$$P_{yp} = 1427 \text{ psi}$$

Thus, for the material used, yielding would begin at about 1427 psi, and catastrophic plastic blowout would occur when the pressure reached 1455 psi, by (5-112). It is also of interest to note that if the tensile ultimate strength had been used in an equation like (5-120), with no account being taken of the change in D and t because of plastic deformation, an ultimate pressure of about 1572 psi would have been calculated. This would have been on the unsafe side by over 100 psi.

QUESTIONS

1. Equation (5-32) represents Hooke's law relationships for a triaxial state of stress. Based on these equations, write
(a) the Hooke's law relationships for a biaxial state of stress and
(b) the Hooke's law relationships for a uniaxial state of stress.
(c) Does a uniaxial state of stress imply a uniaxial state of strain? Explain.

2. Write the expressions for the first, second, and third invariants of strain.

3. It has been calculated that the critical point in a 4340 steel part is subjected to a state of stress in which $\sigma_x = 6000$ psi, $\tau_{xy} = 4000$ psi, and the remaining stress components are all zero. For this state of stress determine the sum of the total normal strains in the x, y, and z directions; that is, determine the magnitude of $\varepsilon_x + \varepsilon_y + \varepsilon_z$.

4. Solve Hooke's law equations of (5-32) explicitly for σ_x, σ_y, and σ_z.

5. Consider the case of an elastic material sandwiched between two perfectly rigid plates and bonded to them. Normal compressive loading on the plates then is imposed to give compressive stress σ_z in the material normal to the plates. Assuming that the bond attaching the material to the plates completely prevents all lateral strain, that is, $\varepsilon_x = \varepsilon_y = 0$, determine the *apparent* modulus of elasticity σ_z / ε_z in the z direction in terms of E and ν for the material. Show that it may be many times greater than E if the material in the compressed layer is made nearly incompressible under conditions of hydrostatic pressure.

6. For the case of pure biaxial shear, that is, τ_{xy} is the only nonzero component of stress, write expressions for the principal normal strains. Is this a biaxial state of strain? Explain.

7. A standard 0.505-inch-diameter tensile specimen, when subjected to a tensile test in the laboratory, exhibited a maximum load of 26,000 lb and total elongation of 0.52 inch at fracture, using a 2.0-inch gage length. Determine the following:
(a) ultimate strength,
(b) percent elongation, and
(c) toughness merit number.

8. In a tension test of a steel test specimen 0.25 inch by 1 inch rectangular cross section, a gage length of 8.0 inches was used. Test data included the following observations: Load at onset of yielding was 8500 lb, ultimate load was 14,700 lb, fracture load was 11,700 lb, elongation at fracture was 2.25 inches, and total deformation in the gage length at 4000-lb load was 0.0044 inch. Determine the following:
(a) lower yield point,
(b) ultimate strength,
(c) modulus of elasticity,
(d) percent elongation,
(e) modulus of resilience, and
(f) toughness merit number.

9. A tensile test on a specimen of circular cross section of original diameter 0.364 inch was performed to obtain stress-strain data for a steel alloy. In the plastic range the loads and corresponding diameters were measured as shown in the chart.

Load (lb)	Diameter (in.)
3500	0.352
4300	0.350
4800	0.347
5175	0.344
5400	0.340
5650	0.334
5775	0.324
5800	0.316
5700	0.302
5600	0.290
5300	0.270
5050	0.260
4950	0.255
4650	0.235
Fracture	

(a) Determine the true stress and true strain values.

(b) Determine the engineering stress and engineering strain values.

(c) Plot true stress true strain on a log-log plot and determine the strength coefficient k and strain-hardening exponent n.

(d) Plot a true stress-strain curve and an engineering stress-strain curve on the same plot and compare the curves.

10.   For some materials the true-stress-true-strain curve may be well approximated by the expression $\delta = S'/E + k(S')^b$. Derive an expression for true strain at the instability point, δ_u, for such a material.

11.   An annealed aluminum alloy has a strength coefficient of 55,900 psi and a strain-hardening exponent of 0.211. Utilizing the results of the development of the simple deformation theory, equations (5-83), (5-84), and (5-85), plot the $\sigma_1' - \delta_1$ stress-strain relationships for each of the following principal stress ratios, interpret the corresponding states of stress, and make any observations you think pertinent:

(a) $\dfrac{\sigma_2'}{\sigma_1'} = -0.5;$ $\dfrac{\sigma_3'}{\sigma_1'} = 0$ (f) $\dfrac{\sigma_2'}{\sigma_1'} = \dfrac{\sigma_3'}{\sigma_1'} = +0.5$

(b) $\dfrac{\sigma_2'}{\sigma_1'} = +0.5;$ $\dfrac{\sigma_3'}{\sigma_1'} = 0$ (g) $\dfrac{\sigma_2'}{\sigma_1'} = \dfrac{\sigma_3'}{\sigma_1'} = -1.0$

(c) $\dfrac{\sigma_2'}{\sigma_1'} = -1.0;$ $\dfrac{\sigma_3'}{\sigma_1'} = 0$ (h) $\dfrac{\sigma_2'}{\sigma_1'} = \dfrac{\sigma_3'}{\sigma_1'} = +1.0$

(d) $\dfrac{\sigma_2'}{\sigma_1'} = +1.0;$ $\dfrac{\sigma_3'}{\sigma_1'} = 0$ (i) $\dfrac{\sigma_2'}{\sigma_1'} = \dfrac{\sigma_3'}{\sigma_1'} = 0$

(e) $\dfrac{\sigma_2'}{\sigma_1'} = \dfrac{\sigma_3'}{\sigma_1'} = -0.5$

12.   A thin-walled cylindrical pressure vessel of initial diameter D_o and initial thickness t_o is subjected to both an internal pressure p and an axial force P. Derive an expression for the maximum principal strain at instability for the case in which circumferential stress component σ_1' is greater than axial stress component σ_2'.

13.   For the thin-walled pressure vessel of problem 12, derive an expression for the maximum principal strain at instability for the case where circumferential stress component σ_1' is less than axial stress component σ_2'.

14.   A thin-walled spherical pressure vessel of inside diameter equal to 50 inches is to be made from a type 430 annealed stainless steel. The vessel is to be designed so that it will just reach the point of incipient local blowout (instability point) as the pressure reaches a value of 2000 psi. Of what initial thickness should the pressure vessel wall be made?

15.   The pinned-joint structure shown in Figure Q5.15 is to be used to support the load P, which is essentially a static load. However, on some occasions an overload condition may exist. If the structural members are solid cylindrical bars of 4130 steel in the hot-rolled and annealed condition, do the following (use strength coefficient of 169,400 and strain-hardening exponent of 0.118):
(a) Derive in detail the equation for the load P_{yp} at which first yielding would be predicted.
(b) Calculate a numerical value for P_{yp}.
(c) Determine at what load all bars are just fully plastic. (Assume elastic perfectly plastic material behavior.)

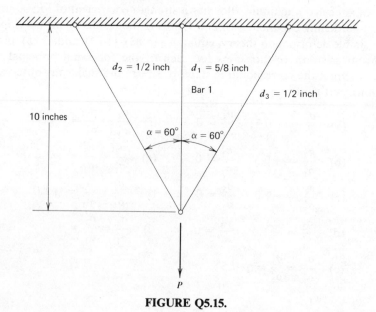

FIGURE Q5.15.

(d) Derive in detail the equation for the load P_i at which plastic instability would be predicted.

(e) Calculate a numerical value for P_i.

(f) Determine the total elongation of bar 1 when plastic instability occurs.

REFERENCES

1. Marin, J., *Mechanical Behavior of Engineering Materials*, Prentice-Hall, Englewood Cliffs, N. J. 1962.

2. Nadai, A., *Theory of Flow and Fracture of Solids*, Vol. 1, McGraw-Hill, New York, 1950.

3. Polakowski, N. H., and Ripling, E. J., *Strength and Structure of Engineering Materials*, Prentice-Hall, Englewood Cliffs, N. J., 1966.

4. Guy, A. G., *Essentials of Materials Science*, McGraw-Hill, New York, 1976.

CHAPTER 6

Combined Stress Theories of Failure and Their Use in Design

6.1 INTRODUCTION

Predicting failure and establishing a geometry that will avert failure is a relatively simple matter if the machine part is subjected to a static uniaxial state of stress. It is necessary only to have available the simple uniaxial stress-strain curve for the material of interest, which can be readily obtained from one or a few simple tension and compression experiments. For example, if yielding has been established as the governing failure mode for a uniaxially stressed machine part under consideration, failure would be predicted to occur when the maximum normal stress in the part reaches the level of the yield point determined from the stress-strain curve established in the simple tension test experiment.

If the machine part under consideration is subjected to a biaxial or a triaxial state of stress, however, the prediction of failure is far more difficult. No longer can one predict yielding, for example, when the maximum normal stress reaches the tensile yield point because the other normal stress components may also influence yielding. Furthermore, there is not one or a few simple experiments that can be performed to characterize failure in the multiaxial state of stress. A large number of complex multiaxial tests would be required in which all of the stress components would be varied over their entire range of values in all possible combinations, and even then it would be difficult to assess the influence of outside factors such as stress concentration, temperature, and environment. Such a testing program would be prohibitively costly and time-consuming, and perhaps not even possible for certain states of stress. When engineers face a problem of this complexity, their inclination is always to attempt to develop a theory that relates behavior in the complex situation to behavior in a simple easily evaluated test through some characteristic "modulus." Specifically, when one desires to predict failure in a machine part subjected to a multiaxial state of stress, it is usual to utilize a theory that relates failure in the multiaxial state of stress to failure by the same mode in a simple tension test through a well chosen "modulus" such as stress, strain, or energy. To be useful such moduli must be calculable in the multiaxial state of stress and readily measurable in the simple uniaxial evaluation test.

The basic assumption that constitutes the framework for all combined stress failure theories is that *failure is predicted to occur when the maximum value of the selected mechanical modulus in the multiaxial state of stress becomes equal to or exceeds the value of the same modulus that produces failure in a simple uniaxial stress test using the same material.*

To further illustrate the concept, consider again a thin-walled pressure vessel as seen in Figure 5.6. The internal pressure produces an essentially biaxial state of stress with a component σ_1 in the hoop direction and a component σ_2 in the longitudinal direction. From elementary stress analyses such as those resulting in (5-89) and (5-91), one could readily calculate the magnitude of the maximum normal stress, $(\sigma_n)_{max}$, as a function of internal pressure. One could also select a coupon of the same material, subject it to a simple tensile test, and document experimentally the magnitude of maximum normal stress, say σ_f, that causes failure in the tensile test. Then, utilizing the general framework upon which all failure theories are based, one might devise a failure theory for this case of biaxial stress in the pressure vessel wall as follows:

Failure is predicted to occur if the calculated value of $(\sigma_n)_{max}$ for the biaxial state of stress in the pressure vessel wall becomes equal to or exceeds σ_f, the failure value of the maximum normal stress in the simple tension test.

Experimentation with a variety of materials would show that the theory works well for certain materials but not very well for others. Such attempts to devise failure theories and experimentally verify them has led to many failure theory proposals, six of which are presented in this chapter.

To summarize, the development of any useful combined stress failure theory must contain three essential ingredients:

1. It must provide an applicable model, describable by explicit mathematical relationships, that relates the external loading to the stresses, strains, or other calculable mechanical moduli at the critical point in the multiaxial state of stress.
2. It must be based on critical physical properties of the material that are *measurable*.
3. It must relate the calculable mechanical modulus in the multiaxial state of stress to a measurable criterion of failure based on the critical physical properties determined in a simple uniaxial test.

Many combined stress failure theories that contain these essential ingredients have been postulated. Six of these theories are presented, in detail, in the following sections, including

1. Maximum normal stress theory,
2. Maximum shearing stress theory,
3. Maximum normal strain theory,
4. Total strain energy theory,
5. Distortion energy theory, and
6. Mohr's failure theory.

Some of these are presented for their historical interest and to indicate limitations that inspired the development of other more accurate combined stress theories of failure. Others are presented because they represent the best theories currently available.

6.2 MAXIMUM NORMAL STRESS THEORY (RANKINE'S THEORY)

In words, the maximum normal stress theory, proposed by Rankine, may be expressed as follows:

> Failure is predicted to occur in the multiaxial state of stress when the maximum principal normal stress becomes equal to or exceeds the maximum normal stress at the time of failure in a simple uniaxial stress test using a specimen of the same material.

The general triaxial state of stress at a point may be fully described by the three principal normal stresses and their directions, as was shown in Chapter 4. For a uniaxial stress test the only nonzero normal stress component is a principal stress in the direction of the applied force. Failure in a uniaxial stress test depends on the failure mode of interest and the way in which the material responds to the loading applied. Thus the pertinent failure strength σ_f at the time of incipient uniaxial failure might be the fracture strength, yield strength, proportional limit, or something else, depending on the failure mode that governs failure at the critical point of interest in the multiaxially stressed machine part. Further, for some materials the failure strength in tensile loading may be different from the failure strength in compressive loading for the same mode of failure. For example, the compressive ultimate strength for cast iron is significantly greater than the tensile ultimate strength. With all these factors in mind, the maximum normal stress theory may be mathematically formulated from the word statement above as follows:

Failure is predicted by the maximum normal stress theory to occur if

$$\sigma_1 \geq \sigma_t$$
$$\sigma_2 \geq \sigma_t$$
$$\sigma_3 \geq \sigma_t$$
$$\sigma_1 \leq \sigma_c$$
$$\sigma_2 \leq \sigma_c$$
$$\sigma_3 \leq \sigma_c \tag{6-1}$$

where σ_1, σ_2, and σ_3 are the principal normal stresses; σ_t is the uniaxial failure strength in tension; and σ_c is the uniaxial failure strength in compression. It is important to note that failure is predicted to occur if any one expression of (6-1) is satisfied.

Failure prediction in the triaxial state of stress by the maximum normal stress theory may be represented graphically, as shown in Figure 6.1. The surfaces of the cubical element represent the boundary of incipient failure. All states of stress that lie outside the cube would be predicted to result in failure, whereas all points that lie within the cube would safely tolerate the applied loads without failure. If the tensile failure strength σ_t is equal to the compressive failure strength σ_c, the cube is symmetrical about the coordinate origin. If the tensile and compressive failure strengths differ, the cubical failure surface is displaced so that the center of the cube no longer coincides with the σ_1-σ_2-σ_3 coordinate origin.

Evaluation of the proposed maximum normal stress theory leads to the observation that the prediction of failure is based solely on the magnitude of the maximum normal component of stress, regardless of the magnitude or direction of the other two principal stresses. For example, if one examines the

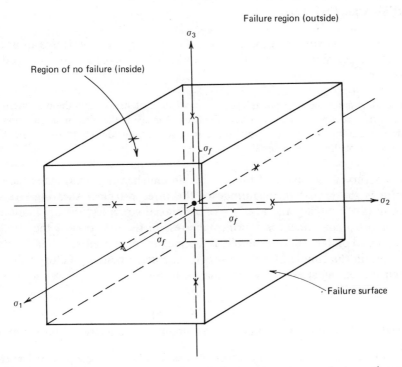

FIGURE 6.1. Graphical representation of the maximum normal stress theory of failure for the general multiaxial state of stress.

case of hydrostatic stress,* either tension or compression, this theory predicts yielding failure when the magnitude of the principal stress $\sigma = \sigma_1 = \sigma_2 = \sigma_3$ becomes equal to the simple tensile yield point. Thus the maximum normal stress theory predicts the onset of yielding in the case of hydrostatic tension or compression when the principal stress σ reaches the simple tensile yield point σ_{yp} for the material. Much experimental evidence exists that indicates that pure hydrostatic compression does not produce any plastic yielding at all in a compact crystalline or amorphous solid, but only a small reversible elastic contraction. Further, it has also been shown experimentally that pure hydrostatic tension does not produce any plastic flow or yielding either. Thus, as a theory for predicting the onset of yielding, the maximum normal stress theory is generally poor. For materials that behave in a ductile fashion, one would not use the maximum normal stress theory.

On the other hand, for brittle materials the maximum normal stress theory is probably the best available failure theory, though it may yield conservative results for some states of stress. Supporting data for this conclusion are shown in Figure 6.8 for certain biaxial stress conditions.

6.3 MAXIMUM SHEARING STRESS THEORY (TRESCA-GUEST THEORY)

In words, the maximum shearing stress theory, proposed by Tresca in about 1865 and experimentally supported by Guest about 1900, may be expressed as follows:

> Failure is predicted to occur in the multiaxial state of stress when the maximum shearing stress magnitude becomes equal to or exceeds the maximum shearing stress magnitude at the time of failure in a simple uniaxial stress test using a specimen of the same material.

It was observed in Chapter 4 that the maximum shearing stress magnitude at a point is the largest in magnitude of the three principal shearing stresses given in (4-55), (4-56), and (4-57). For a uniaxial stress test the only nonzero normal stress component is a principal stress in the direction of the applied force. Based on (4-57), for example, if the maximum normal stress at the time of failure in the simple uniaxial test is σ_f, the corresponding failure value of principal shearing stress τ_f for the uniaxial test is

$$\tau_f = \frac{\sigma_f}{2} \tag{6-2}$$

With these observations in mind the maximum shearing stress theory may be

---

*Hydrostatic stress is defined as the state of stress in which all three principal stresses are equal.

mathematically formulated from the word statement as follows:

Failure is predicted by the maximum shearing stress theory to occur if

$$|\tau_1| \geq |\tau_f|$$
$$|\tau_2| \geq |\tau_f|$$
$$|\tau_3| \geq |\tau_f| \tag{6-3}$$

Utilizing (4-55), (4-56), and (4-57) together with (6-2), we can formulate the maximum shearing stress theory in terms of principal normal stresses as follows:

Failure is predicted by the maximum shearing stress theory to occur if

$$|\sigma_1 - \sigma_2| \geq |\sigma_f|$$
$$|\sigma_2 - \sigma_3| \geq |\sigma_f|$$
$$|\sigma_3 - \sigma_1| \geq |\sigma_f| \tag{6-4}$$

where σ_1, σ_2, and σ_3 are the principal normal stress and σ_f is the uniaxial failure strength in tension. It is important to note that failure is predicted to occur if any one expression of (6-4) is satisfied.

Failure prediction in the triaxial state of stress by the maximum shearing stress theory may be graphically represented by the sketch of Figure 6.2. In this case the failure surface is a hexagonal cylinder whose axis makes equal angles with the three principal axes. It may be observed that all states of stress that lie within the cylinder are points of no failure, whereas points outside the cylinder are predicted to cause failure. In contrast with the maximum normal stress theory, it may be noted that the maximum shearing stress theory is more successful in predicting the experimentally observed behavior of ductile metals under hydrostatic states of stress. This is true because hydrostatic tension or compression lies on the cylinder axis and therefore is always in the region of no failure; hence, yielding would never be predicted to occur under a hydrostatic state of stress. Experimentally, triaxial hydrostatic compression of engineering metals to very high stress levels has been observed to produce no yielding and no fracture. Under hydrostatic tension, fracture is experimentally observed but at values of the principal stress nearly twice the simple uniaxial rupture stress (9). Thus the maximum shearing stress theory seems to model the behavior of engineering metals under hydrostatic states of stress, especially with respect to yielding and ductile behavior. Experimental evidence for other states of stress verifies that the maximum shearing stress theory is in general a good theory for predicting failure of ductile materials, as illustrated in Figure 6.8. It has been observed that only one other theory, the distortion energy theory, gives better agreement with experimental data for ductile behavior under multiaxial states of stress.

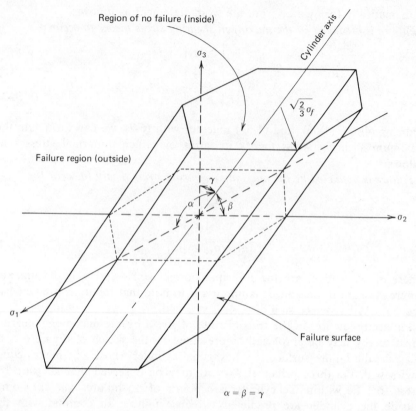

Region of no failure (inside)

Cylinder axis

σ_3

$\sqrt{\frac{2}{3}}\sigma_f$

Failure region (outside)

γ

α β

σ_2

σ_1

Failure surface

$\alpha = \beta = \gamma$

FIGURE 6.2. Graphical representation of the maximum shearing stress theory of failure for the general multiaxial state of stress.

6.4   MAXIMUM NORMAL STRAIN THEORY (ST. VENANT'S THEORY)

In words, the maximum normal strain theory, proposed by St. Venant, may be expressed as follows:

Failure is predicted to occur in the multiaxial state of stress when the maximum principal normal strain becomes equal to or exceeds the maximum normal strain at the time of failure in a simple uniaxial stress test using a specimen of the same material.

The generalized Hooke's law equations of (5-33), which may be summarized as

$$\varepsilon_i = \frac{1}{E}\left[\sigma_i - \nu(\sigma_j + \sigma_k)\right] \qquad (6\text{-}5)$$

may be evaluated for the simple tension test by noting that σ_i corresponds to the uniaxial failure strength, whereas the principal stresses σ_j and σ_k are both zero. Thus ε_f, the principal strain at the time of failure in a uniaxial stress test, becomes

$$\varepsilon_f = \frac{\sigma_f}{E} \tag{6-6}$$

Utilizing these expressions, we may mathematically formulate the maximum normal strain theory from the word statement as follows.

Failure is predicted by the maximum normal strain theory to occur if

$$\varepsilon_1 \geq \varepsilon_f$$
$$\varepsilon_2 \geq \varepsilon_f$$
$$\varepsilon_3 \geq \varepsilon_f \tag{6-7}$$
$$\varepsilon_1 \leq -\varepsilon_f$$
$$\varepsilon_2 \leq -\varepsilon_f$$
$$\varepsilon_3 \leq -\varepsilon_f$$

Expressing the strains of (6-7) in terms of stresses, the maximum normal strain theory becomes:

Failure is predicted by the maximum normal strain theory to occur if

$$\sigma_1 - \nu(\sigma_2 + \sigma_3) \geq \sigma_f$$
$$\sigma_2 \quad \nu(\sigma_1 + \sigma_3) \geq \sigma_f$$
$$\sigma_3 - \nu(\sigma_1 + \sigma_2) \geq \sigma_f \tag{6-8}$$
$$\sigma_1 - \nu(\sigma_2 + \sigma_3) \leq -\sigma_f$$
$$\sigma_2 - \nu(\sigma_1 + \sigma_3) \leq -\sigma_f$$
$$\sigma_3 - \nu(\sigma_1 + \sigma_2) \leq -\sigma_f$$

where again σ_1, σ_2, and σ_3 are principal stresses in the multiaxial state of stress and σ_f is the uniaxial failure strength. It is important to note that failure is predicted to occur if any one expression of (6-8) is satisfied.

Failure prediction in the triaxial state of stress by the maximum normal strain theory is represented geometrically in Figure 6.3. In this case the failure surface consists of the bounding surfaces of four straight tetrahedral pyramids symmetrically stacked along a space diagonal. One of the pyramids is designated as *a-b-c-d* in Figure 6.3. As for the case of the maximum normal stress theory, the closed failure surface predicts failure in hydrostatic states of stress, which is not supported by experimental evidence, especially for prediction of yielding and ductile behavior.

Further, the theory has been found to be less than acceptable for brittle material failures as well. The result is, therefore, that this theory has historical importance but is not currently used by designers, since other better theories

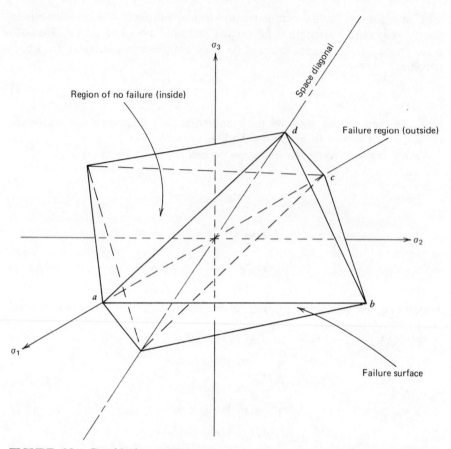

FIGURE 6.3. Graphical representation of the maximum normal strain theory of failure for the general multiaxial state of stress.

are available. Interestingly, however, some renewed interest is being shown for the maximum normal strain theory by investigators working with reinforced composite materials in multiaxial states of stress.

6.5 TOTAL STRAIN ENERGY THEORY (BELTRAMI THEORY)

In words, the total strain energy theory, proposed by Beltrami about 1885, may be expressed as follows:

> Failure is predicted to occur in the multiaxial state of stress when the total strain energy per unit volume becomes equal to or exceeds the total strain energy per unit volume at the time of failure in a simple uniaxial stress test using a specimen of the same material.

To express the total strain energy theory mathematically it is necessary to develop an expression for the total strain energy per unit volume in a multiaxial state of stress, evaluate the total strain energy per unit volume at the time of failure in a uniaxial stress test, and combine these strain energy expressions in a failure theory. The total strain energy stored in a small element of volume dx-dy-dz under the influence of principal stresses σ_1-σ_2-σ_3 and their associated strains may be computed by employing the loss-free general energy equation, which asserts that the total strain energy U_T stored in the volume element must equal the work W done on the element; that is,

$$U_T = W \qquad (6\text{-}9)$$

The work W may be expressed as the mean force multiplied by the distance through which it acts, if the relationship between force and deflection is linear. Thus the work may be written as

$$W = \left(\frac{F_f + F_o}{2} \right) d \qquad (6\text{-}10)$$

when F_o is the original applied force, F_f is the final applied force, and d is the distance through which the force moves. Recognizing that the final force in the σ_1 direction is the stress σ_1 multiplied by the area on which it acts, that the original force is zero, and that the distance through which the force moves is the total deformation of the volume element in the σ_1 direction, the work associated with the application of the stress σ_1 is

$$W_1 = \left(\frac{\sigma_1\, dy\, dz}{2} \right) (\varepsilon_1\, dx) \qquad (6\text{-}11)$$

which, by utilizing (6-9), may be equated to total strain energy stored as a result of the application of σ_1 alone. Thus,

$$U_{T_1} = \frac{\sigma_1 \varepsilon_1}{2}\, dx\, dy\, dz \qquad (6\text{-}12)$$

The total strain energy per unit volume attributable to the application of σ_1 alone may then be obtained by dividing (6-12) by the volume to yield

$$u_1 = \frac{U_{T_1}}{dx\, dy\, dz} = \frac{\sigma_1 \varepsilon_1}{2} \qquad (6\text{-}13)$$

where u_1 is total strain energy per unit volume stored as a result of the application of σ_1. Similar expressions for strain energy per unit volume stored as a result of the application of σ_2 and σ_3, respectively, may be written as

$$u_2 = \frac{\sigma_2 \varepsilon_2}{2} \qquad (6\text{-}14)$$

and

$$u_3 = \frac{\sigma_3 \varepsilon_3}{2} \qquad (6\text{-}15)$$

Neglecting higher order terms, we find the total strain energy per unit volume stored as a result of the simultaneous application of σ_1, σ_2, and σ_3 may be obtained by summing (6-13), (6-14), and (6-15) to yield

$$u_T = u_1 + u_2 + u_3 \qquad (6\text{-}16)$$

or

$$u_T = \tfrac{1}{2}\left[\sigma_1\varepsilon_1 + \sigma_2\varepsilon_2 + \sigma_3\varepsilon_3\right] \qquad (6\text{-}17)$$

The strains ε_i in (6-17) may be expressed as functions of the three principal stresses by employing the Hooke's law equations of (6-5). Thus the complete expression for total strain energy per unit volume may be written as

$$u_T = \frac{1}{2E}\left[\sigma_1^2 + \sigma_2^2 + \sigma_3^2 - 2\nu(\sigma_1\sigma_2 + \sigma_2\sigma_3 + \sigma_3\sigma_1)\right] \qquad (6\text{-}18)$$

For a uniaxial stress test the only nonzero principal stress is, at the time of

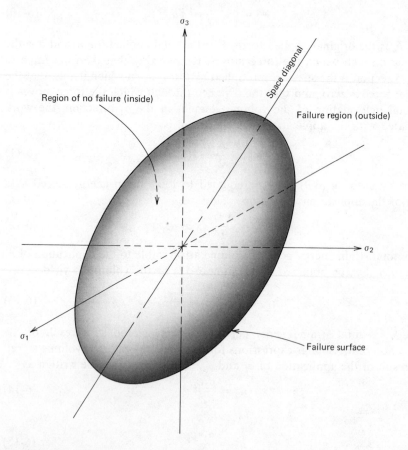

FIGURE 6.4. Graphical representation of the total strain energy theory of failure for the general multiaxial state of stress.

failure, equal to the failure strength σ_f, and the corresponding value of total strain energy u_{Tf} becomes

$$u_{Tf} = \frac{1}{2E}\left[\sigma_f^2\right] \tag{6-19}$$

Utilizing these last two expressions, we may mathematically formulate the total strain energy theory of failure from the word statement as follows:

Failure is predicted by the total strain energy theory to occur if

$$u_T \geq u_{Tf} \tag{6-20}$$

or if

$$\left[\sigma_1^2 + \sigma_2^2 + \sigma_3^2 - 2\nu(\sigma_1\sigma_2 + \sigma_2\sigma_3 + \sigma_3\sigma_1)\right] \geq \sigma_f^2 \tag{6-21}$$

Since Hooke's law has been used, this theory is limited in validity to the linear elastic range, just as the maximum normal strain theory of St. Venant was. Equation (6-21) is the equation of an ellipsoid symmetrical about a space diagonal, as illustrated in Figure 6.4. As for the other failure theories, the region enclosed within the ellipsoidal surface represents states of stress that would be predicted by this theory to result in nonfailure, whereas points outside the failure surface would be predicted to result in failure. It may be noted that the closed failure surface predicts failure for hydrostatic states of stress and, to this extent at least, is in disagreement with experimental observations. This theory of failure is not used by designers but is of historical significance in the development of the widely used distortion energy theory described in Section 6.6.

6.6 DISTORTION ENERGY THEORY (HUBER-VON MISES-HENCKY THEORY)

In words, the distortion energy theory, proposed first in 1904 by Huber with later contributions by Von Mises and Hencky, may be expressed as follows:

Failure is predicted to occur in the multiaxial state of stress when the distortion energy per unit volume becomes equal to or exceeds the distortion energy per unit volume at the time of failure in a simple uniaxial stress test using a specimen of the same material.

The distortion energy theory was developed as an improvement over the total strain energy theory, accounting for the experimental observation that hydrostatic states of stress are not properly assessed by the total strain energy theory. In developing the distortion energy theory, the postulate was made that the total strain energy can be divided into two parts: the energy associated solely with change in volume, termed dilatation energy, and the energy associated solely with change in shape, termed distortion energy. It was further postulated that failure, particularly under conditions of ductile behavior, is related only to the distortion energy, with no contribution from the dilatation energy.

To develop an expression for the distortion energy, a volume element subjected to principal stresses σ_1, σ_2, and σ_3 may be represented as shown in Figure 6.5, where the principal stress vectors are divided into two parallel components in each direction. One of these components, S, is equal in magnitude for all three coordinate directions. We may write

$$S + \sigma_1'' = \sigma_1$$
$$S + \sigma_2'' = \sigma_2$$
$$S + \sigma_3'' = \sigma_3 \qquad (6\text{-}22)$$

In these expressions the stress component S is equal in all three directions and is, therefore, properly described as a hydrostatic stress component. The hydrostatic stress components act to change the volume and are the sole contributors to dilatation energy as just defined.

Again referring to Figure 6.5, we find the total volume change produced by the σ_1-σ_2-σ_3 state of stress depicted at the left of the figure may be calculated by considering a differential volume element of dimensions dx-dy-dz. For such an element the original unstrained volume V_o would be

$$V_o = dx\, dy\, dz \qquad (6\text{-}23)$$

After subjecting the element to the σ_1-σ_2-σ_3 stress field, strains of ε_1, ε_2, and ε_3 would be produced in the x, y, and z directions respectively. The final volume V_f of the strained element, neglecting higher order terms, would be

$$V_f = dx\, dy\, dz + \varepsilon_1\, dx(dy\, dz) + \varepsilon_2\, dy(dx\, dz) + \varepsilon_3\, dz(dx\, dy) \qquad (6\text{-}24)$$

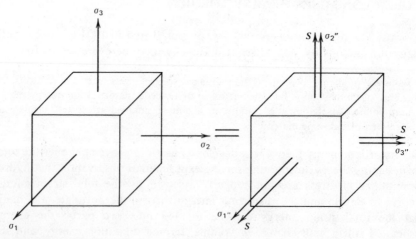

FIGURE 6.5. Principal stresses on an element represented as dilatation and distortion components.

whence

$$V_f = dx\,dy\,dz(1 + \varepsilon_1 + \varepsilon_2 + \varepsilon_3) \tag{6-25}$$

The change in volume ΔV then would be

$$\Delta V = V_f - V_o = dx\,dy\,dz\big[(1 + \varepsilon_1 + \varepsilon_2 + \varepsilon_3) - 1\big] \tag{6-26}$$

To express the volume change on a per unit volume basis, the expression of (6-26) is divided by dx-dy-dz to give the change in volume per unit volume Δv as

$$\Delta v = \varepsilon_1 + \varepsilon_2 + \varepsilon_3 \tag{6-27}$$

Employing Hooke's law to express the strains in terms of stresses, we may write (6-27) as

$$\Delta v = \frac{1}{E}\big[\sigma_1 + \sigma_2 + \sigma_3 - 2\nu(\sigma_1 + \sigma_2 + \sigma_3)\big] \tag{6-28}$$

or

$$\Delta v = \frac{1 - 2\nu}{E}(\sigma_1 + \sigma_2 + \sigma_3) \tag{6-29}$$

Referring now to the state of stress as depicted at the right side of Figure 6.5, by the development leading to (6-27), we write the change in volume per unit volume Δv_s as a result of S acting alone as

$$\Delta v_s = \varepsilon_s + \varepsilon_s + \varepsilon_s = 3\varepsilon_s \tag{6-30}$$

Again utilizing Hooke's law, this expression may be written in terms of stress as

$$\Delta v_s = 3\left\{\frac{1}{E}\big[S - \nu(S + S)\big]\right\} = \frac{1 - 2\nu}{E}(3S) \tag{6-31}$$

Since the stress component S has been postulated to be the only contributor to volume change on the right-hand side of the equality in Figure 6.5, and since the entire volume change for the left-hand side of Figure 6.5 is given in (6-29), the expressions of (6-29) and (6-31) must be equal; hence,

$$\frac{1 - 2\nu}{E}(\sigma_1 + \sigma_2 + \sigma_3) = \frac{1 - 2\nu}{E}(3S) \tag{6-32}$$

Solving for the value of S to make the original postulate true, we get

$$S = \frac{\sigma_1 + \sigma_2 + \sigma_3}{3} \tag{6-33}$$

Thus, if the stress component S is to account for the volume-changing portion of the stress field, it must be equal in magnitude to the arithmetic mean of the three principal stresses.

With this evaluation of S completed, the next task is to write an expression for total strain energy per unit volume under the influence of stress field σ_1-σ_2-σ_3 and another expression for strain energy per unit volume associated with volume change only, as a result of S acting alone. The difference between these two expressions will necessarily be the energy associated solely with distortion. The expression for total strain energy per unit volume has already been developed in (6-18) as

$$u_T = \frac{1}{2E}\left[\sigma_1^2 + \sigma_2^2 + \sigma_3^2 - 2\nu(\sigma_1\sigma_2 + \sigma_2\sigma_3 + \sigma_1\sigma_3)\right] \qquad (6\text{-}34)$$

To obtain an expression for strain energy associated with volume change alone, u_V, it is necessary only to substitute the volume-changing stress component S for each of the stresses σ_1, σ_2, and σ_3 in (6-34) to give

$$u_V = \frac{1}{2E}\left[3S^2 - 2\nu(3S^2)\right] \qquad (6\text{-}35)$$

or

$$u_V = \frac{3(1-2\nu)}{2E}S^2 \qquad (6\text{-}36)$$

Utilizing equation (6-33), we may write the volume-changing strain energy u_V in terms of the principal stresses as

$$u_V = \frac{3(1-2\nu)}{2E}\left[\frac{\sigma_1 + \sigma_2 + \sigma_3}{3}\right]^2 \qquad (6\text{-}37)$$

As anticipated, then, the distortion energy per unit volume u_d may be written as

$$u_d = u_T - u_V \qquad (6\text{-}38)$$

or, from (6-34) and (6-37),

$$u_d = \frac{1}{2E}\left[\sigma_1^2 + \sigma_2^2 + \sigma_3^2 - 2\nu(\sigma_1\sigma_2 + \sigma_2\sigma_3 + \sigma_1\sigma_3)\right]$$
$$- \frac{3(1-2\nu)}{2E}\left[\frac{\sigma_1 + \sigma_2 + \sigma_3}{3}\right]^2 \qquad (6\text{-}39)$$

which may be written more compactly as

$$u_d = \frac{1}{2}\left[\frac{1+\nu}{3E}\right]\left[(\sigma_1 - \sigma_2)^2 + (\sigma_2 - \sigma_3)^2 + (\sigma_3 - \sigma_1)^2\right] \qquad (6\text{-}40)$$

With this expression for distortion energy per unit volume in the multiaxial state of stress, the uniaxial distortion energy per unit volume at the time of failure u_{df} may be obtained as usual by setting σ_2 and σ_3 equal to zero and σ_1 equal to the uniaxial failure strength σ_f. Thus,

$$u_{df} = \frac{1+\nu}{3E}\sigma_f^2 \qquad (6\text{-}41)$$

With all of these expressions at hand, the distortion energy theory of failure may be mathematically formulated from the word statement of the theory as follows:

Failure is predicted by the distortion energy theory to occur if

$$\tfrac{1}{2}\left[(\sigma_1 - \sigma_2)^2 + (\sigma_2 - \sigma_3)^2 + (\sigma_3 - \sigma_1)^2\right] \geq \sigma_f^2 \qquad (6\text{-}42)$$

It is worth noting that exactly the same result is obtained by a different logic if one develops an octahedral shearing stress theory. The octahedral planes are four pairs of parallel planes whose normals make equal angles with the principal axes. These planes, therefore, form a regular octahedron whose corners lie on the principal axes. The resultant shearing stress on an octahedral plane is called the *octahedral shearing stress* and may be calculated* as

$$\tau_o = \tfrac{1}{3}\sqrt{(\sigma_1 - \sigma_2)^2 + (\sigma_2 - \sigma_3)^2 + (\sigma_3 - \sigma_1)^2} \qquad (6\text{-}43)$$

where τ_o is the octahedral shearing stress. Evaluation of the octahedral shearing stress at the time of failure in a uniaxial stress test, τ_{o_f} is accomplished as usual by noting that σ_2 and σ_3 are zero and σ_1 is equal to the uniaxial failure strength σ_f. Thus,

$$\tau_{o_f} = \tfrac{1}{3}\sqrt{2\sigma_f^2} \qquad (6\text{-}44)$$

Utilizing (6-43) and (6-44), then, we may express the octahedral shearing stress theory of failure mathematically as follows:
Failure is predicted to occur if

$$(\sigma_1 - \sigma_2)^2 + (\sigma_2 - \sigma_3)^2 + (\sigma_3 - \sigma_1)^2 \geq 2\sigma_f^2 \qquad (6\text{-}45)$$

This octahedral shearing stress theory is exactly equivalent to the distortion energy theory of failure expressed in (6-42). Therefore, the theory is referred to in the literature by either name. In this text the name *distortion energy theory* will be preferred.

Failure prediction in the triaxial state of stress by the distortion energy theory may be graphically represented by the sketch of Figure 6.6. In this case the failure surface is a circular cylinder whose axis makes equal angles with the three principal axes. All states of stress that lie within the cylinder are points of no failure, whereas points outside the cylinder are predicted to result in failure. It may be observed that the distortion energy theory, like the maximum shearing stress theory, has the potential for properly assessing the hydrostatic states of stress, since these states of stress always lie on the cylinder axis in the region of no failure. Of all the theories dealing with ductile behavior, the distortion energy theory agrees best with experimental data available, as shown, for example, from Figure 6.8, and is widely used in predicting failure of ductile metals. Further, although it was derived for behavior in the elastic range, data indicate that the theory is valid well into the plastic range.

*See pp. 99–105 of ref. 1.

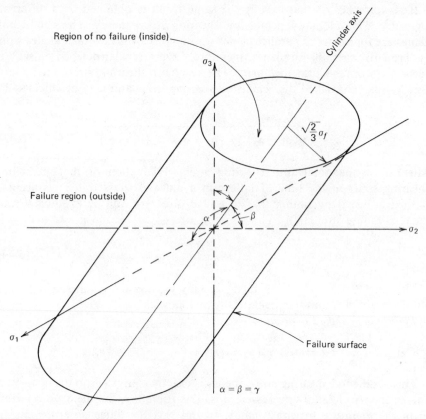

FIGURE 6.6. Graphical representation of the distortion energy theory of failure for the general multiaxial state of stress.

6.7 FAILURE THEORY COMPARISON IN BIAXIAL STATE OF STRESS

Predictions of failure by the five theories discussed thus far in this chapter may be compared for the biaxial state of stress by setting σ_3 equal to zero. Figure 6.7 depicts such a comparison, where the failure envelopes are reduced to plane figures as shown. Where necessary, the value of Possion's ratio has been taken as 0.35, and the axes have been normalized by dividing principal stresses σ_1 and σ_2 by the uniaxial failure strength σ_f. It is worth noting that the distortion energy theory and the maximum shear stress theory are very similar, with the maximum shearing stress theory always more conservative. Also, the maximum normal stress theory is seen to be in agreement with the maximum shearing stress theory in the first and third quadrants where the signs of the principal stresses are the same, but major disagreements appear in the second and fourth quadrants where the principal stresses are of opposite

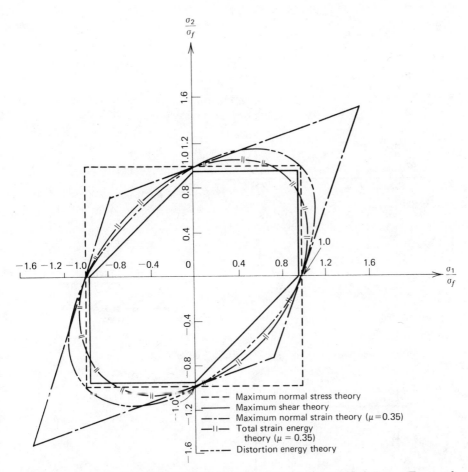

FIGURE 6.7. Comparison of failure theories for a biaxial state of stress. (From ref. 2; Joseph Marin, *Mechanical Behavior of Engineering Materials*, © 1962, p. 123. Adapted by permission of Prentice-Hall, Inc., Englewood Cliffs, New Jersey.)

sign. Experimental investigations support the prediction of the maximum normal stress theory for brittle behavior and the distortion energy theory or maximum shearing stress theory for ductile behavior. The other theories are less accurate and seldom used. Some supporting data for these conclusions are shown in Figure 6.8.

6.8   MOHR'S FAILURE THEORY

The Mohr theory of failure, proposed by Otto Mohr in 1900, is an extension of the maximum shearing stress theory of failure based on an interpretation

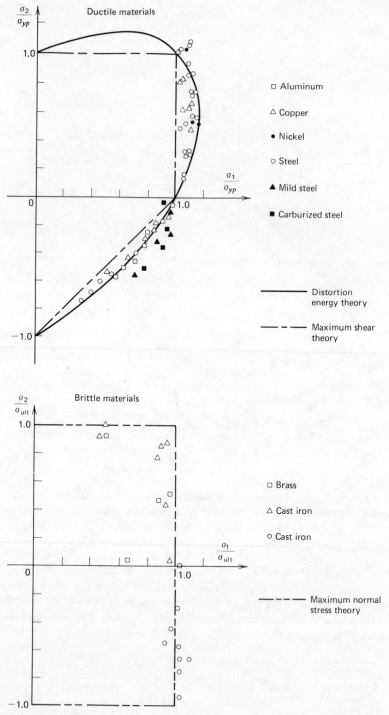

FIGURE 6.8. Comparison of biaxial strength data with theories of failure for a variety of ductile and brittle materials.

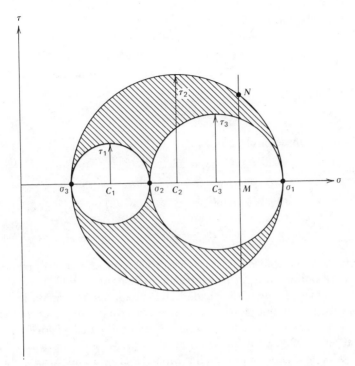

FIGURE 6.9.   Mohr's stress circles for a triaxial state of stress.

of the "three-dimensional" Mohr's circle. Further, the theory is conveniently able to account for materials with uniaxial failure strength properties in compression that differ from the uniaxial tensile failure strength properties. Before stating the Mohr theory of failure, it is necessary to examine the Mohr's circle analogy for the general triaxial state of stress. Figure 6.9 shows a plot on the τ-σ plane, where shearing stress τ and normal stress σ are plotted on orthogonal axes as is prescribed in the Mohr's circle analogy.* In Figure 6.9 it may be noted that three Mohr's circles are plotted, one corresponding to each of the three biaxial principal states of stress viewed by looking along the 1, 2, and 3 axes, one at a time. Thus, viewing the multiaxial state of stress along principal axis number 1 yields a Mohr's circle centered at C_1, with the circle intersecting the σ axis at σ_2 and σ_3. The other circles centered at C_2 and C_3 are drawn in like manner by viewing the state of stress along the other two principal axes. In Figure 6.9 it will be found that the normal and shearing stress components acting on any plane through the point will lie in the shaded area of the τ-σ plane, including its boundaries. The center positions along the

*Described, for example, in *Strength of Materials*, Part I, by S. Timoshenko, D. Van Nostrand, New York, 1958, p. 40.

σ axis of Figure 6.9 are

$$C_1 = \frac{\sigma_2 + \sigma_3}{2}$$

$$C_2 = \frac{\sigma_1 + \sigma_3}{2} \qquad \text{(6-46)}$$

$$C_3 = \frac{\sigma_1 + \sigma_2}{2}$$

The radii of the three circles are

$$R_1 = \frac{\sigma_2 - \sigma_3}{2}$$

$$R_2 = \frac{\sigma_1 - \sigma_3}{2} \qquad \text{(6-47)}$$

$$R_3 = \frac{\sigma_1 - \sigma_2}{2}$$

Referring again to the τ-σ plot of Figure 6.9, consider an arbitrary vertical line M-N. All points on this line represent a constant normal stress σ, and all admissable values of τ lie in the shaded area, including its boundaries. For the chosen σ position of line M-N, the maximum τ that can occur at the point must lie on the outer circle at N. If one assumes failure to be governed by the maximum value of τ, then for any line such as M-N the critical value of τ would always lie on the largest circle. Mohr asserted in his theory, therefore, that the outer or largest circle is sufficient to determine the failure condition, without regard for the intermediate principal stress value.

To proceed with the development, consider a material whose yield point in compression differs from its tensile yield point. Suppose that sample specimens of this material were tested in a uniaxial tension test, a uniaxial compression test, and a torsional shear test, noting yielding in each test. (The torsional shear test is, of course, not a uniaxial stress test.) Suppose then that the results of these three simple tests were plotted on a τ-σ plane as shown in Figure 6.10a. The circle 0-σ_{ypt} is obtained for yielding in the tensile test, the circle 0-σ_{ypc} for yielding in the compression test, and 0-τ_{yp} for yielding in the torsion test. With these circles an envelope circle is constructed to lie tangent to these three test-based circles, above and below. By so doing, a failure region is defined to lie outside the envelope circles on the τ-σ plane as shown in Figure 6.10b. The Mohr theory of failure may now be stated as follows:

Failure is predicted to occur in the multiaxial state of stress when the largest Mohr's circle associated with the state of stress at a given critical point becomes tangent to or exceeds the bounds of the failure envelope determined from the conditions of failure in simple tensile, compressive, and torsion tests using specimens of the same material.

If the Mohr theory were plotted for the multiaxial state of stress, the failure

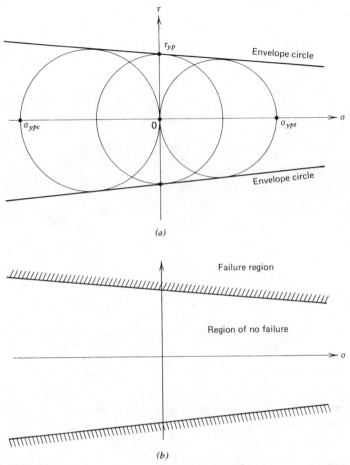

FIGURE 6.10. Failure region for the Mohr theory based on results of a uniaxial tension test, uniaxial compression test, and a torsional shear test.

surface would be very similar to the hexagonal cylinder of Figure 6.2 for the maximum shearing stress theory. However, in the case of the Mohr theory, the legs of the hexagon would be of unequal length. Figure 6.11 illustrates this comparison in the σ_1-σ_2 plane for the biaxial state of stress.

It may be observed that the Mohr theory, like the maximum shearing stress theory, satisfactorily predicts the no-yielding response of materials under hydrostatic states of stress. Mathematical formulation of the Mohr theory will not be pursued here since it can be only approximated. If compressive and tensile properties of a material of interest are significantly different, and if yielding is the failure mode, Mohr's theory might well be used, either in graphical form or in the form of a computer-augmented numerical solution.

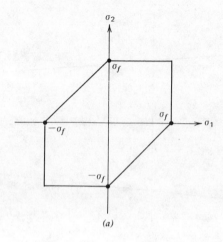

(a)

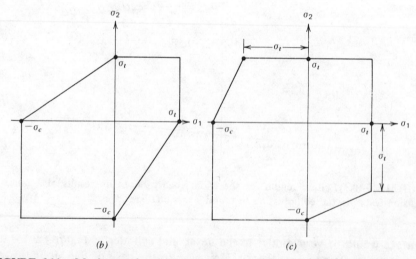

(b)   (c)

FIGURE 6.11. Maximum shearing stress theory compared to Mohr's Theory for a biaxial state of stress. (*a*) Maximum shearing stress theory (ductile). (*b*) Mohr's theory (ductile). (*c*) Modified Mohr's theory (brittle).

Further, if the material behavior is brittle, and the compressive and tensile properties of the material are significantly different, a modified Mohr's theory seems to give the best general agreement with experimental data. The empirical modification, shown in Figure 6.11, involves a linear extension of the failure boundary into the second and fourth quadrants, to a value of $-\sigma_t$, with a linear connection then made with $-\sigma_c$ as shown.

6.9 SUMMARY OF FAILURE THEORY EVALUATION

Evaluation of the six failure theories discussed in this chapter in light of experimental evidence leads to the following observations:

1. For isotropic materials that fail by brittle fracture, the maximum normal stress theory is the best theory to use.

2. For materials that fail by brittle fracture but exhibit a compressive ultimate strength that is significantly different from the tensile ultimate strength, the modified Mohr's theory is the best theory to use.

3. For isotropic materials that fail by yielding or ductile rupture, the distortion energy theory is the best theory to use.

4. For isotropic materials that fail by yielding or ductile rupture, the maximum shearing stress theory is almost as good as the distortion energy theory.

5. For materials that fail by yielding but exhibit a compressive yield strength that is significantly different from the tensile yield strength, Mohr's theory is a good theory to use.

6. As a rule of thumb, the maximum normal stress theory would be used for isotropic materials that exhibit a ductility of less than 5 percent elongation in 2 inches, and either the distortion energy theory or maximum shearing stress theory would be used for isotropic materials that exhibit a ductility of 5 percent or more in a 2-inch gage length. Where possible, a fracture mechanics analysis should be performed.

6.10 COMBINED STRESS FAILURE THEORIES AS DESIGN TOOLS

Having now examined several combined stress failure theories, it is pertinent to consider how one would use such failure theories in the design process. Designers are interested in averting failure but at the same time are interested in the efficient use of materials. The designer is therefore constantly faced with deciding how close to incipient failure he should design the part to operate. In making such a decision it must be recognized that some parts should be designed for infinite life, some parts should be designed for a finite life, and sometimes the design is time independent. It is also important to note that the margin between operating stress levels and failure strengths must be decided with due regard for the many uncertainties associated with material properties; loading configurations; accuracy of assumed physical and mathematical models relating load, material, and failure response; and other factors. A designer might integrate these considerations by proceeding through the following steps:

1. Examine the design specifications and proposed configuration to determine which failure mode or failure modes probably govern the design.

2. Determine the material strength properties that are directly related to the applicable failure modes, and with this information select one or more materials suitable for the task. Cost and availability are always important considerations at this stage.

3. Obtain uniaxial failure strength data for the selected materials in the applicable failure modes. These data may be obtained from available literature or, if necessary, by performing experimental laboratory tests.

4. Select a factor of safety consistent with all constraints on the design.

5. Calculate a design stress by dividing the failure strength by the factor of safety.

6. Transform the applicable failure theory into a design theory by using the equality and inserting the design stress in lieu of failure strength.

7. Determine the geometry of the machine part by solving for dimensions from the design theory. Frequently this step is an iterative or cut-and-try procedure.

8. Fabricate, test, modify, and experimentally develop the configuration of the machine part until it performs its function reliably and is compatible with all design constraints, including such considerations as, weight, cost, aesthetic appeal, etc.

While most of these steps are self-explanatory, the process of selecting a safety factor consistent with the constraints on the design is important enough to discuss in a little more detail. Unacceptable consequences are associated with a poor choice for the safety factor, either on the high side or on the low side. If the safety factor is picked too small, the machine part will have an unduly high probability of failure. If the safety factor is picked too large, the weight, size, and cost may be prohibitive. Judgment in the proper selection of a safety factor is enhanced by experience and a good working knowledge of the limitations on the models and mathematical techniques employed. Assessment of assumptions used in the stress analyses associated with the design, as well as the ability to anticipate which failure modes govern the design, also influences selection of the safety factor. There is no substitute for experience in the selection of a safety factor, but a good selection can be made if one gives due consideration to each of the following eight factors:

1. The accuracy with which the loads, forces, or other failure-inducing agents can be determined.

2. The accuracy with which the stresses or other mechanical moduli can be determined from the forces or other failure-inducing agents.

3. The accuracy with which the failure strength or other mechanical failure modulus can be determined for the material in the appropriate failure mode.

4. The need to conserve material, weight, space, or dollars.

5. The seriousness of the consequences of failure in terms of human life and/or property damage.
6. The quality of workmanship in manufacture.
7. The conditions of operation.
8. The quality of maintenance available or possible during operation.

An average value for the safety factor is about 2. This value would be modified, either up or down, depending on the assessment of each of the eight factors listed.

Selecting the safety factor, predicting the failure mode, and determining the associated failure strength, and calculating from these the design stress, are important steps in making design tools of combined stress failure theories. Substituting design stress for failure strength and using the equality in any applicable failure theory turns it into a design tool by which critical dimensions may be established. Failure theories properly evaluated are, therefore, among the most important links in the design process.

6.11 USING THE IDEAS

To illustrate the use of some of the combined stress failure theories, consider the following problem: It is desired to support a boundary fence across a river to exclude animals from an industrial property site. The proposed design calls for the fence to be supported by a pretensioned steel cable between two torsion bar support members, one on each river bank, buried in a fixed foundation as illustrated in Figure 6.12. The proposed dimensions of the cable support bracket are shown in Figure 6.12a, and the cable loading is estimated to be a static load of 100,000 lb, as indicated in the sketch. Materials under consideration for this cable support bracket are ASTM Class 60 gray cast iron, Grade 35018 ferritic malleable iron, and AISI 1020 steel. Properties for these materials taken from standard reference sources are summarized in Table 6.1.

Table 6.1. Properties for three candidate materials

Property	Class 60 Gray Iron	Grade 35018 Malleable Iron	AISI 1020 Steel
Ultimate tensile strength, psi	60,000	53,000	60,000
Ultimate compressive strength, psi	170,000	220,000	
Ultimate shear strength, psi		48,000	
Tensile yield strength, psi		35,000	43,000
Compr. yield strength, psi		45,000	
Shear yield strength, psi		20,500	
Elong. in 2 inches, percent	< 0.5	18	38

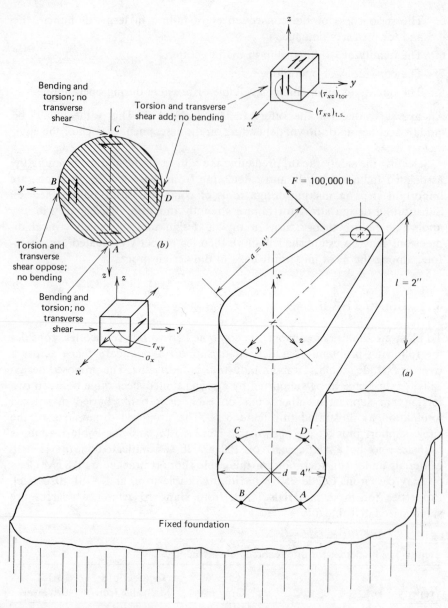

Bending and torsion; no transverse shear

Torsion and transverse shear add; no bending

$(\tau_{xz})_{tor}$

$(\tau_{xz})_{t.s.}$

$F = 100,000$ lb

Torsion and transverse shear oppose; no bending

Bending and torsion; no transverse shear

τ_{xy}

σ_x

$a = 4''$

$l = 2''$

(a)

(b)

$d = 4''$

Fixed foundation

FIGURE 6.12. Cable support bracket problem.

152

Considering only the cylindrical portion of the cable support bracket and neglecting any stress concentration effects, determine which, if any, of the three candidate materials are acceptable for the design as proposed if either yielding or fracture is to be regarded as failure.

A brief study of Figure 6.12 indicates that the combination of loading and geometry will generate multiaxial states of stress in the bracket. Surveying the material properties in Table 6.1, we may note that, based on the rule of thumb on ductility given in Section 6.9, the class 60 gray iron is to be regarded as brittle, whereas the other two materials are to be regarded as ductile. Further, the 35018 malleable iron properties in compression differ significantly from those in tension. Based on these observations and the combined stress failure theory summary of Section 6.9, the following procedures are indicated:

1. For failure analyses involving the class 60 gray iron, use the maximum normal stress theory, (6-1).
2. For failure analysis involving the 1020 steel, use the distortion energy theory as a first choice or the maximum shearing stress theory as a good alternative. That is, use (6-42) or (6-4).
3. For failure analyses involving the grade 35018 malleable iron, use the Mohr graphical theory as discussed in Section 6.8.

Before any of these failure theories may be utilized, however, the bracket must be carefully stress analyzed. Since attention is to be given only to the cylindrical portion of the bracket, a brief assessment of the loading indicates that torsion, bending, and transverse shear are produced by load F on the cylindrical member. Torsional shear stresses will reach maximum value over the entire outer surface of the cylinder. Bending stresses will reach maximum values at the wall cross section where the largest bending moment occurs and at points farthest from the neutral axis of bending. Transverse shear stresses will have the same pattern at all sections along the cylinder, with maximum transverse shear stress occurring along the neutral axis of bending, diminishing to zero at the points farthest from the neutral axis. Based on these observations it may be concluded that the most critical section is at the wall, and the potential critical points within that cross section will be at A, B, C, and/or D, as shown in Figure 6.12b. As indicated, at critical point A bending and torsional stresses combine effects, but the transverse shear stresses are zero. At critical point C the same state of stress is generated, except that the bending stress is compressive rather than tensile. At critical points B and D torsional shear and transverse shear add their effects, but bending stresses are zero. Careful analysis shows, however, that torsion and transverse shear add at D whereas they subtract at C. From the preceding discussion, then, it may be concluded that only critical points A and D need be investigated. Referring to the state of stress at A shown in Figure 6.12, it may be calculated that

$$\sigma_x = \frac{M_b c}{I} = \frac{Flc}{I} = \frac{(100,000)(2)(2)}{\left(\dfrac{\pi (4)^4}{64}\right)} = 31,830 \text{ psi} \qquad (6\text{-}48)$$

$$\tau_{xy} = \frac{Tc}{J} = \frac{Fac}{J} = \frac{(100,000)(4)(2)}{\left(\dfrac{\pi (4)^4}{32}\right)} = 31,830 \text{ psi} \qquad (6\text{-}49)$$

At critical point D of Figure 6.12, it may be calculated that

$$(\tau_{xz})_{\text{tr.sh.}} = \frac{4}{3}\frac{F}{A} = \frac{4}{3}\frac{(100,000)}{\left(\dfrac{\pi (4)^2}{4}\right)} = 10,610 \text{ psi} \qquad (6\text{-}50)$$

$$(\tau_{xz})_{\text{tor}} = \frac{Fac}{J} = \frac{100,000(4)(2)}{\left(\dfrac{\pi (4)^4}{32}\right)} = 31,830 \text{ psi} \qquad (6\text{-}51)$$

Since these shear components may be added directly,

$$\tau_{xz} = 10,610 + 31,830 = 42,440 \text{ psi} \qquad (6\text{-}52)$$

To implement use of any of the combined stress failure theories it is necessary to determine the principal normal stresses, which may be done by utilizing the general stress cubic equation (4-23) at each critical point of interest, that is, at A and D.

For critical point A the only nonzero stress components are σ_x and τ_{xy}, so the stress cubic reduces to

$$\sigma^3 - \sigma^2 \sigma_x + \sigma \tau_{xy}^2 = 0 \qquad (6\text{-}53)$$

or

$$\sigma\left(\sigma^2 - \sigma \sigma_x + \tau_{xy}^2\right) = 0 \qquad (6\text{-}54)$$

Utilizing the quadratic formula, then, the three solutions are

$$
\begin{aligned}
\sigma_1 &= \frac{\sigma_x}{2} + \sqrt{\left(\frac{\sigma_x}{2}\right)^2 + \tau_{xy}^2} \\[4pt]
\sigma_2 &= 0 \\[4pt]
\sigma_3 &= \frac{\sigma_x}{2} - \sqrt{\left(\frac{\sigma_x}{2}\right)^2 + \tau_{xy}^2}
\end{aligned}
\qquad (6\text{-}55)
$$

Substituting numerical values from (6-48) and (6-45), these principal stresses then become

$$\sigma_1 = 51,500 \text{ psi}$$
$$\sigma_2 = 0 \qquad (6\text{-}56)$$
$$\sigma_3 = -19,670 \text{ psi}$$

For critical point D the only nonzero stress component is τ_{xz}, so the stress cubic reduces to

$$\sigma^3 + \sigma(-\tau_{xz}^2) = 0 \qquad (6\text{-}57)$$

or

$$\sigma(\sigma^2 - \tau_{xz}^2) = 0 \qquad (6\text{-}58)$$

from which the three solutions are

$$\sigma_1 = \tau_{xz}$$
$$\sigma_2 = 0 \qquad (6\text{-}59)$$
$$\sigma_3 = -\tau_{xz}$$

and numerical values for the principal stresses at point D become

$$\sigma_1 = 42,440 \text{ psi}$$
$$\sigma_2 = 0 \qquad (6\text{-}60)$$
$$\sigma_3 = -42,440 \text{ psi}$$

With these stress analyses complete, the three candidate materials may be investigated using the appropriate theories of failure. It is interesting to observe before proceeding further that brittle materials will be more critical at point A (location of maximum principal normal stres), whereas ductile materials will be more critical at point D (location of maximum shearing stress, that is, maximum difference between principal normal stresses).

Considering the class 60 gray cast iron, a brittle material, we find that the maximum normal stress theory of (6-1) states, using numerical values just calculated, that *failure is predicted to occur by brittle fracture at critical point A if*

$$51,500 \geq 60,000$$
$$0 \geq 60,000 \qquad (6\text{-}61)$$
$$-19,670 \geq 60,000$$

or if

$$51,500 \leq -170,000$$
$$0 \leq -170,000 \qquad (6\text{-}62)$$
$$-19,670 \leq -170,000$$

Of these equations, the first one of (6-61) is clearly the governing equation, and it does not predict failure. Also from (6-60), the maximum normal stress is lower at critical point D than for critical point A. Thus, failure would not be predicted if the class 60 gray iron were to be used. On the other hand, the margin of safety might be questioned for this material, since the safety factor in such a solution would be, from the first equation of (6-61),

$$n_{C.I.} = \frac{60,000}{51,500} = 1.17 \tag{6-63}$$

As stated in 6.10, an average value of safety factor is about 2. It is unlikely that the calculated value of safety factor of 1.17 would be acceptable in this application; and if the gray iron were used, it would be recommended that the part be redesigned to give a larger factor of safety.

Considering next the 1020 steel, a ductile isotropic material, the distortion energy theory of (6-42) states, using numerical values calculated, that *failure is predicted to occur by yielding at critical point D if*

$$\frac{1}{2}\left[(42,440 - 0)^2 + (0 + 42,440)^2 + (-42,440 - 42,440)^2\right] \geq (43,000)^2 \tag{6-64}$$

or if

$$5.40 \times 10^9 \geq 1.85 \times 10^9 \tag{6-65}$$

and the conclusion is that failure by yielding is clearly predicted. If the maximum shear theory of (6-4) is used here, it would state that *failure is predicted to occur by yielding at critical point D if*

$$|42,440 - 0| \geq 43,000$$
$$|0 + 42,440| \geq 43,000 \tag{6-66}$$
$$|-42,440 - 42,440| \geq 43,000$$

Thus, from the third of these equations failure is clearly predicted, which is not unexpected since the distortion energy theory predicted failure and the maximum shear stress theory is always on the "conservative" side compared to the distortion energy theory. A redesign is clearly necessary if 1020 steel is to be used in this application.

Finally, considering the grade 35018 malleable iron, a ductile material whose tensile and compressive yield strength are significantly different, we find that the Mohr theory of Section 6.8 would be most applicable. To proceed with the graphical solution, the failure envelope is first constructed, as indicated in Figure 6.13a, by drawing to scale the Mohr's circles representing tensile yielding, shear yielding, and compressive yielding on the τ-σ plane and then constructing the tangent circle envelope to these three circles as shown. Transferring the failure envelope to Figure 6.13b, we next construct on the τ-σ plane, the Mohr's circles representing the states of stress at critical points A and D, using the standard techniques of Mohr's circle construction* referring to the states of stress, as depicted in Figure 6.12b. Since the Mohr's

*See, for example, p. 135 of ref. 8.

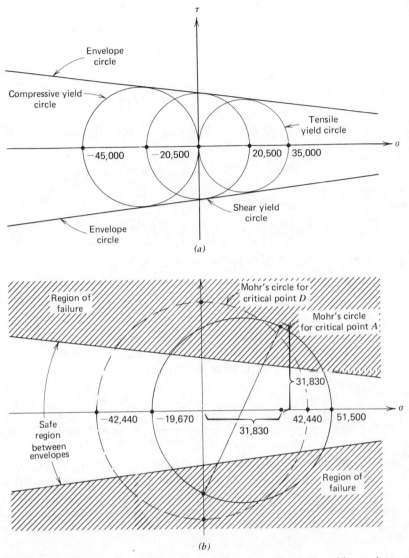

FIGURE 6.13. Mohr's theory solution to cable support bracket problem using grade 35018 malleable iron.

circles of stress for both critical point A and critical point D exceed the failure boundary in Figure 6.13b, failure is predicted and a redesign is necessary if the grade 35018 malleable iron is to be used.

Surveying all the results, it might well be concluded that the class 60 gray iron is the best material selection for this application, if a section size increase is specified to give the desired safety factor. To pursue this, it may be deduced that, for a safety factor of 2, the part must be sized to give

$$\sigma_d = \sigma_1 = \frac{\sigma_u}{n_{C.I.}} \tag{6-67}$$

or

$$\sigma_d = \sigma_1 = \frac{60,000}{2} = 30,000 \text{ psi} \tag{6-68}$$

where σ_d is the design stress desired. Carrying this result back to (6-55), then,

$$\sigma_d = \sigma_1 = \frac{\sigma_x}{2} + \sqrt{\left(\frac{\sigma_x}{2}\right)^2 + \tau_{xy}^2} \tag{6-69}$$

and substituting from (6-48) and (6-49) for σ_x and τ_{xy},

$$\sigma_d = \frac{Flc}{2I} + \sqrt{\left(\frac{Flc}{2I}\right)^2 + \left(\frac{Fac}{J}\right)^2} \tag{6-70}$$

Since $J = 2I$ and $a = 2l$, this may be rewritten as

$$\sigma_d = \frac{Flc}{2I}\left[1 + \sqrt{1 + 2^2}\right] = \left(\frac{1 + \sqrt{5}}{2}\right)\frac{Flc}{I} \tag{6-71}$$

or

$$\sigma_d = \frac{1.62\,Fl(d/2)}{(\pi d^4/64)} = \frac{51.78\,Fl}{\pi d^3} \tag{6-72}$$

or

$$d = \sqrt[3]{\frac{51.78\,Fl}{\pi\sigma_d}} = \sqrt[3]{\frac{51.78(100,000)(2)}{\pi(30,000)}} = 4.79 \text{ inches} \tag{6-73}$$

Thus the design recommendation would be to make the cable support bracket of class 60 gray iron with a 4.79-inch diameter for the cylindrical portion. As a final caution, recall that stress concentration effects have been neglected and should be included, as discussed later in Chapter 12. Further, critical points in locations other than the cylindrical portion of the bracket should also be identified and investigated.

QUESTIONS

1. Explain why it is often necessary for a designer to utilize a failure theory.
2. What are the essential attributes of any useful failure theory?
3. What is the basic assumption that constitutes the framework for all failure theories?
4. State, in words only and also mathematically, the definition of failure according to the maximum normal stress theory of failure.

5.   State, in words only and also mathematically, the definition of failure according to the maximum shear stress theory of failure.

6.   Define or describe what is meant by the term *distortion energy*.

7.   Derive an equation by means of which distortion energy may be calculated from the principal normal stresses.

8.   State, in words only and also mathematically, the definition of failure according to the distortion energy theory of failure.

9.   Summarize the conditions under which each of the three theories of failure noted in problems 4, 5, and 8 has been found to be most applicable.

10.   (a) Write in *words* a "first strain invariant" theory of failure. Be complete and precise.
(b) Derive a complete mathematical expression for your "first strain invariant" theory of failure, expressing the final result in terms of *principal stresses and material properties*. Recall that

$$I_1 = \text{first strain invariant} = \varepsilon_i + \varepsilon_j + \varepsilon_k$$

(c) How could you establish the validity of this theory of failure?

11.   A solid shaft of circular cross section is subjected to a pure torsional moment M_t. Predict the diameter of the shaft corresponding to incipient failure under the torque M_t by (a) the maximum normal stress theory, (b) maximum shearing stress theory, and (c) the distortion energy theory. (d) Find the ratio of the diameters predicted by the maximum shearing stress theory and distortion energy theory to that predicted by the maximum normal stress theory.

12.   The solid cylindrical cantilever bar shown in Figure Q6.12 is subjected to *pure* torsional moment T about the z axis, *pure* bending moment M_b about the y axis, and pure tensile force P along the x axis, all at the same time. The material is a ductile aluminum alloy.

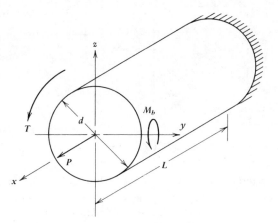

FIGURE Q6.12.

(a) Carefully identify the most critical point(s), neglecting stress concentration. Give detailed reasoning for your selection(s).

(b) At the critical point(s) clearly draw an element of volume, carefully labeled, showing all stress vectors.

(c) Carefully explain how you would determine whether to expect yielding at the critical point.

13.   Transform the principal normal stress theory of failure, the maximum shearing stress theory of failure, and the distortion energy theory of failure into appropriate *design* theories involving the safety factor concept. Explain the conditions under which a designer should use each of these design theories.

14.   Pick three machine parts with which you are familiar and decide, on the basis of the eight factors to be considered in selecting a safety factor, what an appropriate safety factor might be in each case.

15.   In the triaxial state of stress shown in Figure Q6.15, determine whether or not failure should occur. Use the maximum normal stress theory for brittle material and both the distortion energy theory and the maximum shear stress theory for ductile materials.

(a) For an element stressed as shown made of 319-T6 aluminum. (σ_{yp} = 24,000 psi, σ_u = 36,000 psi, e = 2.0 percent in 2 inches.)

(b) For an element stressed as shown but made of 220-T4 aluminum. (σ_{yp} = 25,000 psi, σ_u = 46,000 psi, e = 14 percent in 2 inches.)

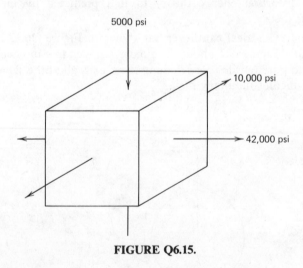

5000 psi

10,000 psi

42,000 psi

FIGURE Q6.15.

16.   A critical machine part in an aircraft flap actuator may be adequately modeled as a solid cylindrical bar subjected to an axial force P of 10,000 lbs, a bending moment M_b = 1500 in-lb, and a torsional moment T = 4000 in-lb. If it is proposed to use a 1-inch-diameter bar of 7075-T6 aluminum (σ_{yp} =

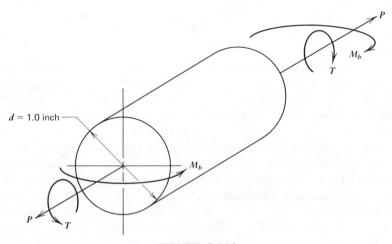

FIGURE Q6.16.

72,000 psi, σ_u = 82,000 psi, e = 11 percent in 2 inches), what would be the safety factor on yielding for this member? Give a clear and complete analysis, stating what you are doing at every step.

17.   An aircraft wing flap actuator housing is made of cast magnesium alloy AZ63A-T4 (σ_{yp} = 14,000 psi, σ_u = 40,000 psi, e = 12 percent in 2 inches). At the suspected critical point it has been calculated that the state of stress is as shown in Figure Q6.17. Would you predict failure of the part by yielding?

18.   A 4340 steel bar of 1-inch-diameter solid circular cross section is heat-treated and drawn at 1000°F to give a Brinell Hardness Number (BHN)

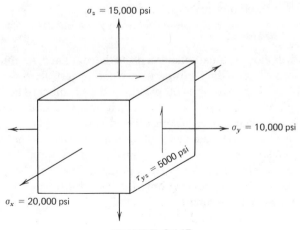

FIGURE Q6.17.

of 377. If this bar is simultaneously subjected to a pure torsional moment of 12,000 in-lb and a bending moment of 10,000 in-lb, utilize the Mohr theory of failure to predict whether or not failure should occur.

19. A 2.5-inch-diameter shaft is made of 4340 steel, oil quenched and tempered at 800°F. The shaft is subjected to a pure static torsional moment of 70,000 in-lb and a pure static bending moment of 50,000 in-lb under critical loading conditions. Determine the factor of safety on yielding for this shaft.

20. A horizontal solid metal bar is 20 inches long, has a cylindrical cross section of 1-inch diameter, and is fixed at one end as a cantilever beam. At the free end the bar is subjected to a vertical downward bending force of 250 lb and a torsional moment about the cylinder axis of 4700 in-lb. Neglect stress concentration.

(a) If the bar material were brittle, would you predict failure under the specified loading if $\sigma_u = 69,000$ psi? Explain.

(b) If the bar were ductile, would you predict failure under the specified loading if $\sigma_{yp} = 63,000$ psi? Explain.

21. A 1-inch-diameter solid cylindrical bar of 6061-T6 aluminum is 27 inches long. The bar is simultaneously subjected to an axial force of $P = 25,000$ lb and a pure torsional moment about the axis of symmetry of $T = 3000$ in-lb. The material has an ultimate strength of 45,000 psi, yield point of 40,000 psi, and elongation in 2 inches of 12 percent.

(a) Establish the location(s) of the critical point(s).

(b) Using a coordinate system of your choice and the usual elemental volume, draw a sketch to completely define the state of stress at the critical point.

(c) Calculate the principal stresses. *Do not use the Mohr's circle technique for doing this.*

(d) Determine the direction cosines for the principal planes upon which the *largest* principal stress acts.

(e) Calculate the maximum principal shearing stress.

(f) Determine whether or not failure by yielding would take place under the specified loading.

22. A round, hollow tubular member is to be designed to fulfill the dual function of a shaft providing power to a rotating test chamber and also a pressure supply pipe providing 5000 psi air to the chamber. The hollow shaft will be configured and supported so that it may be considered to be a closed-end pressure vessel. Because of other design specifications, the hollow shaft must have an inside diameter of exactly 5 inches. The steady drive torque on the shaft is estimated to be a constant 60,000 in-lb. The material to be used is 4130 steel in the annealed condition. The yield strength of this material is 114,000 psi. Considering only the cylindrical wall:

(a) Determine the outside diameter of the shaft/vessel that would just prevent but be on the verge of yielding.

(b) Give your best estimate of how much overpressure could be tolerated before a local blowout would be expected.

(c) Recommend a wall thickness that you would consider to provide a *safe* design for this machine part.

REFERENCES

1. Nadai, A., *Theory of Flow and Fracture of Solids*, Vol. 1, McGraw-Hill, New York, 1950.

2. Marin, J., *Mechanical Behavior of Engineering Materials*, Prentice-Hall, Englewood Cliffs, N.J., 1962.

3. Polakowski, N. H., and Ripling, E. J., *Strength and Structure of Engineering Materials*, Prentice-Hall, Englewood Cliffs, N.J., 1966.

4. Juvinall, R. C., *Engineering Considerations of Stress, Strain, and Strength*, McGraw-Hill, New York, 1967.

5. Timoshenko, S., *Strength of Materials*, Part II, Van Nostrand, New York, 1952.

6. Drucker, D. C., *Introduction to Mechanics of Deformable Solids*, McGraw-Hill, New York, 1967.

7. Mendelson, A., *Plasticity: Theory and Application*, Macmillan Co., New York, 1968.

8. Crandall, S. H., and Dahl, N. C., *An Introduction to the Mechanics of Solids*, McGraw-Hill, New York, 1959.

9. McAdam, J. J., Jr., Geil, G. M., and Jenkins, W. H., "Influence of Plastic Extension and Compression on the Fracture Stress of Metals," ASTM 1947 preprint; McAdam, J. J., Jr., Geil, G. M., and Mebs, R. W., "Influence of Plastic Deformation, Combined Stresses and Low Temperatures on the Breaking Stress of Ferritic Steels," *Metals Technology* (August 1947); McAdam, J. J., Jr., "The Technical Cohesive Strength of Metals in Terms of Principal Stresses," *Metals Technology* (December 1944).

CHAPTER 7

High-Cycle Fatigue

7.1 INTRODUCTION

Static or quasistatic loading is rarely observed in modern engineering practice, making it essential for the designer to address himself to the implications of repeated loads, fluctuating loads, and rapidly applied loads. By far, the majority of engineering design projects involve machine parts subjected to fluctuating or cyclic loads. Such loading induces fluctuating or cyclic stresses that often result in failure by fatigue. It might be observed at the outset that the term *fatigue* adopted well over a century ago, may not be the best choice of terminology since many aspects of the phenomenon are distinctly different from the processes of biological fatigue. For example, it is difficult to detect the progressive changes in material properties that occur during fatigue stressing, and failure may therefore occur with little or no warning. Also, periods of rest, with the fatigue stress removed, do not lead to any measurable healing or recovery from the effects of the prior cyclic stressing. Thus, the damage done during the fatigue process is *cumulative*, and generally unrecoverable.

Fatigue, although a complex subject, has not been neglected by researchers. Estimates have been made (43) that indicate that if one wished to keep up with the literature by reading one report per working day, he would fall behind more than a year for every year he read, and catching up on the existing backlog of literature would be virtually impossible. Yet the designer is increasingly challenged by the demands for higher performance, higher speed, higher temperature, lighter weight, and longer life—and all this at a reasonable cost and in a reasonable production time. To successfully achieve these objectives the designer must squarely face the problems of preventing failure by fatigue. Some of these problems may be characterized as follows:

1. Calculations of life are generally less accurate and less dependable than strength calculations. Order of magnitude errors in life estimates are not unusual.
2. Fatigue characteristics of a material cannot be deduced from other mechanical properties. They must be measured directly.
3. Full-scale prototype testing is usually necessary to assure an acceptable service life.

4. Results of different but "identical" tests may differ widely, requiring, therefore, a statistical interpretation by the designer.
5. Materials and design configurations must often be selected to provide slow crack propagation and, if possible, detection of cracks before they become dangerous.
6. The concepts of "fail-safe" design must often be implemented to achieve acceptable reliability. That is, even if a given structural element fails, the overall structure must remain intact and capable of supporting the load on a short-term emergency basis.

Fatigue failure investigations over the years have led to the observation that the fatigue process actually embraces two domains of cyclic stressing or straining that are significantly different in character, and in each of which failure is probably produced by different physical mechanisms. One domain of cyclic loading is that for which significant plastic strain occurs during each cycle. This domain is associated with high loads and short lives, or low numbers of cycles to produce fatigue failure, and is commonly referred to as *low-cycle fatigue*. The other domain of cyclic loading is that for which the strain cycles are largely confined to the elastic range. This domain is associated with lower loads and long lives, or high numbers of cycles to produce fatigue failure, and is commonly referred to as *high-cycle fatigue*. Low-cycle fatigue is typically associated with cycle lives from one up to about 10^4 or 10^5 cycles, and high-cycle fatigue for lives greater than about 10^4 or 10^5 cycles. This chapter addresses itself to high-cycle fatigue. Chapter 11 discusses the problems associated with low-cycle fatigue.

7.2 HISTORICAL REMARKS

For centuries man has been aware that he could break wood or metal by repeatedly bending it back and forth with a large amplitude. It came as something of a surprise, however, when he found that repeated stressing would produce fracture even with the stress amplitude held well within the elastic range of the material. The first fatigue investigations seem to have been reported by a German mining engineer, W. A. S. Albert, who in 1829 performed some repeated loading tests on iron chain. Some of the earliest fatigue failures in service occurred in the axles of stage coaches. When railway systems began to develop rapidly about the middle of the nineteenth century, fatigue failures in railway axles became a widespread problem that began to draw the first serious attention to cyclic loading effects. This was the first time that many similar parts of machines had been subjected to millions of cycles at stress levels well below the yield point, with documented service failures appearing with disturbing regularity. As is often done in the case of unexplained service failures, attempts were made to reproduce the failures in the laboratory. Between 1852 and 1870 the German railway engineer, August Wöhler (44) set up and conducted the first systematic fatigue investigation. He conducted tests on full-scale railway axles and also small-scale bending, torsion, and axial cyclic loading tests for several different materials. Some of Wöhler's original data were reported as shown in Figure 7.1. These data for Krupp axle steel are plotted on what has become very well known as the *S-N* diagram, and fatigue data today are often presented in much the same way.

At about the same time, other engineers began to concern themselves with the problems of failure associated with repeated fluctuating loads. With the rapidly developing railway system, wrought iron bridges were beginning to replace brick and stone structures, and questions were raised regarding the suitability of iron bridges for railway structures. Full-scale tests of riveted girders were conducted. Some girders 22 feet long and 16 inches deep were subjected to millions of cycles of loading. By 1900 over 80 papers on the subject of fatigue had been published, reporting fatigue failures not only in railway axles and bridge structures, but also in chains, crank shafts, marine propeller shafts, and wire rope.

As the "age of technology" began to accelerate in the early twentieth century with higher speed machinery, the advent of high-speed turbines, and the development of the aviation industry, much more attention was turned toward trying to understand the fatigue problem. By the midtwentieth century widespread activity in fatigue investigation was worldwide at both the

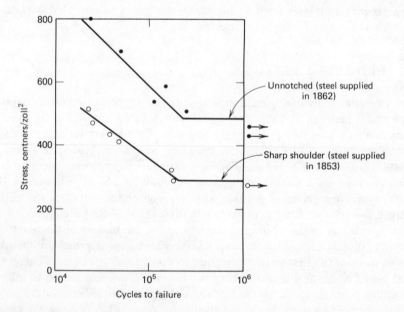

FIGURE 7.1. Fatigue data reported by Wöhler in an early investigation on fatigue of railway axle steel. Note: 1 centner = 50 kg, 1 zoll = 1 inch, 1 centner/zoll² ≑ 110 psi.

microscopic level where physicists and metallurgists were attempting to explain the basic phenomena and at the macroscopic level where engineers were attempting to design components and systems by employing simple laboratory test data together with semiempirical design theories. The development of dislocation theory during this period contributed much to the understanding of fatigue at the microscopic level. Development of the electron microscope with its superior resolution greatly contributed to the direct observation of fatigue processes. No less important a contribution has been made at the macroscopic level by the development of the high-speed computer, which gives designers a powerful computational tool for making better estimates of fatigue life and strength. Finally, fracture mechanics concepts have in the past decade added an important measure of insight into crack propagation behavior, as well as providing the basis for a practical new failure prediction tool useful to a designer when faced with fatigue loading conditions.

7.3   THE NATURE OF FATIGUE

Fatigue may be characterized as a progressive failure phenomenon that proceeds by the *initiation* and *propagation* of cracks to an unstable size. Although there is not complete agreement on the microscopic details of the initiation and propagation of the cracks, some of the better explanations offered are as follows:*

Fatigue crack nuclei, from which cracks grow and often propagate to failure, are thought to be formed through the movements of dislocations that produce fine slip bands at the crystal surfaces. It may be recalled by referring to Figure 3.7 that the application of a static shearing stress produces slip steps at the surface that are on the order of 10^{-4} to 10^{-5} cm in height. These slip bands are commonly characterized as *coarse* slip bands. Under cyclic loading it is more usual to observe *fine* slip bands of about 10^{-7} cm in height, which ultimately turn out to be the regions in which the fatigue cracks were initiated. Further, the fatigue slip bands give rise to surface grooves and ridges (45), shown schematically in Figure 7.2, as a result of the reversed slip on adjacent slip planes caused by load reversal. These grooves and ridges may be either sharply saw-toothed in shape or smoothly rounded corrugations. If many planes slip, the resulting striations are shallow and undulating, whereas if only a few closely spaced planes are operative, sharply defined crevices and walls are formed. In certain cases clearly defined extrusions have been found (46) as a result of reversed slip, and detectable slip band cracks are always found at these extrusions, which may be in fact intrusions generated by reversed slip. Once formed, these intrusions grow in depth by

*See pp. 34–100 of ref. 19.

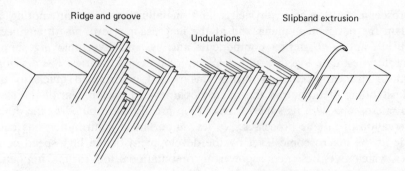

FIGURE 7.2. Grooves and ridges formed by reversed loading and associated regions of fine slip.

the reversed slip process, and their growth may well constitute the major portion of the fatigue life of the metal.

Another proposed explanation for the initiation of fatigue nuclei is based on an observation that many dislocation loops are produced by cyclic stresses. It is postulated that the interaction of these loops produces many vacancies that condense at the operative slip planes in sufficient quantities to form stable holes in the lattice structure. Certain holes have been found (47) along slip bands and grain boundaries, but the mechanism of their formation has not been clearly established. These holes then constitute fatigue nuclei from which cracks grow.

The extension of the fatigue nucleus by reversed slip takes place generally along the slip planes most closely aligned with the direction of maximum shearing stress. As long as the crack continues to grow along the active slip plane, no change in the basic mechanism of growth has been observed. This type of crack growth has been arbitrarily designated as Stage I crack growth. Stage I crack growth, which may occupy a large or small portion of the total fatigue life, seems to be favored by low applied stresses and conditions that lead to slow crack growth. If periodic high stress cycles, notches, or conditions that lead to a high ratio of tensile stress to shear stress component exist, then Stage I crack growth may be expected to give way to Stage II crack growth.

Stage II crack growth is governed not by the local shearing stress but by the maximum principal normal stress in the neighborhood of the crack tip. Thus the crack tip is caused to deviate from its slip path and propagate in a direction roughly perpendicular to the direction of the maximum normal stress. The fracture surface during Stage II growth is characterized by striations and *beach marks* that can be related in their density and width to the applied stress level (48). The fracture surface produced during the Stage II growth is relatively smooth.

Finally the crack length reaches a critical dimension and one additional cycle then causes complete failure. The final failure region will typically show evidence of plastic deformation produced just prior to final separation. For ductile materials the final fracture area often appears as a shear lip produced by crack propagation along the planes of maximum shear. Several features of a typical fatigue failure surface are illustrated schematically in Figure 7.3.

Although designers find these basic observations of great interest, they must be even more interested in the macroscopic phenomenological aspects of fatigue failure and in avoiding fatigue failure during the design life. Some of the macroscopic effects and basic data requiring consideration in designing under fatigue loading include:

1. The effects of a simple, completely reversed alternating stress on the strength and properties of engineering materials.
2. The effects of a steady stress with superposed alternating component, that is, the effects of cyclic stresses with a nonzero mean.
3. The effects of alternating stresses in a multiaxial state of stress.
4. The effects of stress gradients and residual stresses, such as imposed by shot-peening or cold-rolling, for example.
5. The effects of stress raisers, such as notches, fillets, holes, threads, riveted joints, and welds.

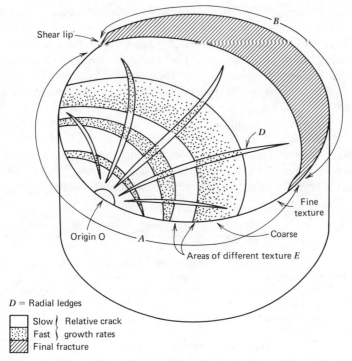

FIGURE 7.3. Features of the failure surface of a typical ductile metal subjected to alternating cyclic loads. (From ref. 19; Reproduced by kind permission of Blackie and Son, Limited, Glasgow.)

6. The effects of surface finish, including the effects of machining, cladding, electroplating, and coating.
7. The effects of temperature on fatigue behavior of engineering materials.
8. The effects of size of the structural element.
9. The effects of accumulating cycles at various stress levels and the permanence of the effect.
10. The extent of the variation in fatigue properties to be expected for a given material.
11. The effects of humidity, corrosive media, and other environmental factors.
12. The effects of interaction between fatigue and other modes of failure, such as creep, corrosion, and fretting.

These important factors and others will be considered in the remaining pages of this chapter. The questions of importance will also include "How does one obtain useful data?" and "What is done with it once it is obtained?"

7.4   FATIGUE LOADING

Faced with the design of a fatigue sensitive element in a machine or structure, a designer is critically interested in the fatigue response of engineering materials to various loadings that might occur throughout the design life of the machine under consideration. That is, he is interested in the effects of various *loading spectra* and associated *stress spectra*, which will in general be a function of the design configuration and the operational use of the machine.

Perhaps the simplest fatigue stress spectrum to which an element may be subjected is a zero-mean sinusoidal stress-time pattern of constant amplitude and fixed frequency, applied for a specified number of cycles. Such a stress-time pattern, often referred to as a completely reversed cyclic stress, is illustrated in Figure 7.4a. Utilizing the sketch of Figure 7.4, we can conveniently define several useful terms and symbols; these include:

$$\sigma_{max} = \text{maximum stress in the cycle}$$

$$\sigma_m = \text{mean stress} = \frac{\sigma_{max} + \sigma_{min}}{2}$$

$$\sigma_{min} = \text{minimum stress in the cycle}$$

$$\sigma_a = \text{alternating stress amplitude} = \frac{\sigma_{max} - \sigma_{min}}{2}$$

$$\Delta\sigma = \text{range of stress} = \sigma_{max} - \sigma_{min}$$

$$R = \text{stress ratio} = \frac{\sigma_{min}}{\sigma_{max}}$$

$$A = \text{amplitude ratio} = \frac{\sigma_a}{\sigma_m} = \frac{1 - R}{1 + R}$$

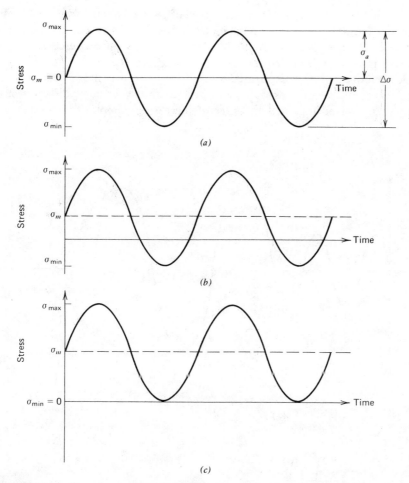

FIGURE 7.4. Several constant-amplitude stress time patterns of interest. (*a*) Completely reversed, $R = -1$. (*b*) Nonzero mean stress. (*c*) Released tension, $R = 0$.

Any two of the quantities just defined, except the combination σ_a and $\Delta\sigma$ or the combination A and R, are sufficient to completely describe the stress-time pattern shown.

A second type of stress-time pattern often encountered is the nonzero mean spectrum shown in Figure 7.4*b*. This pattern is very similar to the completely reversed case except that the mean stress is either tensile or compressive, in any event different from zero. The nonzero mean case may be thought of as a static stress equal in magnitude to the mean σ_m, with a superposed completely reversed cyclic stress of amplitude σ_a.

A special case of nonzero mean stress, illustrated in Figure 7.4*c* is often encountered in practice. In this special case the minimum stress σ_{min} is zero. That is, the stress ranges from zero up to some tensile maximum and then back to zero. This type of stressing is often called *released tension*. For released tension it may be noted that the mean stress is half the maximum stress, or $\sigma_m = \sigma_{max}/2$. A similar but less frequently encountered stress-time pattern is called *released compression*, where $\sigma_{max} = 0$ and $\sigma_m = \sigma_{min}/2$.

A somewhat more complicated stress-time pattern is illustrated in Figure 7.5a where the mean stress is zero but there are two or more different stress amplitudes mixed together. In order of increasing complexity, the sketch of

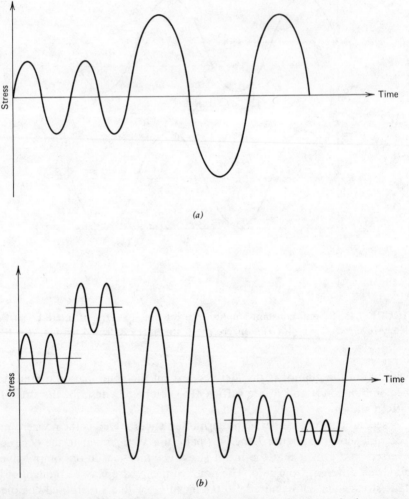

(a)

(b)

FIGURE 7.5.   Stress-time patterns in which the amplitude changes or both mean and amplitude change to produce a more complicated stress spectrum. (a) Zero mean, changing amplitude. (b) Changing mean and amplitude.

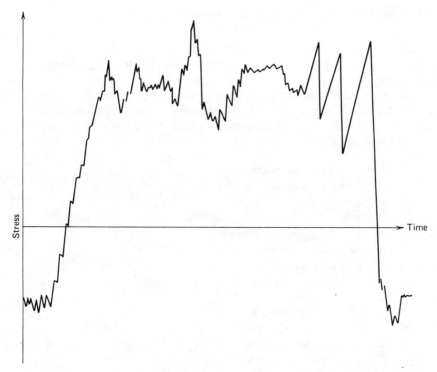

FIGURE 7.6. A quasi-random stress-time pattern that might be typical of an operational aircraft during any given mission.

Figure 7.5*b* illustrates the case in which not only the stress amplitude varies but also the magnitude of the mean stress periodically changes. It may be noted that this stress-time spectrum is beginning to approach a degree of realism. Finally, in Figure 7.6 a sketch of a realistic stress spectrum is given. This type of quasi-random stress-time pattern might be encountered in an airframe structural member during a typical mission including refueling, taxi, takeoff, gusts, maneuvers, and landing. The obtaining of useful, realistic data is a challenging task in itself. Instrumentation of existing machines, such as operational aircraft, provide some useful information to the designer if his mission is similar to the one performed by the instrumented machine. Recorded data from accelerometers, strain gauges, and other transducers may in any event provide a basis from which a statistical representation can be developed and extrapolated to future needs if the fatigue processes are understood. Having a good estimate of the loading spectrum is a good beginning, but even with such information available, the fatigue designer's task in developing geometry and selecting materials to successfully avert failure by fatigue is not an easy one.

7.5 LABORATORY FATIGUE TESTING

Designing machine parts or structures that are subjected to fatigue loading is usually based on the results of laboratory fatigue tests using specimens of the material of interest. Because laboratory fatigue data are so useful to the designer who knows the merits and limitations of such data, it will be useful to consider briefly some of the more common testing machines found in a fatigue testing laboratory. Laboratory fatigue testing machines might be classified in the following way:

    I.   Rotating-bending machines
        A. Constant bending moment type
        B. Cantilever bending type
   II.   Reciprocating-bending machines
  III.   Axial direct-stress machines
        A. Brute-force type
        B. Resonant type
  IV.   Vibrating shaker machines
        A. Mechanical type
        B. Electromagnetic type
   V.   Repeated torsion machines
  VI.   Multiaxial stress machines
 VII.   Computer-controlled closed loop machines
VIII.   Component testing machines for special applications
  IX.   Full-scale or prototype fatigue testing systems

Rotating-bending machines of the constant bending moment type may be described schematically as shown in Figure 7.7. With this type of device the region of the rotating beam between the inboard bearings is subjected to a constant bending moment all along its length. While under the influence of this constant moment, the specimen is caused to rotate with the drive spindles about a longitudinal axis. Any point on the surface is thereby subjected to a completely reversed stress-time pattern, as can be deduced by following the stress history of a point as it rotates from maximum compression at the top position down through zero stress when at the side, through maximum tension at the bottom, and then back through zero to maximum compression at the top again. The stress-time pattern for a typical point at the surface of the critical section is shown below the sketch of the machine in Figure 7.7. This machine is basically of the constant load variety and is not well adaptable to nonzero mean fatigue testing.

The cantilever type of rotating-bending fatigue testing machine is shown schematically in Figure 7.8. It is very similar to the rotating-bending machine of Figure 7.7 except that the bending moment varies along the beam, making the axial location of the critical section important in applying the proper stress level. The stress-time pattern is again completely reversed unless some

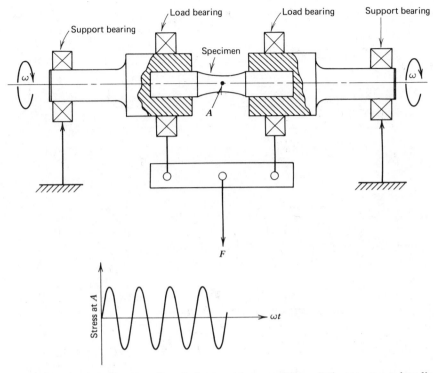

FIGURE 7.7. Rotating-bending fatigue testing machine of the constant bending moment type.

special device is added to produce an axial stress component. The machine is basically of the constant load variety.

Reciprocating-bending machines, often called flat-plate machines because of the typical flat specimens usually used, may be characterized by the sketch of Figure 7.9. The reciprocating-bending machine is capable of producing either completely reversed stresses or nonzero mean cyclic stresses by judiciously positioning the specimen clamping vise with respect to the mean displacement position of the eccentric crank drive. This machine is a constant deflection machine, in contrast to the rotating-bending machines described earlier. With a suitable feedback control system, the device can be made to behave like a constant force machine.

Direct-stress machines, often called push-pull machines, impose a direct tensile or compressive stress on a specimen by pushing and pulling in the direction of the specimen axis. In the brute-force type of direct-stress machine, as represented schematically in Figure 7.10, the stress-time pattern is produced by an eccentric crank driving a lever system attached to one end of the specimen while the other end of the system is held fixed to the frame.

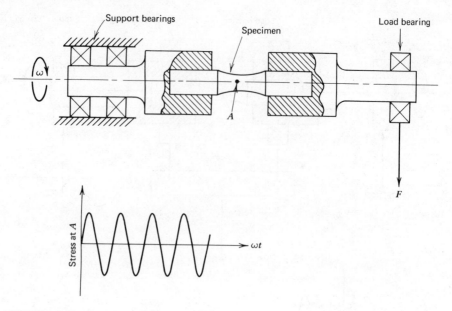

FIGURE 7.8. Rotating-bending fatigue testing machine of the cantilever bending type.

Special precautions are usually taken to provide parallel grip alignment and straight line motion of the driven end. This machine is basically of the constant deflection type, but often embodies a control system that converts it to what is essentially a constant force machine. Nonzero mean stress testing is relatively easy with this machine because the mean stress can be altered by simply moving the position of the fixed end.

Resonant-type direct-stress fatigue testing machines are basically spring-coupled two-mass vibrating systems with the fatigue specimen placed at the node. Such a system is shown schematically in Figure 7.11. Excitation of one mass at the proper frequency causes the system to resonate, and by properly controlling the amplitude of oscillation, the specimen may be subjected to any desired stress level. Nonzero mean stresses can be produced in these machines by addition of an auxiliary spring in parallel with the specimen, but this introduces practical problems of measuring and controlling stress in the specimen. The machine is best suited to producing completely reversed stress-time patterns.

Vibrating shakers are often used for fatigue testing components and subassemblies. These shakers may be driven mechanically, either by eccentric-crank brute-force drives or by a system of rotating unbalanced masses, or they may be driven electromagnetically by supplying an alternating current to drive an electromagnetic pole piece. In some applications small shakers may be

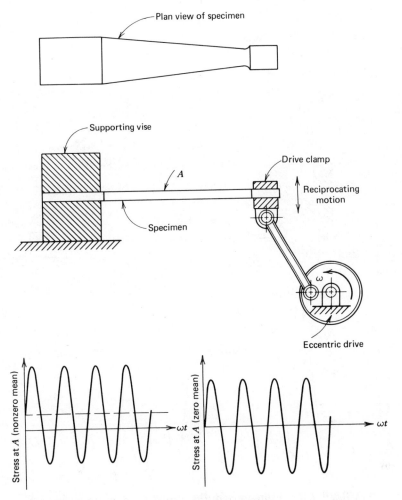

FIGURE 7.9.   Reciprocating-bending fatigue testing machine.

mounted on the structure to be tested and in other cases the shaker is used as
a testing machine upon which the test piece is mounted. Repeated torsion
machines usually utilize a specimen with circular cross section, subjecting it
to torsional stresses first in one direction and then the other. Repeated torsion
data are not uniaxial and therefore are more difficult to interpret and use.
Repeated torsion machines are for this reason not widely used. Other multiaxial testing machines utilize thin-walled pressure vessels or flat plates as
fatigue test specimens. The thin-walled pressure vessels may be pressurized,
either statically or cyclically, and simultaneously subjected to bending, torsion, or direct tension or compression, either statically or cyclically. Flat

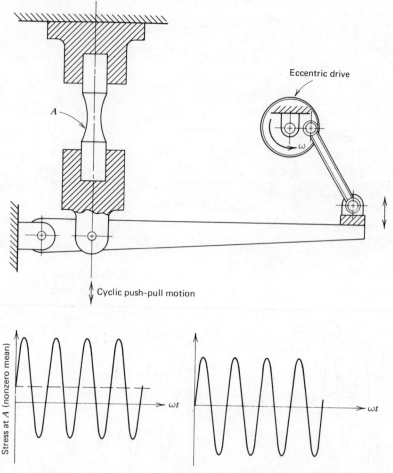

FIGURE 7.10.   Brute-force direct-stress fatigue testing machine.

plates may be subjected to cyclic anticlastic bending to produce certain multiaxial states of stress.

Computer-controlled fatigue testing machines are widely used in all modern fatigue testing laboratories. Usually such machines take the form of precisely controlled hydraulic systems with feedback to electronic controlling devices capable of producing and controlling virtually any strain-time, load-time, or displacement-time pattern desired. A schematic diagram of such a system is shown in Figure 7.12.

Special testing machines for component testing and full-scale prototype testing systems are not found in the general fatigue laboratory. These systems

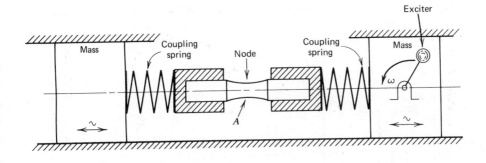

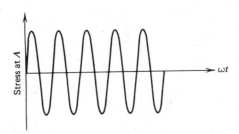

FIGURE 7.11.   Resonant-type direct-stress fatigue testing machine.

are built up especially to suit a particular need, for example, to perform a full-scale fatigue test of a commercial jet aircraft.

It may be observed that fatigue testing machines range from very simple to very complex. The very complex testing systems, used for example to test a full-scale prototype, produce very specialized data applicable only to the particular prototype and test conditions used; thus, for the particular prototype and test conditions the results are very accurate, but extrapolation to other test conditions and other pieces of hardware is difficult, if not impossible. On the other hand, simple smooth-specimen laboratory fatigue data are very general and can be utilized in designing virtually any piece of hardware made of the specimen material. However, to use such data in practice requires a quantitative knowledge of many pertinent differences between the laboratory and the application, including the effects of nonzero mean stress, varying stress amplitude, environment, size, temperature, surface finish, residual stress pattern, and others. Fatigue testing is performed at the extremely simple level of smooth specimen testing, the extremely complex level of full-scale prototype testing, and everywhere in the spectrum between. Valid arguments can be made for testing at all levels.

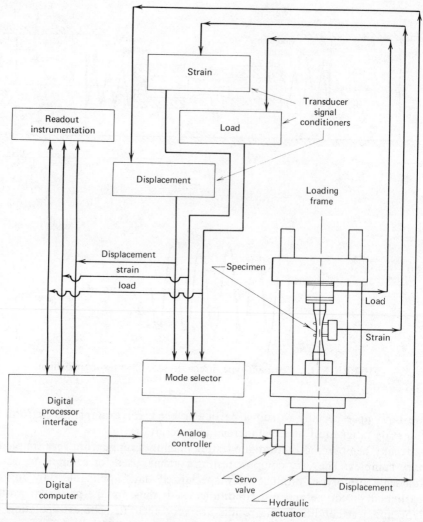

FIGURE 7.12. Schematic diagram of a computer-controlled closed-loop fatigue testing machine.

7.6 THE *S-N-P* CURVES—A BASIC DESIGN TOOL

Basic fatigue data in the high-cycle life range can be conveniently displayed on a plot of cyclic stress level versus the logarithm of life, or alternatively, on a log-log plot of stress versus life. These plots, called *S-N* curves, constitute design information of fundamental importance for machine parts subjected to repeated loading. Because of the scatter of fatigue life data at any given stress level, it must be recognized that there is not only one *S-N* curve for a given material, but a family of *S-N* curves with probability of failure as the parameter. These curves are called the *S-N-P* curves, or curves of constant probability of failure on a stress versus life plot.

To develop an *S-N-P* plot in the fatigue laboratory by "standard" methods, one would proceed in the following way:

1. Select a large group of carefully prepared, polished fatigue specimens of the material of interest and subdivide them into four or five smaller groups of at least 15 specimens each.

2. Select four or five stress levels, perhaps judged by a few exploratory tests, that span the stress range of the *S-N* curve.

3. Run an entire subgroup at each of the selected stress levels following the procedures to be outlined here.

4. To make each test run, mount a specimen in the testing machine, using due care to avoid spurious stresses. Set the machine for the desired stress amplitude, with cycle counter set on zero.

5. Start the machine and run at constant stress amplitude until the specimen fails or the machine reaches a predetermined *runout* criterion, for example, 5×10^7 cycles.

6. Record the stress amplitude used and the cycle count at the time of failure or runout.

7. Using a new specimen, repeat the procedure, again recording the stress level and life at failure or runout. Continue to repeat this procedure until all specimens designated for the selected stress level have been tested.

8. Change to a new stress level and repeat the preceding procedure until all specimens designated for this second stress level have been tested. Repeat this procedure until all selected stress levels have been tested. Note that the entire output from a complete fatigue test is a single point on the *S-N* plot.

9. Plot all data collected on a stress versus log-life coordinate system as shown in Figure 7.13. Runouts, or points for which fatigue failure was not observed during the test, are indicated by a small arrow to the right.

Considering the data plotted in Figure 7.13, one could simply construct a visual mean curve through the data. Doing this, it becomes clear that a substantial scatter of data about the mean clouds the design usefulness of such a mean curve. A better approach would be to construct for each stress level a *histogram*, such as the one shown in Figure 7.14, which shows the *distribution* of failures as a function of the log life for the sample tested. Computation of the sample mean and variance permits the estimation of population mean and variance if the form of the distribution is known for fatigue tests at a constant stress level. Details of these ideas and procedures are discussed much more fully in Chapter 9. Extensive testing of large samples has indicated that a log-normal distribution of life at a constant

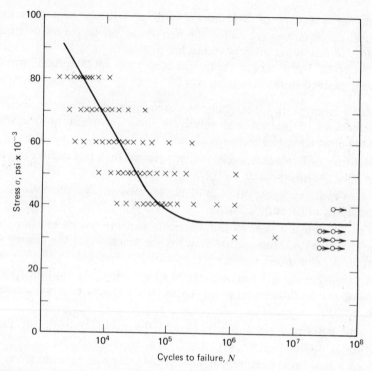

FIGURE 7.13. Plot of stress-cycle (S-N) data as it might be collected by laboratory fatigue testing of a new alloy.

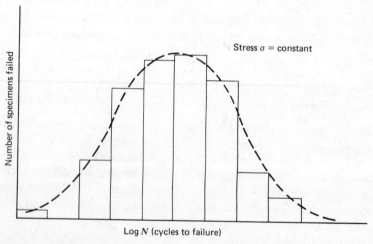

FIGURE 7.14. Distribution of fatigue specimens failed at a constant stress level as a function of logarithm of life.

stress level is a good estimate. Assuming the life distribution to be log normal, the sample mean and variance can be used to specify any desired probability of failure. Repeating the analysis at all test stress levels, we can connect points of equal probability of failure to obtain curves of constant probability of failure on the *S-N* plot. Such a family of *S-N-P* curves is shown in Figure 7.15. It is also of interest to note that the "reliability" R is defined to be 1 minus the probability of failure; hence, $R = (1 - P)$. Thus, the 5 percent probability of failure curve may alternatively be designated as the 95 percent reliability curve ($R = 0.95$), for example. Thus, some reference will be found in the literature to these so-called *R-S-N* curves. It should also be noted that references to the "*S-N* curve" in the literature generally refer to the *mean* curve unless otherwise specified.

The mean *S-N* curves sketched in Figure 7.16 distinguish two types of material response to cyclic loading commonly observed. The ferrous alloys and titanium exhibit a steep branch in the relatively short life range, leveling off to approach a stress asymptote at longer lives. This stress asymptote is called the *fatigue limit* (formerly called endurance limit) and is the stress level below which an infinite number of cycles can be sustained without failure. The nonferrous alloys do not exhibit an asymptote, and the curve of stress versus life continues to drop off indefinitely. For such alloys there is no fatigue limit, and failure as a result of cyclic load is only a matter of applying

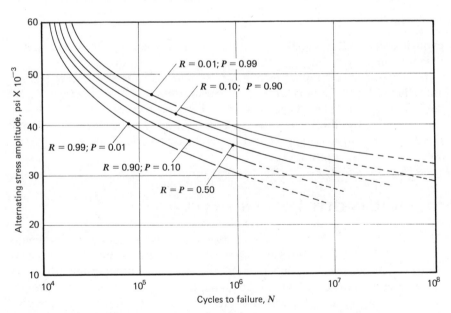

FIGURE 7.15. Family of *S-N-P* curves, or *R-S-N* curves, for 7075-T6 aluminum alloy. Note: P = probability of failure; R = reliability = $1 - P$. (Adapted from ref. 16, p. 117; with permission from John Wiley & Sons, Inc.)

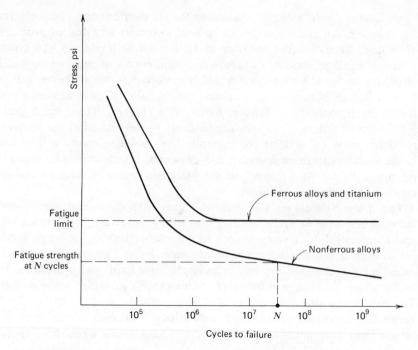

FIGURE 7.16. Two types of material response to cyclic loading.

enough cycles. All materials, however, exhibit a relatively flat curve in the long life range.

To characterize the failure response of nonferrous materials, and of ferrous alloys in the finite life range, the term *fatigue strength at a specified life*, S_N, is used. The term fatigue strength identifies the stress level at which failure will occur at the specified life. The specification of *fatigue strength* without specifying the corresponding life is meaningless. The specification of a *fatigue limit* always implies infinite life.

7.7   FACTORS THAT AFFECT *S-N-P* CURVES

The difference in failure response of machine parts or laboratory specimens because of different metallurgical, configurational, environmental, and operational influences may be conveniently indicated by examining the effects on the *S-N* curve brought about by changing these factors. Qualitative evaluations of many of these effects are presented in the following paragraphs. These factors will be discussed in detail on a quantitative basis where possible. The references at the end of this chapter virtually all contain extensive discussions on various factors affecting *S-N* curves.

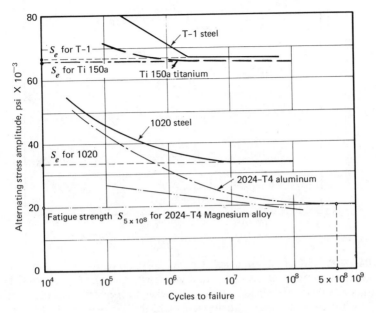

FIGURE 7.17. Effect of material composition on the S-N curve. Note that ferrous and titanium alloys exhibit a well-defined fatigue limit, whereas other alloy compositions do not. (Data from refs. 6 and 21)

Material Composition

Materials divide themselves into two broad groups with respect to *S-N* failure response, as has been indicated in Figure 7.16. The ferrous alloys and titanium exhibit a rather well-defined fatigue limit, which is well established by the time 10^7 cycles of stress have been applied. The other nonferrous alloys do not exhibit a fatigue limit at all, with their *S-N* curves continuing to fall off at lives of 10^8, 10^9, and larger numbers of cycles. Data that illustrate these characteristics for several different alloys are shown in Figure 7.17.

Grain Size and Grain Direction

Fine-grained materials exhibit fatigue properties that are superior to the fatigue properties of coarse-grained materials of the same composition.* Although grain size in ferritic steels seems to have only a minor effect, for austenitic steels and many nonferrous alloys the degradation of fatigue strength with larger grain size becomes significant, as illustrated for example in Figures 7.18 and 7.19. The superiority of fine grain size becomes less

*See p. 252 of ref. 3.

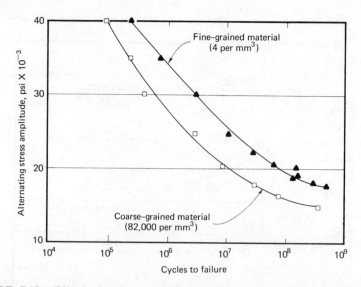

FIGURE 7.18. Effect of grain size on the *S-N* curve for 18S aluminum alloy. Average diameter ratio of coarse to fine grains is approximately 27 to 1. Nominal composition: 4.0 percent copper, 2.0 percent nickel, 0.6 percent magnesium. Note that at a life of 10^8 cycles the mean fatigue strength of the coarse-grained material is about 3000 psi lower than for fine-grained material. (Data from ref. 3; adapted from *Fatigue and Fracture of Metals*, by W. M. Murray, by permission of The MIT Press, Cambridge, Massachusetts, copyright 1952)

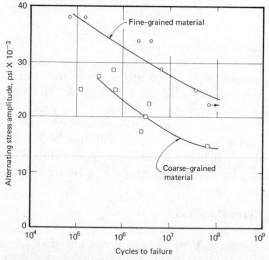

FIGURE 7.19. Effect of grain size on the *S-N* curve for an extruded British aluminum alloy. Average diameter ratio of coarse to fine grains is approximately 15 to 1. Nominal composition: 2.5 percent copper, 0.7 percent manganese, 0.8 percent magnesium, 1.0 percent silicon, 0.5 percent iron. Note that a life of 10^8 cycles the mean fatigue strength of the coarse-grained material is about 8000 psi lower than for the fine-grained material. (Data from ref. 3; adapted from *Fatigue and Fracture of Metals*, by W. M. Murray, by permission of The MIT Press, Cambridge, Massachusetts, copyright 1952)

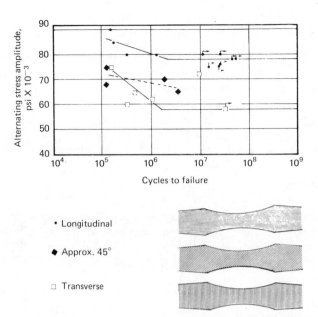

FIGURE 7.20. Effect on the *S-N* curve of grain flow direction relative to longitudinal loading direction for specimens machined from crankshaft forgings. Nominal composition: 0.41 percent carbon, 0.47 percent manganese, 0.01 percent silicon, 0.04 percent phosphorus, 1.8 percent nickel. $S_u = 139,000$ psi, $S_{yp} = 115,000$ psi, $e(2.0$ inches$) = 20$ percent. (Data from ref. 22)

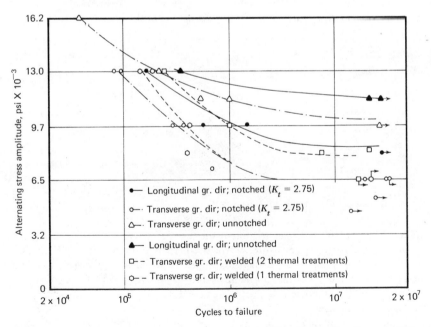

FIGURE 7.21. Effect on the *S-N* curve of grain direction under several different conditions for LA141A alloy sheet tested in completely reversed bending. (Data from ref. 23)

significant at elevated temperatures when the characteristic transgranular room temperature fatigue cracking gives way to intergranular cracking paths.

Grain flow direction in forged or rolled sheet products relative to the direction of loading also has a significant effect on fatigue strength, with cyclic loading across the grain direction (transverse) giving fatigue properties that are inferior to cyclic loading along the grain direction (longitudinal). Examples of this behavior are illustrated in Figures 7.20 and 7.21.

Heat Treatment

Fatigue strength properties, like static strength properties, are, for many alloys, strongly affected by heat treatment procedure. The effects of various heat treatment procedures on the *S-N* curves for several different ferrous alloys are illustrated in Figures 7.22, 7.23, 7.24, and 7.25.

Welding

No matter what welding technique is used, the weld joint must be regarded as a metallurgically nonhomogeneous region ranging from unheated parent metal through the heat-affected zone, to the weld metal zone, which may be

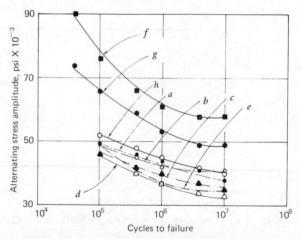

FIGURE 7.22. Effects of heat treatment on the *S-N* curve of cast steel. (*a*) Cast steel, acid open hearth, large casting, normalized, 1650°F, drawn at 750°F. (*b*) Cast steel, acid electric, 1650°F, air cooled, hydrogen removed at 750°F. (*c*) Cast steel, basic open hearth, large casting, normalized, 1650°F, drawn at 750°F. (*d*) Cast steel, converter, 1650°F, air cooled, hydrogen removed at 750°F. (*e*) Cast steel, triplex, large casting, normalized, 1650°F, drawn at 750°F. (*f*) Copper cast steel, 1650°F, air cooled, copper precipitated at 930°F. (*g*) Copper cast steel, 1650°F, air cooled, copper in solution. (*h*) Cast steel, basic electric, large casting, normalized, 1650°F, drawn at 750°F. (Data from ref. 6)

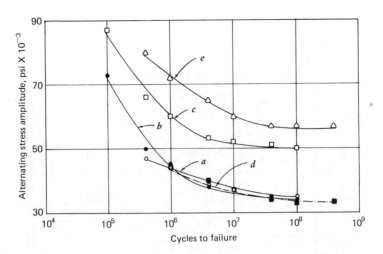

FIGURE 7.23. Effects of heat treatment on the *S-N* curve of carbon steel. (*a*) Carbon steel (0.35 C, 0.55 Mn, 0.19 Si), annealed. (*b*) Carbon steel (0.36 C, 0.61 Mn, 0.019 Si), 1550°F, furnace cooled. (*c*) Carbon steel (0.36 C, 0.61 Mn, 0.019 Si), 1550°F, water quenched at 900°F, furnace cooled. (*d*) Carbon steel (0.37 C; 0.58 Mn, 0.16 Si), 4 inch billet, normalized, 1495°F, furnace cooled. (*e*) Carbon steel (0.37 C, 0.38 Mn, 0.16 Si), 4 inch billet, 1550°F, water quenched at 1050°F, air cooled. (Data from ref. 6)

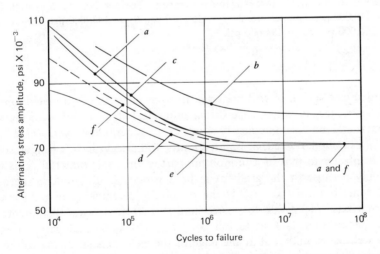

FIGURE 7.24. Effects of heat treatment on the *S-N* curve of SAE 2340 steel. All data from rotating bending tests. Heat treatment and (test specimen diameter, inch): (*a*) 1450°F, oil quenched, tempered 1200°F, (0.12). (*b*) 1450°F, lead bath 755°F, no draw, (0.212). (*c*) $\frac{7}{8}$-inch bar, 1450°F, oil quenched, 60 min., 1100°F, (0.212). (*d*) $\frac{7}{8}$-inch bar, 1450°F, air blast, no draw, (0.235). (*e*) $\frac{7}{8}$-inch bar, 1450°F, air blast, 60 min., 700°F, (0.235). (*f*) $\frac{7}{8}$-inch bar, 1450°F, oil quenched, 60 min., 1200°F, (0.235). (Data from ref. 6)

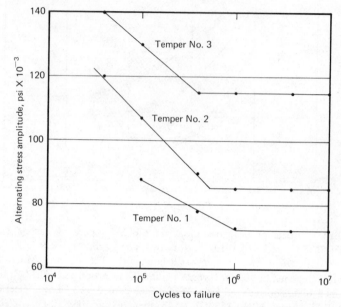

FIGURE 7.25.   Effects of heat treatment on the S-N curve of SAE 4130 steel, using 0.19 inch diameter rotating bending specimens cut from $\frac{3}{8}$ inch plate, 1625°F, oil quenched, followed by three different tempers. Temper No. 1: $S_u = 129{,}000$ psi, $S_{yp} = 118{,}000$ psi. Temper No. 2: $S_u = 150{,}000$ psi, $S_{yp} = 143{,}000$ psi. Temper No. 3: $S_u = 206{,}000$ psi, $S_{yp} = 194{,}000$ psi.

regarded as a region of cast material. In some cases the entire weld joint may be heat treated, in which case the metallurgical structure of weld metal and parent metal may become nearly identical. In general, however, welded joints, as well as bolted, riveted, or bonded joints, will always have a fatigue strength that is inferior to that of a monolithic part of the same material. This is true not only because of the gradient in homogeneity of the material across the weld zone, but also because of the possibility of cracking in weld metal or adjacent base metal due to postcooling shrinkage stresses, incomplete weld metal penetration, lack of fusion between weld metal and parent metal on prior weld runs, undercut at the edge of the weld deposit owing to incorrect welding technique, overlap of weld metal due to overflowing beyond the fusion zone, slag inclusions or porosity due to faulty welding technique, misshapen welds, or welds with surface defects that produce geometrical stress concentrations. Some of the effects of welding on fatigue properties are illustrated in Figures 7.26, 7.27, and 7.28.

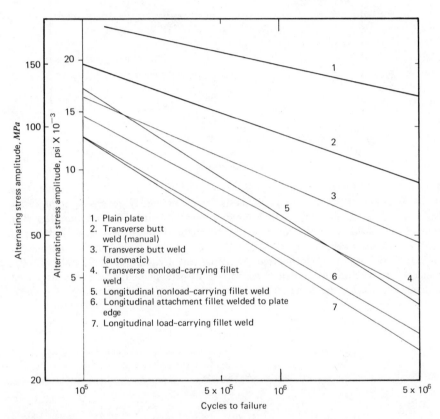

FIGURE 7.26. Effects of welding detail on the *S-N* curve of structural steel, with yield strength in the range 30–52,000 psi. Tests were released tension ($\sigma_{min} = 0$). (Data from ref. 24)

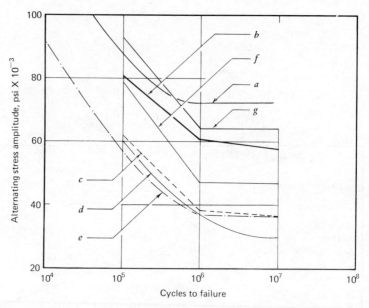

FIGURE 7.27. Effects of flash welded butt joints, with various heat treatments, on the *S-N* curve of SAE 4130 steel. Tests were completely reversed reciprocating bending. All specimens were $\frac{1}{4}$ inch plate ground to $\frac{1}{8}$ inch after welding. (*a*) Heat treated, unwelded ($\sigma_u = 192,000$ psi). (*b*) Normalized, unwelded ($\sigma_u = 121,000$ psi). (*c*) Normalized, welded ($\sigma_u = 124,000$ psi). (*d*) Normalized, welded, normalized ($\sigma_u = 113,000$ psi). (*e*) Normalized, welded, heat treated ($\sigma_u = 172,000$ psi). (*f*) Normalized, welded, heat treated, reinforced 10 percent one side. (*g*) Normalized, welded, heat treated, reinforced 10 percent both sides. (Data from ref. 6)

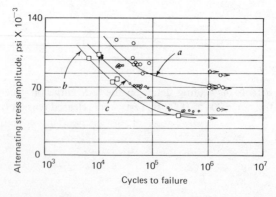

FIGURE 7.28. Effects of welding on the *S-N* curve of Titanium 6A1-4V alloy at room temperature under tension-tension loading with $R = 0.05$. (*a*) Parent metal coupons, vacuum annealed. (*b*) As-welded coupons. (*c*) Diffusion-bonded coupons. (Data from ref. 25)

192

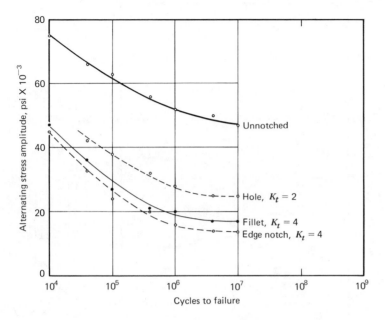

FIGURE 7.29. Effects of geometrical discontinuities on the *S-N* curve of SAE 4130 steel sheet, normalized, tested in completely reversed axial fatigue test. Specimen dimensions (t = thickness, w = width, r = notch radius): Unnotched: t = 0.075 inch, w = 1.5 inches. Hole: t = 0.075 inch, w = 4.5 inches, r = 1.5 inches. Fillet: t = 0.075 inch, w_{net} = 1.5 inches, w_{gross} = 2.25 inches, r = 0.0195 inch. Edge notch: t = 0.075 inch, w_{net} = 1.5 inches, w_{gross} = 2.25 inches, r = 0.057 inch. (Data from ref. 6)

Geometrical Discontinuities

Stress concentration effects due to changes in shape, geometrical discontinuities, or joints may strongly affect the fatigue strength of a machine part, even if the part is made of a ductile material. The seriousness of notches, holes, fillets, joints, and other stress raisers depends upon the relative dimensions, type of loading, and notch sensitivity of the material. Several examples of the influence of geometrical discontinuities upon fatigue are illustrated in Figures 7.29, 7.30, 7.31, and 7.32. A more detailed discussion of stress concentration is presented in Chapter 9.

Surface Conditions

A very high proportion of all fatigue failures nucleate at the surface of the affected part; hence, surface conditions become an extremely important factor influencing fatigue strength. The usual standard by which various surface conditions are judged is the polished laboratory specimen. Irregular

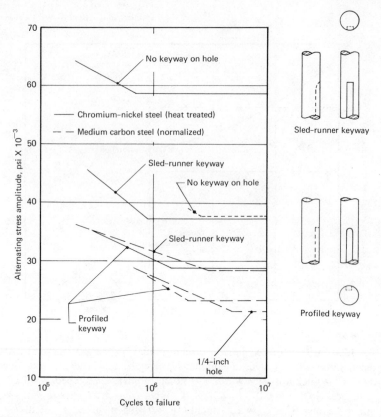

FIGURE 7.30. Effects of keyways and holes on the *S-N* curves of two types of steel. (Data from ref. 26), copyright ASTM; adapted with permission.)

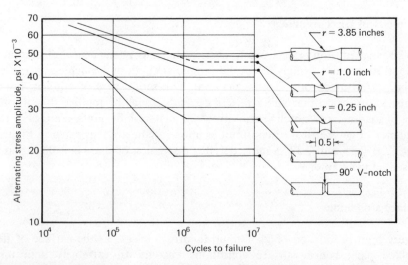

FIGURE 7.31. Effects of various geometrical discontinuities on the *S-N* curve of a 0.49 percent carbon steel, water quenched and tempered at 1200°F. Stock size was 0.40 inch, machined to nominal specimen size of 0.275 inch diameter. (Data from ref. 27)

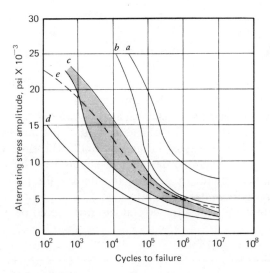

FIGURE 7.32. Effect of bonded lap joints and riveted joints on the *S-N* curve of 2024-T3 Alclad Aluminum sheet, tested in released tension loading. (Note: t = sheet thickness, L = overlap length) (*a*) Unnotched sheet. (*b*) Lap joint, $\sqrt{t}/L = 0.14$. (*c*) Lap joint, $\sqrt{t}/L = 0.28$. (*d*) Lap joint, $\sqrt{t}/L = 0.36$. (*e*) Upper limit for riveted joint (approx.). (Data from ref. 28)

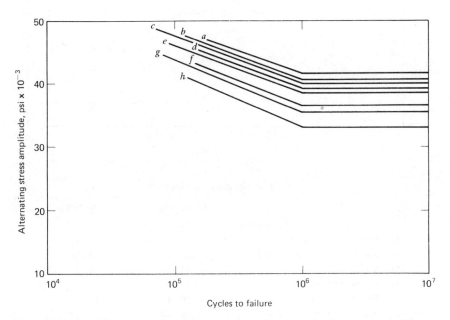

FIGURE 7.33. Effect of surface finish on the *S-N* curve of 0.33 percent carbon steel specimens, tested in a rotating cantilever beam machine. (*a*) High polish, longitudinal direction. (*b*) FF emery finish. (*c*) No. 1 emery finish. (*d*) Coarse emery finish. (*e*) Smooth file. (*f*) As turned. (*g*) Bastard file. (*h*) Coarse file. (Data from ref. 29)

and rough surfaces generally exhibit inferior fatigue properties compared to smooth surfaces. Figures 7.33, 7.34, and 7.35 illustrate these observations. Cladding, plating, and coating also very often reduce the fatigue strength of the plated member. Fatigue nuclei, initiated in the coating, continue to propagate on into the base metal, causing a reduction in fatigue strength of 10 to 50 percent in some cases. Zinc and cadmium plating seem to have a relatively small effect in reducing fatigue strength, whereas nickel and chromium plating have a substantial detrimental effect. The effects of anodizing seem to be relatively small. In general, the thicker the layer of plating or coating, the more detrimental is the effect on fatigue properties. Figures 7.36 and 7.37 illustrate these effects for several cases. It should be noted, however,

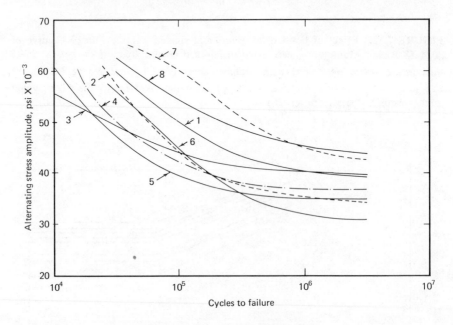

Surface finish, microinches rms

		Longitudinal	Transverse
(1)	Longitudinal hand polish	11–14	15–19
(2)	Smooth mill cut	21–26	9–13
(3)	Sand blast—condition 1	55–60	55–56
(4)	Vapor hone	54–59	48–55
(5)	Hand burnish	35–45	85–100
(6)	Rough mill cut	110–140	13–35
(7)	Sand blast—condition 2	44–49	44–49
(8)	Sand blast and hand polish	16–24	24–29

FIGURE 7.34. Effects of surface finish on the *S-N* curve of 7075-T6 aluminum alloy extrusions. (Data from ref. 30)

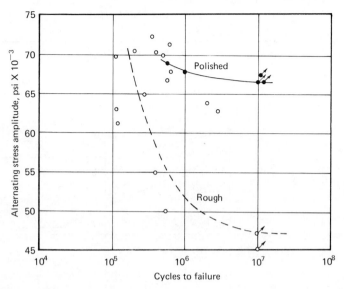

FIGURE 7.35. Effects of polishing die marks from the surface on the *S-N* curve of heat-treated chromium-molybdenum aircraft tubing. Tube dimensions: 0.5-inch O.D. with 0.065-inch wall. Heat treatment: oil quench 1625°F, draw 650°F.

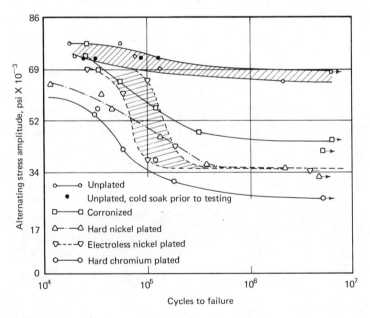

FIGURE 7.36. Effects of several electrodeposited coatings on the *S-N* curve of low alloy steel at room temperature under axial tension-tension loading with $R = 0.02$. Static ultimate strengths: unplated; 172,300 psi; corronized, 177,700 psi; hard nickel plated, 176,100 psi; electroless nickel plated, 182,100 psi; hard aluminum plated; 162,400 psi. (Data from ref. 16; adapted with permission from John Wiley & Sons, Inc.)

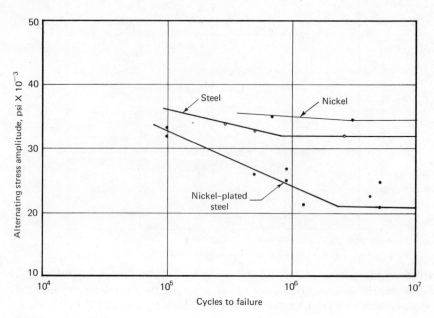

FIGURE 7.37. Effect of electrodeposited nickel plating on the *S-N* curve of steel. (Data from ref. 32)

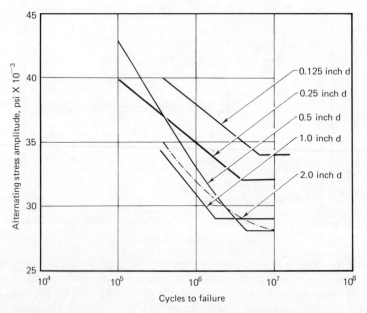

FIGURE 7.38. Size effects on the *S-N* curve of SAE 1020 steel specimens cut from a $3\frac{1}{2}$ inch diameter hot rolled bar, tested in rotating bending. (Data from ref. 33)

198

that the corrosion protection afforded by most plating processes more than offsets the intrinsic strength loss due to the plating.

Size Effect

Larger specimens and machine parts are observed to exhibit poorer fatigue strength than smaller specimens or machine parts, especially when subjected to cyclic bending stresses. One explanation offered for this observation is that larger specimens have a higher probability of containing a more serious stress concentration or fatigue nucleus simply because they have a greater volume and greater surface area. The size effect in comparing the fatigue limit of a one-half-inch-diameter bar with a 6-inch-diameter bar might be as much as 15 to 20 percent. An illustration of the size effect is shown in Figure 7.38.

Residual Surface Stresses

Residual stresses in the surface layer of specimens or machine parts, whether induced intentionally or accidentally, play an important role in the overall fatigue properties of the part.

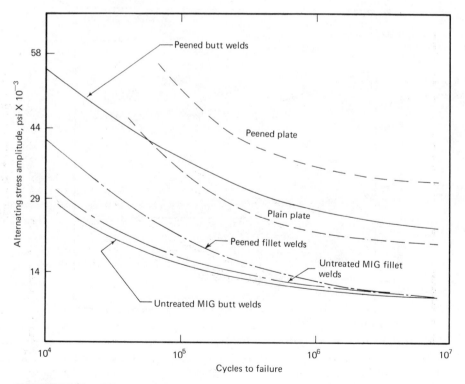

FIGURE 7.39. Effect of shot peening on the *S-N* curves for welded and unwelded steel plate. (Data from ref. 24)

If the induced residual stresses at the surface are tensile, the fatigue strength is diminished. If the induced residual stresses at the surface are compressive, the fatigue strength is improved. Three common means of producing residual surface stresses are by shot-peening, by cold-rolling, and by presetting or prestressing. The reason for the beneficial effects of residual compressive stresses at the surface is that fatigue cracks find it much more difficult to propagate through a compressive stress field. Residual stress effects are shown in Figures 7.39 through 7.44 and in Figure 14.7 for a variety of different shot-peening, cold-rolling, and prestressing conditions. It may also be observed that certain surface treatments, such as nitriding or carburizing, produce favorable residual compressive stresses at the surface. Chromium plating, on the other hand, produces residual tensile surface stresses and therefore degrades the fatigue strength of the plated member.

Another important observation of interest to the designer is that not only the mean fatigue strength of a specimen or machine part is improved by surface treatments such as shot-peening and cold-rolling, but the scatter of failure data is notably diminished. That is, the standard deviation of the

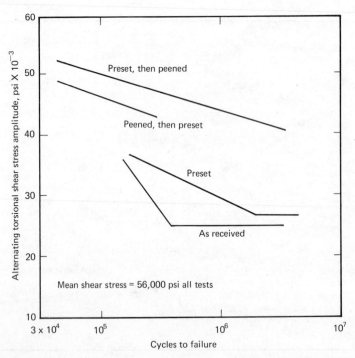

FIGURE 7.40. Effects of shot peening and/or presetting on the *S-N* curve of hot wound helical coil springs made of 0.9 percent carbon steel. Spring details: hardness = Vickers DPH 550, wire diameter = $\frac{1}{2}$ inch, mean coil diameter = $2\frac{5}{8}$ inches, number of turns = 6, free length = $5\frac{1}{16}$ or 6 inches. (Data from ref. 8)

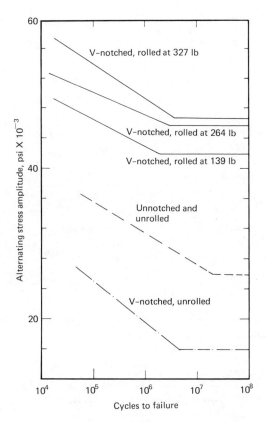

FIGURE 7.41. Effects of surface cold-rolling on the *S-N* curve of ferritic nodular iron. Note: Rolling accomplished by three hardened steel contoured rollers, spring loaded into V-notch to roll notch root only, prior to fatigue test. (Data from ref. 34)

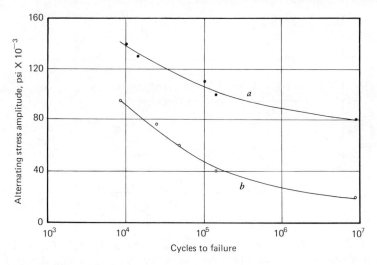

FIGURE 7.42. Effects of cold-rolling threads before and after heat treatment on *S-N* curve for 220,000 psi ultimate strength bolts. (*a*) Rolled after heat treatment. (*b*) Rolled before heat treatment. (Data from ref. 16; with permission from John Wiley & Sons, Inc.)

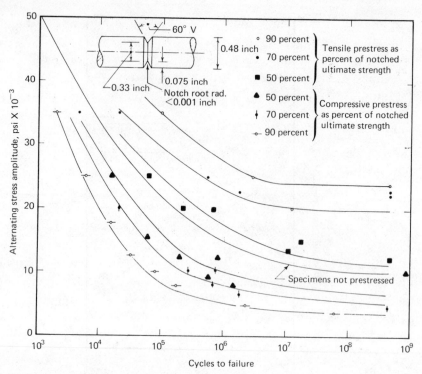

FIGURE 7.43. Effects of axial static prestress on the *S-N* curve of 7075-T6 specimens tested in rotating bending fatigue tests. (Data from ref. 35)

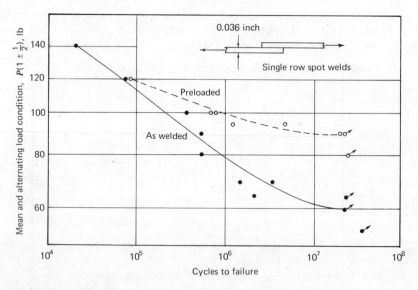

FIGURE 7.44. Effects of axial static preload on the *S-N* curve of spot-welded lap joints in aluminum alloy sheet tested in tension-tension loading with $R = 0.33$. Note: Specimens were statically preloaded to two-thirds of ultimate strength of joint. (Data from ref. 36; reprinted with permission from McGraw Hill Book Company.)

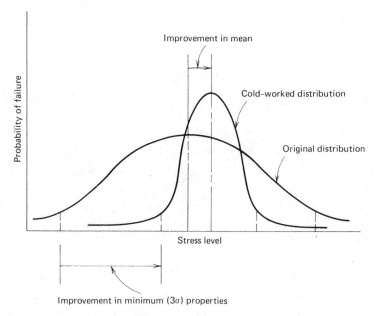

FIGURE 7.45. Illustration of improvement in minimum fatigue properties brought about by reduction in scatter through shot-peening or cold-rolling.

distribution is significantly decreased by the induced compressive stress layer at the surface of the machine part. As can be seen in Figure 7.45, the "minimum" fatigue strength may be greatly improved even if the mean improvement is small. Designers are usually interested in the "minimum" properties. Because of the practical importance of this observation, designers often specify shot-peening in critical regions of highly stressed fatigue-sensitive machine parts.

Operating Temperature

The temperature of operation may have a significant influence on the fatigue strength. Generally speaking, the fatigue strength is enhanced at temperatures below room temperature and diminished at temperatures above room temperature, although exceptions may be found. In the range of temperature from around zero on the Fahrenheit scale to about one-half the absolute melting temperature, where creep processes begin to become important, the effect of temperature would be classified as slight in most cases. At higher temperatures the fatigue strength diminishes significantly, as shown, for example, in Figures 7.46 and 7.47. A further significant observation is that the alloys that exhibit a fatigue limit at room temperature lose this characteristic at elevated temperature, as shown in Figure 7.46.

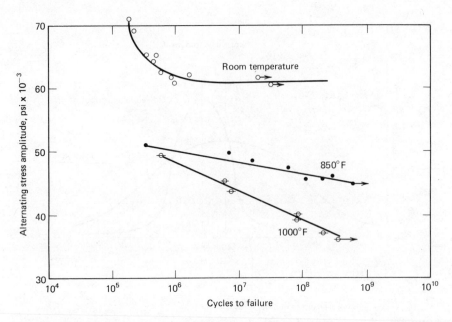

FIGURE 7.46. Effect of operating temperature on the S-N curve of a 12 percent chromium steel alloy. Alloy composition = 0.10 C, 0.45 Mn, 0.21 Ni, 12.3 Cr, and 0.38 Mo. (Data from ref. 37)

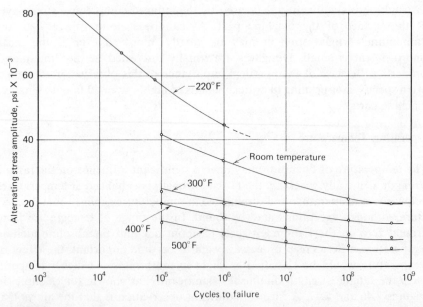

FIGURE 7.47. Effect of operating temperature on the S-N curve of 2024-T4 aluminum alloy tested in rotating bending. (Data from refs. 6 and 38)

204

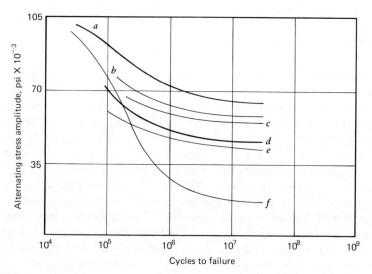

FIGURE 7.48. Effects of corrosion on the *S-N* curve of alloy steel tested in rotating bending. (*a*) Fatigue strength in air. (*b*) Precorroded for 1 day in tap water. (*c*) Precorroded for 2 days in tap water. (*d*) Precorroded for 6 days in tap water. (*e*) Precorroded for 10 days in tap water. (*f*) Bending fatigue test carried out in tap water. (Data from ref. 39; adapted with permission of Pergamon Press, Inc.)

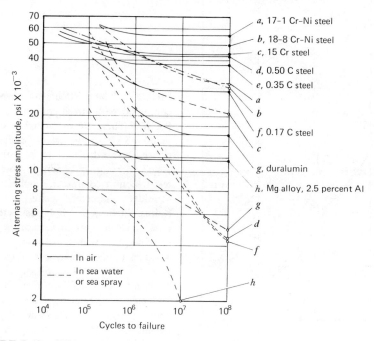

FIGURE 7.49. Effects of corrosion on the *S-N* curves of various aircraft materials tested in push-pull loading in sea water or sea spray. (Data from ref. 40)

Corrosion

A corrosive environment tends to lower the fatigue strength of engineering materials, often by a large amount. For many materials, including the ordinary structural steels, plain distilled water on the surface may reduce the fatigue strength to less than two-thirds of its value in a dry environment. As shown in Figures 7.48 and 7.49, a tap water or salt spray environment may reduce fatigue strength even more drastically for some materials. Even the use of certain solvents and cleaning agents sometimes used to clean the surfaces of test specimens may have a deleterious effect on fatigue strength. For example, carbon tetrachloride used to clean titanium specimens will result in a lowering of fatigue strength, especially if the specimens are subsequently operated at elevated temperatures. Combined effects of corrosion and stress concentration are illustrated for a few cases in Figure 7.50. It should be noted also that, similar to the case of elevated temperature influence on fatigue properties, materials that exhibit a fatigue limit in dry air lose this characteristic in sustained corrosive environments.

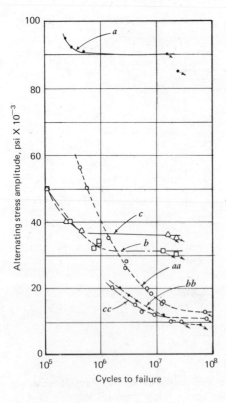

FIGURE 7.50. Effects of corrosion combined with stress concentration on the S-N curve of quenched and tempered SAE 3140 steel specimens, polished with drilled hole, or with fillet, tested in rotating bending. Note: $d =$ test section diameter, inches; $D =$ shoulder diameter, inches; $r =$ fillet radius, inches; $a =$ hole diameter, inches. (a) Solid specimen, in air, $d = 0.3$. (b) Specimen with hole, in air, $d = 0.4$, $a = 0.04$. (c) Specimen with fillet, in air, $D = 0.5$, $r = 0.022$. (aa) Solid specimen, in water. (bb) Specimen with hole, in water. (cc) Specimen with fillet, in water. (Data from ref. 2; adapted with permission from John Wiley & Sons, Inc.)

Fretting

Under certain circumstances fretting action may lead to a drastic reduction in the fatigue strength of a machine part. Based on the definition of fretting in Section 2.3, it may be noted that fretting is a surface disturbance that arises in joints and connections where surfaces in contact undergo small amplitude cyclic relative motion. Under certain fretting conditions the fatigue strength may be reduced to only one-third or less of the fatigue strength without fretting. A more detailed discussion of fretting damage is presented in Chapter 14. Some illustrations of the effect of fretting on fatigue strength are shown in Figures 7.51, 7.52, and 7.53, with additional data given in Figures 14.4, 14.5, and 14.7.

Operating Speed

In the range of cyclic speeds from about 200 cycles per minute up to about 7000 cycles per minute, little, if any, effect of operating speed on fatigue strength has been observed for most materials. This observation assumes, however, that the temperature of the specimen is not significantly elevated during the test. There is some evidence that a small decrease in fatigue strength is observed at speeds slower than about 200 cycles per minute. In the speed range from about 7000 cycles per minute up to about 60,000 to 90,000 cycles per minute most materials exhibit a significant improvement in fatigue

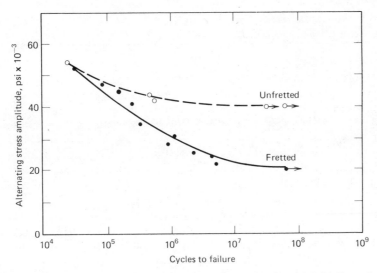

FIGURE 7.51. Effect of fretting on the *S-N* curve of a forged 0.24 C steel. (Data from ref. 36; reprinted with permission from McGraw Hill Book Company.)

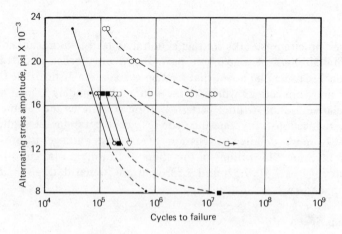

○ No fretting
• Fretting with clamping pressure of 16,000 psi
○ Fretting with clamping pressure of 16,000 psi
■ Fretting with clamping pressure of 500 psi

▽ Fretting with clamping pressure of 120 psi
□ Fretting with clamping pressure of 20 psi

FIGURE 7.52. Effects of fretting on the *S-N* curve of L-65 aluminum alloy, using various clamping pressures. Note: All tests were tension-tension loading with a tensile mean stress of 25,000 psi. (Data from ref. 8, p. 386)

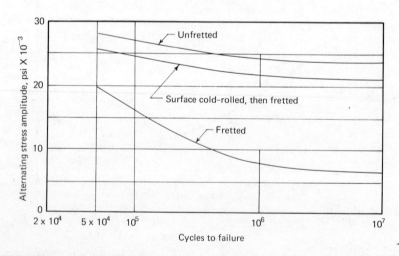

FIGURE 7.53. Effects of fretting on the *S-N* curve of magnesium alloys, showing the effect of surface cold-rolling. (Data from ref. 36; reprinted with permission from McGraw Hill Book Company.)

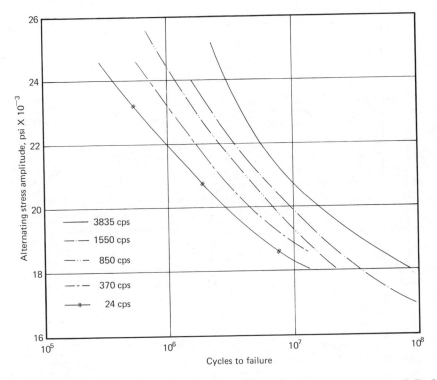

FIGURE 7.54. Effects of operating speed on the *S-N* curve of Hiduminium R.R. 56 aluminum alloy. (Data from ref. 8, p. 368)

strength, as shown, for example, in Figure 7.54; but above these speeds many materials exhibit a sharp decrease in fatigue strength, as indicated in Figure 7.55 for several materials. It has also been noted that rest periods between cycles or between blocks of cycles have no measurable effect on fatigue strength.

Configuration of Stress-Time Pattern

In actual operation the shape and configuration of the stress-time pattern takes many forms. Laboratory studies and in-service studies have led to several observations of interest with regard to the configuration of the stress-time pattern. A very common stress-time pattern used in laboratory fatigue testing is the completely reversed sinusoidal wave form, illustrated in Figure 7.56*a*. If a completely reversed ramp loading is used to produce the same peak values, as shown in Figure 7.56*b*, the fatigue behavior is essentially unchanged from the results of the sinusoidal pattern. Thus, the fatigue life seems to be relatively insensitive to a change in wave form as long as the peak values are the same.

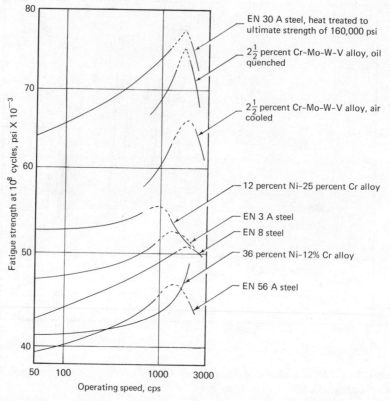

FIGURE 7.55. Effect of operating speed on the fatigue strength at 10^8 cycles for several different ferrous alloys. (Data from ref. 8, p. 381)

If a very small amplitude ripple of cyclic stress is superposed upon a large amplitude lower frequency wave, as shown in Figure 7.56c, it has been noted that the failure in many cases proceeds as if the tiny ripple were not present.

However, it should be pointed out that a small ripple superposed on a static mean stress might be very significant since it transforms a static loading into a fatigue loading situation, with all of the added problems of progressive failure caused by cyclic stresses. Further, if the amplitude of the ripple becomes large enough, additional damage may be contributed by the ripple.

If the cyclic wave form is more complex, for example the resultant of adding the first harmonic to the fundamental frequency of a sinusoidal wave form, certain observations may be made. The complex wave form shown in Figure 7.56d contains primary and secondary peaks. As long as the secondary peaks are small compared to the primary peak values, they contribute little to failure, and the cycle can be approximated by a sine wave of amplitude σ_{max} and period T, as shown. If the wave form is distorted as shown in Figure

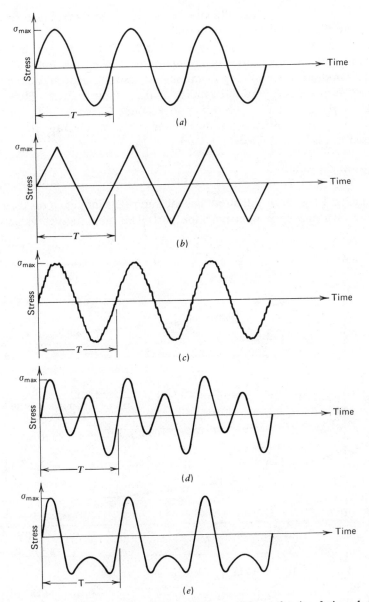

FIGURE 7.56. A variety of stress-time patterns used in evaluating fatigue behavior. (*a*) Completely reversed sinusoid. (*b*) Completely reversed ramp. (*c*) Superposed ripple. (*d*) Secondary peaks. (*e*) Distorted peaks.

7.56e, it may also be successfully approximated in many cases by a sinusoidal wave form of amplitude σ_{max} and period T.

The overall observation that might be made is that fatigue behavior is rather insensitive to the shape and configuration of the wave form. Therefore, a certain amount of smoothing of random stress-time data may be done for analysis without destroying the validity of the fatigue analysis. However, to the extent possible it is much better to account for each and every stress reversal by a good cycle counting method, such as the rain flow method discussed in detail in Section 8.5.

Nonzero Mean Stress

The magnitude of the mean stress has an important influence on the fatigue behavior of a specimen or a machine part. The quantitative evaluation of

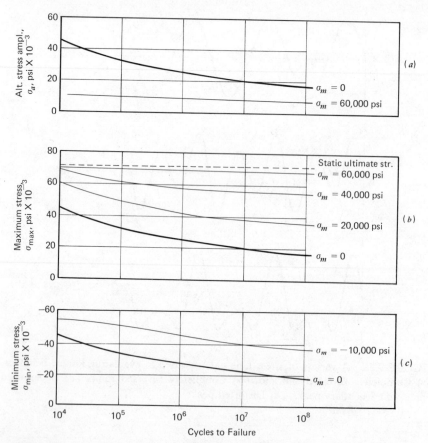

FIGURE 7.57. Various ways of presenting the influence of nonzero mean stress on the fatigue behavior of 2014-T6 aluminum alloy. (Adapted from ref. 41)

mean stress effects is discussed in detail in Section 7.9. Qualitatively, it may simply be observed that for a given cyclic stress amplitude, the addition of a mean stress will result in diminished fatigue life. This point is illustrated in Figure 7.57a. However, the same point is illustrated in another way by the plots of Figures 7.57b and 7.57c. Fatigue failure is usually governed by the maximum stress σ_{max} in the cycle if a tensile mean stress σ_m is present. For the tensile mean stress case of Figure 7.57b it may be noted that, for a given value of σ_{max}, increasing the mean stress σ_m will yield longer life. For the compressive mean stress case of Figure 7.57c it may likewise be noted that for a given value of σ_{min}, increasing the absolute value of compressive mean stress will yield longer life, though the effect is small. These conclusions are consistent with the plot of Figure 7.57a, since increasing the mean stress σ_m for a constant value of maximum stress σ_{max} implies a decrease in alternating stress amplitude σ_a, and correspondingly longer life.

It may be observed from the preceding discussion that for a given fixed value of peak stress the worst possible situation is to have a completely reversed stress-time cycle. Often the concept of preloading can be used to advantage in fatigue design to provide longer life by inducing a nonzero mean stress.

Damage Accumulation

The damage caused by the application of cyclic stresses accumulates with the addition of each damaging cycle. The concept of cumulative damage will be

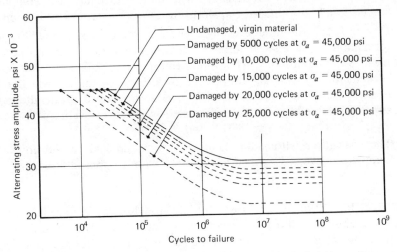

FIGURE 7.58. Illustration of the influence of accumulated fatigue damage on subsequent fatigue behavior of carbon steel. Note: Life of virgin material at $\sigma_a = 45,000$ psi is approximately 30,000 cycles. (Data from ref. 2; reprinted with permission from John Wiley & Sons, Inc.)

discussed in detail in Chapter 8, but the idea is depicted in Figure 7.58. The basic concept of importance is that the material is changed after the accumulation of fatigue damage caused by cyclic loading and actually exhibits a new and different S-N curve. Thus, the S-N curve constantly changes with the addition of damaging cycles of stress, as illustrated in Figure 7.58.

7.8 USING THE FACTORS IN DESIGN

All the factors discussed in the preceding section must be evaluated quantitatively for the specific materials and conditions of interest if they are to be of any real use to the designer. Although the data presented in Figures 7.17 through 7.57 will be of some help, it will often be necessary for the designer to search the literature or perform special tests to satisfy his own special needs for such information. After numerical strength modification factors are found for each of the important effects discussed in Section 7.7, the procedure used to obtain a fatigue design stress would typically include the following steps:

1. For the material of interest, obtain the basic completely reversed ($R = -1$) experimental laboratory data for polished specimens from statistically significant testing programs. The fatigue strength plot would be extracted directly from S-N-P curves at the desired probability level, such as those shown in Figure 7.15. If available data are other than completely reversed ($R \neq -1$), the procedures of Step 2 (following) would be applied twice, once to transform the data into equivalent completely reversed data and then to transform the stresses as discussed in Step 2.

2. Having found the fatigue strength curve at the desired probability level for completely reversed loading, we next modify it to account for nonzero mean stress conditions. This modification would utilize a relationship such as the Goodman, Gerber, or Soderberg equations, which relate nonzero mean alternating stresses to equivalent zero mean alternating stress. These relationships are discussed in detail in Section 7.9 and are given in another form in (8-114) and (8-115).

3. The strength modification factors are next applied to the equivalent completely reversed S-N curve, together with an appropriate safety factor, to arrive finally at a design S-N curve.

The procedure outlined here is not entirely straightforward since interactions among strength modification factors may occur in some instances. For example, the temperature effect and the residual stress effect may interact, or other effects may interact to produce unknown influences on the design stress. For these reasons it is always advisable to subject the final design to full-scale prototype testing under simulated service conditions.

7.9   THE INFLUENCE OF NONZERO MEAN STRESS

Most basic fatigue data collected in the laboratory are for completely reversed alternating stresses, that is, zero mean cyclic stresses. Most service applications involve nonzero mean cyclic stresses. It is, therefore, very important to a designer to know the influence of mean stress on fatigue behavior so that he can utilize basic completely reversed laboratory data in designing machine parts subjected to nonzero mean cyclic stresses.

High-cycle fatigue data collected from a series of experiments devised to investigate combinations of alternating stress amplitude σ_a and mean stress σ_m are characterized in Figure 7.59 for a failure life of N cycles. A different σ_a versus σ_m plot would be obtained for every different failure life N. By definition, the fatigue strength σ_N at N cycles is plotted on the σ_a axis, where σ_m is zero. As shown, the failure data points tend to follow a curve that passes through the ultimate strength value on the σ_m axis. The influence of mean stress on fatigue failure is different for compressive mean stress values than for tensile mean stress values. In the tensile mean stress region failure is very sensitive to the magnitude of the mean stress, whereas in the compressive mean stress region failure is rather insensitive to the magnitude of the mean stress. The influence of mean stress in the compressive region is greater for shorter lives than for longer lives, as shown in Figure 7.60.

If a designer is fortunate enough to find data for his proposed material under the mean stress conditions and design life of interest to him, he should, of course, use these data. Such data are typically presented on so-called *master diagrams* for the material. Three such master diagrams are shown in Figure 7.61 for three different alloys.

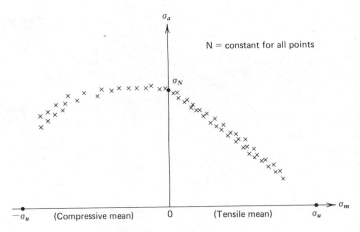

FIGURE 7.59.   Simulated high-cycle fatigue failure data showing the influence of mean stress.

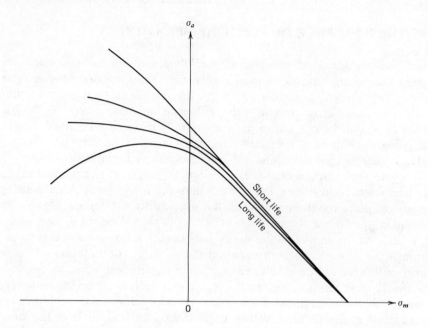

FIGURE 7.60. Comparison of mean stress effect on fatigue failure for high-cycle and low-cycle fatigue.

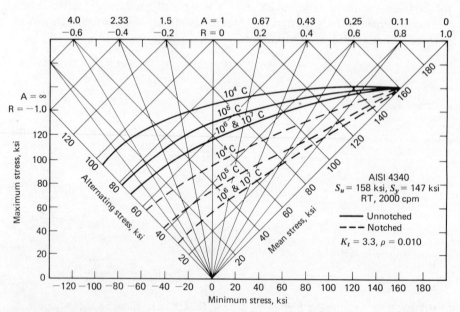

FIGURE 7.61. Master diagrams for alloys of steel, aluminum, and titanium. (From ref. 14, pp. 317, 322).

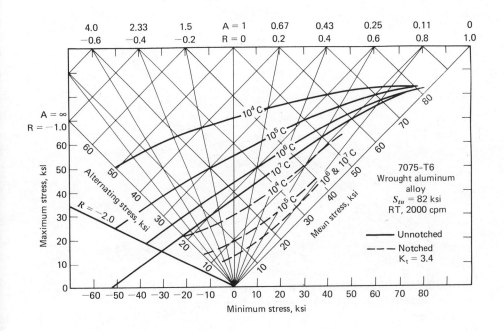

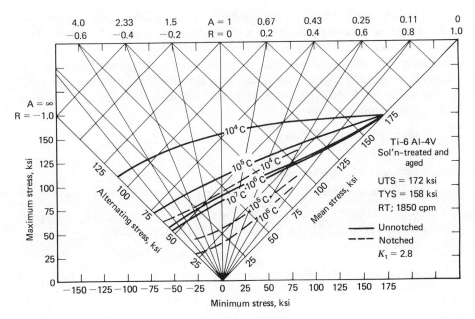

FIGURE 7.61. (*Continued*)

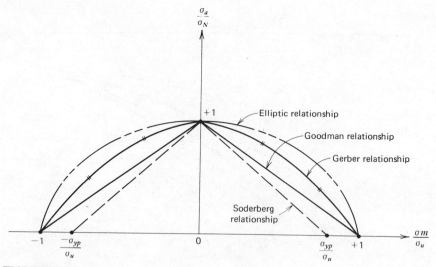

FIGURE 7.62. Illustration of the form of several empirical relationships for estimating the influence of nonzero-mean stress on fatigue failure.

If data are not available to the designer, he may estimate the influence of nonzero mean stress by any one of several empirical relationships that relate failure at a given life under nonzero mean conditions to failure at the same life under zero mean cyclic stresses. Historically, the plot of alternating stress amplitude σ_a versus mean stress σ_m has been the object of numerous empirical curve-fitting attempts. The more successful attempts have resulted in four different relationships, namely:

1. Goodman's linear relationship
2. Gerber's parabolic relationship
3. Soderberg's linear relationship
4. The elliptic relationship

These four relationships may best be illustrated on a normalized $\sigma_a - \sigma_m$ plot as shown in Figure 7.62. The equations for these four relationships may be written as follows:

Goodman's Linear Relationship

$$\frac{\sigma_a}{\sigma_N} + \frac{\sigma_m}{\sigma_u} = 1 \tag{7-1}$$

Gerber's Parabolic Relationship

$$\frac{\sigma_a}{\sigma_N} + \left(\frac{\sigma_m}{\sigma_u}\right)^2 = 1 \tag{7-2}$$

Soderberg's Linear Relationship

$$\frac{\sigma_a}{\sigma_N} + \frac{\sigma_m}{\sigma_{yp}} = 1 \qquad (7\text{-}3)$$

Elliptic Relationship

$$\left(\frac{\sigma_a}{\sigma_N}\right)^2 + \left(\frac{\sigma_m}{\sigma_u}\right)^2 = 1 \qquad (7\text{-}4)$$

A modified form of the Goodman relationship is recommended for general use under conditions of high-cycle fatigue. To develop working equations for the modified Goodman relationships a range-of-stress diagram is shown in Figure 7.63, in which selected stresses are plotted on the vertical axis versus

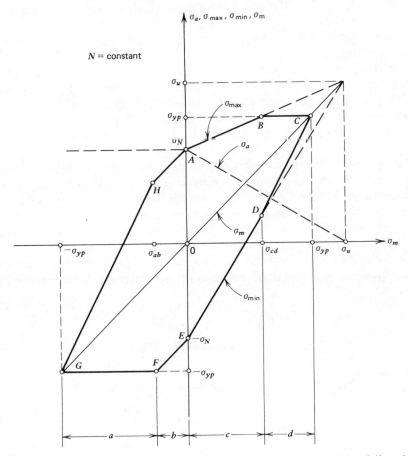

FIGURE 7.63.   Modified Goodman range-of-stress diagram for fatigue failure in N cycles.

mean stress on the horizontal axis. The entire range-of-stress diagram in Figure 7.63 is constructed for a single fatigue lifetime of N cycles.

The modified Goodman range-of-stress diagram is constructed in the following way: Using information from Figure 7.59, a dashed straight-line approximation passes through the point σ_N on the vertical σ_a axis and through the point σ_u on the horizontal σ_m axis. In the compressive mean stress region, the failure data are approximated by a horizontal straight line passing through the point σ_N on the vertical σ_a axis. These dashed lines are based on experimental results.

Next, a line representing the mean stress σ_m is plotted. If the scale of the vertical axis is equal to the scale of the horizontal axis, the σ_m line will make a 45 degree angle with the coordinate axes. The σ_m line is used as a reference for plotting the maximum stress σ_{max} and minimum stress σ_{min} on the range-of-stress diagram. The line representing σ_{max} is obtained by adding the σ_a line to the σ_m line. The line representing σ_{min} is obtained by subtracting the σ_a line from the σ_m line. In the compressive mean stress region this results in two parallel 45 degree lines, one passing through $+\sigma_N$ and the other one passing through $-\sigma_N$.

The diagram is next truncated at the yield point, both in tension and in compression, and the diagram is completed. It may be noted that the lines C-D and G-H are generated by the yield point truncation and the fact that stress cycles must be symmetrical about the mean stress line. By definition of mean stress, the σ_{min} value must always be the same distance below σ_m as the σ_{max} value is above σ_m.

The modified Goodman range-of-stress diagram shown in Figure 7.63 is a failure locus for the case of *uniaxial* fatigue stressing. Any cyclic loading that produces a stress amplitude that exceeds the bounds of the locus will cause failure in fewer than N cycles. Any stress amplitude that lies within the locus will result in more than N cycles without failure. Stress amplitudes that just touch the locus produce failure in exactly N cycles.

To facilitate further analysis, the diagram of Figure 7.63 is divided into four regions: a, b, c, and d. The bounds of these four regions are defined in Table 7-1.

Table 7.1. Bounds on Regions in Modified Goodman Range-of-Stress Diagram

Region	Limiting Value of σ_m
a	$-\sigma_{yp} \leq \sigma_m \leq \sigma_{ab}$
b	$\sigma_{ab} \leq \sigma_m \leq 0$
c	$0 \leq \sigma_m \leq \sigma_{cd}$
d	$\sigma_{cd} \leq \sigma_m \leq \sigma_{yp}$

From geometrical considerations in Figure 7.63, it may be determined that

$$\sigma_{ab} = \sigma_N - \sigma_{yp} \tag{7-5}$$

where σ_N is the fatigue strength at N cycles and σ_{yp} is the yield point strength for the material. Also, from geometry of the diagram it may be written that

$$\sigma_{cd} = \frac{\sigma_{yp} - \sigma_N}{1 - r} \tag{7-6}$$

where

$$r \equiv \frac{\sigma_N}{\sigma_u} \tag{7-7}$$

The symbol σ_u is used to denote the ultimate strength of the material. The stresses σ_{ab} and σ_{cd} are particular values of mean stress corresponding to the region interfaces a-b and c-d, respectively.

It is now possible to write failure equations for each of the four regions just defined. For line GF in Region a,

$$\sigma_{\min} = -\sigma_{yp} \tag{7-8}$$

For line FE in Region b, recalling that the slope of the line is unity,

$$\sigma_{\min} = \sigma_m - \sigma_N \tag{7-9}$$

Next, for line AB in Region c

$$\sigma_{\max} = \sigma_N + \sigma_m\left(\frac{\sigma_u - \sigma_N}{\sigma_u}\right) \tag{7-10}$$

or utilizing the definition (7-7)

$$\sigma_{\max} = \sigma_N + \sigma_m(1 - r) \tag{7-11}$$

Finally, for line BC in Region d

$$\sigma_{\max} = \sigma_{yp} \tag{7-12}$$

These four equations represent conditions of incipient fatigue failure in N cycles for each of their respective regions of validity. To put these four failure equations in a more useful form, the minimum stress $\sigma_{\min}$ may be eliminated from (7-8) and (7-9) by recalling that

$$\sigma_m = \frac{\sigma_{\min} + \sigma_{\max}}{2} \tag{7-13}$$

whence

$$\sigma_{\min} = 2\sigma_m - \sigma_{\max} \tag{7-14}$$

Making this substituting for $\sigma_{\min}$ into (7-8) and writing it as a failure equation valid in Region a, we may state that

*Failure is predicted to occur in N cycles or less when operating in Region **a** if*

$$\sigma_{\max} - 2\sigma_m \geq \sigma_{yp} \tag{7-15}$$

Table 7.2.   Failure Prediction Equations for Nonzero Mean Cyclic Stressing

Region	Failure Equation (*failure is predicted to occur if:*)	Limits on Validity of Equation
a	$\sigma_{max} - 2\sigma_m \geq \sigma_{yp}$	$-\sigma_{yp} \leq \sigma_m \leq (\sigma_N - \sigma_{yp})$
b	$\sigma_{max} - \sigma_m \geq \sigma_N$	$(\sigma_N - \sigma_{yp}) \leq \sigma_m \leq 0$
c	$\sigma_{max} - (1 - r)\sigma_m \geq \sigma_N$	$0 \leq \sigma_m \leq \left(\dfrac{\sigma_{yp} - \sigma_N}{1 - r}\right)$
d	$\sigma_{max} \geq \sigma_{yp}$	$\left(\dfrac{\sigma_{yp} - \sigma_N}{1 - r}\right) \leq \sigma_m \leq \sigma_{yp}$
	where $r \equiv \dfrac{\sigma_N}{\sigma_u}$	

Similarly, (7-14) may be combined with (7-9) and the result may be written as a failure equation valid in Region *b*:

> *Failure is predicted to occur in N cycles or less when operating in Region **b** if*

$$\sigma_{max} - \sigma_m \geq \sigma_N \qquad (7\text{-}16)$$

Writing (7-11) as a failure equation valid in Region *c*,

> *Failure is predicted to occur in N cycles or less when operating in Region **c** if*

$$\sigma_{max} - (1 - r)\sigma_m \geq \sigma_N \qquad (7\text{-}17)$$

Finally, writing (7-12) as a failure equation valid in Region *d*,

> *Failure is predicted to occur in N cycles or less when operating in Region **d** if*

$$\sigma_{max} \geq \sigma_{yp} \qquad (7\text{-}18)$$

These resulting failure equations and their regions of validity are summarized in Table 7.2.

Utilizing the results of Table 7.2, a designer may estimate whether failure will occur under any nonzero mean stress condition at the design life if he knows the ultimate strength of the material, the yield strength, and the simple completely reversed fatigue strength for the material at the design life. All these properties are usually available.

7.10   USING THE IDEAS

Consider the following example of a simple fatigue design problem: A 4140 steel linkage bar of solid circular cross section is to be subjected to an axial cyclic force that ranges from a maximum of 80,000 lb tension to a minimum of 30,000 lb compression. The material has an ultimate strength of 158,000 psi and a yield strength of 133,000 psi. The mean fatigue limit for the material has been determined to be 72,000 psi. Using a factor of safety of 2, calculate the required bar diameter to provide infinite life.

To solve this problem, the failure equations of Table 7.2 are examined. To determine which equation is applicable, it is necessary to know in which region of mean stress the bar will be operating. Since this is a case of axial loading, the mean stress is proportional to the mean load and may be written as

$$\sigma_m = \frac{\sigma_{max} + \sigma_{min}}{2} = \frac{(P_{max}/A) + (P_{min}/A)}{2} \tag{7-19}$$

From the problem statement the maximum load P_{max} is 80,000 lbs (tension), the minimum load P_{min} is $-30,000$ lbs (compression), and the area A is unknown since the dimensions are yet to be established. From (7-19), then, the mean stress σ_m may be written as

$$\sigma_m = \frac{80,000 + (-30,000)}{2A} = \frac{25,000}{A} \tag{7-20}$$

Since the area A will always be positive, it is clear from (7-20) that the mean stress will lie in either Region c or Region d of Figure 7.63. For the 4140 material the range of Region c is

$$0 \le \sigma_m \le \frac{\sigma_{yp} - \sigma_N}{1 - r} \tag{7-21}$$

or

$$0 \le \sigma_m \le \frac{133,000 - 72,000}{1 - \dfrac{72,000}{158,000}} \tag{7-22}$$

or

$$0 \le \sigma_m \le 112,000 \text{ psi} \tag{7-23}$$

and the range of Region d is

$$112,200 \le \sigma_m \le 133,000 \text{ psi} \tag{7-24}$$

As a first trial it may be assumed that the link will be operated in Region c, and therefore the Region c failure equation in Table 7.2 is valid. For this case, in which the link is loaded in uniaxial direct stress, the proper relationship from Table 7.2 is therefore

$$\frac{P_{max}}{A_f} - (1 - r)\frac{P_m}{A_f} = \sigma_e \tag{7-25}$$

where P_{max} is the maximum load in the cycle, P_m is the mean load, A_f is the required cross-sectional area of the link to produce failure in N cycles, and σ_e is the fatigue limit for the material. Solving (7-25) for A_f yields

$$A_f = \frac{P_{max} - (1 - r)P_m}{\sigma_e} \tag{7-26}$$

or

$$A_f = \frac{80,000 - (0.544)25,000}{72,000} \tag{7-27}$$

whence

$$A_f = 0.93 \text{ in.}^2 \tag{7-28}$$

With this trial value of A_f the mean stress may be calculated from (7-20) as

$$\sigma_m = \frac{25,000}{0.93} = 26,900 \text{ psi} \tag{7-29}$$

which does in fact lie in Region c as defined in (7-23). Thus, the assumption of operation in Region c was correct and use of (7-25) was valid. With the validity of (7-25) established, the relationship (7-26) for required cross-sectional area must be converted from a failure equation to a design equation by introducing the safety factor to give

$$A_d = \frac{P_{\max} - (1 - r)P_m}{\sigma_e/n} \tag{7-30}$$

where A_d is the design area and n is the safety factor. From (7-30), then, with a safety factor of 2,

$$A_d = 1.86 \text{ in.}^2 \tag{7.31}$$

Since the bar was to be of solid circular cross section, the diameter is given by

$$D = \sqrt{\frac{4A_d}{\pi}} = \sqrt{\frac{4(1.86)}{\pi}} \tag{7-32}$$

or

$$D = 1.54 \text{ inches} \tag{7-33}$$

7.11 MULTIAXIAL FATIGUE STRESSES

Uniaxial cyclic loading has been implicit in all of the discussions presented thus far in Chapter 7. Most real design situations, including rotating shafts, connecting links, turbine buckets, aircraft structures, automotive parts, and many others, involve a multiaxial state of cyclic stress. Although verifying data are relatively sparse, an acceptable design approach for machine parts subjected to a multiaxial state of cyclic stress is to make the following *fundamental assertion for multiaxial stress fatigue*:

Failure for the multiaxial state of cyclic stress is predicted to occur, according to the theory associated with a particular mechanical "modulus," if and when any range of that "modulus" that has been induced by the cyclic loading is of sufficiently critical nature that failure would occur in the uniaxial state of cyclic stress for an identical range of the same "modulus."

The mechanical "modulus" is, as usual, a measurable quantity such as principal normal stress, principal shearing stress, or distortion energy. The range of the mechanical modulus may be defined by any two appropriate and independent values of the modulus, such as σ_{max} and σ_m or σ_{max} and σ_{min}. The three most useful mechanical moduli for designing under conditions of multiaxial fatigue are maximum normal stress, maximum shearing stress, and distortion energy. The use of each of these three moduli under conditions of multiaxial cyclic stresses with nonzero means is described in later paragraphs. Throughout the discussions, it will be assumed that the loads producing the multiaxial components of stress are always in phase. This assumption is usually accurate, is usually conservative in the design sense, and is widely used by designers because it affords a significant simplication in the analysis. Therefore, if the loads are known to be out of phase, the designer might choose to analyze carefully the cyclic state of stress actually produced by the out of phase loads, or he might choose to perform his analysis assuming the loads to be in phase, recognizing that the error he makes will be on the conservative side. Some investigators (42) have developed useful procedures for the case of multiaxial cyclic stresses that are not synchronous.

Maximum Normal Stress Multiaxial Fatigue Failure Theory

The maximum normal stress multiaxial fatigue failure theory represents a combination of the maximum normal stress theory for static stresses, developed in Section 6.2, and the modified Goodman relationships, developed in Section 7.9. Writing the results of this combination in complete detail results in a total of 12 failure equations, which, when written together with their regions of validity, are as follows:

Failure is predicted to occur if:

$$\sigma_{1max} - 2\sigma_{1m} - \sigma_{yp} \geq 0; \; -\sigma_{yp} \leq \sigma_{1m} \leq (\sigma_N - \sigma_{yp}) \tag{7-34}$$

$$\sigma_{1max} - \sigma_{1m} - \sigma_N \geq 0; \; (\sigma_N - \sigma_{yp}) \leq \sigma_{1m} \leq 0 \tag{7-35}$$

$$\sigma_{1max} - \sigma_{1m}(1 - r) - \sigma_N \geq 0; \; 0 \leq \sigma_{1m} \leq \left(\frac{\sigma_{yp} - \sigma_N}{1 - r}\right) \tag{7-36}$$

$$\sigma_{1max} - \sigma_{yp} \geq 0; \; \left(\frac{\sigma_{yp} - \sigma_N}{1 - r}\right) \leq \sigma_{1m} \leq \sigma_{yp} \tag{7-37}$$

or if

$$\sigma_{2max} - 2\sigma_{2m} - \sigma_{yp} \geq 0; \; -\sigma_{yp} \leq \sigma_{2m} \leq (\sigma_N - \sigma_{yp}) \tag{7-38}$$

$$\sigma_{2max} - \sigma_{2m} - \sigma_N \geq 0; \; (\sigma_N - \sigma_{yp}) \leq \sigma_{2m} \leq 0 \tag{7-39}$$

$$\sigma_{2max} - \sigma_{2m}(1 - r) - \sigma_N \geq 0; \; 0 \leq \sigma_{2m} \leq \frac{(\sigma_{yp} - \sigma_N)}{1 - r} \tag{7-40}$$

$$\sigma_{2max} - \sigma_{yp} \geq 0; \; \frac{(\sigma_{yp} - \sigma_N)}{1 - r} \leq \sigma_{2m} \leq \sigma_{yp} \tag{7-41}$$

or if

$$\sigma_{3max} - 2\sigma_{3m} - \sigma_{yp} \geq 0; \quad -\sigma_{yp} \leq \sigma_{3m} \leq (\sigma_N - \sigma_{yp}) \qquad (7\text{-}42)$$

$$\sigma_{3max} - \sigma_{3m} - \sigma_N \geq 0; \quad (\sigma_N - \sigma_{yp}) \leq \sigma_{3m} \leq 0 \qquad (7\text{-}43)$$

$$\sigma_{3max} - \sigma_{3m}(1 - r) - \sigma_N \geq 0; \quad 0 \leq \sigma_{3m} \leq \frac{(\sigma_{yp} - \sigma_N)}{1 - r} \qquad (7\text{-}44)$$

$$\sigma_{3max} - \sigma_{yp} \geq 0; \quad \frac{(\sigma_{yp} - \sigma_N)}{1 - r} \leq \sigma_{3m} \leq \sigma_{yp} \qquad (7\text{-}45)$$

Whichever one of these 12 equations proves to be most critical will govern the design, and geometry is determined in accordance with this most critical equation. It should also be noted that the subscripts 1, 2, and 3 in these equations denote the three principal normal stresses.

Maximum Shearing Stress Multiaxial Fatigue Failure Theory

The maximum shearing stress multiaxial fatigue theory represents a combination of the maximum shearing stress theory for static stresses, developed in Section 6.3, and the modified Goodman relationships, developed in Section 7.9. It may be recalled from (6-2) that for a uniaxial state of stress the largest principal shearing stress τ is equal to one-half the largest principal normal stress σ. This relationship is utilized by substituting 2τ for σ everywhere that it appears in the modified Goodman relationships. Writing the results of this substitution for the first principal shearing stress τ_1 yields the following set of equations:

Failure is predicted to occur if:

$$2\tau_{1max} - 4\tau_{1m} - \sigma_{yp} \geq 0; \quad -\sigma_{yp} \leq 2\tau_{1m} \leq (\sigma_N - \sigma_{yp}) \qquad (7\text{-}46)$$

$$2\tau_{1max} - 2\tau_{1m} - \sigma_N \geq 0; \quad (\sigma_N - \sigma_{yp}) \leq 2\tau_{1m} \leq 0 \qquad (7\text{-}47)$$

$$2\tau_{1max} - 2\tau_{1m}(1 - r) - \sigma_N \geq 0; \quad 0 \leq 2\tau_{1m} \leq \frac{(\sigma_{yp} - \sigma_N)}{1 - r} \qquad (7\text{-}48)$$

$$2\tau_{1max} - \sigma_{yp} \geq 0; \quad \frac{(\sigma_{yp} - \sigma_N)}{1 - r} \leq 2\tau_{1m} \leq \sigma_{yp} \qquad (7\text{-}49)$$

In addition to this set of equations for τ_1, there are two other sets of equations, identical in structure, that must be written for τ_2 and τ_3. This results in a total of 12 failure equations that must be examined to determine the most critical equation that governs the design. The values of τ_1, τ_2, and τ_3 to be used in these failure equations are the multiaxial principal shearing stresses as defined earlier in (4-55), (4-56), and (4-57), specifically

$$\tau_1 = \pm \left(\frac{\sigma_2 - \sigma_3}{2} \right) \tag{7-50}$$

$$\tau_2 = \pm \left(\frac{\sigma_3 - \sigma_1}{2} \right) \tag{7-51}$$

$$\tau_3 = \pm \left(\frac{\sigma_1 - \sigma_2}{2} \right) \tag{7-52}$$

where σ_1, σ_2, and σ_3 are the principal normal stresses. The geometry of the machine part would then be determined in accordance with the most critical of the 12 equations.

Distortion Energy Multiaxial Fatigue Failure Theory

The distortion energy multiaxial fatigue failure theory represents a combination of the distortion energy theory for a static state of stress, discussed in Section 6.6, and the modified Goodman relationships, developed in Section 7.9. The expression for distortion energy per unit volume was developed and presented for the static stress situation in (6-40) as

$$u_d = \left[\frac{1 + \nu}{3E} \right] \left[\frac{(\sigma_1 - \sigma_2)^2}{2} + \frac{(\sigma_2 - \sigma_3)^2}{2} + \frac{(\sigma_3 - \sigma_1)^2}{2} \right] \tag{7-53}$$

Using this basic expression for distortion energy per unit volume, we can make three definitions to implement the use of this theory; these are

$$u_{dmax} = \left[\frac{1 + \nu}{3E} \right]$$
$$\times \left[\frac{(\sigma_{1max} - \sigma_{2max})^2}{2} + \frac{(\sigma_{2max} - \sigma_{3max})^2}{2} + \frac{(\sigma_{3max} - \sigma_{1max})^2}{2} \right] \tag{7-54}$$

$$u_{dm} = \left[\frac{1 + \nu}{3E} \right] \left[\frac{(\sigma_{1m} - \sigma_{2m})^2}{2} + \frac{(\sigma_{2m} - \sigma_{3m})^2}{2} + \frac{(\sigma_{3m} - \sigma_{1m})^2}{2} \right] \tag{7-55}$$

$$u_{dmin} = \left[\frac{1 + \nu}{3E} \right]$$
$$\times \left[\frac{(\sigma_{1min} - \sigma_{2min})^2}{2} + \frac{(\sigma_{2min} - \sigma_{3min})^2}{2} + \frac{(\sigma_{3min} - \sigma_{1min})^2}{2} \right] \tag{7-56}$$

It should be carefully noted that under these definitions

$$u_{dm} \neq \frac{u_{dmax} + u_{dmin}}{2} \tag{7-57}$$

To obtain the proper value of u_{dm}, the mean principal stresses σ_{1m}, σ_{2m}, and σ_{3m} must be computed and substituted into (7-55).

From equations (7-54), (7-55), and (7-56) it may be observed that for a uniaxial state of cyclic stress the expressions for $u_{d\max}$, u_{dm}, and $u_{d\min}$ become

$$u_{d\max} = \frac{1 + \nu}{3E}(\sigma_{\max})^2 \tag{7-58}$$

$$u_{dm} = \frac{1 + \nu}{3E}(\sigma_m)^2 \tag{7-59}$$

$$u_{d\min} = \frac{1 + \nu}{3E}(\sigma_{\min})^2 \tag{7-60}$$

Solving these equations for $\sigma_{\max}$, σ_m, and $\sigma_{\min}$ yields

$$\sigma_{\max} = \sqrt{\frac{3E}{1 + \nu}}\sqrt{u_{d\max}} \tag{7-61}$$

$$\sigma_m = \sqrt{\frac{3E}{1 + \nu}}\sqrt{u_{dm}} \tag{7-62}$$

$$\sigma_{\min} = \sqrt{\frac{3E}{1 + \nu}}\sqrt{u_{d\min}} \tag{7-63}$$

Now, substituting the expressions of (7-61), (7-62), and (7-63) into the modified Goodman relationships results in the following failure prediction equations:

Failure is predicted to occur if:

$$\left[\frac{3E}{1 + \nu}\right]^{1/2}\left[(u_{d\max})^{1/2} - 2(u_{dm})^{1/2}\right] - \sigma_{yp} \geq 0 \tag{7-64}$$

$$\left[\frac{3E}{1 + \nu}\right]^{1/2}\left[(u_{d\max})^{1/2} - (u_{dm})^{1/2}\right] - \sigma_N \geq 0 \tag{7-65}$$

$$\left[\frac{3E}{1 + \nu}\right]^{1/2}\left[(u_{d\max})^{1/2} - (u_{dm})^{1/2}(1 - r)\right] - \sigma_N \geq 0 \tag{7-66}$$

$$\left[\frac{3E}{1 + \nu}\right]^{1/2}\left[u_{d\max}\right]^{1/2} - \sigma_{yp} \geq 0 \tag{7-67}$$

where, as defined in (7-7), $r \equiv \sigma_N/\sigma_u$.

Since distortion energy is always a positive quantity in the nontrivial case, it cannot be used in specifying the region of validity for (7-64) through (7-67). Instead, all four equations must be evaluated and the design based on the most critical equation, with the knowledge that certain of these four equations may have to be discarded because they do not make sense physically. This will become clear in the example cited in Section 7.13.

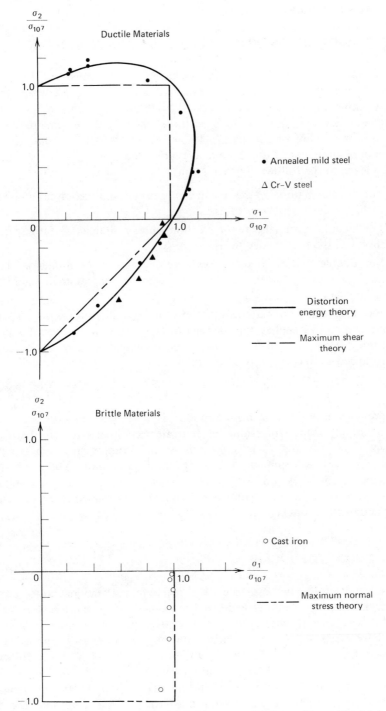

FIGURE 7.64. Comparison of biaxial fatigue strength data with multiaxial fatigue failure theories for ductile and brittle materials. (See Chap. 7 of ref. 9)

7.12 USE OF MULTIAXIAL FATIGUE FAILURE THEORIES

Three multiaxial fatigue failure theories have been described in the preceding section. Each of these failure theories provides a means of predicting failure, or designing to prevent failure, under conditions of nonzero mean cyclic multiaxial states of stress as long as the stress amplitude remains constant throughout the life of the part. Although there is only a relatively small amount of multiaxial fatigue data available, the existing experimental evidence leads to the following observations:

1. For brittle materials, the maximum normal multiaxial fatigue failure theory is the best theory to use.
2. For ductile materials, the distortion energy multiaxial fatigue failure theory is the best theory to use.
3. For ductile materials, the maximum shearing stress multiaxial fatigue failure theory is almost as good as the distortion energy multiaxial fatigue failure theory.
4. As a rule of thumb, materials that exhibit a ductility of less than 5 percent elongation in 2 inches may be regarded as brittle, whereas materials that exhibit a ductility of 5 percent or more elongation in 2 inches may be regarded as ductile.

Some data illustrating the validity of these observations are plotted in Figure 7.64.

Another viewpoint in using the modified Goodman equation is presented in Section 8.5 where expressions for equivalent completely reversed stress are shown in (8-114) and (8-115) for tensile and compressive nonzero mean conditions with a specified alternating stress amplitude. These equations, developed from (7-9) and (7-11), are very useful in converting any nonzero mean cyclic stress into an "equivalent" completely reversed stress in terms of fatigue damage produced.

7.13 USING THE IDEAS

Consider the following example of a machine part subjected to a cyclic multiaxial state of stress: It is desired to design a solid circular torsion bar, fixed at one end, to withstand $N = 5 \times 10^8$ cycles of released torsion produced by a released cyclic moment with $M_{t\max} = 1500$ in-lb applied at the free end. Calculate the bar diameter d to provide a life of $N = 5 \times 10^8$ cycles if the material is 2024-T4 aluminum alloy with $\sigma_u = 68,000$ psi, $\sigma_{yp} = 48,000$ psi, elongation of 19 percent in 2 inches, and S-N properties as shown in Figure 7.17. The first step in solving this problem is to determine the three principal stresses for the case of pure torsion by employing the stress cubic equation (4-23). As has been shown in (4-60), (4-61), and (4-62), the three principal normal stresses for pure torsion are, together with the corresponding expressions for maximum, mean, and minimum values as illustrated in Figure 7.65,

$$\sigma_1 = |\tau| = \frac{16|M_{tmax}|}{\pi d^3} \; ; \; \sigma_{1max} = |\tau_{max}|, \; \sigma_{1m} = \frac{1}{2}|\tau_{max}|, \; \sigma_{1min} = 0 \qquad (7\text{-}68)$$

$$\sigma_2 = 0; \; \sigma_{2max} = 0, \; \sigma_{2m} = 0, \; \sigma_{2min} = 0 \qquad (7\text{-}69)$$

$$\sigma_3 = -|\tau| = \frac{-16|M_{tmax}|}{\pi d^3} \; ; \; \sigma_{3max} = 0, \; \sigma_{3m} = -\frac{1}{2}|\tau_{max}|, \; \sigma_{3min} = -|\tau_{max}|$$

$$(7\text{-}70)$$

The material is ductile, so the distortion energy fatigue failure theory is selected. Using the values of principal stresses in (7-54) and (7-55), we may determine the magnitudes of u_{dmax} and u_{dm} to be

$$u_{dmax} = \frac{1+\nu}{3E}\left[\frac{(|\tau_{max}| - 0)^2}{2} + \frac{(0-0)^2}{2} + \frac{(0 - |\tau_{max}|)^2}{2}\right] \qquad (7\text{-}71)$$

or

$$u_{dmax} = \frac{1+\nu}{3E}\left[|\tau_{max}|^2\right] \qquad (7\text{-}72)$$

and, similarly,

$$u_{dm} = \left(\frac{1+\nu}{3E}\right)\left(\frac{1}{2}\right)\left[\left(\frac{|\tau_{max}| - 0}{2}\right)^2 + \left(\frac{0 + |\tau_{max}|}{2}\right)^2 + \left(\frac{-|\tau_{max}|}{2} - \frac{|\tau_{max}|}{2}\right)^2\right]$$

$$(7\text{-}73)$$

or

$$u_{dm} = \frac{1+\nu}{3E}\left[\frac{3}{4}|\tau_{max}|^2\right] \qquad (7\text{-}74)$$

These values are next substituted into (7-64) to obtain
Failure is predicted to occur if:

$$\left[\frac{3E}{1+\nu}\right]^{1/2}\left[\left(\frac{1+\nu}{3E}\right)^{1/2}|\tau_{max}| - 2\left(\frac{1+\nu}{3E}\right)^{1/2}\left(\frac{3}{4}\right)^{1/2}|\tau_{max}|\right] - \sigma_{yp} \geq 0$$

$$(7\text{-}75)$$

or if

$$(1 - \sqrt{3})|\tau_{max}| - \sigma_{yp} \geq 0 \qquad (7\text{-}76)$$

Similarly, from (7-65) through (7-67),
Failure is predicted to occur if:

$$\left[1 - \sqrt{\frac{3}{2}}\right]|\tau_{max}| - \sigma_N \geq 0 \qquad (7\text{-}77)$$

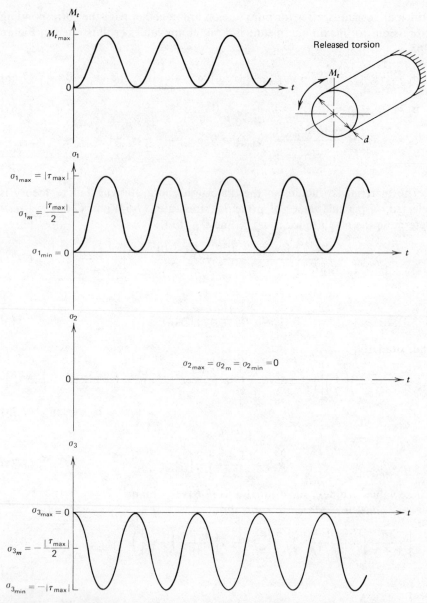

FIGURE 7.65. Principal stresses as a function of time for released cyclic torsion example of Section 7.13.

or

$$\left[1 - \frac{(1-r)\sqrt{3}}{2} \right] |\tau_{max}| - \sigma_N \geq 0 \qquad (7\text{-}78)$$

or

$$|\tau_{max}| - \sigma_{yp} \geq 0 \qquad (7\text{-}79)$$

Equations (7-76) through (7-79) are the governing failure equations and, from (7-68), the magnitude of $|\tau_{max}|$ in all these equations is given by

$$|\tau_{max}| = \frac{16|M_{tmax}|}{\pi d^3} \qquad (7\text{-}80)$$

where M_{tmax} is the maximum value of cyclic torque, d is the bar diameter, σ_N is the completely reversed fatigue strength corresponding to a life of N cycles, and σ_{yp} is the yield strength of the material.

The expression (7-76) is meaningless since it can never exceed zero. Expressions (7-77), (7-78), and (7-79) can be solved for diameter d, dropping the inequality, to give

$$d = \sqrt[3]{\frac{16\left[1 - \sqrt{\frac{3}{2}} \right]|M_{tmax}|}{\pi\sigma_N}} \qquad (7\text{-}81)$$

or

$$d = \sqrt[3]{\frac{16\left[1 - \frac{(1-r)\sqrt{3}}{2} \right]|M_{tmax}|}{\pi\sigma_N}} \qquad (7\text{-}82)$$

or

$$d = \sqrt[3]{\frac{16|M_{tmax}|}{\pi\sigma_{yp}}} \qquad (7\text{-}83)$$

Whichever one gives the most critical (largest) value for required diameter d will govern the design. To make a useful design equation of the governing expression, an appropriate safety factor must be introduced by dividing the failure strengths, σ_N and σ_{yp}, by the safety factor.

Selecting a safety factor of 2, reading $\sigma_N = 20{,}000$ psi at 5×10^8 cycles from Figure 7.17, and using the properties just given, (7-81), (7-82), and (7-83) become

$$d = \sqrt[3]{\frac{16\left[1 - \sqrt{\frac{3}{2}}\right][1500]}{\pi(20,000/2)}} = 0.47 \text{ inch} \qquad (7\text{-}84)$$

$$d = \sqrt[3]{\frac{16\left[1 - \frac{(1 - [20/68])\sqrt{3}}{2}\right][1500]}{\pi(20,000/2)}} = 0.79 \text{ inch} \qquad (7\text{-}85)$$

$$d = \sqrt[3]{\frac{16(1500)}{\pi(48,000/2)}} = 0.68 \text{ inch} \qquad (7\text{-}86)$$

The design specification, then, should be a shaft of diameter $d = 0.79$ inch.

QUESTIONS

1.   Distinguish the differences between high-cycle fatigue and low-cycle fatigue.

2.   Identify several problems a designer must recognize when dealing with fatigue loading as compared with static loading.

3.   Describe at least two hypotheses for the initiation and propagation of fatigue nuclei.

4.   Describe the characteristic appearance of a fatigue fracture surface and explain why it has such an appearance.

5.   Sketch a family of S-N-P curves, explain the meaning and utility of these curves, and explain in detail how such a family of curves would be produced in the laboratory.

6.   Make a list of factors that may influence the S-N curve and indicate briefly what the influences might be in each case.

7.   Explain how a designer might utilize a master diagram, such as the ones shown in Figure 7.61.

8.   State and explain the basic concepts behind the *fundamental assertion for multiaxial stress fatigue*.

9.   (a)   An aluminum bar of solid cylindrical cross section is to be subjected to a cyclic axial loading that ranges from 5000-lb tension to 10,000-lb tension. The material has an ultimate strength of 100,000 psi, a yield strength of 80,000 psi, a mean fatigue strength of 40,000 psi at 10^5 cycles, and an elongation of 8% in 2 inches. Calculate the bar diameter that should be used to just produce failure in 10^5 cycles, on the average.

(b)   If, instead of the loading specified in part (a), the cyclic axial loading ranged from 15,000-lb tension to 20,000-lb tension, calculate the bar diameter that should be used to just produce failure in 10^5 cycles, on the average.

(c)   Compare the results of parts (a) and (b), making any observations you think appropriate.

10.   The S-N data for a completely reversed bending fatigue test are as shown in the chart.

S (psi)	N (cycles)
17×10^4	2×10^4
15.1×10^4	5×10^4
14.1×10^4	1×10^5
12.7×10^4	2×10^5
12.5×10^4	5×10^5
12.3×10^4	1×10^6
12.1×10^4	$2 \times 10^6 \to \infty$

The ultimate strength is 218,000 psi, and the yield point is 200,000 psi. Determine and plot the S-N curve for a mean stress of 40,000 psi tension for the material.

11.   The σ_{max}-N data for direct stress fatigue tests in which the mean stress σ_m was 25,000 psi tension for all tests are as shown in the chart.

σ_{max} (psi)	N (cycles)
150,000	2×10^4
131,000	5×10^4
121,000	1×10^5
107,000	2×10^5
105,000	5×10^5
103,000	1×10^6
102,000	2×10^6

The ultimate strength is 240,000 psi, and the yield point is 225,000 psi.

(a)   Determine and plot the σ_{max}-N curve for this material for a mean stress of 50,000 psi tension.

(b)   Determine and plot on the same graph sheet the σ_{max}-N for this material for a mean stress of 50,000 psi compression.

12.   A thick-walled cylindrical pressure vessel, closed at both ends, must have an inside diameter of 3.00 inches to meet design requirements. The material to be used is a steel alloy with the following properties:

$$S_u = 250{,}000 \text{ psi}$$

$$S_{ypt} = 200{,}000 \text{ psi}$$

$$S_{ypc} = 200{,}000 \text{ psi}$$

$$S_e = 100{,}000 \text{ psi}$$

elongation is 2 inches = 4 percent

The vessel is to be pressurized cyclically from 0 to 15,000 psi once a minute continuously for 10 years. Using a safety factor of 1.5, determine the outside diameter required according to the (a) distortion energy multiaxial fatigue failure theory, and (b) maximum normal stress multiaxial fatigue failure theory.

13.   A hollow tubular steel bar is used as a torsion spring that is subjected to a cyclic pure torque ranging from -5000 in-lb to $+15,000$ in-lb. It is desired to use a tube with wall thickness of 10% of the outside diameter. The material has an ultimate strength of 200,000 psi and a yield point of 180,000 psi. The fatigue limit is 95,000 psi. Find the tube dimensions that should provide infinite life by using (a) the maximum normal stress multiaxial fatigue failure theory, (b) the maximum shearing stress multiaxial fatigue failure theory, and (c) the distortion energy multiaxial fatigue failure theory.

14.   A power transmission shaft of a solid cylindrical shape is to be made of hot-rolled 1020 steel with $\sigma_u = 65,000$ psi, $\sigma_{yp} = 43,000$ psi, $\sigma_e = 32,000$ psi, and $e = 36$ percent elongation in 2 inches. The shaft is to transmit 85 horsepower steadily at a rotational speed of 1800 rpm with no fluctuations in torque or speed. At the critical location midspan between bearings, the rotating shaft is also subjected to a pure bending moment of 2000 in-lb fixed in a vertical plane. Using the most accurate methods known to you, determine the required shaft diameter to provide infinite life, assuming a safety factor of unity. Give a clear description of what you are doing at each step.

15.   The support bracket, made of alloy Y and shown in Figure Q7.15a, consists of a solid circular bar, fixed at one end, with an arm at the free end. A cyclic load F is applied at the offset end or the arm. The load F alternates sinusoidally from a maximum value of 10,000 lb to a minimum value of 5,000

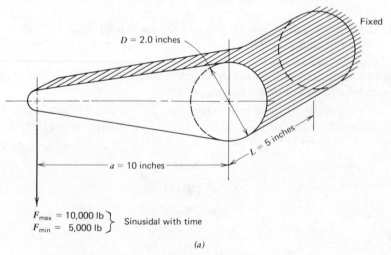

(a)

FIGURE Q7.15.

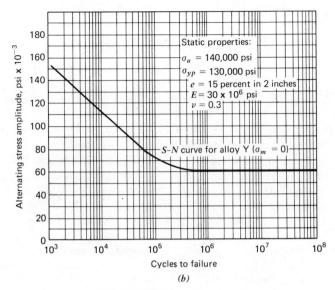

FIGURE Q7.15. (*Continued*)

1b. Considering only the cylindrical portion of the bracket, and neglecting any stress concentration effects, estimate the number of cycles that can be applied to the bracket before fatigue failure occurs. Static and fatigue properties for alloy *Y* are shown in Figure Q7.15b.

REFERENCES

1. Cazaud, R., *La Fatigue des Métaux* (French), translated by A.J. Fenner as *Fatigue of Metals*, Chapman and Hall, Ltd., London, 1953.

2. Battelle Memorial Institute, *Prevention of Fatigue in Metals*, John Wiley & Sons, New York, 1941.

3. Murray, W.M. (ed.), *Fatigue and Fracture of Metals*, John Wiley & Sons, New York, 1952.

4. *Manual on Fatigue Testing*, STP-91 and STP-91-A, American Society for Testing and Materials, Philadelphia, 1949 and 1963.

5. *Basic Mechanisms of Fatigue*, STP-237, American Society for Testing and Materials, Philadelphia, 1958.

6. Grover, H.J., Gordon, S.A., and Jackson, L.R., *Fatigue of Metals and Structures*, Government Printing Office, Washington, D.C., 1954.

7. Freudenthal, A.M., *Fatigue in Aircraft Structures*, Academic Press, New York, 1956.

8. *Proceedings of International Conference on Fatigue*, American Society of Mechanical Engineers (jointly with Institution of Mechanical Engineers), New York 1956.

9. Sines, G., and Waisman, J.L., *Metal Fatigue*, McGraw-Hill, New York, 1959.

10. Heywood, R.B., *Designing Against Fatigue of Metals*, Reinhold, New York, 1962.

11. Kennedy, A.J., *Processes of Creep and Fatigue in Metals*, John Wiley & Sons, New York, 1963.

12. Averback, B.L., Felbeck, D.K., Hahn, G.T., and Thomas, H.A., *Fracture*, MIT Press, Cambridge, 1959.

13. McClintock, F.A., and Argon, A.S., *Mechanical Behavior of Materials*, Addison-Wesley, Reading, Mass., 1966.

14. Grover, H.J., *Fatigue of Aircraft Structures*, Government Printing Office, Washington, D.C., 1966.

15. *Fatigue Crack Propagation*, STP-415, American Society for Testing and Materials, Philadelphia, 1967.

16. Madayag, A.F., *Metal Fatigue, Theory and Design*, John Wiley & Sons, New York, 1969.

17. Forrest, P.G., *Fatigue of Metals*, Addison-Wesley, Reading, Mass., 1962.

18. Mann, J.Y., *Fatigue of Materials*, Melbourne University Press, Melbourne, 1967.

19. Forsyth, P.J.E., *The Physical Basis of Metal Fatigue*, Blackie and Son, Ltd., London, 1969.

20. Manson, S.S. (ed.), *Metal Fatigue Damage*, STP-495, American Society for Testing and Materials, Philadelphia, 1971.

21. Higdon, A., Ohlsen, E.H., Stiles, W.B., and Weese, J.A., *Mechanics of Materials*, John Wiley & Sons, New York, 1967.

22. Johnson, J.B., "Aircraft Engine Material," *SAE Journal*, **40** (March 1937): 153–162.

23. Jackson, R.J., and Frost, P.D., *Properties and Current Applications of Magnesium-Lithium Alloys*, SP-5068, NASA, Washington, D.C., 1967.

24. *Proceedings of the Conference on Welded Structures*, Vols. I and II, The Welding Institute, Cambridge, England, 1971.

25. "Materials and Processes for the '70s," *Science of Advanced Materials and Process Engineering, Proceedings*, **15**, Western Periodicals, Azusa, Calif., 1969.

26. Peterson, R.E., "Fatigue of Shafts Having Keyways." *ASTM Proceedings*, **32**, p. 2 (1932): 413–420.

27. Keysers, C.A., *Material Science in Engineering*, Charles E. Merrill Co., Columbus, Ohio, 1968.

28. *Welding Handbook*, Vol. 3, American Welding Society, New York, 1961.

29. Thomas, W.N., "Effect of Scratches and Various Workshop Finishes upon the Fatigue Strength of Steel," *Engineering*, **116** (1923): 449ff.

30. Hooker, R.N., "Surface Finish vs. Fatigue Life for 75S-T6 Spar Cap Material," *Tech Note Dev-950*, Douglas Aircraft Co., Long Beach, Calif.

31. Johnson, J.B., "Dependence of Aviation on Metallurgy," *Metal and Alloys*, **1** (1930): 450.

32. Almen, J.O., "Fatigue Loss and Gain by Electro-plating," *Product Engineering*, **22**, No. 6 (1951).

33. Moore, H.F., "A Study of Size Effect and Notch Sensitivity in Fatigue Tests of Steel," *ASTM Proceedings*, **45** (1945): 507.

34. Pope, J.A., *Metal Fatigue*, Chapman & Hall, Ltd., London, 1959.

35. Lyst, J.O., "The Effect of Residual Strains Upon the Rotating Beam Fatigue Properties of Some Aluminum Alloys," *Technical Report No. 9-60-34*, Alcoa, Pittsburgh, 1960.

36. Horger, O.J., *Metals Engineering Design* (ASME Handbook), American Society of Mechanical Engineers, McGraw-Hill, New York, 1953.

37. Smith, G.V., *Properties of Metals at Elevated Temperatures*, McGraw-Hill, New York, 1950.

38. Pousopa, J., *Low Temperature Fatigue Properties and Cumulative Damage Response of 20204-T4 Aluminum Alloy*, M.S. Thesis, Ohio State University, Columbus, 1977.

39. Sors, L., *Fatigue Design of Machine Components*, Pergamon Press, New York, 1971.

40. Gough, H.J., and Sopwith, D.G., *Journal of Iron and Steel Institute*, **127** (1933): 301.

41. Howell, F.M., and Miller, J.L., "Axial Stress Fatigue Strengths of Several Structural Aluminum Alloys," *ASTM Proceedings*, 55 (1955): 955.

42. Miller, W.R., Ohji, K., and Marin, J., "Rotating Principal Stress Axes in High Cycle Fatigue," ASME Paper No. 66-WA/Met 9, American Society of Mechanical Engineers, New York, 1966.

43. Manson, S.S., "Fatigue: A Complex Subject—Some Simple Approximations," *Experimental Mechanics*, July 1965.

44. Wöhler, A., "Versuche über die Festigkeit der Eisenbahnwagen-Achsen," *Zeitschrift für Bauwesen*, 1860.

45. Forsyth, P.J.E., *Proceedings of the Royal Society* (London), A242 (1957): 198.

46. Forsyth, P.J.E., and Stubbington, C.A., "The Slip Band Extrusion Effect Observed in Some Aluminum Alloys Subjected to Cyclic Stresses," *Journal of the Institute of Metals*, 83 (1954): 395.

47. Polmear, I.J., and Bainbridge, I.F., *Philosophical Magazine*, 4, No. 48 (1959): 1296.

48. Forsythe, P.J.E., and Ryder, D.A., *Aircraft Engineering*, April 1, 1960.

CHAPTER 8

Concepts of Cumulative Damage, Life Prediction, and Fracture Control

8.1 INTRODUCTION

In virtually every engineering application where fatigue is an important failure mode, the alternating stress amplitude may be expected to vary or change in some way during the service life. Such variations and changes in load amplitude, often referred to as *spectrum loading*, make the direct use of standard *S-N* curves inapplicable because these curves are developed and presented for constant stress amplitude operation. Therefore, it becomes important to a designer to have available a theory or hypothesis, verified by experimental observations, that will permit good design estimates to be made for operation under conditions of spectrum loading using the standard constant-amplitude *S-N* curves that are more readily available.

The basic postulate adopted by all fatigue investigators working with spectrum loading is that operation at any given cyclic stress amplitude will produce *fatigue damage*, the seriousness of which will be related to the number of cycles of operation at that stress amplitude and also related to the total number of cycles that would be required to produce failure of an undamaged specimen at that stress amplitude. It is further postulated that the damage incurred is permanent and operation at several different stress amplitudes in sequence will result in an accumulation of total damage equal to the sum of the damage increments accrued at each individual stress level. When the total accumulated damage reaches a critical value, fatigue failure occurs. Although the concept is simple in principle, much difficulty is encountered in practice because the proper assessment of the amount of damage incurred by operation at any given stress level S_i for a specified number of cycles n_i is not straightforward. Many different *cumulative damage* theories have been proposed for the purposes of assessing fatigue damage caused by operation at any given stress level and the addition of damage increments to properly predict failure under conditions of spectrum loading.

8.2   THE LINEAR DAMAGE THEORY

The first cumulative damage theory was proposed by Palmgren in 1924 and later developed by Miner in 1945. This linear theory, which is still widely used, is referred to as the *Palmgren-Miner hypothesis* or the *linear damage rule*. The theory may be described using the *S-N* plot shown in Figure 8.1.

By definition of the *S-N* curve, operation at a constant stress amplitude S_1 will produce complete damage, or failure, in N_1 cycles. Operation at stress amplitude S_1 for a number of cycles n_1 smaller than N_1 will produce a smaller fraction of damage, say D_1. D_1 is usually termed the *damage fraction*. Operation over a spectrum of different stress levels results in a damage fraction D_i for each of the different stress levels S_i in the spectrum. When these damage fractions sum to unity, failure is predicted; that is,

Failure is predicted to occur if:

$$D_1 + D_2 + \cdots + D_{i-1} + D_i \geq 1 \tag{8-1}$$

The Palmgren-Miner hypothesis asserts that the damage fraction at any stress level S_i is linearly proportional to the ratio of number of cycles of operation to the total number of cycles that would produce failure at that

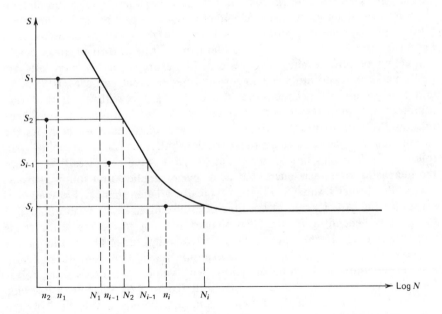

FIGURE 8.1. Illustration of spectrum loading where n_i cycles of operation are accrued at each of the different corresponding stress levels S_i, and the N_i are cycles to failure at each S_i.

stress level; that is

$$D_i = \frac{n_i}{N_i} \tag{8-2}$$

By the Palmgren-Miner hypothesis, then, utilizing (8-2), we may write (8-1) as

Failure is predicted to occur if:

$$\frac{n_1}{N_1} + \frac{n_2}{N_2} + \cdots + \frac{n_{i-1}}{N_{i-1}} + \frac{n_i}{N_i} \geq 1 \tag{8-3}$$

or

Failure is predicted to occur if:

$$\sum_{j=1}^{i} \frac{n_j}{N_j} \geq 1 \tag{8-4}$$

This is a complete statement of the Palmgren-Miner hypothesis, or the linear damage rule. It has one sterling virtue, namely, *simplicity*; and for this reason it is widely used. It must be recognized, however, that in its simplicity certain significant influences are unaccounted for, and failure prediction errors may, therefore, be expected. Perhaps the most significant shortcomings of the linear theory are that no influence of the order of application of various stress levels is recognized, and damage is assumed to accumulate at the same rate at a given stress level without regard to past history. Experimental data indicate that the order in which various stress levels are applied does have a significant influence and also that damage rate at a given stress level is a function of prior cyclic stress history. For example, if two-step tests are performed using laboratory specimens subjected to cyclic stress levels $S_1 > S_2$, and two groups of specimens are tested, one group with S_1 applied first followed by S_2 and the second group with S_2 applied first followed by S_1, the resulting Miner's sums at the time of failure are significantly different for the two groups. For the decreasing stress sequence, that is, S_1 cycles applied first, followed by S_2 cycles, the Miner's sum $\Sigma(n/N)$ is typically less than 1. For the increasing stress sequence, that is, S_2 cycles applied first followed by S_1 cycles, the Miner's sum $\Sigma(n/N)$ is typically greater than 1. Experimental values for the Miner's sum at the time of failure often range from about 1/4 to about 4, depending on the type of decreasing or increasing cyclic stress amplitudes used. If the various cyclic stress amplitudes are mixed in the sequence in a quasi-random way, the experimental Miner's sum more nearly approaches unity at the time of failure, with values of Miner's sums corresponding to failure in the range of about 0.6 to 1.6. Since many service applications involve quasi-random fluctuating stresses, the use of the Palmgren-Miner linear damage rule is often satisfactory for failure prediction.

To cloud the laboratory specimen results that show experimental values of $\Sigma(n/N)$ greater than unity for increasing stress amplitude sequences and values less than unity for decreasing stress amplitude sequences, results reported for spectrum loading tests of components and structures consistently seem to show a directly opposite trend. That is, for components and structures it appears that increasing stress amplitude sequences result in more damage than the same cycle blocks of stress amplitude in decreasing sequence. This seeming paradox has not been clearly explained, but it is undoubtedly related to the residual stresses introduced in the components and structures by virtue of stress concentrations in the structural joints and discontinuities. High stresses applied early and then relaxed result in a residual compressive stress field in regions of local stress concentration, and later application of lower level cyclic stresses produce less damage than if the residual compression had never been induced. Even a single cycle of a very high stress level may have an important influence on fatigue failure.

In spite of all the problems cited, the Palmgren-Miner linear damage rule is frequently used because of its simplicity and the experimental fact that other much more complex cumulative damage theories do not always yield a significant improvement in failure prediction reliability. A better understanding of the cumulative damage process may be obtained, however, by examining several other proposed cumulative damage theories.

8.3 CUMULATIVE DAMAGE THEORIES

If the Pamgren-Miner linear damage theory is plotted with damage fraction D as a function of cycle ratio n/N, the result is the straight line shown in Figure 8.2 as curve 2. Based on a survey of experimental results, however, the fatigue damage often accumulates nonlinearly, as indicated by curves 1 and 3 of Figure 8.2. Furthermore, experimental evidence indicates that damage curves on a plot such as Figure 8.2 are a function of cyclic stress amplitude levels, with lower curves corresponding to lower stress levels. In Figure 8.2, then, curve 1 would correspond to a higher stress level than curve 2 and curve 2 would correspond to a higher stress level than curve 3. To emphasize the implication of curves 1 and 3, it may be noted that a damage fraction of 0.4 would be predicted by the Palmgren-Miner Theory for a cycle ratio of $n/N = 0.4$, whereas the same cycle ratio would actually produce a damage fraction of 0.78 under conditions of curve 1 or a damage fraction of 0.08 under conditions of curve 3. Several theories have been proposed to approximate the nonlinear relationship between damage and cycle ratio.

Marco-Starkey Cumulative Damage Theory

One of the first nonlinear cumulative damage theories was proposed by Marco and Starkey (1). A similar hypothesis was also proposed by Richart and Newmark (2). The Marco-Starkey Theory is based on the following

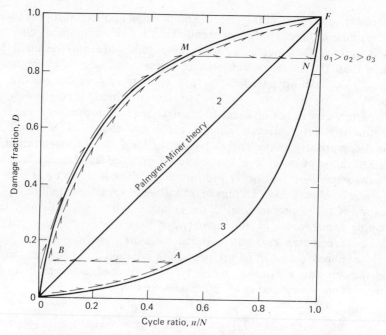

FIGURE 8.2. Fatigue damage as a function of cycle ratio.

postulates:

1. Damage curves for each level of completely reversed sinusoidal stress amplitude may be defined by the relationship

$$D = \left(\frac{n}{N}\right)^{m_i} \tag{8-5}$$

where m_i is a function of the stress level.

2. A specimen subjected to any sequence of completely reversed sinusoidal stresses will fail when D reaches unity.

3. Failure, or 100 percent damage, will be reached when $\Sigma(n/N)$ reaches a critical value, which may be approximated from the expression

$$\Sigma \frac{n}{N} \doteq \int_0^1 \frac{\left[1 + \dfrac{N_1}{N_2} + \dfrac{N_1}{N_3} + \cdots + \dfrac{N_1}{N_i}\right] dD}{\left[1 + \dfrac{N_1}{N_2} r_2 D\left(\dfrac{r_2 - 1}{r_2}\right) + \cdots + \dfrac{N_1}{N_i} r_i D\left(\dfrac{r_i - 1}{r_i}\right)\right]} \tag{8-6}$$

where $N_1, N_2, \cdots N_i =$ the numbers of cycles of completely reversed stresses $S_1, S_2, \cdots S_i$ to produce failure, with subscripts $1, 2, \cdots, i$ the order of application of the stress levels.

$D =$ damage

$r_i =$ ratio of exponents m_i/m_1 in the damage equations representing the two stress levels S_i and S_1.

$m_i =$ exponent in the damage equation associated with the ith stress level.

Use of (8-6) is very cumbersome, and many experimental data are required to properly evaluate the constants in the equation. Qualitatively, however, the use of this theory may be illustrated by referring again to Figure 8.2 where damage D is plotted versus cycle ratio n/N for three different stress levels. Consider, for example, two sequences of stress σ_1 and σ_3. For the first sequence, σ_1 is applied first for a cycle ratio of $(n/N) = 0.5$; then σ_3 is applied until failure takes place at $D = 1.0$. For the second sequence, σ_3 is applied first for a cycle ratio of $(n/N) = 0.5$, followed by operation at stress level σ_1 until failure occurs.

From Figure 8.2 for the first sequence path, O-M-N-F represents the process; and the cycle ratio to produce failure may be computed from the curve as

$$\Sigma \frac{n}{N} = \left(\frac{n}{N}\right)_{\sigma_1} + \left(\frac{n}{N}\right)_{\sigma_3} \tag{8-7}$$

or

$$\Sigma \frac{n}{N} = 0.5 + (1 - 0.98) \tag{8-8}$$

or

$$\Sigma \frac{n}{N} = 0.52 \tag{8-9}$$

For the second sequence of the same two stresses, path O-A-B-F represents the process; and the cycle ratio to produce failure may be computed from the curve as

$$\Sigma \frac{n}{N} = \left(\frac{n}{N}\right)_{\sigma_3} + \left(\frac{n}{N}\right)_{\sigma_1} \tag{8-10}$$

or

$$\Sigma \frac{n}{N} = 0.5 + (1 - 0.03) \tag{8-11}$$

or

$$\Sigma \frac{n}{N} = 1.47 \tag{8-12}$$

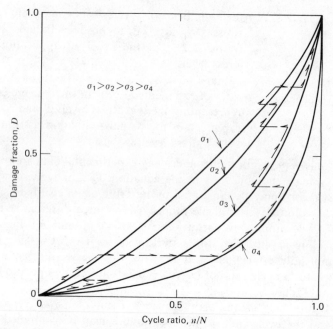

FIGURE 8.3. Damage history plotted for a sequence of many different cyclic stresses. Loading sequence: $\sigma_3 - \sigma_1 - \sigma_4 - \sigma_3 - \sigma_2 - \sigma_1 - \sigma_3$.

Recognizing that σ_1 is greater than σ_3, we may observe that this analysis indicates a summation of cycle ratios less than unity when the higher stress level is applied first and a cycle ratio greater than unity when the lower stress level is applied first in the sequence. This is in accord with experimental results from laboratory tests.

To summarize the concept of this theory, then, one obtains an estimate of the cycle ratio corresponding to failure by first plotting curves of damage fraction versus cycle ratio as a function of stress level, as shown in Figure 8.3. Each of these curves may be represented by an empirical relationship of the form of (8-5). The sequence of operating stress levels is then established and the damage history is plotted, as in Figure 8.3, by proceeding along the proper curves, in order, until a damage fraction of unity is reached. One moves from curve to curve along lines of constant damage (horizontal path lines).

Henry Cumulative Damage Theory

The cumulative damage theory proposed by Henry (3) is based on the concept that the S-N curve is shifted as fatigue damage accumulates and that fatigue damage may be defined as the ratio of the reduction in fatigue limit to the original fatigue limit of virgin material; that is,

$$D = \frac{E_o - E}{E_o} \qquad (8\text{-}13)$$

where D = damage
E_o = original fatigue limit
E = fatigue limit after damage

In the development of the Henry Theory, it was further assumed that the virgin S-N curve could be represented by the equation of an equilateral hyperbola referred to the stress axis and a line passing through E_o parallel to the cycle axis as the asymptotes of the hyperbola. Thus the equation assumed for the S-N curve has the form

$$N = \frac{k_o}{S - E_o} \qquad (8\text{-}14)$$

where N = number of cycles to failure at stress amplitude S
S = completely reversed amplitude of applied stress
k_o = material constant
E_o = original fatigue limit

It is implied in this development that *no damage* is accrued by operation at cyclic stress levels below the fatigue limit. Henry further assumed that the S-N curve after damage could also be represented by the equation of an equilateral hyperbola, whence

$$N_r = \frac{k}{S - E} \qquad (8\text{-}15)$$

where N_r = number of remaining cycles to failure at stress amplitude S
S = completely reversed amplitude of applied stress
k = material constant
E = damaged fatigue limit (reduced from E_o)

Based on the examination of some data, and making some heuristic arguments, Henry further asserted that it is approximately true that

$$\frac{k}{k_o} = \frac{E}{E_o} \qquad (8\text{-}16)$$

Based on the foregoing assumptions, the damage relationship proposed by Henry was developed as follows: If n cycles of stress amplitude S are applied to a specimen, the remaining life N_r at that stress amplitude is given by

$$N_r = N - n \qquad (8\text{-}17)$$

where N is the total number of cycles required to produce failure of the virgin material when subjected to stress amplitude S. Combining (8-17) with (8-15) gives

$$N - n = \left(\frac{k}{S - E} \right) \qquad (8\text{-}18)$$

Then, dividing through by N yields

$$1 - \frac{n}{N} = \frac{1}{N} \left(\frac{k}{S - E} \right) \qquad (8\text{-}19)$$

Now, substituting for N in (8-19) its value from (8-14) we obtain

$$1 - \frac{n}{N} = \frac{S - E_o}{k_o} \left(\frac{k}{S - E} \right) \qquad (8\text{-}20)$$

or

$$1 - \frac{n}{N} = \frac{k}{k_o} \frac{(S - E_o)}{(S - E)} \qquad (8\text{-}21)$$

Then, utilizing the assumption formulated in (8-15),

$$1 - \frac{n}{N} = \frac{E}{E_o} \frac{(S - E_o)}{(S - E)} \qquad (8\text{-}22)$$

which may be solved for E to give

$$E = \frac{S\left(1 - \dfrac{n}{N}\right)}{\left(\dfrac{S - E_o}{E_o}\right) + \left(1 - \dfrac{n}{N}\right)} \qquad (8\text{-}23)$$

Equation (8-23) is one useful form of the Henry theory. It gives an expression for the current value E of the fatigue limit after n cycles of stress amplitude S have been applied, if the total number of cycles to failure was originally N at stress level S and the original fatigue limit was E_o. Equation (8-23) may be readily converted to a damage equation by utilizing (8-13) to give

$$D = \frac{E_o - E}{E_o} = 1 - \frac{E}{E_o} = 1 - \frac{S\left(1 - \dfrac{n}{N}\right)}{(S - E_o) + E_o\left(1 - \dfrac{n}{N}\right)} \qquad (8\text{-}24)$$

which may be rearranged to give

$$D = \frac{\left(\dfrac{n}{N}\right)}{1 + \left(\dfrac{E_o}{S - E_o}\right)\left(1 - \dfrac{n}{N}\right)} \qquad (8\text{-}25)$$

where D = damage fraction
$\quad\quad n$ = number of cycles applied at stress amplitude S
$\quad\quad N$ = number of cycles to failure
$\quad\quad E_o$ = original fatigue limit
$\quad\quad S$ = applied stress amplitude

The Henry theory may be extended to a sequence of different stress levels by applying (8-23) or (8-24) successively in the order of applied stress levels. In this sequential procedure, the value of E_o must be updated after the application of each stress amplitude. Thus, a sequence of values for fatigue limit would be obtained, say $E_o, E_1, E_2, \cdots$, where E_o is the original fatigue limit, E_1 is the fatigue limit after applying n_1 cycles of stress level S_1, and so on. Such a procedure allows one to estimate the diminishing fatigue limit as damage accrues. A useful modification of the Henry theory involves passing all the S-N curves through the point S_u at 1 cycle, connecting this point to the computed value of E from (8-23) at 10^6 cycles by a straight line on a semilog S-N plot to obtain the complete S-N curve at each damage level.

Gatts Cumulative Damage Theory

Many similarities may be noted between the cumulative damage theory postulated by Gatts (4) and the Henry theory described previously. Gatts, however, postulated that the fatigue strength and fatigue limit change *continuously* with the application of stress cycles, and that the change is proportional to some function of the stress amplitude. These postulates give rise to the equation

$$\frac{dS_q}{dn} = -kD(S) \tag{8-26}$$

where S_q = instantaneous value of strength
$\quad\quad n$ = number of stress cycles applied
$\quad\quad k$ = proportionality constant
$\quad D(S)$ = damage function, a function of stress level

To utilize (8-26) it is necessary to explicitly define an appropriate damage function $D(S)$. Gatts proposed that a power function of the form $D(S) = (S - S_e)^\rho$ be assumed, which leads to an expression of (8-26) as

$$\frac{dS_q}{dn} = -k(S - S_e)^\rho \tag{8-27}$$

where S_e = fatigue limit
$\quad\quad k, \rho$ = empirical constants

To evaluate the constants in this expression, Gatts proposed that the damage function be related to strain energy associated with strains and stresses that exceed the level of the fatigue limit. Referring to Figure 8.4, then, (8-27) may be written as

$$\frac{dS_q}{dn} = -k' \int_{\epsilon_e}^{\epsilon_s} (\sigma_\epsilon - S_e) \, d\epsilon \tag{8-28}$$

Then making the assumption that $\sigma_\epsilon = E'\epsilon$, (8-28) may be rewritten as

$$\frac{dS_q}{dn} = -k_1 \int_{S_e}^{S} (\sigma - S_e) \, d\sigma \tag{8-29}$$

Thus the assumption has been made that the damage during the nth cycle is proportional to the cross-hatched area in Figure 8.4. Next, treating S_e as a constant, since it changes only a small amount during the nth cycle, the indicated integration of (8-29) is performed to give

$$\frac{dS_q}{dn} = -k(S - S_e)^2 \tag{8-30}$$

Next, it is assumed that

$$S_e = CS_q \tag{8-31}$$

where C is a material constant. Substituting this expression for S_e into (8-30) yields

$$\frac{dS_q}{dn} = -k(S - CS_q)^2 \tag{8-32}$$

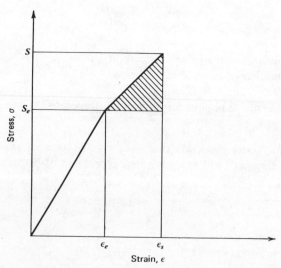

FIGURE 8.4. Stress-strain plot for nth cycle of stress as proposed by Gatts in developing a cumulative damage theory.

or, alternatively,

$$\frac{dS_e}{dn} = -k(S - S_e)^2 \qquad (8\text{-}33)$$

Defining the original fatigue limit to be S_{eo}, we integrate (8-33) to give

$$kn = \frac{1}{S - S_{eo}} - \frac{1}{S - S_e} \qquad (8\text{-}34)$$

where S_{eo} = fatigue limit when $n = 0$
  S_e = fatigue limit, a function of cyclic stress history, not a constant
  S = amplitude of applied cyclic stress
  n = number of cycles of stress applied

It was also asserted that the boundary conditions could be established as

$$S_q = S_u \text{ for } n = 0$$
$$S_q = S \text{ for } n = N \qquad (8\text{-}35)$$

Finally, then, the equation for the S-N curve was established from (8-34) and (8-35) as

$$kN = \frac{1}{S - S_{eo}} - \frac{1}{S(1 - C)} \qquad (8\text{-}36)$$

This is the equation for an ogee-shaped curve, as shown in Figure 8.5, which is in qualitative agreement with the experimentally established shapes of most S-N curves.

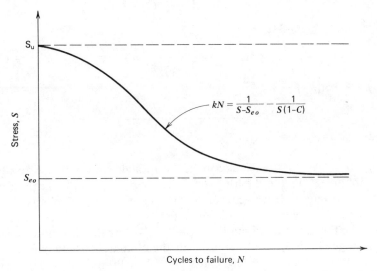

FIGURE 8.5. S-N curve approximation proposed by Gatts. (After ref. 4)

Gatts next step was to develop a normalized version of the S-N curve. To proceed with this, the following definitions were established.

$\gamma = \dfrac{S}{S_{eo}}$ = *stress amplitude ratio*, the ratio of the stress amplitude to the original value of the fatigue limit.

$\beta = \dfrac{n}{N}$ = *cycle ratio*, the ratio of number of cycles applied to number of cycles to failure at stress amplitude S.

$\gamma_e = \dfrac{S_e}{S_{eo}}$ = *fatigue limit ratio*, the ratio of current value of fatigue limit to original value of fatigue limit.

$L = \dfrac{N}{N^*}$ = *life ratio*, the ratio of number of cycles N required to cause failure at any given stress amplitude ratio γ to the number of cycles N^* required to cause failure at an arbitrary reference value of γ, namely γ^*.

With these definitions, (8-36) may be rewritten, with $K = kS_e$, as

$$KN = \frac{1}{\gamma - 1} - \frac{1}{\gamma(1 - C)} \tag{8-37}$$

Equation (8-37) may be evaluated for any arbitrary reference value of stress amplitude ratio γ^* as

$$KN^* = \frac{1}{\gamma^* - 1} - \frac{1}{\gamma^*(1 - C)} \tag{8-38}$$

Dividing (8-37) by (8-38), then,

$$K^*L = \frac{1}{\gamma - 1} - \frac{1}{\gamma(1 - C)} \tag{8-39}$$

where K^* is defined to be

$$K^* = \frac{1}{\gamma^* - 1} - \frac{1}{\gamma^*(1 - C)} \tag{8-40}$$

It may be noted that (8-30) is a nondimensionalized equation relating the stress amplitude ratio γ to the life ratio L. This equation allows the plotting of fatigue data from a wide range of materials on a common γ-L plot, as long as the constant C is the same for all the materials and the reference value γ^* is picked to be the same for all materials plotted. Noting that for most steels the original fatigue limit S_{eo} is about one-half the ultimate strength, we find that (8-31) yields a value of $C = 0.5$. Gatts plotted (8-39) and data from several sources for steel material as shown in Figure 8.6. It may be noted that the

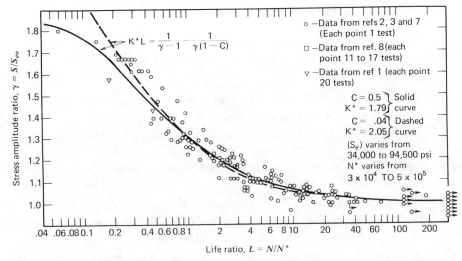

FIGURE 8.6. Theoretical γ-L curves (nondimensionalized S-N curves) and composite data from constant stress amplitude tests of steels (After ref. 4)

agreement is good between (8-39) and the data, even though a variety of different steels is plotted.

The useful result from the development of expression (8-39), then, is that the entire S-N curve for a new alloy may be quickly established to a good approximation by experimentally evaluating the two quantities S_{eo} and N^* and then utilizing the γ-L expression of (8-39) to evaluate the entire S-N curve for the material.

Finally, the damage expression proposed by Gatts is written by combining the results of (8-34) and (8-36), using the definitions already given for β and γ, as

$$\gamma_e = \gamma \left[1 - \frac{1}{\dfrac{\beta}{1-C} + \dfrac{\gamma}{\gamma-1}(1-\beta)} \right] \qquad (8\text{-}41)$$

Comparing this expression with (8-23), it may be observed that the Gatts damage equation and the Henry damage equation are very similar. In fact, if in (8-41) the assumption were made that $(S_e)_N = 0$ rather than $(S_e)_N = CS$, (8-41) would reduce to

$$\gamma_e = \frac{\gamma(1-\beta)}{\gamma-\beta} \qquad (8\text{-}42)$$

which is equivalent to the Henry theory expression of (8-23).

Corten-Dolan Cumulative Damage Theory

The Corten-Dolan (5) theory of cumulative damage is based on six assumptions that are, at least qualitatively, well supported by data. These assumptions are

1. A nucleation period (possibly a small number of cycles) may be required to initiate permanent fatigue damage.
2. The number of damage nuclei (submicroscopic voids) that form throughout the member increases as the stress is increased.
3. Damage at a given stress amplitude propagates at an increasing rate with increased numbers of cycles.
4. The rate of damage per cycle increases as the stress is increased.
5. The total damage that constitutes failure in a given member is a constant for all possible stress histories that could be applied.
6. Damage will continue to be *propagated* at stress levels that are lower than the minimum stress required to *initiate* damage.

To simplify development of their hypothesis, Corten and Dolan initially assumed the nucleation period to be zero, or

$$N' = 0 \tag{8-43}$$

where N' = number of cycles of stress level S required to nucleate damage

A power-law relationship was next assumed to exist between damage per nucleus and applied cycles to give

$$D' = rN^a \tag{8-44}$$

where D' = damage per nucleus
r = coefficient of damage propagation rate, a function of stress level
N = number of cycles of stressing, corresponding to the damage
a = damage propagation exponent

From (8-44), if the total number of fatigue damage nuclei is defined to be m, the total fatigue damage D is given by

$$D = mD' = mrN^a \tag{8-45}$$

For example, then, for two different constant stress amplitudes S_1 and S_2 damage at failure D_f may be expressed as

$$D_f = m_1 r_1 N_1^{a_1} = m_2 r_2 N_2^{a_2} \tag{8-46}$$

Based on this expression, the cumulative damage fraction D may be plotted as a function of cycles, as shown in Figure 8.7, where the two curves represent the two cyclic stress amplitudes S_1 and S_2. To illustrate the development of this theory, a simple two-level cyclic stress history is applied in blocks of cycles, as shown in Figure 8.8. Note that the following definitions may be

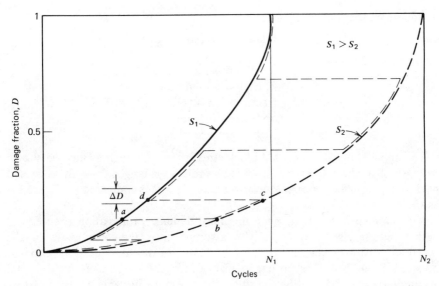

FIGURE 8.7. Plot of damage versus cycles for two different stress levels according to the Corten-Dolan theory.

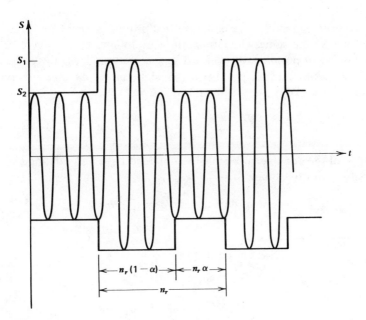

FIGURE 8.8. A simple two-level cyclic stress spectrum used to develop the concepts of the Corten-Dolan cumulative damage theory.

255

written from the sketch of Figure 8.8:

$$S_1 = \text{higher stress amplitude}$$
$$S_2 = \text{lower stress amplitude}$$
$$n_r = \text{total number of cycles}$$
$$\text{in each repeated block}$$
$$\alpha = \text{fraction of } n_r \text{ cycles that}$$
$$\text{are incurred at the}$$
$$\text{higher stress level } S_1$$

The application of a repeated block stress spectrum, such as the one shown in Figure 8.8 where the stress level is periodically changed from S_1 to S_2 and back to S_1, may be represented on a damage plot, as indicated in Figure 8.7 as the dashed line.

It should be noted that, once initiated, damage propagation proceeds at both stress levels S_1 and S_2. The assumption is therefore made in the remainder of this development that the number of damage nuclei m_1 initiated by stress level S_1 will be propagated at both stress levels. Further, the total accumulated damage will be the sum of the damage increments produced by operation at stress level S_1 plus the damage increments produced by operation at stress level S_2. One convenient way to sum the damage increments is to find the equivalent number of cycles at S_1 that corresponds to identical damage due to the actual number of cycles $n_r (1 - \alpha)$ at stress level S_2. If this is successfully accomplished, it is possible to work exclusively along the S_1 curve of Figure 8.7.

Referring to Figure 8.7, we can note that a damage increment ΔD may be defined that is the same whether the process follows the S_1 curve or the S_2 curve. For small numbers of cycles ΔN, a corresponding damage increment ΔD may be written as the product of the slope of the damage curve and the cycle increment. Thus, for small values of ΔN

$$\frac{\Delta D}{\Delta N} = \text{slope of damage curve} \tag{8-47}$$

The slope of the damage curve may be obtained by differentiating (8-45) with respect to cycles to obtain

$$\text{slope of damage curve} = \frac{d}{dN}(mrN^a) \tag{8-48}$$

or

$$\text{slope of damage curve} = mraN^{a-1} \tag{8-49}$$

Combining (8-47) with (8-49), then,

$$\frac{\Delta D}{\Delta N} = mraN^{a-1} \tag{8-50}$$

For point d on the S_1 curve in Figure 8.7, based on (8-50), the damage increment may be written as

$$\Delta D = m_1 a_1 r_1 N_d^{a_1 - 1} \Delta N_1 \tag{8-51}$$

Similarly, for point C on the S_2 curve the damage increment may be written as

$$\Delta D = m_1 a_2 r_2 N_c^{a-1} N_2 \tag{8-52}$$

where it should be noted that m_1 is used as the number of propagating damage nuclei rather than m_2, since m_1 nuclei have already been initiated by operation at stress level S_1 and all are assumed to continue to propagate at S_2. Now, since the damage increment ΔD is the same whether the process follows the S_1 curve or the S_2 curve, the expressions of (8-51) and (8-52) may be equated to yield

$$m_1 a_1 r_1 N_d^{a_1 - 1} \Delta N_1 = m_1 a_2 r_2 N_c^{a_2 - 1} \Delta N_2 \tag{8-53}$$

This expression may then be solved for ΔN_1 to obtain

$$\Delta N_1 = \left(\frac{a_2}{a_1} \right) \left(\frac{r_2}{r_1} \right) \frac{N_c^{a_2 - 1}}{N_d^{a_1 - 1}} \Delta N_2 \tag{8-54}$$

Also, from (8-45), since the damage at c and d in Figure 8.7 is the same,

$$D = m_1 r_1 N_d^{a_1} = m_1 r_2 N_c^{a_2} \tag{8-55}$$

from which

$$N_d = \left(\frac{r_2}{r_1} \right)^{1/a_1} = N_c^{a_2/a_1} \tag{8-56}$$

Substituting the result of (8-56) into (8-54) then gives

$$\Delta N_1 = \left(\frac{a_2}{a_1} \right) \left(\frac{r_2}{r_1} \right) \frac{N_c^{a_2 - 1}}{\left[\left(\frac{r_2}{r_1} \right)^{1/a_1} N_c^{a_2/a_1} \right]^{a_1 - 1}} \Delta N_2 \tag{8-57}$$

which may be reduced to

$$\Delta N_1 = \left(\frac{a_2}{a_1} \right) \left(\frac{r_2}{r_1} \right)^{1/a_1} N_c^{(a_2/a_1) - 1} \Delta N_2 \tag{8-58}$$

To simplify this expression the following two quantities are next defined:

$$A \equiv \frac{a_2}{a_1} \tag{8-59}$$

$$R \equiv \frac{r_2}{r_1} \tag{8-60}$$

With these definitions, expression (8-58) may be written as

$$\Delta N_1 = A R^{1/a_1} N_c^{A - 1} \Delta N_2 \tag{8-61}$$

Using (8-61) together with (8-45), then, we may write an expression for the summation of all damage increments referred to the S_1 damage curve in Figure 8.7:

$$D = \sum_{i=1}^{g} \Delta D_i = m_1 r_1 \Big[N_g \alpha + AR^{1/a_1}(1 - \alpha)n_r \{ n_r^{A-1} + (2n_r)^{A-1} + (3n_r)^{A-1} \\ + \cdots + (gn_r)^{A-1} \} \Big]^{a_1}$$
(8-62)

where N_g = total number of stress cycles in all "g" repeated blocks applied to the member

g = number of repeated blocks in the spectrum

$n_r, 2n_r, 3n_r, \cdots$ = appropriate values of N_c (8-60) for each repeated block in the summation

The series in (8-62) may be evaluated and simplified to yield

$$D = m_1 r_1 \Big[N_g \alpha + R^{1/a_1}(1 - \alpha)N_g^A \Big]^{a_1}$$
(8-63)

Recall now that one of the six basic assumptions was that the damage corresponding to failure is constant no matter what stress spectrum is imposed. Therefore, damage at failure may be evaluated by utilizing a constant stress amplitude S_1 and the associated number of cycles to failure, N_1. By hypothesis, then, the damage at failure calculated in this way will be equal to the damage at failure calculated by (8-63). Evaluating both (8-45) and (8-63) at the time of failure and equating the results then gives

$$D_f = m_1 r_1 N_1^{a_1} = m_1 r_1 \Big[(N_g)_f \alpha + R^{1/a_1}(1 - \alpha)(N_g)_f^A \Big]^{a_1}$$
(8-64)

where $(N_g)_f$ = total number of cycles to failure for all repeated blocks

Simplification of (8-64) then yields

$$N_1 = \alpha(N_g)_f + R^{1/a_1}(1 - \alpha)(N_g)_f^A$$
(8-65)

It may be observed that if a_1 and a_2 were approximately equal, the magnitude of A would be approximately 1, and (8-65) would become

$$(N_g)_f = \frac{N_1}{\alpha + R^{1/a}(1 - \alpha)}$$
(8-66)

For a wide range of stress amplitudes, especially for steel materials, it does turn out that the assumption $a_1 \doteq a_2 \doteq a$ is well verified experimentally (5).

At the outset of the Corten-Dolan development, the nucleation period N' was assumed to be nil. However, it may be true that a substantial number of cycles will be required to initiate the fatigue damage nuclei. To account for this possibility, the term $(N_i - N_i')$ is substituted for N_i in (8-46) to give

$$D_f = m_1 r_1 (N_1 - N_1')^{a_1} = m_2 r_2 (N_2 - N_2')^{a_2}$$
(8-67)

If it is now hypothesized that the damage nucleation period under a spectrum of fluctuating stress amplitudes is the same as the nucleation period under the highest constant stress amplitude S_1 (somewhat justified because the number of damage nuclei is postulated to be a function of the highest stress amplitude S_1), and if $A = 1$, (8-64) may be rewritten as

$$(N_g)_f = \frac{(1 - \beta)N_1}{\alpha + R^{1/a}(1 - \alpha)} + \left(\frac{\beta}{\alpha}\right)N_1 \qquad (8\text{-}68)$$

where

$$\beta \equiv \frac{N_1'}{N_1} \qquad (8\text{-}69)$$

Equation (8-68) is developed by substituting $(N_g - N_1'/\alpha)$ for N_g in (8-64) and solving for $(N_g)_f$. This substitution is justified because a total of N_1'/α cycles at stresses S_1 and S_2 is required to produce N_1' cycles at S_1, and it is postulated that N_1' cycles at S_1 are required to initiate the damage nuclei. It may be noted that (8-68) is very similar to (8-66) except that (8-68) provides for a nucleation period.

Employing a substantial body of experimental data, using the up-and-down testing method and 20 specimens per data point to establish 95 percent confidence limits on the mean, and using steel wire specimens, Corten and Dolan established very good agreement (5) between theory and experiment with the assumptions that $A = 1$ and $\beta = 0$. That is, they established that over a wide range of stress levels for the steel material tested, (8-66) was very accurate. Similar conclusions were reached by Dolan later in testing samples of 7075-T6 aluminum alloy*

It was further suggested by Corten and Dolan that the stress-dependent ratio R might be equivalent to another ratio involving stresses S_1 and S_2 that could be more easily evaluated. Pursuing this thought, they investigated the relationship

$$R^{1/a} = \left(\frac{S_2}{S_1}\right)^d \qquad (8\text{-}70)$$

The data obtained from their experimental tests on steel material are presented in Figure 8.9, and some later data on aluminum material are shown in Figure 8.10. From these plots it was concluded that a single value of d in (8-70) could be established for any given material. Since (8-70) may be written as

$$\log R^{1/a} = d\log\left(\frac{S_2}{S_1}\right) \qquad (8\text{-}71)$$

the slopes of the lines in Figures 8.9 and 8.10 yield the value of d for the

*See p. 438 of ref. 14.

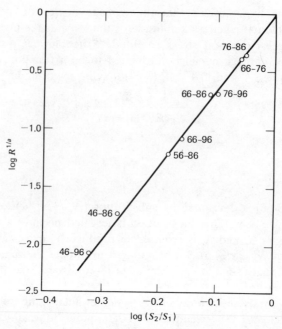

FIGURE 8.9. Correlation between $\log(S_2/S_1)$ and $\log R^{1/a}$ in Corten-Dolan theory for steel alloy wire (Brite Basic), 0.05 inch diameter. Numbers adjacent to points indicate magnitudes in thousands of psi of the fluctuating stresses used in the experiments. (After ref. 5)

materials represented. It may be noted that a mean value of $d = 6.57$ represents the steel material with a range on d of 6.2 to 6.9. The mean value for the aluminum alloy is 6.0.

Using the same concepts that led to (8-66) and the result of (8-70) Corten and Dolan developed an expression for estimating the number of cycles to failure for repeated blocks of many different stress levels:

$$(N_g)_f = \frac{N_1}{\alpha_1 + \alpha_2(S_2/S_1)^d + \alpha_3(S_3/S_1)^d + \cdots + \alpha_i(S_i/S_1)^d} \quad (8\text{-}72)$$

where $(N_g)_f$ = total number of cycles to failure under the conditions of imposed fluctuating stress amplitude history

d = material constant

N_1 = cycles to failure at the highest stress amplitude S_1.

$\alpha_1, \alpha_2, \cdots, \alpha_i$ = the fraction of cycles imposed at stresses $S_1, S_2, \cdots, S_i$, respectively.

The assumption has been made in this expression that $A = 1$ and $\beta = 0$. In utilizing this theory, an experimental value of d must be obtained for each material of interest.

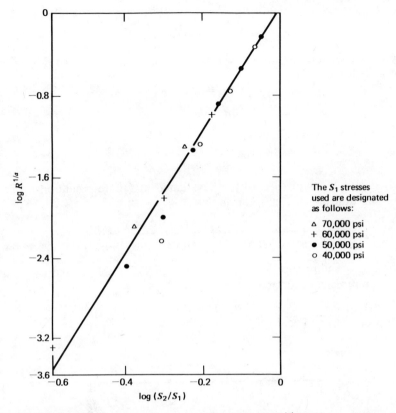

FIGURE 8.10. Correlation between $\log(S_2/S_1)$ and $\log R^{1/a}$ in Corten-Dolan theory for aluminum alloy 7075-T6. (After ref. 14, p. 438)

For purposes of comparison with the Marin theory,* it will be convenient to express (8-72) in a different form. To do this it may be noted that

$$\alpha_1 n_r + \alpha_2 n_r + \alpha_3 n_r + \cdots + \alpha_i n_r = n_r \qquad (8\text{-}73)$$

where n_r is the number of cycles in one repeated block and the α_i's are the fractions of the number of cycles associated with the stress levels S_i. Also

$$g n_r = N_g \qquad (8\text{-}74)$$

where g is the number of repeated blocks and N_g is the total number of cycles imposed under the entire fluctuating stress history. Multiplying (8-73) by g yields

$$\alpha_1 g n_r + \alpha_2 g n_r + \cdots + \alpha_i g n_r = g n_r \qquad (8\text{-}75)$$

*See p. 200 of ref. 6.

whence, by (8-74),

$$\alpha_1 N_g + \alpha_2 N_g + \cdots + \alpha_i N_g = N_g \tag{8-76}$$

Also, if n_i is the total number of cycles imposed at each stress level S_i, it may be observed that

$$n_1 = \alpha_1 N_g$$
$$n_2 = \alpha_2 N_g$$
$$\vdots \tag{8-77}$$
$$n_i = \alpha_i N_g$$

Dividing both sides of (8-72) by N_g evaluated at failure conditions, that is, $(N_g)_f$, and utilizing (8-77), it is true that

$$1 = \frac{N_1}{N_1 + n_2(S_2/S_1)^d + n_3(S_3/S_1)^d + \cdots + n_i(S_i/S_1)^d} \tag{8-78}$$

Further, if the numerator and denominator of (8-78) are divided by N_1 and the equation rearranged, one obtains

$$\left(\frac{n_1}{N_1}\right) + \left(\frac{n_2}{N_1}\right)\left(\frac{S_2}{S_1}\right)^d + \left(\frac{n_3}{N_1}\right)\left(\frac{S_3}{S_1}\right)^d + \cdots + \left(\frac{n_i}{N_1}\right)\left(\frac{S_i}{S_1}\right)^d = 1 \tag{8-79}$$

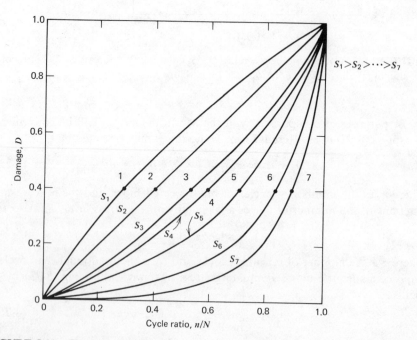

FIGURE 8.11.   Damage as a function of cycle ratio for many different stress levels.

This form of the Corten-Dolan cumulative damage relationship will be used later for comparison with the Marin cumulative damage theory.

Marin Cumulative Damage Theory

Marin's cumulative damage theory (10) is based on a consideration of the relationship between damage as a function of cycle ratio and changes in the *S-N* curve due to damage accumulation. If experimentally determined damage curves for several different stress levels were plotted as shown in Figure 8.11, a line of constant damage could be selected, for example $D = 0.4$, and a distinct point could be identified on each of the curves, every point corresponding to a damage $D = 0.4$. These points are noted in Figure 8.11 as 1 through 7. These seven points could then be plotted on an *S-N* plot at the seven stress levels corresponding to the seven damage curves shown in Figure 8.11. Such an *S-N* plot is illustrated in Figure 8.12 by the curve labeled $D = 0.4$. The *S-N* curve plotted in this way is therefore a curve of constant damage $D = 0.4$. The original *S-N* curve for the material may be considered to be a line of constant damage $D = 1.0$. Other lines of constant damage may be established by noting the intersections of the seven curves with the selected damage level in Figure 8.11 and plotting the results in Figure 8.12. A family of constant damage lines may be established in this way and plotted as shown in Figure 8.12.

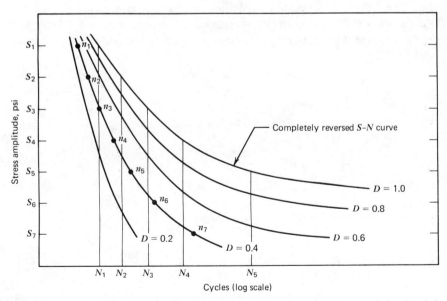

FIGURE 8.12. *S-N* plot showing lines of constant damage developed by Marin's theory.

The concept of constant damage lines on the S-N plot leads directly to the observation that the damage produced by n_i cycles of operation at any stress level S_i is exactly equivalent to n_1 cycles of stress level S_1, where $n_1, n_2, n_3, \cdots n_i$ are all points on the constant damage curve. It is possible, therefore, to find an equivalent number of cycles at a reference stress level that would produce the same damage as n_i cycles of operation at the actual stress level S_i. Using this concept, and picking the highest stress level as the reference stress level S_1, Marin writes,

$$n_{2e} = n_2\left(\frac{S_2}{S_1}\right)^y$$

$$n_{3e} = n_3\left(\frac{S_3}{S_1}\right)^y$$

$$\vdots$$

$$n_{ie} = n_i\left(\frac{S_i}{S_1}\right)^y \tag{8-80}$$

where n_{ie} = number of cycles of operation at reference stress S_1 to produce damage equivalent to n_i actual cycles at stress level S_i
n_i = actual number of cycles of operation at stress level S_i
S_i = actual operating stress level at which damage is produced

It may be observed in (8-80) that if y is greater than or equal to 1, n_{ie} will always be less than n_i because by definition of the reference stress level S_1 it is always larger than any other stress level S_i. Based on (8-80), Marin next defined a sequence of damage ratios corresponding to operation at each stress level as

$$R_1 = \frac{n_1}{N_1}$$

$$R_2 = \frac{n_{2e}}{N_1}$$

$$R_3 = \frac{n_{3e}}{N_1}$$

$$\vdots \tag{8-81}$$

$$R_i = \frac{n_{ie}}{N_1}$$

It is asserted, then, that the summation of these life ratios must equal unity at the time of failure. Hence, failure is predicted to occur when

$$R_1 + R_2 + R_3 + \cdots + R_i = 1 \tag{8-82}$$

Substituting the expression of (8-81) into (8-82) then yields

$$\frac{n_1}{N_1} + \frac{n_{2e}}{N_1} = \frac{n_{3e}}{N_1} + \cdots + \frac{n_{ie}}{N_1} = 1 \tag{8-83}$$

and finally, replacing the values of n_{ie} by their expressions form (8-80) gives a prediction of failure by the Marin theory when

$$\left(\frac{n_1}{N_1}\right) + \left(\frac{n_2}{N_1}\right)\left(\frac{S_2}{S_1}\right)^y + \left(\frac{n_3}{N_1}\right)\left(\frac{S_3}{S_1}\right)^y + \cdots + \left(\frac{n_i}{N_1}\right)\left(\frac{S_i}{S_1}\right)^y = 1 \tag{8-84}$$

Comparing (8-84) with the result of the Corten-Dolan development shown in (8-79) we may observe that the results are exactly the same if the exponent y in the Marin equation is made equal to the exponent d in the Corten-Dolan expression.

The Marin expression was developed further by assuming that the S-N curve could be adequately approximated by an equation of the form

$$S^x N = k \tag{8-85}$$

Based on this assumption, it is true that

$$S_2^x N_2 = S_1^x N_1$$
$$S_3^x N_3 = S_1^x N_1$$
$$\vdots \tag{8-86}$$
$$S_i^x N_i = S_1^x N_1$$

Solving these expressions for the values of N_i and substituting the results appropriately for the N_1 values in (8-84) gives

$$\left(\frac{n_1}{N_1}\right) + \left(\frac{n_2}{N_2}\right)\left(\frac{S_2}{S_1}\right)^q + \left(\frac{n_3}{N_3}\right)\left(\frac{S_3}{S_1}\right)^q + \cdots + \left(\frac{n_i}{N_i}\right)\left(\frac{S_i}{S_1}\right)^q = 1 \tag{8-87}$$

where

$$q = y - x \tag{8-88}$$

It is interesting to note that (8-87) reduces to Miner's hypothesis if $q = 0$, that is, if $y = x$.

From (8-87), then, failure may be predicted under conditions of spectrum loading if one knows the material constant q and the completely reversed virgin S-N curve for the material. If one is interested in the *remaining life* at a given stress level S_i after having operated under some known loading spectrum, it may be obtained by solving (8-87) for n_i to give

$$n_{ir} = N_i\left[\left(1 - \frac{n_1}{N_1}\right)\left(\frac{S_1}{S_i}\right)^q - \left(\frac{n_2}{N_2}\right)\left(\frac{S_2}{S_i}\right)^q - \cdots - \left(\frac{n_{i-1}}{N_{i-1}}\right)\left(\frac{S_{i-1}}{S_i}\right)^q\right] \tag{8-89}$$

where n_{ir} = remaining life in cycles at stress level S_i after having applied n_1
    cycles at S_1, n_2 cycles at $S_2, \cdots$, and n_{i-1} cycles at S_{i-1}

Manson Double Linear Damage Rule

In 1960 it was proposed by Grover (11) that cumulative damage estimates might be improved by breaking the fatigue process down into a crack initiation phase and a crack propagation phase and applying a linear damage rule to each phase separately. No basis was suggested for quantitatively defining the ranges for these two phases, however, until Manson (12, 13) presented an empirical technique for establishing the ranges and the damage equations for these two phases. Manson suggested that the crack propagation period could be expressed as

$$N_p = PN_f^p \tag{8-90}$$

where N_p = number of cycles to propagate a crack to failure *after* it has been
    initiated
N_f = total number of cycles to failure
P = propagation coefficient to be determined experimentally
p = propagation exponent to be determined experimentally

It was next noted that the crack initiation period N' could then be written as

$$N' = N_f - N_p \tag{8-91}$$

or, using (8-90),

$$N' = N_f - PN_f^p \tag{8-92}$$

The propagation exponent p was selected by Manson to be 0.6, based on an integrated consideration of data for many different materials. The propagation coefficient P was experimentally determined to best fit experimental data from two-stress-level spectrum testing (12, 13) and found to have a best-fit value of 14. Thus, (8-90) and (8-92) become

$$N_p = 14N_f^{0.6} \tag{8-93}$$

and

$$N' = N_f - 14N_f^{0.6} \tag{8-94}$$

Additional experimentation led Manson to restrict these equations to a range of failure lives N_f that exceed about 730 cycles. At total lives to failure less than 730 cycles, the crack initiation seemed to occur during the first stress cycle because of the high stress level associated with such short lives, and the entire life N_f seemed to be associated with the propagation phase. To summarize,

$$\left. \begin{array}{l} N' = N_f - 14N_f^{0.6} \\ N_p = 14N_f^{0.6} \end{array} \right\} \text{ for } N_f > 730 \text{ cycles} \tag{8-95}$$

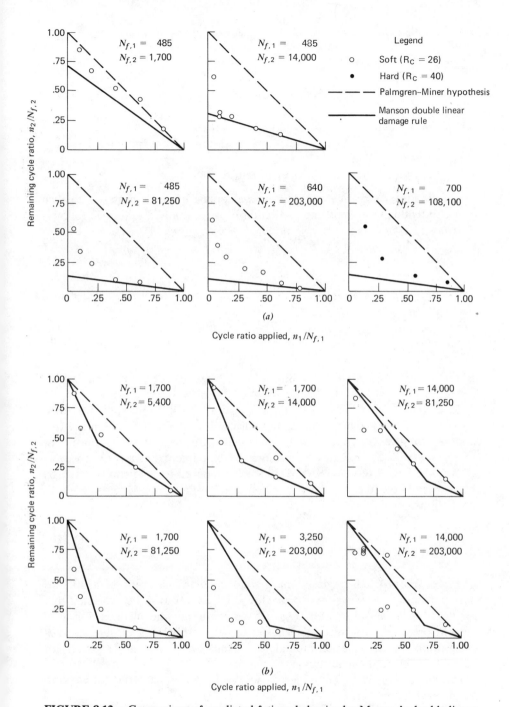

FIGURE 8.13. Comparison of predicted fatigue behavior by Manson's double linear damage rule and Palmgren-Miner linear damage rule with experimental data for two-stress-level tests on SAE 4130 steel. R. R. Moore rotating-bending: (*a*) high to low stress with low initial life; (*b*) high to low stress with relatively high initial life. (After ref. 13, copyright ASTM; adapted with permission)

and

$$\left.\begin{array}{l} N' = 0 \\ N_p = N_f \end{array}\right\} \text{ for } N_f \le 730 \text{ cycles} \qquad (8\text{-}96)$$

where N' = number of cycles to initiate a crack

N_p = number of cycles to propagate a crack from the initiation stage to failure

N_f = total number of cycles to failure

Utilizing these empirical expressions, we apply a linear damage rule to each phase individually to yield the prediction of initiation and failure as follows: *Fatigue nuclei of critical size are initiated when*

$$\sum_{i=1}^{m} \frac{n_i}{N_i'} = 1 \qquad (8\text{-}97)$$

and *fatigue cracks are propagated to failure if cracks of critical size have been initiated and then*

$$\sum_{j=1}^{q} \frac{n_j}{(N_p)_j} = 1 \qquad (8\text{-}98)$$

where in each case n is the number of cycles applied at the ith or j^{th} stress level.

Thus, (8-97) must be applied until a critical crack is initiated, and then (8-98) is used until failure is predicted. Use of this double linear damage rule appears to give relatively good agreement with two-stress-level tests for several materials (13). Figure 8.13 shows the experimental results for a variety of two-stress-level tests on SAE 4130 steel compared to predictions made by Manson's double linear damage rule and also the Palmgren-Miner hypothesis predictions.

8.4   USING THE IDEAS

Consider the problem of designing a solid circular link made of 4340 steel heat treated to a hardness of Rockwell C-35. The link is to be subjected to a spectrum of axial loads, and it is desired to design the member for a 99 percent probability of survival. The 99 percent probability of survival *S-N* design data based on experimental test results are shown in Table 8.1 for completely reversed cyclic stresses.

The actual link is to be subjected to the following spectrum of loading during each duty cycle:

$$P_A = 22,000 \text{ lb for } 1200 \text{ cycles}$$

$$P_B = 12,000 \text{ lb for } 7,000 \text{ cycles}$$

$$P_C = 6500 \text{ lb for } 50,000 \text{ cycles} \qquad (8\text{-}99)$$

Table 8.1. Fatigue S-N Data for 99 Percent Probability of Survival (4340 steel; Rockwell C-35)

Stress Amplitude, S (psi)	N, cycles
168,000	100
160,000	1,350
150,000	3,500
140,000	7,100
130,000	14,200
120,000	28,000
110,000	55,500
100,000	110,000
90,000	216,000
80,000	440,000
70,000	1,980,000
68,000	∞

This duty cycle is to be repeated three times during the life of the bar. It is desired to find the required cross-sectional area for 99 percent probability of survival using the Palmgren-Miner theory, Henry theory, Gatts theory, Corten-Dolan theory, Marin theory, and Manson theory.

To proceed with the solution, the 99 percent probability of survival S-N curve is plotted as shown in Figure 8.14. The procedure used in determining the area required for the cross section of the link is an iterative process for each of the six theories. To estimate the required area based on the Palmgren-Miner theory, the cross-sectional area is arbitrarily assumed to be 0.2 inch^2 as a first guess. This permits a first-cut calculation of stresses for use with (8-98) as

$$S_{A_1} = \frac{P_A}{A_1} = \frac{22,000}{0.2} = 110,000 \text{ psi for 1200 cycles}$$

$$S_{B_1} = \frac{P_B}{A_1} = \frac{12,000}{0.2} = 60,000 \text{ psi for 7000 cycles}$$

$$S_{C_1} = \frac{P_C}{A_1} = \frac{6500}{0.2} = 32,500 \text{ psi for 50,000 cycles} \qquad (8\text{-}100)$$

Three repetitions of this block of stresses are to be imposed during the life of the link. Utilizing (8-100) and the S-N curve of Figure 8.14, we may make the tabulation shown in Table 8.2 for the first iteration, and from the last column of Table 8.2,

$$\sum \frac{n}{N} = 0.065 \qquad (8\text{-}101)$$

Since the sum of cycle ratios is less than 1, the selected area was too large and a new smaller area must be tried.

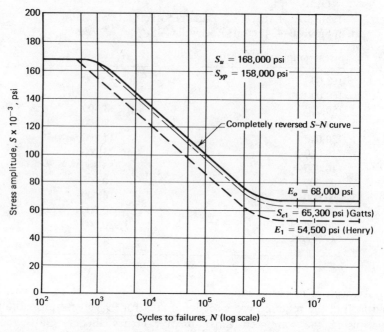

FIGURE 8.14. *S-N* curve plotted for 99 percent probability of survival for 4340 steel material used in illustrative link design. The first-step "damaged" *S-N* curves are shown for both the Henry and Gatts damage theories.

Table 8.2. First Iteration Values for Palmgren-Miner Solution

Load Level	P (lb)	A_1 (in.^{2})	S_1 (psi)	N_1 (cycles)	n_{r1} (cycles)	n_1 (cycles)	n_1/N_1
A	22,000	0.2	110,000	55,500	1,200	3,600	0.065
B	12,000	0.2	60,000	∞	7,000	21,000	0
C	6,500	0.2	32,500	∞	50,000	150,000	0

Try next $A_2 = 0.15$ inch^2. The data then are tabulated as shown in Table 8.3, and from the last column

$$\sum \frac{n}{N} = 0.85 \qquad (8\text{-}102)$$

This sum is much closer to unity but the area is still slightly large.

Next try $A_3 = 0.148$ and tabulate the data as in Table 8.4. From the last column of Table 8.4,

$$\sum \frac{n}{N} = 1.00 \qquad (8\text{-}103)$$

Based on the Palmgren-Miner theory, therefore, the design area would be 0.148 inch^2.

Table 8.3. Second Iteration Values for Palmgren-Miner Solution

Load Level	P (lb)	A_2 (in.^{2})	S_2 (psi)	N_2 (cycles)	n_{r_2} (cycles)	n_2 (cycles)	n_2/N_2
A	22,000	0.15	146,700	4,500	1,200	3,600	0.80
B	12,000	0.15	80,000	445,000	7,000	21,000	0.05
C	6,500	0.15	43,300	∞	50,000	150,000	0

Table 8.4. Third Iteration Values for Palmgren-Miner Solution

Load Level	P (lb)	A_3 (in.^{2})	S_3 (psi)	N_3 (cycles)	n_{r3} (cycles)	n_3 (cycles)	n_3/N_3
A	22,000	0.148	148,700	3,800	1,200	3,600	0.948
B	12,000	0.148	81,200	400,000	7,000	21,000	0.053
C	6,500	0.148	44,000	∞	50,000	150,000	0

To estimate the required area for the link based on the Henry theory, (8-23) and (8-13) are repetitively employed in conjunction with the S-N curve. As a first estimate of area, the Palmgren-Miner result of $A = 0.148$ will be utilized. We then construct Table 8.5.

It is clear that failure occurs at the end of load level C in loading block 2 because the number of cycles to failure at that stress level is 50,000 and 50,000 cycles are imposed. Thus the residual fatigue limit is zero and by (8-13) the damage is unity, which implies failure. The conclusion is that the selected area is too small and a new larger area must be tried for the next iteration. The design objective is to pick the area so that the cycle ratio n/N reaches a value of unity just after the final cycle of load level C during block 3 has been applied.

To estimate the required area for the link based on the Gatts theory the procedure is very similar to that just described for the Henry theory. Failure is predicted when γ_e becomes zero in the equation of Gatts (8-41). Table 8-6 may be constructed for a first iteration using an area of 0.5 inch^2 and a value of $C = S_{eo}/S_u = 0.4$

The procedure indicated in the table is carried out sequentially for all three blocks of loading. The design objective is to achieve a value of $\beta = 1$ just as the final cycle of load level C during block 3 is completed. Under these circumstances, (8-41) evaluated at failure becomes $\gamma_e = C\gamma$, or failure is predicted to occur when the residual value of fatigue limit is reduced to CS in magnitude.

To obtain the required area for the link based on the Corten-Dolan theory (8-79) is utilized. Selecting a trial area of 0.15 in.^{2} and using the mean value of

Table 8.5. First Iteration Values for Henry Solution

E (psi)	Load Level	Block Number	P (lb)	A_1 (in.^{2})	S_1 (psi)	N_1 (cycles)	n_1	n_1/N_1	$(S-E_o)/E_o$
$E_o = 68,000$	A	1	22,000	0.148	148,700	3,800	1,200	0.316	1.19
$E_1 = 54,500$	B	1	12,000	0.148	81,200	160,000	7,000	0.044	0.49
$E_2 = 53,200$	C	1	6,500	0.148	44,000	∞	50,000	0	—
$E_3 = 53,200$	A	2	22,000	0.148	148,700	1,400	1,200	0.857	1.80
$E_4 = 10,950$	B	2	12,000	0.148	81,200	8,000	7,000	0.875	6.40
$E_5 = 1,550$	C	2	6,500	0.148	44,000	50,000	50,000	1	(failure)
$E_6 = 0$									

Table 8.6. First Iteration Values for Gatts Solution

S_e	Load Level	Block number	P	A_1	S_1	N_1	n_1	β_1	γ_1	γ_e
$S_{eo} = 68,000$	A	1	22,000	0.15	146,700	4,500	1,200	0.266	2.16	0.960
$S_{e1} = 65,300$	B	1	12,000	0.15	80,000	350,000	7,000	0.02	1.22	0.998
$S_{e2} = 65,200$	C	1	6,500	0.15	43,300	∞	50,000	0	0.66	1.0
$S_{e3} = 65,200$	A	2	22,000	0.15	146,700	3,600	1,200	0.33	2.25	0.96
$S_{e4} = 62,500$	B	2	.	.	.					

$d = 6.57$, the Corten-Dolan equation for the first iteration becomes

$$D = \left(\frac{3600}{4500}\right) + \left(\frac{21,000}{4500}\right)\left(\frac{80,000}{146,700}\right)^{6.57} + \left[\left(\frac{150,000}{4500}\right)\right.$$

$$\times \left.\left(\frac{43,300}{146,700}\right)^{6.57}\right] \overset{?}{=} 1 \tag{8-104}$$

from which

$$D = 0.89 \tag{8-105}$$

Since this is less than 1, the area selected was a little too large. Additional trials must be made until the selected area yields a value of unity from (8-79).

To utilize the Marin theory as expressed in (8-87), the S-N data are first plotted on logarithmic coordinates, as shown in Figure 8.15, to determine the magnitude of the exponent x, which for this material is 8.05. Assuming a value of $y = 6.57$ based on Corten-Dolan data, the Marin expression of (8-87) may be evaluated for an assumed area of 0.15 inch^2 as

$$D = \left(\frac{3600}{4500}\right) + \left(\frac{21,000}{445,000}\right)\left(\frac{80,000}{146,700}\right)^{-1.48} + \left[\frac{150,000}{\infty}\right.$$

$$\left.\left(\frac{43,300}{146,700}\right)^{-1.48}\right] \overset{?}{=} 1 \tag{8-106}$$

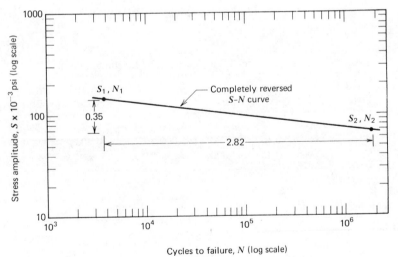

$$x = \frac{\log N_2 - \log N_1}{\log S_1 - \log N_2} = \frac{2.82}{0.35} = 8.05$$

FIGURE 8.15. Logarithmic plot of S-N data used to determine exponent x in Marin's equation for the S-N curve. Note that x must be the negative reciprocal of the slope of the S-N curve shown to satisfy (8-85).

from which

$$D = 0.92 \tag{8-107}$$

Since this value of damage D is less than 1, the area selected is a little too large. Additional trials are necessary to adjust the area until the calculated value of D becomes unity, at which time the selected area meets the design requirements.

Finally, the Manson double linear damage rule may be applied by selecting a trial area of 0.15 inch^2 and utilizing (8-95) or (8-96) together with (8-97) and (8-98) in sequence. It may be observed from the S-N curve that for the highest stress level in the spectrum, the total life N_f exceeds 730 cycles so the equations of (8-95) are to be used. Using (8-95) for each of the three stress levels S_A, S_B, and S_C, one obtains the results shown in Table 8.7. Next, (8-97) is applied by accumulating initiation damage until (8-97) reaches a value of unity. This procedure is tabulated Table 8.8, where the subscript 1 refers to the first iteration on area.

Thus it is concluded that a critical nucleus is created during load level A of block 2 since $\Sigma n/N$ exceeded unity during this block of cycles. To determine exactly when the $\Sigma n/N$ reaches unity during block 2, load level A, it may be noted that

$$0.534 + \frac{n'_{2A}}{2320} = 1$$

or

$$n'_{2A} = 2320(1 - 0.534) = 1080 \text{ cycles} \tag{8-108}$$

Table 8.7. Basic Data for First Iteration of Manson Solution

Load Level	P (lb)	A_1 (in.^{2})	S_1 (psi)	N_{f1} (cycles)	$N_{f1}^{0.6}$	N_{p1} (cycles)	N'_1 (cycles)
A	22,000	0.15	146,700	4,500	156	2,180	2,320
B	12,000	0.15	80,000	445,000	2,450	34,300	410,700
C	6,500	0.15	43,300	∞	∞	∞	∞

Table 8.8. First Iteration on Crack Initiation Using Manson Solution

Load Level	Block Number	P	A_1	S_1	n_1	N'_1	$\frac{n_1}{N'_1}$	$\Sigma \frac{n_1}{N_1}$
A	1	22,000	0.15	146,700	1,200	2,320	0.517	0.517
B	1	12,000	0.15	80,000	7,000	410,700	0.017	0.534
C	1	6,500	0.15	43,300	50,000	∞	0	0.534
A	2	22,000	0.15	146,700	1,200	2,320	0.517	1.051

Table 8.9. First Iteration on Crack Propagation to Failure Using Manson Solution

Load Level	Block Number	P	A_1	S_1	n_1	N_{p1}	$\dfrac{n_1}{N_{p1}}$	$\sum \dfrac{n_1}{N_{p1}}$
A	2	22,000	0.15	146,700	120	2,180	0.06	0.06
B	2	12,000	0.15	80,000	7,000	34,300	0.20	0.26
C	2	6,500	0.15	43,300	50,000	∞	0	0.26
A	3	22,000	0.15	146,700	1,200	2,180	0.55	0.81
B	3	12,000	0.15	80,000	7,000	34,300	0.20	1.01

where n'_{2_A} is the number of cycles of operation during load level A of block 2 that have passed when the critical nucleus is initiated. The remaining 120 cycles at this load level, and all other remaining cycles in the design spectrum, contribute to propagation. For the propagation phase, (8-98) is applied as indicated in Table 8.9. Here it may be noted that failure is barely predicted during load level B of the third block of loading. This indicates that the area selected was slightly too small.

Although this result is close to satisfactory since load level C will produce no damage, in principle it is necessary to assume a slightly larger area and repeat the entire calculation, with the design objective of reaching a value $\sum n/N_p = 1$ just after the application of the final cycle at the final load level of the final block in the loading spectrum.

8.5 LIFE PREDICTION BASED ON LOCAL STRESS-STRAIN AND FRACTURE MECHANICS CONCEPTS

In recent years it has been recognized that the fatigue failure process involves three phases. A crack initiation phase occurs first, followed by a crack propagation phase; finally, when the crack reaches a critical size, the final phase of unstable rapid crack growth to fracture completes the failure process. The modeling of each of these phases has been under intense scrutiny, but the models have not yet been developed in a coordinated way to provide a widely accepted engineering design tool. Nevertheless, it is worthwhile to examine recent developments in these modeling efforts since great progress has been made in recent years, especially in the crack propagation modeling and in modeling the final fracture phase, as discussed in Sections 3.7 through 3.10.

Local Stress-Strain Approach to Crack Initiation

Although progress has been much slower in modeling the crack initiation phase, the most promising approach to the prediction of crack initiation seems to be the local stress-strain concept. The basic premise of the local stress-strain approach is that the local fatigue response of the material at the critical point, that is, the site of crack initiation, is analogous to the fatigue response of a small, smooth specimen subjected to the same cyclic strains and stresses (15). This concept is illustrated schematically in Figure 8.16 for a simple notched plate under cyclic loading. The cyclic stress-strain response of the critical material may be determined from the characterizing smooth specimen through appropriate laboratory testing. To properly perform such laboratory tests, the local cyclic stress-strain history at the critical point in the structure must be determined, either by analytical or experimental means. Thus, valid stress analysis procedures, finite element modeling, or experimental strain measurements are necessary, and the ability to properly account for plastic behavior must be included. In performing smooth specimen tests of this type it must be recognized that the phenomena of cyclic hardening, cyclic softening, and cycle-dependent stress relaxation, as well as sequential loading effects and residual stress effects, may be experienced by the specimen as it accumulates fatigue damage presumed to be the same as at the critical point in the structural member being simulated. Some data have been accumulated to support the validity of this postulate (16–18).

Digital computer simulation of the smooth specimen simulation has also been shown to be feasible (18–20). To successfully utilize this computer simulation technique, it is necessary to have access to both monotonic and

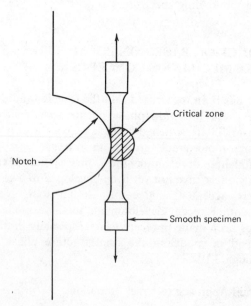

FIGURE 8.16.   Smooth specimen analog of material at critical point in the structure. (See ref. 15)

cyclic material properties, since the stress-strain response of most materials changes significantly with cyclic straining into the plastic range. Several examples of this are shown in Figure 8.17. If cyclic materials response data are not available from the literature or an accessible data bank, it is necessary to perform enough smooth specimen testing to characterize the cyclic stress-strain response and fracture resistance of the material. With cyclic materials properties available, the computer simulation model for prediction of crack initiation must contain the ability to

1. compute local stresses and strains, including means and ranges, from the applied loads and geometry of the structure.
2. count cycles and associate mean and range values of stress and strain with each cycle.
3. convert nonzero mean cycles to equivalent completely reversed cycles.
4. compute fatigue damage in each cycle from stress and/or strain amplitudes and cyclic materials properties.
5. compute damage cycle by cycle and sum the damage to give desired prediction of crack initiation.

To compute local stresses and strains from the external loading and geometry, Neuber's rule is often used in conjunction with the cyclic stress-strain properties and fatigue stress concentration factor. Neuber's rule may be written as (21)

$$K_t = (K_\varepsilon K_\sigma)^{1/2} \tag{8-109}$$

where $K_\varepsilon = \dfrac{\varepsilon}{e} = $ local strain concentration factor

$K_\sigma = \dfrac{\sigma}{S} = $ local stress concentration factor

$K_t = $ theoretical stress concentration factor
$\varepsilon = $ local strain
$e = $ nominal strain
$\sigma = $ local stress
$S = $ nominal stress

This relationship was modified for application to fatigue loading (22) by utilizing the fatigue stress concentration factor together with nominal stress *range* ΔS, nominal strain *range* Δe, local stress *range* $\Delta \sigma$, and local strain *range* $\Delta \varepsilon$ to transform (8-109) into

$$K_f (\Delta S \Delta e)^{1/2} = (\Delta \sigma \Delta \varepsilon)^{1/2} \tag{8-110}$$

If loads are small enough that the material behavior of the overall member is nominally elastic, then $\Delta S / \Delta e = E$ and (8-110) becomes

$$\frac{(K_f \Delta S)^2}{E} = \Delta \sigma \Delta \varepsilon \tag{8-111}$$

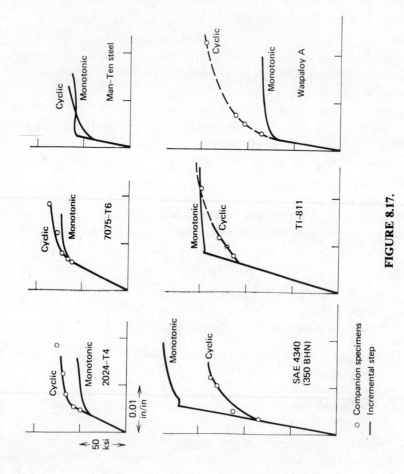

FIGURE 8.17.

○ Companion specimens
— Incremental step

Man-Ten steel

Cyclic
Monotonic

Waspaloy A

Cyclic
Monotonic

7075-T6

Cyclic
Monotonic

TI-811

Monotonic
Cyclic

2024-T4

Cyclic
Monotonic

SAE 4340
(350 BHN)

Monotonic
Cyclic

← 50 ksi →

0.01
in/in

278

Material	Condition	0.2 percent yield strength, monotonic σ_{yp}/cyclic σ_{yp} ksi	Strain-hardening exponent n(monotonic)/n'(cyclic)	Cyclic behavior
OFHC copper	Annealed	3/20	0.40/0.15	Hardens
	Partial annealed	37/29	0.13/0.16	Stable
	Cold worked	50/34	0.10/0.12	Softens
2024 aluminum alloy	T4	44/65	0.20/0.11	Hardens
7075 aluminum alloy	T6	68/75	0.11/0.11	Hardens
Man-Ten steel	As-received	55/50	0.15/0.16	Softens and hardens
SAE 4340 steel	Quenched and tempered, 350 BHN	170/110	0.066/0.14	Softens
Ti-8Al-1Mo-1V	Duplex annealed	145/115	0.078/0.14	Softens and hardens
Waspaloy		79/102	0.11/0.17	Hardens
SAE 1045 steel	Quenched and tempered, 595 BHN	270/250	0.071/0.14	Stable
	Quenched and tempered, 500 BHN	245/185	0.047/0.12	Softens
	Quenched and tempered, 450 BHN	220/140	0.041/0.15	Softens
	Quenched and tempered, 390 BHN	185/110	0.044/0.17	Softens
SAE 4142 steel	As-quenched, 670 BHN	235/...	0.14/....	Hardens
	Quenched and tempered, 560 BHN	245/250	0.092/0.13	Stable
	Quenched and tempered, 475 BHN	250/195	0.048/0.12	Softens
	Quenched and tempered, 450 BHN	230/155	0.040/0.17	Softens
	Quenched and tempered, 380 BHN	200/120	0.051/0.18	Softens

FIGURE 8.17. Cyclic stress-strain behavior of several materials. (From ref. 74, copyright ASTM; reprinted with permission.)

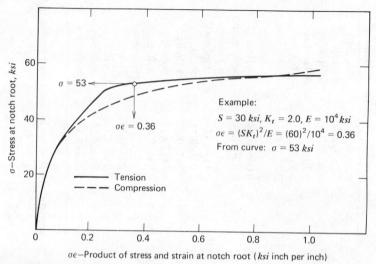

FIGURE 8.18. Stress versus product of stress and strain for 2024-T3 aluminum. (After ref. 23, copyright ASTM; reprinted with permission.)

All the terms on the left side of (8-111) are known from the geometry and from loading and material properties of the structure. To resolve the right-hand term, curves may be prepared* from the stable cyclic stress-strain curve to give σ versus the product $\sigma\epsilon$, as shown in Figure 8.18. This technique provides one means for computing local stresses and strains, if the stable cyclic stress-strain curve is available.

For most engineering metals the stable cyclic stress-strain curve is significantly different from the static stress-strain curve, as shown in Figure 8.17. The cyclic stress-strain curve may be obtained in a variety of ways. Basically, it involves passing a curve through the tips of stable hysteresis loops of several "identical" companion specimens tested at different strain ranges. A stable hysteresis loop is one that does not change size or shape significantly upon the application of additional stress-strain cycles. Other techniques involve passing a curve through the stable hysteresis loops generated for a specimen subjected to incremental steps in strain range or performing a static test, recording the stress-strain curve after a stable hysteresis loop has been achieved through cyclic straining. An empirical expression for the cyclic stress-strain curve, which is satisfactory for most engineering metals, may be obtained by separating the cyclic strain amplitude $\Delta\varepsilon/2$ into elastic and plastic components to yield[†]

$$\frac{\Delta\varepsilon}{2} = \frac{\Delta\varepsilon_e}{2} + \frac{\Delta\varepsilon_p}{2} = \frac{\Delta\sigma}{2E} + \left[\frac{\Delta\sigma}{2k'}\right]^{1/n'} \qquad (8\text{-}112)$$

*See, for example, ref. 23.
[†]From p. 7 of ref. 15.

where k' and n' are the *cyclic* strength coefficient and *cyclic* strain-hardening exponent, respectively, determined from the intercept and slope of a log-log plot of cyclic stress amplitude versus cyclic plastic strain amplitude. The static counterpart of this technique was demonstrated in Figure 5.2 and (5-18). Some values for n' are shown in Figure 8.17. The use of (8-112), then, provides another means for computing local stresses and strains, when used in conjunction with (8-111). In strain-controlled testing at constant strain range, the stress range may increase with the application of more cycles, remain unchanged, or decrease with the application of more cycles. If the stress range increases under strain-controlled cycling, the material is said to cyclically harden; and if the stress range decreases, the material is said to cyclically soften. Some materials may harden under certain conditions and soften under others, as shown in Figure 8.17. The need to include cyclic hardening and softening in the prediction model is dependent on the accuracy of other parts of the model, and in some cases these transient phenomena may be regarded as second-order effects.

Cycle-dependent stress relaxation may be very important in cases in which occasional large overload cycles produce large residual stresses in the structure that relax with additional cycles of local plastic strain. An example of cycle-dependent relaxation of mean stress is shown in Figure 8.19. Like cyclic hardening and softening, cyclic stress relaxation is difficult to model. However, since the presence or absence of a mean stress at the critical point in a

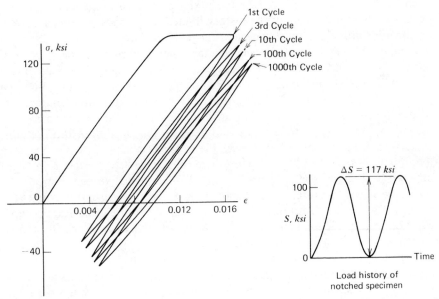

FIGURE 8.19. Cyclic softening and relaxation of mean stress under Neuber control ($\sigma \varepsilon = K$) for Ti-8Al-1Mo-1V with $K_f = 1.75$. (From ref. 16, copyright ASTM, reprinted with permission.)

structure may influence the life by several orders of magnitude, it is important to model cyclic stress relaxation. One such empirical model (23) to represent the cyclic relaxation of residual stress σ_R is

$$\frac{\sigma_R}{\sigma_o} = e^{\frac{-an(K_f \Delta S)^2}{E\sigma_{yp}}} \tag{8-113}$$

where σ_R is the local residual mean stress after n cycles following application of the overload nominal stress that produced the original local mean stress value σ_o. The material constant a for one case (23) was found to be about 0.005.

To properly interpret complex load, stress, or strain versus time histories requires that an appropriate cycle counting method be used. The six cycle counting methods shown in Table 8.10 have each been shown to be deficient in some respect (24). Two additional cycle counting methods, the *range pair method* and the *rain flow method*, have gained wide acceptance as counting methods (24).

The range pair counting method counts a strain range as a cycle if it can be paired with a subsequent straining of equal magnitude in the opposite direction. For a complicated history, some of the ranges counted as cycles will be simple ranges during which the strain does not change direction, but others will be interrupted by smaller ranges that will also be counted as cycles. Cycle counting by the range pair method is illustrated in Figure 8.20. The counted ranges are marked with solid lines and the paired ranges with dashed lines.

Each peak is taken in order as the initial peak of a range, except that a peak is skipped if the part of the history immediately following it has already been paired with a previously counted range. If the initial peak of a range is a minimum, a cycle is counted between this minimum and the most positive maximum that occurs before the strain becomes more negative than the initial peak of the range. For example, in Figure 8.20 a cycle is counted between peak 1 and peak 8, peak 8 being the most positive maximum that occurs before the strain becomes more negative than peak 1. If the initial peak of a range is a maximum, a cycle is counted between this maximum and the most negative minimum that occurs before the strain becomes more positive than the initial peak of the range. For example, in Figure 8.20 a cycle is counted between peak 2 and peak 3, peak 3 being the most negative minimum before the strain becomes more positive than peak 2. Each range that is counted is paired with the next straining of equal magnitude in the opposite direction. For example, in Figure 8.20 part of the range between peaks 8 and 9 is paired with the range counted between peaks 1 and 8.

The *rain flow cycle counting method*, illustrated in Figure 8.21, is probably more widely used than any other method. The strain-time history is plotted so that the time axis is vertically downward, and the lines connecting the strain peaks are imagined to be a series of pagoda roofs. Several rules are imposed on rain dripping down these roofs so that cycles and half cycles are defined.

Table 8.10. Cycle Counting Methods (24)

Name	Example	Description
Peak count	Stress or strain / Mean / Time	All maximums above the mean and all minimums below the mean are counted
Peak-between-mean crossing count	Stress or strain / Mean / Time	Only the largest peak between successive crossings of the mean is counted
Level crossing count	Stress or strain / 3, 2, 1, Mean / Time	All positive slope level crossings above the mean and negative slope level crossings below the mean are counted
Fatigue meter count	Stress or strain / 3, 2, 1, Mean / 3', 2', 1'	Similar to level crossing except that only one count is made between successive crossings of a lower level associated with each counting level
Range count	Stress or strain / r_1, r_2, r_3, r_4, r_5, r_6, r_7, r_8 / 0 / Time	Each range, i.e., the difference between successive peak values, is counted as 1/2 cycle, the amplitude of which is half the range.
Range-mean count	Stress or strain / m_1, m_2, m_3, m_4, m_5, m_6, m_7, m_8 / 0 / Time	Ranges are counted as above, and the mean value of each range is also considered

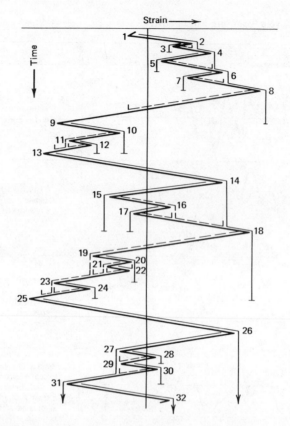

FIGURE 8.20. Example of range pair cycle counting method. (After ref. 24.)

Rain flow begins successively at the inside of each strain peak. The rain flow initiating at each peak is allowed to drip down and continue except that, if it initiates at a minimum, it must stop when it comes opposite a minimum more negative than the minimum from which it initiated. For example, in Figure 8.21 begin at peak 1 and stop opposite peak 9, peak 9 being more negative than peak 1. A half cycle is thus counted between peaks 1 and 8. Similarly, if the rain flow initiates at a maximum, it must stop when it comes opposite a maximum more positive than the maximum from which it initiated. For example, in Figure 8.21 begin at peak 2 and stop opposite peak 4, thus counting a half cycle between peaks 2 and 3. A rain flow must also stop if it meets the rain from a roof above. For example, in Figure 8.21, the half cycle beginning at peak 3 ends beneath peak 2. Note that every part of the strain-time history is counted once and only once.

When this procedure is applied to a strain history, a half cycle is counted between the most positive maximum and the most negative minimum. Assume that of these two the most positive maximum occurs first. Half cycles

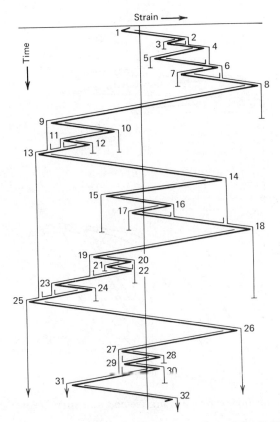

FIGURE 8.21. Example of rain flow cycle counting method. (After ref. 24.)

are also counted between the most positive maximum and the most negative minimum that occur before it in the history, between this minimum and the most positive maximum occurring previous to it, and so on to the beginning of the history. After the most negative minimum in the history, half cycles are counted that terminate at the most positive maximum occurring subsequently in the history, the most negative minimum occurring after this maximum, and so on to the end of the history. The strain ranges counted as half cycles therefore increase in magnitude to the maximum and then decrease.

All other strainings are counted as interruptions of these half cycles, or as interruptions of the interruptions, etc., and will always occur in pairs of equal magnitude to form full cycles. The rain flow counting method corresponds to the stable cyclic stress-strain behavior of a metal in that all strain ranges counted as cycles will form closed stress-strain hysteresis loops, and those counted as half cycles will not. This is illustrated in Figure 8.22.

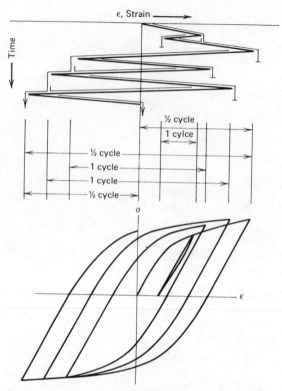

FIGURE 8.22. Rain flow cycle counting method and stress-strain hysteresis loops. (After ref. 24.)

If cycles are to be counted for a "duty cycle" or a "mission profile" that is repeated until failure occurs, one complete strain cycle should be counted between the most positive and most negative peaks in the sequence, and other smaller complete cycles that are interruptions of this largest cycle should also be counted. This will be accomplished by either the rain flow or the range pair cycle counting methods if the cycle counting is started at either the most positive or most negative peak in the sequence. Under these circumstances the rain flow method will count no half cycles and will give a cycle count identical to that obtained using the range pair method. The rain flow and range pair cycle counting methods can be considered equivalent for most practical situations, and from either of these counting methods the range and mean may be tabulated for each cycle in the stress-strain history.

The nonzero mean stress cycles may be converted to equivalent completely reversed cycles by utilizing either the modified Goodman equations of (7-9) or (7-11) or some empirical expression based on specific material data. For a tensile mean stress, then, from (7-11), the equivalent completely reversed stress σ_{eqC-R} is

$$\sigma_{eqC-R} = \frac{\sigma_a}{1 - \sigma_m/\sigma_u} ; \qquad \sigma_m \geq 0 \tag{8-114}$$

and for a compressive mean stress, from (7-9),

$$\sigma_{eqC-R} = \sigma_a ; \qquad \sigma_m \leq 0 \tag{8-115}$$

To compute the fatigue damage in each cycle associated with the equivalent completely reversed stress and strain range, it is necessary to have available data for strain amplitude versus cycles to failure, N_f, (or reversals to failure, $2N_f$), as illustrated in Figure 8.23.

Noting that the total strain amplitude may be expressed as the sum of elastic strain amplitude plus plastic strain amplitude, each linear with cycles to failure on a log-log plot, an empirical expression for the total strain amplitude versus life may be written as

$$\frac{\Delta\varepsilon}{2} = \frac{\Delta\varepsilon_e}{2} + \frac{\Delta\varepsilon_p}{2} = \frac{\sigma_f'}{E}(2N_f)^b + \varepsilon_f'(2N_f)^c \tag{8-116}$$

The constants b and σ_f'/E are the slope and one-reversal intercept of the elastic strain amplitude versus reversals to failure line in Figure 8.23, and the constant c and ε_f' are the slope and one-reversal intercept of the plastic strain amplitude versus life line in Figure 8.23. Although these constants are best

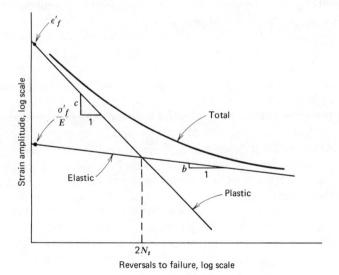

FIGURE 8.23. Schematic representation of elastic, plastic, and total strain amplitude versus fatigue life. (After ref. 24.)

evaluated from cyclic testing, they may be approximated from static properties, if fatigue data are unavailable. This may be done by taking σ_f' equal to true fracture strength σ_f, ε_f' equal to true fracture ductility ε_f, c equal to -0.6 and b equal to $-0.16 \log 2\sigma_f/\sigma_u$. However, actual fatigue data should be used where available.

Finally, damage is summed by utilizing an appropriate cumulative damage theory. Using the procedure described in this section for prediction of crack initiation, the Palmgren-Miner linear damage hypothesis of (8-4) gives results that are as good as any other proposed technique. Thus, when the sum of cycle ratios becomes equal to unity, it is predicted that crack initiation has occurred. It must be emphasized again that the prediction techniques described in this section are practical only with the help of a digital computer program designed to carry forth the tedious cycle-by-cycle analyses involved, and the validity of the method remains to be proven, even for uniaxial loading. Another practical difficulty lies in the definition and detection of an "initiated" crack. Thus it must be cautioned that the state of the art in prediction of fatigue crack initiation has not yet progressed far enough to provide a designer with proven tools for such predictions.

8.6   FRACTURE MECHANICS APPROACH TO CRACK PROPAGATION

As discussed in Section 3.7, the concepts of linear elastic fracture mechanics may be employed to predict the size of a crack in a given structure that will, under specified loadings, propagate spontaneously to failure. This critical crack size is determined from the critical stress intensity factor, as defined, for example, in (3-39). A fatigue crack that has been initiated by cyclic loading, or any other preexisting flaw in the structure or the material, may be expected to grow under sustained cyclic loading until it reaches the critical size from which it will propagate rapidly to catastrophic failure in accordance with the laws of fracture mechanics. Typically, the time for a fatigue-initiated crack or a preexisting flaw to grow to critical size is a significant portion of the useful life of a structure. Thus, not only is it necessary to understand the crack initiation phase and the definition of critical crack size, but an understanding of the growth of a crack from a known subcritical size is also essential.

Many different models have been proposed for predicting the rate at which a fatigue crack grows from a specified initial flaw size to the final critical crack size. One study (25) lists 33 proposed crack growth "laws." In a critical analysis of fatigue crack-growth-rate behavior reported by several investigators, Paris and Erdogan (38) determined that crack growth rate could be approximated by an expression of the form

$$\frac{da}{dN} = f(\Delta\sigma, a, C) \tag{8-117}$$

where $\Delta\sigma$ is the alternating nominal stress range, a is crack length, and C is a parameter that depends upon mean load, material properties, and secondary variables.

Equation (8-117) indicates that fatigue crack growth depends upon magnitude of alternating stress range and crack length. Equation (3-38) indicates that the stress intensity factor K depends upon applied stress and crack size. Thus, it was concluded that fatigue crack growth rate might be related to stress intensity factor; and since crack growth rate seemed to be related to *range* of alternating stress, it was thought that da/dN might be related to *range* of stress intensity factor ΔK, or

$$\frac{da}{dN} = g(\Delta K) \tag{8-118}$$

where $\Delta K = C\Delta\sigma\sqrt{\pi a}$, and not only is the stress fluctuating but the crack is growing. This observation has been supported by many subsequent investigations, and most crack propagation data produced have been characterized in terms of ΔK and plotted either as a log-log or log-linear function of ΔK. For example, Figure 8.24 illustrates the dependence of fatigue crack growth on

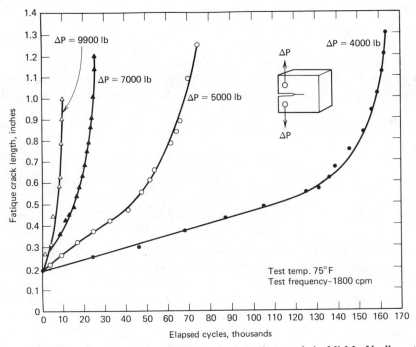

FIGURE 8.24. Effect of cyclic-load range on crack growth in Ni-Mo-V alloy steel for released tension loading. (From ref. 66, copyright Society for Experimental Stress Analysis, 1971; reprinted with permission.)

stress intensity factor. The crack growth rate da/dN, indicated by the slope of the a versus N curves, increases with both applied load and crack length. Since the crack-tip stress intensity factor range ΔK also increases with applied load and crack length, it is clear that crack growth rate is related to the applied stress intensity factor range.

To plot the data of Figure 8.24 in terms of stress intensity factor range and crack growth rate, the crack growth rate is determined from the slope of the a versus N curves between successive data points. Corresponding values of ΔK are computed from applied load range and mean crack length for each interval, using the proper stress intensity factor expression for the geometry of the specimen being tested. The results of this procedure are shown in Figure 8.25 for the data presented in Figure 8.24. It should be noted that all the curves of Figure 8.24 incorporate themselves into a single curve in Figure 8.25 through use of the stress intensity factor concept, and the curve of Figure 8.25 is therefore applicable to any combination of cyclic stress range and crack length for released loading on specimens of this geometry.

Fatigue crack-growth-rate data similar to that shown in Figure 8.25 have been reported for a wide variety of engineering metals. Thus, it was concluded (38) that the fatigue crack-growth-rate for engineering metals could be generalized to the form

$$\frac{da}{dN} = C_{PE}(\Delta K)^n \qquad (8\text{-}119)$$

where n is the slope of the log da/dN versus log ΔK plot, such as shown in Figure 8.25, and C_{PE} is an empirical parameter that depends upon material properties, frequency, mean load, and perhaps other secondary variables. Thus, if the parameters C_{PE} and n are known for a particular application, the crack length a_N, after application of N cycles of loading, may be computed from the expression

$$a_N = a_i + \sum_{i=1}^{N} C_{PE} \Delta K^n \qquad (8\text{-}120)$$

or

$$a_N = a_i + \int_{1}^{N} C_{PE} \Delta K^n \, dN \qquad (8\text{-}121)$$

where a_i is the initial crack length and N is the total number of loading cycles. It must be emphasized that (8-119), (8-120), and (8-121) are applicable only to the linear portion of the da/dN versus ΔK curve. That is, if the generalized crack growth curve is represented (67) as shown in Figure 8.26, (8-119), (8-120), and (8-121) are applicable only to Region II crack growth. Region I of Figure 8.26 corresponds to the nucleation period, and Region III corresponds to the transition into the unstable regime of rapid crack extension.

Although (8-119) has been widely quoted, the exponent has been shown to vary significantly, and the expression does not take into account either the

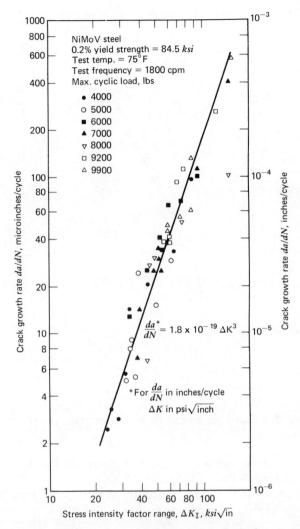

FIGURE 8.25. Crack growth rate as a function of stress-intensity range for Ni-Mo-V steel. (From ref. 66, copyright Society for Experimental Stress Analysis, 1971; reprinted with permission.)

influence of magnitude of peak stress or the existence of a threshold value of ΔK as indicated in Figure 8.26 at the boundary between Regions I and II. Several expressions have been developed to account for the peak stress effect, but the most widely used seems to be (59)

$$\frac{da}{dN} = \frac{C_F \Delta K^m}{(1 - R)K_c - \Delta K} \qquad (8\text{-}122)$$

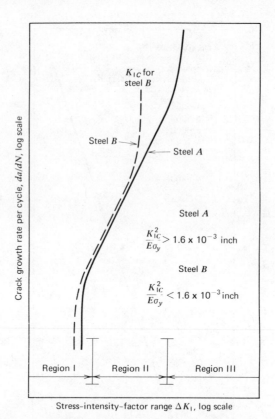

FIGURE 8.26. Schematic representation of fatigue crack growth in steel. (See ref. 67.)

where R is the stress ratio $\sigma_{min}/\sigma_{max}$ and K_c is the fracture toughness for unstable crack growth during monotonic loading.

A further modification of this expression to account for apparent threshold stress intensity levels for crack propagation yields (68)

$$\frac{da}{dN} = \frac{C_{HS}(\Delta K - \Delta K_{th})^m}{(1 - R)K_c - \Delta K}$$ (8-123)

where ΔK_{th} is the threshold stress intensity factor range for fatigue crack propagation.

Crack growth rates determined from constant-amplitude cyclic loading tests are approximately the same as for random loading tests in which the maximum stress is held constant but mean and range of stress vary randomly. However, in random loading tests where the maximum stress is also allowed to vary, the sequence of loading cycles may have a marked effect on crack growth rate, with the overall crack growth being significantly higher for random loading spectra (69).

Several investigations have shown that in some circumstances a significant delay in crack propagation occurs following intermittent application of high stresses. That is, fatigue damage and crack extension are dependent upon preceding cyclic load history. This dependence of crack extension on preceding history and the effects upon future damage increments are referred to as *interaction* effects. Most of the interaction studies conducted thus far have dealt with *retardation* of crack growth as a result of the application of occasional tensile overload cycles. Retardation can be characterized as a period of reduced crack growth rate following the application of a peak load or loads higher and in the same direction as those peaks that follow, as shown schematically in Figure 8.27 for a single overload followed by constant-level cyclic loading. The *retardation* effect has been modeled by considering the size of the plastic zone ahead of the crack tip relative to the size of the plastic zone produced by the overload, as shown in Figure 8.28. In the application of an overload such as P, in Figure 8.27, if its direction tends to open a crack such as the one shown in Figure 8.28, it will first produce local tensile stresses at the crack tip that are below the yield strength of the material but that then exceed the yield strength to produce a zone of local plastic deformation. As the loading is increased to its final value P_1, the size of the yield zone grows because of the redistribution of local stresses brought about by the relatively

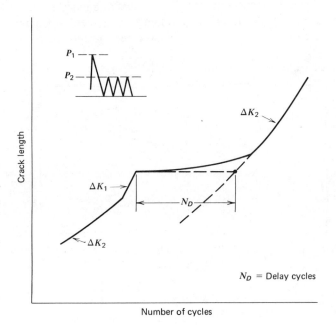

FIGURE 8.27. Schematic illustration of delay in crack growth following the application of single overload. (From ref. 70; reprinted with permission of Sijthoff & Noordhoff, International Publishers b.v.)

large plastic flow near the crack tip. For a given material and geometry, the magnitude of load applied will dictate the equilibrium location of the elastic-plastic interface between the yield zone and the surrounding elastic material. Such a boundary for overload P_1 is indicated by the dashed line in Figure 8.28. When the overload is removed, the surrounding material forces the yield zone into a state of residual compression that tends to retard subsequent crack growth due to crack-opening loads. Therefore, when loads of smaller magnitude are applied in the same direction as P_1, for example, loads P_2 in Figure 8.27, the associated crack-tip yield zone is smaller and growth is relatively slow until the crack penetrates all the way across the overload yield zone. Unless another overload is applied, the rate of crack propagation then returns to its original value as shown in Figure 8.27 following the delay cycles N_D. One way to estimate the retardation of crack growth rate in such a case is by use of a retardation factor, which may be expressed as a correction factor C_p on the crack growth rate da/dN. One such proposal (71) defines the correction factor to be

$$C_p = \left[\frac{R_y}{a_p - a} \right]^m \qquad \text{for } (R_y + a) < a_p$$

$$C_p = 1 \qquad \text{for } (R_y + a) \geq a_p \qquad (8\text{-}124)$$

where R_y and a_p are plastic-zone size parameters as shown in Figure 8.28, a is crack size, and m is an empirical parameter dependent upon material and stress history. Other retardation models have been proposed, based on devel-

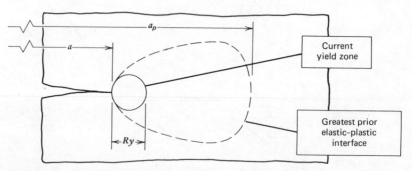

FIGURE 8.28. Description of Wheeler model for crack growth retardation. (See ref. 71.)

$$a_N = a_0 + \sum_{i=1}^{n} C_{pi} \left[\frac{da}{dN} = f(\Delta K_i) \right]$$

$$C_p = \left(\frac{R_y}{a_p - a} \right)^m \qquad \text{for } (a + R_y) < a_p$$

or

$$C_p = 1 \qquad \text{for } (a + R_y) \geq a_p$$

opment and decay of a residual stress field ahead of the crack tip (72) or on the concept that after plastic deformation has occurred in the material behind the crack tip a crack-opening load must be applied before further crack growth can take place (73). None of these models admit the possibility of crack-growth acceleration, however. *Acceleration* can be characterized as a period of increased crack growth rate owing to the application, for example, of occasional crack-closing overload cycles, which produce residual tension caused by overload yield zone formation, followed by a sequence of crack-opening load cycles of smaller magnitude. This is a matter that must be given further attention.

Although much progress has been made in crack-growth-rate prediction in the laboratory, it must be recognized that load sequence effects, environmental effects, frequency effects, multiaxial state-of-stress effect, and problems in determination of applicable ΔK values still make it essential to conduct full-scale fatigue tests to ensure proper resistance to failure by fatigue.

8.7 SERVICE LOADING SIMULATION AND FULL-SCALE FATIGUE TESTING

As noted in the preceding section, to achieve a reliable fatigue-resistant design configuration, a realistic service loading simulation test or a full-scale fatigue test will generally be required. For service load simulation testing it is essential that both the test specimen and the loading spectrum, including sequence, be representative of actual service conditions. This means that the specimen should be an actual component, a complete structural subassembly, or a complete full-scale machine or structure. It also means that an exact simulation of the service load-time history would be best, if known. Usually, however, an estimated load-time history must be generated on the basis of a statistical representation of similar mission loading spectra obtained from similar instrumented articles already in actual use.

In designing service loading simulation spectra, special attention must be given to the highest load levels to be incorporated because they may exert a major influence on crack propagation and fatigue life. As noted in Section 8.6, the life-enhancing retardation phenomenon associated with large load peaks can be extremely important. It should be recognized that the largest peak loads in service will vary from article to article on a statistical basis, so that some articles will experience the maximum load peak more than once, whereas others will never see this load level. For these reasons it is usual to truncate the design loading spectrum, discarding all load levels that occur fewer than 10 times during the projected design lifetime. This is illustrated in Figure 8.29 for an aircraft application. Although it may seem intuitively wrong to discard the high load peaks to achieve a more critical simulation test spectrum, it must be realized that the crack retardation effect of the occasional high loads gives a longer test life than if the high loads were omitted.

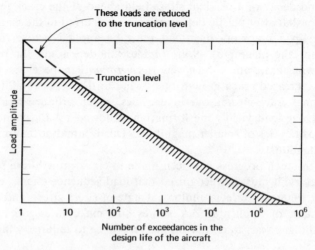

FIGURE 8.29.   Example of truncating the infrequently occurring high amplitudes of a loading spectrum. (After ref. 75.)

Thus it is conservative to omit the 10 highest loads from the simulation test spectrum. Likewise, truncation is utilized in analyses for establishing suitable inspection intervals for aircraft so that propagating cracks will have a higher probability of being detected before they approach critical size. A more detailed discussion of inspection intervals is given in Section 8.8.

The application of fail-safe or limit loads at regular intervals during a flight simulation or full-scale fatigue test *must be avoided*, since they may contribute to crack growth retardation which is not typical of the actual flight spectrum. Limit loading should be applied only at the end of the fatigue test.

Low-amplitude cycles are often omitted from service simulation testing to save time. However, it must be recognized that these cycles may contribute to fatigue crack nucleation through the fretting process, and omitting these low-amplitude cycles, and the fretting they would normally produce, may result in unsafe simulation test predictions of service fatigue life. Various empirical procedures have been developed, however, for establishing effective in-service mission simulation spectra (76).

Full-scale fatigue testing of an article such as a newly designed aircraft is extremely expensive. Such tests generally would be regarded as accelerated tests, where flight simulation testing over a six- to 12-month testing period is designed to represent 10 or more years of actual service history. Flight simulation testing of an aircraft component on a modern closed-loop fatigue testing machine may be accomplished in one or two weeks or less. The benefits of full-scale testing include (a) discovery of fatigue-critical elements and design deficiencies, (b) determination of times to detectable cracking, (c) obtaining data on crack propagation, (d) determination of remaining safe life with cracks, (e) determination of residual strength, (f) establishment of proper inspection intervals, and (g) development of repair methods. Factors that may influence the full-scale fatigue simulation test results include loading rate, environmental factors, statistical scatter, and loading spectrum deviations. Actual service life is usually shorter than the simulation test life, sometimes by factors of as much as 2 to 4. However, in recent years the agreement between full-scale flight simulation testing and in-service experience has improved significantly. Full-scale flight simulation testing is often continued over the long term so that fatigue failures in the test will lead the fleet experience by enough time to redesign and install whatever modifications are required to prevent catastrophic fleet failures in service before they occur.

8.8   DAMAGE TOLERANCE AND FRACTURE CONTROL

The concept of "damage-tolerant" structure, which has developed primarily within the aerospace industry, is characterized by structural configurations that are designed to minimize the loss of aircraft because of the propagation of undetected flaws, cracks, or other similar damage. There are two major design objectives that must be met to produce a damage-tolerant structure. These objectives are controlled safe flaw growth, or safe life with cracks, and positive damage containment, which implies a safe remaining or residual strength. These objectives should not be considered as separate or distinct requirements, however, because it is only by their judicious combination that effective fracture control can be achieved. Furthermore, it must be emphasized that damage-tolerant design is *not* a substitute for a careful fatigue analysis and design as discussed earlier, because the achievement of "fatigue quality" through careful stress analysis, geometry selection, detail design, material selection, surface finish, and workmanship is a necessary *prerequisite* to effective damage-tolerant design and fracture control.

The general goals of damage-tolerant design and fracture control include the selection of fracture-resistant materials and manufacturing processes, the design for inspectability, and the use of damage-tolerant structural configurations such as multiple load paths or crack stoppers, in addition to the usual rules of good design practice.

In the application of the fracture control philosophy, *the basic assumption is made that flaws do exist even in new structures and that they may go undetected.* The first major requirement for damage tolerance, therefore, is that any member in the structure, including each element of a redundant load path group, must have a safe life with assumed cracks present. For any specific application the primary factors influencing the design include the type or class of structure, the quality of the nondestructive inspection (NDI) techniques used in production assembly, the accessibility of the structure to inspection, the assurance that the member will be inspected on schedule when in service, and the probability that a flaw of subcritical size will go undetected even though periodic in-service inspections are made on schedule.

Most structural arrangements may be classified according to load path as Class 1, single load path; Class 2, single primary load path with auxiliary crack arrest features; or Class 3, multiple or redundant load path. Each of these structural classes is illustrated in Figure 8.30. Clearly, for the Class 1 structure it is essential to satisfy the safe-life-with-cracks requirement, because failure is catastrophic. For Class 2 structures, including pressurized cabins and pressure vessels, relatively large amounts of damage may be contained by providing tear straps or stiffeners. There is usually a high probability of damage detection for a Class 2 structure because of fuel or pressure leakage, that is, "leak-before-break" design is characteristic of Class 2 structures. Class 3 structures are usually designed to provide a specified percentage of the original strength, that is, a specified residual strength, during and subsequent to the failure of one element. This is often called "fail-safe" type of structure. However, the preexisting flaw concept requires that all members, including every member of a multiple load path structure, be assumed to contain flaws. It is usual to assume a smaller initial flaw size for Class 3 structures because it is appropriate to take a larger risk of operating with cracks if multiple load paths are available.

In the preceding discussion two basic design philosophies have been mentioned, *safe-life* design and *fail-safe* design. In the case of safe-life design, the technique is to evaluate the expected lifetime by employing the traditional design methods of Chapter 7, often followed by a full-scale test using an

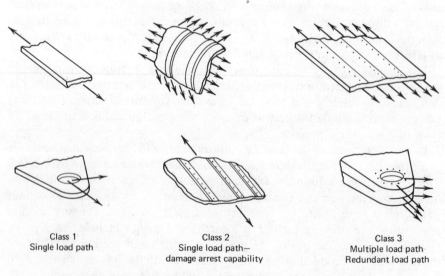

Class 1	Class 2	Class 3
Single load path	Single load path— damage arrest capability	Multiple load path Redundant load path

FIGURE 8.30.   Structural arrangements. (After ref. 77.)

appropriate loading spectrum. A carefully selected safety margin is then required between the calculated and/or test lifetime and the specified design life, to account for uncertainties and scatter. This margin of safety has sometimes been called a *scatter factor*. The fail-safe technique is to provide redundant load paths in the structure so that if failure of a primary structural member occurs, a secondary member is capable of carrying the load on an emergency basis until failure of the primary structure is detected and a repair can be made.

The development of inspection procedures is an important part of any fracture control program. Appropriate inspection procedures must be established for each structural element, and regions within elements may be classified with respect to required nondestructive inspection (NDI) sensitivity. Inspection intervals are established on the basis of crack growth information assuming a specified initial flaw size and a "detectable" flaw size that depends on the NDI procedure. Inspection intervals are established to assure that an undetected flaw will not grow to critical size before the next inspection, with a comfortable margin of safety. The intervals are usually picked so that two inspections will occur before any crack will reach critical size. Cracks larger than the detectable flaw size, a_{det}, are presumed to be discovered and removed. Generally, the establishment of inspection intervals is also based on the assumptions that all critical points are checked at every inspection; that cracks larger than a_{det} are all found, at the latest, by the second inspection; that inspections are performed on schedule, and that inspection techniques are truly nondamaging. Unfortunately, these assumptions are sometimes violated in field practice. For example, some large aircraft may contain as many as 22,000 critical fastener holes in the lower wing surface alone.* Complete inspection of such a large number of sites is not only tedious and time consuming, but subject to error born of the boredom of inspecting 20,000 holes with no serious problems, only to miss one hole with a serious crack (sometimes referred to as a "rogue" crack*). The uncertainties associated with the best current NDI techniques remain significant, and even for large rogue cracks the reliability of detection is only about 80 percent with a high probability that the same crack may be missed on the second inspection.* Furthermore, efforts to improve NDI state of the art have concentrated on lowering the detection threshold to smaller crack sizes but giving insufficient attention to the largest crack size that can be missed. Finally, it must be observed that NDI, in practice, may not be used to check every critical point at every inspection, may miss cracks it is supposed to find, may not always be performed on schedule, and may sometimes be partially damaging in spite of its name. Nonetheless, the use of NDI techniques and the establishment of appropriate inspection intervals represent significant advances in the state of the art and should be implemented in all high-performance fatigue-critical designs.

*See p. 1.5.13 of ref. 79.

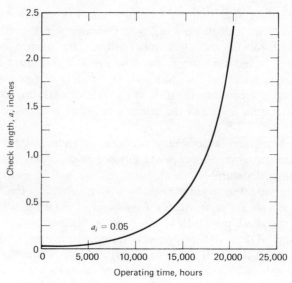

FIGURE 8.31. Crack length as a function of operating time for a hypothetical high performance machine.

In principle, to establish proper inspection intervals requires a knowledge of the initial flaw size a_i in a structural member, the detectable crack length a_{det}, and the critical crack size a_{cr} at which unstable rapid propagation to failure occurs, together with a crack growth versus life relationship or a plot of crack length versus operating hours, such as shown in Figure 8.31. If, for example, $a_i = 0.050$ inch, $a_{det} = 0.15$ inch, and $a_{cr} = 1.5$ inches, we would utilize Figure 8.31 and establish the inspection intervals, with a safety factor of 2, as follows:* The initial inspection interval I_1 would be

$$I_1 = \frac{[\text{time}]_{a_{cr} = 1.5} - [\text{time}]_{a_i = 0.05}}{2} \tag{8-125}$$

or

$$I_1 = \frac{[18{,}000] - [5000]}{2} = 6500 \text{ hours} \tag{8-126}$$

and for the second and subsequent inspection intervals I_2,

$$I_2 = \frac{[\text{time}]_{a_{cr} = 1.5} - [\text{time}]_{a_{det} = 0.15}}{2} \tag{8-127}$$

$$I_2 = \frac{[18{,}000] - [9500]}{2} = 4250 \text{ hours} \tag{8-128}$$

Damage tolerance criteria depend upon the degree of inspectability, frequency of inspection, and class of structure. For structures that are less

*See p. 1.2.17 of ref. 79.

inspectable, less frequently inspected, or less fail safe, the damage tolerance criteria, including initial flaw size, minimum required residual strength, service-induced flaw size, and crack growth rate, must be more conservative.

A good fracture control program should encompass and interact with design, materials selection, fabrication, inspection, and operational phases in the development of any high-performance engineering system. For example, the following is a typical fracture control plan for tasks in each phase of developing a high-performance engineering system (78).*

Design

1. Determine stress and strain distributions.
2. Determine flaw tolerance for regions of greatest fracture hazard.
3. Estimate stable crack growth for typical service periods.
4. Recommend safe operating conditions and specify intervals between inspections.

Materials

1. Determine yield and ultimate strengths.
2. Determine fracture parameters: K_c, K_{Ic}, K_{ISCC}, da/dN.
3. Establish recommended heat treatments.
4. Establish recommended welding methods.

Fabrication

1. Control residual stress, grain growth, and grain direction.
2. Develop or protect strength and fracture properties.
3. Maintain fabrication records.

Inspection

1. Inspect part prior to final fabrication.
2. Inspect fabrication factors such as welding current and speed.
3. Inspect for defects using NDI techniques.
4. Proof test.
5. Estimate largest cracklike defect sizes.

Operation

1. Control stress level and stress fluctuations in service.
2. Protect part from corrosion.
3. Inspect part periodically.

*Reprinted from *Machine Design*, Sept 2, 1971, copyright Penton/IPC, with permission.

8.9   USING THE IDEAS

Suppose that a 0.50-inch-thick plate, 10 inches wide and 30 inches long, used as a tension member in a high-performance structure, undergoes released loading such that the plate experiences released tensile loads fluctuating from 0 to 160,000 lb parallel to the 30-inch dimension. An NDI inspection has discovered a through-the-thickness crack at one edge near midspan of the plate. The crack length is indicated to be 0.075 inch. The material is Ni-Mo-V steel with yield point of 84,500 psi, plane strain fracture toughness of 33,800 psi $\sqrt{\text{in.}}$, and crack growth behavior as depicted in Figure 8.25. An estimate is to be made of the remaining life in cycles before catastrophic fracture of the structural plate occurs.

It may be recalled that three types of information are needed to make a life estimate of a cracked member. These are initial crack size, a_i; rate of crack propagation with applied cycles, da/dN; and critical crack size for unstable growth and fracture, a_{cr}. The methods of Sections 3.7 and 3.8 may be utilized to determine the critical crack size a_{cr}, the approach of Section 8.6 may be used to analyze crack growth, and the initial crack size a_i is provided from the inspection report. The first task is to determine the critical crack size a_{cr}. From (3-45) it may be noted that the minimum thickness for plane strain conditions to prevail is

$$B_{\min} = 2.5 \left(\frac{K_{Ic}}{\sigma_{yp}} \right)^2 = 2.5 \left(\frac{33,800}{84,500} \right)^2 = 0.40 \text{ inch} \qquad (8\text{-}129)$$

and the plane strain assumption for the 0.5-inch plate will therefore be valid. Consequently, the stress intensity factor for a single edge through-the-thickness crack under direct tension, given by (3-50), will be equal to the plane strain fracture toughness when the maximum stress is applied and the crack size is a_{cr}. This may be written as

$$C_I \sigma_{t\,\max} \sqrt{\pi a_{cr}} = K_{Ic} \qquad (8\text{-}130)$$

which may be solved for critical crack size as

$$a_{cr} = \frac{1}{\pi} \left[\frac{K_{Ic}}{C_I \sigma_{t\,\max}} \right]^2 \qquad (8\text{-}131)$$

Figure 3.33 provides information for the evaluation of C_I. Although C_I is a function of crack size, it is not strongly dependent upon crack size. For example, if $a = 0.075$, the initial crack size, $a/b = 0.008$; and from Figure 3.37

$$C_I [1 - (0.008)]^{3/2} = 1.11 \qquad (8\text{-}132)$$

or

$$C_I = 1.12 \qquad (8\text{-}133)$$

If the crack grows to $a = 0.30$ in length, $a/b = 0.30$; and from Figure 3.33

$$C_I = [1 - (0.03)]^{3/2} = 1.08 \qquad (8\text{-}134)$$

or

$$C_I = 1.13 \qquad (8\text{-}135)$$

Finally, the maximum gross section tensile stress under the specified loading is

$$\sigma_{t\,max} = \frac{P}{A} = \frac{160,000}{(0.5)(10)} = 32,000 \text{ psi} \qquad (8\text{-}136)$$

Thus the critical crack size may be computed from (8-131) as

$$a_{cr} = \frac{1}{\pi}\left[\frac{33,800}{1.13(32,000)}\right]^2 = 0.28 \text{ inch} \qquad (8\text{-}137)$$

To estimate the remaining life in cycles, then, it is necessary only to utilize the empirical relationships from Figure 8.25 in the crack growth region between the initial size $a_i = 0.075$ inch and the critical size $a_{cr} = 0.28$ inch. However, it would be desirable to make a plastic-zone correction to the crack size. Since plane strain conditions prevail, (3-42) may be utilized to give

$$r_{Y_\varepsilon} = \frac{1}{6\pi}\left(\frac{33,800}{84,500}\right)^2 = 0.008 \qquad (8\text{-}138)$$

and from (3-43) the effective crack length a' becomes

$$a' = 0.075 + 0.008 = 0.083 \text{ inch} \qquad (8\text{-}139)$$

From Figure 8.25, the governing equation for the stable crack growth region is

$$\frac{da}{dN} = 1.8 \times 10^{-19} \Delta K^3 \qquad (8\text{-}140)$$

or since

$$\Delta K = C_I \Delta\sigma\sqrt{\pi a} \qquad (8\text{-}141)$$

then

$$\frac{da}{dN} = 1.8 \times 10^{-19}\left[C_I \Delta\sigma\sqrt{\pi a}\right]^3 \qquad (8\text{-}142)$$

This equation may be rewritten as

$$\frac{da}{a^{3/2}} = 1.8 \times 10^{-19}\left[C\Delta\sigma\sqrt{\pi}\right]^3 dN \qquad (8\text{-}143)$$

Integrating both sides, and noting that the initial effective crack size is $a' = 0.083$ when $N = 0$ and the critical crack size is $a = 0.28$ inch at failure when $N = N_f$,

$$\int_{0.083}^{0.28} \frac{da}{a^{3/2}} = \int_0^{N_f} 1.8 \times 10^{-19} \left[1.13(32,000)\sqrt{\pi} \right]^3 dN \qquad (8\text{-}144)$$

Performing the integration,

$$\left. \frac{a^{-1/2}}{-1/2} \right]_{0.083}^{0.28} = 4.74 \times 10^{-5} N]_0^{N_f} \qquad (8\text{-}145)$$

or

$$\left. \frac{-2}{\sqrt{a}} \right]_{0.083}^{0.28} = 4.74 \times 10^{-5} N_f \qquad (8\text{-}146)$$

or

$$\frac{-2}{\sqrt{0.28}} + \frac{2}{\sqrt{0.083}} = 4.74 \times 10^{-5} N_f \qquad (8\text{-}147)$$

or

$$N_f = \frac{3.16}{4.74} \times 10^5 = 66,700 \text{ cycles} \qquad (8\text{-}148)$$

Thus, the number of cycles of remaining life after discovery of the 0.075-inch edge flaw is estimated to be 66,700 cycles.

QUESTIONS

1. Discuss the basic assumptions made in using a linear damage rule to assess fatigue damage accumulation and note the major "pitfalls" one might experience in using such a theory. Why, then, is linear damage theory so often used?

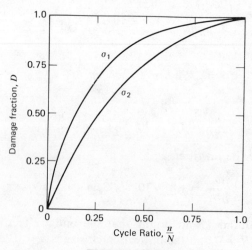

FIGURE Q8.2.

2.   A laboratory testing program has resulted in cumulative damage data for alloy "R" as shown in the plot of Figure Q 8.2 for the two stress levels σ_1 and σ_2 with $\sigma_1 > \sigma_2$. Utilizing this diagram, discuss the effects of order-of-application of the cyclic stresses of amplitude σ_1 and σ_2 and compare your conclusions with the Palmgren-Miner theory prediction.

3.   Compare and contrast the Palmgren-Miner hypothesis with the Manson double linear damage rule.

4.   An alloy steel tension strut for an aircraft application is fabricated from a lot of material that has an ultimate strength of 135,000 psi, yield strength of 120,000 psi, elongation of 20 percent in 2 inches, and fatigue properties based on experiment as shown in the chart.

Stress Amplitude (psi)	Cycles to Failure, N
110,000	6,600
105,000	9,500
100,000	13,500
95,000	19,200
90,000	27,500
85,000	39,000
80,000	55,000
75,000	87,000
73,000	116,000
71,000	170,000
70,000	220,000
69,000	315,000
68,500	400,000
68,000	∞

In service the tension strut is to be subjected to the following spectrum of completely reversed axial loads during each duty cycle:

$$P_a = 11,000 \text{ lb for } 1000 \text{ cycles}$$
$$P_b = 8,300 \text{ lb for } 4000 \text{ cycles}$$
$$P_c = 6,500 \text{ lb for } 50,000 \text{ cycles}$$

This duty cycle is to be repeated three times during the life of the strut. Find the required cross-sectional area for the strut using each of the following cumulative damage theories: Palmgren-Miner theory, Henry theory, Gatts theory, Corten-Dolan theory, Marin theory, and Manson theory.

5.   A special laboratory device is to be shipped by truck from Phoenix to New York City. To protect the device from shock loads it is mounted in a cylindrical container and suspended on a rigid truck bed structure from three special brackets through short lengths of wire rope. The wire ropes are vertical, equally spaced around the cylindrical container, and may be considered to be inextensible. The mass center of the device is located at the

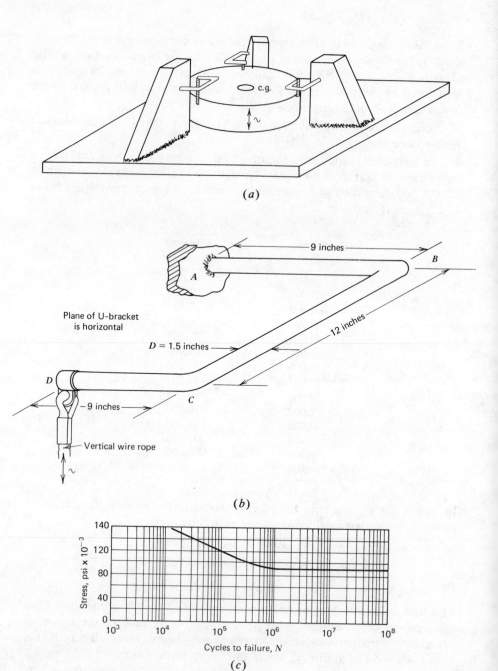

(a)

9 inches

B

A

Plane of U-bracket
is horizontal

D = 1.5 inches

12 inches

D

C

9 inches

Vertical wire rope

(b)

140

120

Stress, psi x 10^{-3}

80

40

0

10^3 10^4 10^5 10^6 10^7 10^8

Cycles to failure, N

(c)

FIGURE Q8.5.

geometrical center of the cylindrical container, and the cylinder is guided so that it may move only vertically (see Figure Q8.5a).

Each of the three brackets is made of a high-strength solid steel bar, 1.5 inches in diameter and bent into a U-shape as shown in Figure Q8.5b.

The plane of the U is horizontal, with one end of the U welded to the rigid structure and the other end supporting the cable. The material has an ultimate strength of 180,000 psi, yield point of 160,000 psi, and an S-N curve as shown in Figure Q8.5c.

Experience with similar devices has resulted in the observation that the cylinder will vibrate in a vertical direction approximately sinusoidally. The maximum amplitude of vibration is estimated to be approximately equal to the static deflection of the 3000-lb mass as it is first loaded onto the supporting U-shaped brackets. It is further estimated that the amplitude distribution may be assumed to be 50 percent at maximum amplitude, 25 percent at three-fourths of maximum amplitude, and 25 percent at half of maximum amplitude. The average truck speed is 50 mph.

You are asked as a consultant to the trucking firm to determine the following:

(a) Where are the potential critical points in the supporting structure?

(b) What failure mode(s) govern(s) at each critical point?

(c) Would you expect failure to occur at the most serious critical point(s)? Support this answer with engineering calculations.

6. A *design* S-N curve for material "x" has been determined by considering all influencing factors and a safety factor; it is plotted in Figure Q8.6. A solid square bar of side dimension "d" is to be designed to withstand the following spectrum of completely reversed axial loading:

One block is

$$1000 \text{ cycles at } 10,000 \text{ lb}$$
$$19,600 \text{ cycles at } 6000 \text{ lb}$$
$$10^7 \text{ cycles at } 1000 \text{ lb}$$

The whole block is to be repeated three times. Determine dimension d.

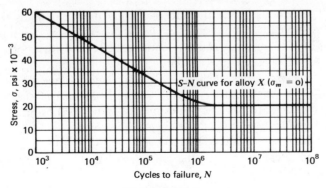

FIGURE Q8.6.

7. The critical point in the main rotor shaft of a new VSTOL aircraft of the ducted-fan type has been instrumented, and during a "typical" mission the equivalent completely reversed stress spectrum found to be 50,000 psi for 15 cycles, 30,000 psi for 100 cycles, 60,000 psi for 3 cycles, and 10,000 psi for 10,000 cycles.

Ten missions of this spectrum have been run. It is desired to overload the shaft to 1.10 times the "typical" loading spectrum.

(a) Estimate the number of additional "overloaded" missions that can be run without failure if the stress spectrum is linearly proportional to the loading spectrum. See S-N data for the shaft material shown in Figure Q8.7.

(b) How many additional runs could have been made under normal loading conditions?

(c) What deficiencies of Miner's theory might give an erroneous prediction in this estimate?

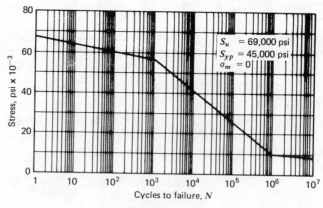

FIGURE Q8.7.

8. A ball joint housing for the end of a wing flap actuator is to be made from a high-strength aluminum alloy for which the S-N curve is given in Figure Q8.7. The critical point is in a solid cylindrical portion and is subjected to direct axial stresses only. The S-N properties given in Figure Q8.7 are mean properties, including all notch and surface effects, etc. The design is to be based on a fatigue strength 3σ below the mean, and the stresswise standard deviation is estimated to be 1000 psi at all lives. The operational *loading* spectrum is as shown in the chart.

The total design lifetime is to be 1000 missions. The $(-)$ is compression and $(+)$ is tension. Determine what the cross-sectional area of the critical cylindrical portion should be.

Step	Segment	Load range (lb)	Cycles per Mission
1	Ground	0	0
2	Post take off	+ 900 to − 300	1
3	Climb, cruise	− 400 to + 200	1
4	Climb, cruise	− 500 to − 175	10
5	Climb, cruise	− 500 to − 225	100
6	Supersonic cruise	+ 2000 to 0	5
7	Terrain following	− 2100 to 0	200
8	Prelanding	− 500 to + 300	2
9	Ground	0	0

9.   The critical section of a special purpose control linkage bar with ball ends is to be made of a solid cylindrical shape. The steel alloy to be used in the cylindrical linkage member has an ultimate strength of 150,000 psi, a yield strength of 132,000 psi, elongation in 2 inches of 14 percent, and completely reversed fatigue properties as shown in the chart for 99.7 percent probability of survival.

Stress Amplitude (psi)	Cycles to Failure, N
150,000	100
145,000	1200
130,000	6500
120,000	15,500
100,000	57,800
80,000	220,000
70,000	473,000
62,000	10^6
60,000	∞

The link is to be designed to withstand a spectrum of axial loads and it is desired to design it for 99.7 percent probability of survival. The loading spectrum for each duty cycle is as follows:

P_A ranges from + 30,000 lb to − 6000 lb for 800 cycles.

P_B ranges from + 18,000 lb to − 18,000 lb for 7500 cycles.

P_C ranges from + 9000 lb to − 9000 lb for 75,000 cycles.

This duty cycle is to be repeated three times during the life of the bar.

Find the required cross-sectional area to provide 99.7 percent probability of survival with a minimum weight of material according to the Palmgren-Miner theory and the Manson theory.

10.   An annealed 1040 steel alloy has an ultimate strength of 54,000 psi, a yield strength of 48,000 psi, elongation in 2 inches of 50 percent, and the

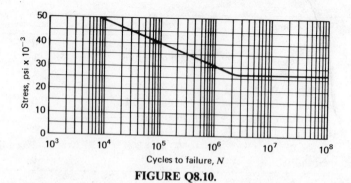

FIGURE Q8.10.

fatigue properties shown in Figure Q8.10. A solid cylindrical bar of 1 inch diameter is made of this material and subjected to the following schedule of completely reversed axial loads: first, 33,770 lb for 5000 cycles; then, 24,350 lb for 3×10^5 cycles; finally, 17,300 lb for 2×10^6 cycles.

Following all of this, the load is changed to 28,700 lbs, completely reversed loading. How many cycles would you predict at this final stress level before failure occurs?

11. In "modern" fatigue analysis and design procedures, three separate phases are defined in the fatigue process. List these three phases and briefly describe the analytical approach to dealing with each of these phases.

12. Experimental values for the properties of an alloy steel have been found to be $\sigma_u = 215,000$ psi, $\sigma_{yp} = 200,000$ psi, $K_{IC} = 74,000$ psi $\sqrt{\text{in.}}$, $e = 20$ percent in 2 inches, $k' = 155,000$ psi, $n' = 0.15$, $\varepsilon_f' = 0.48$, $\sigma_f' = 290,000$ psi, $b = -0.091$, and $c = -0.60$. A direct tension member made of this alloy has a single semicircular edge notch that results in a fatigue stress concentration factor of 1.6. The net cross section of the member at the root of the notch is 0.35 inch thick by 1.43 inches wide. A completely reversed cyclic axial force of 12,500 lbs amplitude is applied to the tension member.

(a) How many cycles would you estimate that it would take to initiate a fatigue crack at the notch root?

(b) What length would you estimate this crack to be at the time it is "initiated" by the calculation of part (a)?

13. For the equation

$$\frac{da}{dN} = C \Delta K^n$$

define each term, describe the physical phenomenon being modeled, and tell what the limiting conditions are on the magnitude of ΔK and the consequences of exceeding the limits of validity.

14. A steel alloy support member for a steam drum in a stationary power station has a rectangular cross section 6 inches thick by 24 inches wide by 100 inches long. Because of dynamic excitations the tensile force in the member

in the direction of the 100-inch dimension ranges from a maximum of $+6.48 \times 10^6$ lbs to a minimum of $+3.60 \times 10^6$ lbs. The material properties for this alloy include $\sigma_{yp} = 100,000$ psi and $K_{Ic} = 150,000$ psi $\sqrt{\text{in}}$. Experimental testing of this alloy has indicated that the effects of nonzero mean stress on crack growth rate are negligible and that crack growth rate for this alloy may be adequately described by the empirical expression

$$\frac{da}{dN} = 1.17 \times 10^{-15}(\Delta K)^{2.25}$$

Inspection has shown that a 0.3-inch-long through-the-thickness edge crack exists at one edge. How many cycles of the loading specified would you predict could be applied before catastrophic fracture would occur?

15. A Ni-Mo-V steel with yield point of 84,500 psi, plane strain fracture toughness of 33,800 psi $\sqrt{\text{in}}$. and crack growth behavior shown in Figure 8.25, is 0.50 inch thick, 10.0 inches wide, and 30.0 inches long. The plate is to be subjected to released tensile load fluctuating from 0 to 160,000 lb applied uniformly in the direction of the 30-inch dimension. A through-the-thickness crack of length 0.075 inch has been detected at one edge. How many cycles of this released tensile loading would you predict could be applied before catastrophic fracture would occur?

16. If the crack growth equation of problem 14 is equally applicable to the tensile member of problem 12 and if a good estimate for problem 12(b) is that a 0.050-inch crack constitutes "initiation," estimate the total life in cycles to fracture of the tensile member in problem 12 from the time the first cycle of loading is applied.

17. The stress-time pattern shown in Figure Q8.17a is to be repeated in blocks until failure of a test component occurs. Using the rain-flow cycle counting method and the S-N curve given in Figure Q.8.17b, estimate the hours of life until failure of this test component. (Hint: See example solution in Section 11.9.)

18. The stress-time pattern shown in Figure Q8.18a is to be repeated in blocks. Using the rain flow cycle counting method and the S-N curve in Figure Q8.18b, estimate the time in hours of testing required to produce failure. (Hint: See example solution in Section 11.9.)

19. A large, ducted two-speed mine ventilation fan (see Figure Q8.19a) consists of eight blades attached to a rotating hub. A critical region has been identified in each fan blade near the hub attachment point. This critical region may be regarded as rectangular in cross section, 2 inches by 4 inches, and may be assumed to experience direct forces only, that is, no bending, torsion, etc.

Because of centripetal acceleration, under high-speed operation, a steady tensile force of 150,000 lb is induced in the critical section. Because of aerodynamically excited vibration, an additional cyclic force of amplitude 100,000 lb is *superposed* upon the centrifugal tensile force. During low-speed

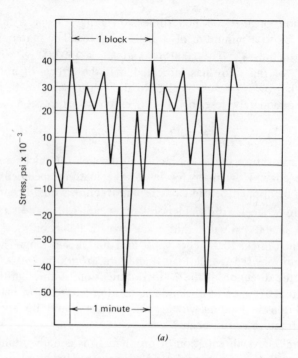

(a)

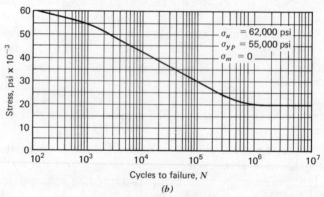

Cycles to failure, N

(b)

FIGURE Q8.17.

operation the steady centrifugal force is reduced to 100,000 lb, but the superposed cyclic force increases to an amplitude of 150,000 lb.

A typical duty cycle consists of 5×10^5 cycles at high speed followed by 10^6 cycles at low speed during one working day. A design life of 18 years is desired.

The material is 7075-T6 forged aluminum alloy with $\sigma_u = 87,000$ psi, $\sigma_{yp} = 78,000$ psi, elongation of 7 percent in 2 inches, and $K_{Ic} = 27,500$ psi $\sqrt{\text{in}}$. The fatigue strength at 10^{10} cycles is 20,000 psi.

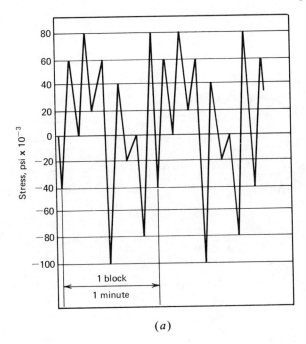

(a)

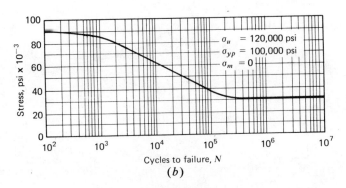

Cycles to failure, N

(b)

FIGURE Q8.18.

Three failure modes of primary interest are yielding, brittle fracture and fatigue.

(a) Would you predict failure by yielding?

(b) Ultrasonic crack inspection has been performed on this rectangular forged region. The inspection equipment has a 99 percent probability of finding cracks 1/8 inch and larger. No cracks have been found in these forgings. Would you predict failure by brittle fracture?

(c) Assuming the fan operates 24 hours per day and 365 days per year, would

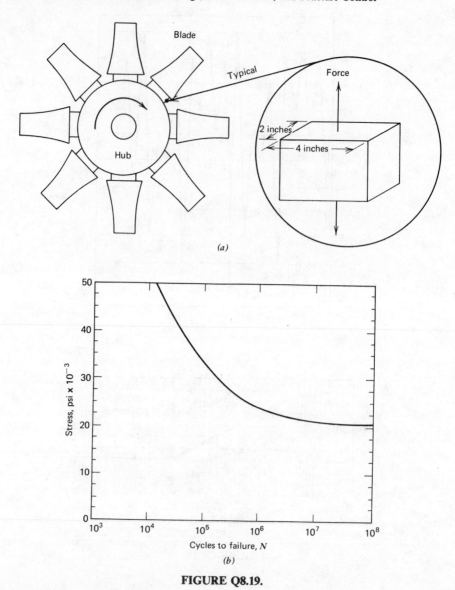

(a)

(b)

FIGURE Q8.19.

you predict failure by fatigue during the 18-year design life? (Hints: Utilize (8-114) and (8-115); see S-N curve in Figure Q8.19b).

(d) If you predict that it will fail by fatigue, about how long will it run before you would expect such a failure?

20. In service loading simulation tests, the test loading spectrum is often truncated to eliminate all tensile load levels that occur fewer than 10 times. Explain in some detail why this is done.

REFERENCES

1. Marco, S. M., and Starkey, W. L., "A Concept of Fatigue Damage," *ASME Transactions*, **76** (1954): 627.

2. Richart, F. E., and Newmark, N. M., "An Hypothesis for the Determination of Cumulative Damage in Fatigue," *ASTM Proceedings* **48** (1946): 767.

3. Henry, D. L., "Theory of Fatigue Damage Accumulation in Steel," *ASME Transactions*, **77** (1955): 913.

4. Gatts, R. R., "Application of a Cumulative Damage Concept to Fatigue," *ASME Transactions*, **83**, Series D, No. 4 (1961): 529.

5. Corten, H. T., and Dolan, T. J., "Cumulative Fatigue Damage," *Proceedings of International Conference on Fatigue of Metals*," *ASME and IME* (1956): 235ff.

6. Kommers, J. B., "The Effect of Overstressing and Understressing in Fatigue," *ASTM Proceedings*, **38** (1938): 249 and **43** (1943): 749.

7. Marco, S. M., Starkey, W. L., and Foster, T. G., "Fatigue Characteristics of 76S-T61 Aluminum Alloy and SAE 4340 Steel," Report No. 2, WADC Contract AF33(038)12393, Ohio State University Research Foundation, Columbus, Ohio, 1951.

8. Bennett, J. A., "A Study of the Damaging Effect of Fatigue Stressing on X4130 Steel," *ASTM Proceedings*, **46** (1946): 693.

9. Freudenthal, A. M., and Heller, R. A., "On Stress Interaction in Fatigue and a Cumulative Damage Rule, Part I," WADC TR58-69, June 1958.

10. Marin, J., *Mechanical Behavior of Engineering Materials*, Prentice-Hall, Englewood Cliffs, N.J., 1962.

11. Grover, H. J., "An Observation Concerning the Cycle Ratio in Cumulative Damage," *Fatigue in Aircraft Structures*, STP-274, American Society for Testing and Materials, Philadelphia, 1960, pp. 120–124.

12. Manson, S. S., "Interfaces Between Fatigue, Creep, and Fracture," *Proceedings of International Conference on Fracture*, Vol. 1, Japanese Society for Strength and Fracture of Metals, Sendai, Japan, September, 1965, and *International Journal of Fracture Mechanics*, March 1966.

13. Manson, S. S., Frecke, J. C., and Ensign, C. R., "Applications of a Double Linear Damage Rule to Cumulative Fatigue," *Fatigue Crack Propagation*, STP-415, American Society for Testing and Materials, Philadelphia, 1967, p. 384.

14. Kennedy, A. J., *Processes of Creep and Fatigue in Metals*, John Wiley & Sons, New York, 1963.

15. Morrow, J. D., Martin, J. F., and Dowling, N. E., "Local Stress-Strain Approach to Cumulative Fatigue Damage Analysis," Final Report, T. & A. M. Report No. 379, Department of Theoretical and Applied Mechanics, University of Illinois, Urbana, January 1974.

16. Stadnick, S. J., and Morrow, J., "Techniques for Smooth Specimen Simulation of the Fatigue Behavior of Notched Members," *Testing for Prediction of Material Performance in Structures and Components*, STP-515, American Society for Testing and Materials, Philadelphia, 1972, pp. 229–252.

17. Stadnick, S. J., "Simulation of Overload Effects in Fatigue Based on Neuber's Analysis," T. & A. M. Report No. 325, Department of Theoretical and Applied Mechanics, University of Illinois, Urbana, 1969.

18. Topper, T. H., and Morrow, J., "Simulation of the Fatigue Behavior at the Notch Root in Spectrum Loaded Notched Members (U)," T. & A. M. Report No. 333, Department of Theoretical and Applied Mechanics, University of Illinois, Urbana, January 1970 (Final Report for Aero Structures Department, Naval Air Development Center).

19. Martin, J. F., Topper, T. H., and Sinclair, G. M., "Computer Based Simulation of Cyclic

Stress-Strain Behavior with Applications to Fatigue," *Materials Research and Standards*, **11**, No. 2 (February 1971): 23.

20.   Martin, J. F., Topper, T. H., and Sinclair, G. M., "Computer Based Simulation of Cyclic Stress-Strain Behavior," T. & A. M. Report No. 326, Department of Theoretical and Applied Mechanics, University of Illinois, Urbana, July 1969.

21.   Neuber, H., "Theory of Stress Concentration for Shear-Strained Prismatical Bodies with Arbitrary Nonlinear Stress-Strain Law," *Journal of Applied Mechanics*, ASME Transactions, **8** (December 1961): 544–550.

22.   Topper, T. H., Wetzel, R. M., and Morrow, J., "Neuber's Rule Applied to Fatigue of Notched Specimens," *Journal of Materials*, **4**, No. 1 (March 1969): 200–209.

23.   Impellizzeri, L. F., "Cumulative Damage Analysis in Structural Fatigue," *Effects of Environment and Complex Load History on Fatigue Life*, STP-462, American Society for Testing and Materials, Philadelphia, 1970, pp. 40–68.

24.   Dowling, N. E., "Fatigue Failure Predictions for Complicated Stress-Strain Histories," *Journal of Materials*, **7**, No. 1 (March 1972): 71–87. See also, Dowling, N. E., "Fatigue Failure Predictions for Complicated Stress-Strain Histories," T. & A. M. Report No. 337, Department of Theoretical and Applied Mechanics, University of Illinois, Urbana, January 1971.

25.   Hoeppner, D. W., and Krupp, W. E., "Prediction of Component Life by Application of Fatigue Crack Growth Knowledge," *Engineering Fracture Mechanics*, **6** (1974): 47–70.

26.   Shanley, F. R., "A Theory of Fatigue Based on Unbonding during Reversed Slip," Rand Corp. Report No. P350 (November 1952), also supplement (May 1963).

27.   Shanley, F. R., "A Proposed Mechanism of Fatigue Failure," *Colloquium on Fatigue*, Stockholm, Sweden, International Union of Theoretical and Applied Mechanics, Springer-Verlag, Berlin, 1956.

28.   Head, A. K., "The Growth of Fatigue Cracks," *Philosophical Magazine*, **44**, Ser. 7 (1953): 925.

29.   Head, A. K., "The Propagation of Fatigue Cracks," *Journal of Applied Mechanics*, **23** (1956): 407.

30.   Weibull, W., *The Propagation of Fatigue Cracks in Light Alloy Plates*, SAAB Aircraft Co., Linkoping, Sweden, 1954.

31.   Frost, N. F., and Dugdale, D. S., "The Propagation of Fatigue Cracks in Sheet Specimens," *Journal of Mechanics and Physics of Solids*, **6** (1958): 92–110.

32.   McEvily, A. J., Jr., and Illg, W., "The Rate of Fatigue Crack Propagation in Two Aluminum Alloys," NACA TN 4394, 1958.

33.   Schijve, J., "Fatigue Crack Propagation in Light Alloy Sheet Material and Structures," NLL Report MP 195, 1960.

34.   Schijve, J., "Fatigue Crack Propagation in Light Alloy Sheet Material and Structures," *Advances in Aeronautical Sciences*, Vol. 3, Pergamon Press, New York, 1962, pp. 287–408.

35.   Paris, P. C., Gomez, M. P., and Anderson, W. E., "A Rational Analytic Theory of Fatigue," *Trend Engineering*, University of Washington, **13**, No. 1 (1961): 9–14.

36.   Paris, P. C., *The Growth of Fatigue Cracks Due to Variations in Load*, Ph.D. Dissertation, Lehigh University, Bethlehem, Pa., 1962.

37.   Paris, P. C., "Crack Propagation Caused by Fluctuating Loads," ASME Paper No. 62-Met 3, ASME, New York, 1962.

38.   Paris, P. C., and Erdogan, F., "A Critical Analysis of Crack Propagation Laws," *Journal of Basic Engineering*, ASME Transactions, Series D, **85**, No. 4 (1963): 528–534.

39.   Paris, P. C., "The Fracture Mechanics Approach to Fatigue," *Fatigue—An Interdisciplinary Approach*, Proceedings of the Tenth Sagamore Army Materials Research Conference, Syracuse University Press, Syracuse, N.Y., 1964, pp. 107–132.

40. Liu, H. W., "Crack Propagation in Thin Sheet Metal Under Repeated Loading," *Journal of Basic Engineering*, ASME Transactions, Series D, **83** (1961).

41. Christensen, R. H., "Cracking and Fracture in Metals and Structures," *Proceedings of Crack Propagation Symposium 2*, Cranfield, England, 1961, pp. 326–374.

42. Weibull, W., "The Effect of Size and Stress History on Fatigue Crack Initiation and Propagation," *Proceedings of Crack Propagation Symposium 2*, Cranfield, England, 1961, pp. 271–286.

43. Valluri, S. R., "A Unified Engineering Theory of High-Stress Level Fatigue," *Aerospace Engineering*, **20** (1961): 18–19, 68–69.

44. Frost, N. E., "The Effect of Mean Stress on the Rate of Growth of Fatigue Cracks in Sheet Materials," *Journal of Mechanical Engineering*, **4**, No. 1 (1962): 22.

45. Stulen, F. B., "The Theoretical Development of the Crack Propagation Formula," TN No. 554, Curtiss-Wright Corp., Caldwell, N.J., 1962.

46. McEvily, A. J. Jr., and Boettner, R. C., "On Fatigue Crack Propagation in F.C.C. Metals, *Acta Metallurgica*, **11** (1963): 725.

47. Weibull, W., "A Theory of Fatigue Crack Propagation in Sheet Specimens," *Acta Metallurgica*, **11** (1963): 745.

48. Liu, H. W., "Fatigue Crack Propagation and Applied Stress Range—An Energy Approach," *Journal of Basic Engineering*, ASME Transcations Series, D, **85** (1963): 116.

49. Liu, H. W., "Size Effects on Fatigue Crack Propagation," Galcit SM 63-7, California Institute of Technology, Air Force Contract No. AF 33(616)-6270, Project No. 7024, Task No. 7066 or ARI Report 64–68, 1964.

50. McClintock, F. A., "On the Plasticity in the Growth of Fatigue Cracks," *Fracture of Solids*, Interscience and AIME, 1963, pp. 65–102.

51. Valluri, S. R., Glassco, J. B., and Bockrath, G. E., "Further Considerations Concerning a Theory of Crack Propagation in Metal Fatigue," SAE Paper No. 752A, 1963.

52. Krafft, J. M., "Correlation of Plane Strain Crack Toughness with Strain Hardening Characteristics of a Low, a Medium, and a High Strength Steel," *Applied Materials Research*, **3**, No. 2 (1964): 88–101.

53. Krafft, J. M., "A Comparison of Cyclic Fatigue Crack Propagation with Single Cycle Crack Toughness and Plastic Flow," Special Report to the ASTM-FTHSM, 1964.

54. Brock, D., and Schijve, J., "The Influence of Mean Stress on the Propagation of Fatigue Cracks in Aluminum Alloy Sheet," NIRIRM 2111, 1961.

55. Manson, S. S., "Interfaces between Fatigue, Creep and Fracture," *Proceedings of First International Conference on Fracture*, Sendai, Japan, 1965.

56. Weertman, J., "Rate of Growth of Fatigue Cracks as Calculated from the Theory of Infinitesimal Dislocations Distributed on a Plane," *Proceedings of International Conference on Fracture*, Vol. 1, Japanese Society for Strength and Fracture of Metals, Sendai, Japan, September 1965.

57. McEvily, Jr., A. J., and Johnston, T. I., "The Role of Cross Slip in Brittle Fracture and Fatigue," *Proceedings of International Conference on Fracture*, Vol. 1, Japanese Society for Strength and Fracture of Metals, Sendai, Japan, September, 1965.

58. Hoeppner, D. W., Pettit, D. F., and Hyler, W. S., "A Study of Fatigue and Other Related Problems Associated with Drill Pipe and Casing Materials for Project Mohole," *Summary Report III*, Federal Clearing House of Scientific Research, Battelle, Columbus, 1967.

59. Forman, R. G., Kearney, V. F., and Engle, R. M., "Numerical Analysis of Crack Propagation in Cyclic-Loaded Structures," *Journal of Basic Engineering*, ASME Transactions, Series D (1967): 89, 459.

60. Rawe, R. A., and Fitman, D. A., "A Parametric Relationship Between Fatigue Life, Cyclic

Stress, and Crack Length in Flat Panels and Cylinders," *Douglas Aircraft Paper 4285*, ASME Winter Annual Meeting, 1967.

61. Walker, E. K., "The Effect of Stress Ratio During Crack Propagation and Fatigue for 2024-T3 and 7075-T6," *The Effects of Environment and Complex Load History on Fatigue Life*, STP-462, American Society for Testing and Materials, Philadelphia, 1970, pp. 1–15.

62. Lehr, K. R., and Lin, H. W., "Fatigue Crack Propagation and Strain Cycling Properties," *International Journal of Fracture Mechanics*, **5**, No. 1 (1969).

63. Donahue, R. J., and McEvily Jr., A. J., *Symposium on Fracture Mechanics*, Carnegie Melon University, Pittsburgh, 1970.

64. Krafft, J. M., "A Comparison of Cyclic Fatigue-Crack Propagation with Single Cycle Crack Toughness and Plastic Flow," presented to ASIMC Committee E24 on Fracture, 1964.

65. Donahue R. J., Clark, H. M., Atanmo, P., Kumble, R., and McEvily Jr, A. J., "Crack Opening Displacement and the Rate of Fatigue Crack Growth," University of Connecticut, Institute of Materials of Science Report, 1971.

66. Clark, W. G., Jr., "Fracture Mechanics in Fatigue," *Experimental Mechanics*, September 1971.

67. Barsom, J. M., "Effect of Cyclic Stress Form on Corrosive Fatigue Crack Propagation Below K_{ISCC} in a High Yield Strength Steel," *Corrosion Fatigue*, 1972, pp. 424–436.

68. Hartman, A., and Schijve, J., "The Effects of Environment and Load Frequency on the Crack Propagation Law for Macro Fatigue Crack Growth in Aluminum Alloys," *Engineering Fracture Mechanics*, **1** (1970): 615.

69. McMillan, J. L., and Pelloux, R. M. N., "Fatigue Crack Propagation under Program and Random Loads," STP-415, American Society for Testing and Materials, Philadelphia, 1967, p. 505.

70. Jonas, D., and Wei, R. P., "An Exploratory Study of Delay in Fatigue Crack Growth," *International Journal of Fracture Mechanics*, **7**, No. 1 (March 1971).

71. Wheeler, D. E., "Spectrum Loading and Crack Growth," ASME Paper No. 71-MET, January 1972.

72. Willenborg, J., Engle, R. M., and Wood, H. A., "A Crack Growth Retardation Model Using an Effective Stress Concept," AFFDL Tech. Memo. 71-1-F1312, Air Force Flight Dynamics Laboratory, Wright-Patterson AFB, Ohio, January 1971.

73. Elber, W., "The Significance of Fatigue Crack Closure," *Damage Tolerance in Aircraft Structures*, STP-486, American Society for Testing and Materials, Philadelphia, 1970, pp. 230–243.

74. Wundt, B. M., *Effects of Notches on Low Cycle Fatigue*, STP-490, American Society for Testing and Materials, Philadelphia, 1972.

75. Schijve, J., "The Accumulation of Fatigue Damage in Aircraft Materials and Structures," AGARD Publication No. AG-157, January 1972.

76. Ekvall, J. C., and Young, L., "Converting Fatigue Loading Spectra for Flight-by Flight Testing of Aircraft and Helicopter Components," Report of Lockheed California Company, Burbank, Calif., 1974.

77. Wood, H. A., "Fracture Control Procedures for Aircraft Structural Integrity," AFFDL Report TR-21-89, Wright-Patterson Air Force Base, Ohio, July 1971.

78. Osgood, C. C., "Assuring Component Life," *Machine Design*, September 2 1971, p. 95.

79. Rich, T. P., and Cartwright, D. J. (eds.), *Case Studies in Fracture Mechanics*, Report No. AMMRC MS77-5, U.S. Army Material Development and Readiness Command, Alexandria, Va., 1977.

CHAPTER 9

Use of Statistics in Fatigue Analysis

9.1 INTRODUCTION

In Section 7.6 it was noted that the scatter in fatigue data makes it necessary to utilize statistical techniques in describing, analyzing, and comparing fatigue failure data so that the designer may achieve the desired reliability on a rational basis. Several uses of statistical tools in fatigue testing and analysis will be described in the following pages, but it will be useful to first consider some of the basic definitions and concepts of statistical analysis.

9.2 DEFINITIONS

A *population* or *universe* is a collection of objects having some common characteristic. The collection may be either finite or infinite. The characteristics of a given population are fixed, though usually unknown.

A *sample* is any subcollection of objects drawn from a population. The characteristics of a sample vary from sample to sample, even though the samples are all drawn from the same population.

The characteristics of a population are called *parameters*. For any given population the parameters are fixed constants. For example, if a population consists of $x_1, x_2, x_3, \ldots, x_n$, where x_i is some measurable characteristic of the population, for example the fatigue life at a specified stress level, then the *mean* life μ is given by

$$\mu = \frac{x_1 + x_2 + x_3, \ldots, x_n}{n} \tag{9-1}$$

The characteristics of samples are called *statistics*. In general, these statistics vary from sample to sample drawn from the same population. For examples, x_1, x_3, and x_5 might be drawn as a first sample and x_2, x_4, x_6, and x_7 drawn as a second sample. The sample means $\bar{x}_1$ and $\bar{x}_2$ for samples 1 and

2, respectively, would be computed as

$$\bar{x}_1 = \frac{x_1 + x_3 + x_5}{3} \tag{9-2}$$

and

$$\bar{x}_2 = \frac{x_2 + x_4 + x_6 + x_7}{4} \tag{9-3}$$

Population parameters are not usually easy to determine because to do so requires that the characteristic of interest must be measured for each and every object in the population. This is difficult if not impossible, and even when it is possible, it would usually be undesirable or impractical. For example, to obtain the fatigue limit for all the specimens in a given heat of 4340 steel would result in damaging or breaking every specimen, and no useable material would remain for making machine parts. Population parameters are, therefore, usually *estimated* on the basis of statistics calculated for *representative* samples drawn from the population. The procedures used to select samples and determine appropriate sample size are often called *statistical design of experiment*.

Several different types of samples may be used to yield statistics useful in the estimation of population parameters. All these sample types have in common some element of *randomness*. By this is meant that any object in the population has as much chance of being selected for the sample as any other object in the population, within the constraints of the sampling procedure. Various sampling procedures may be used to obtain a representative sample with maximum randomness, depending on how much is known about the population and how it may be partitioned. For example, one might use unrestricted random sampling, stratified random sampling, stratified proportional random sampling, or optimum-allocation-of-strata random sampling.

Each of these techniques requires more *a priori* knowledge of the population and its use than the preceding one. Unrestricted random sampling implies random sample selection from the population as a whole. Stratified random sampling implies that one knows how to partition the population into meaningful strata, and random selection is made within each stratum, making sure that all strata are represented. Stratified proportional random sampling implies knowledge of the relative sizes of the strata, with random samples drawn according to sizes of strata. Finally, optimum-allocation-of-strata random sampling implies that one has information about the population strata and the way they will be used so that an optimum sample selection may be made to give the most representative information about the population. In general, the more that is known about a population, the better its parameters may be estimated by an appropriate sampling technique.

Some of the more useful statistics used in estimating population parameters are often classified as *descriptive statistics*. These include measures of location, measures of dispersion, measures of skewness, and measures of kurtosis or peakedness of the distribution of the characteristic of interest. Several such descriptive statistics are listed as follows:

Measure of location

Arithmetic mean	$\bar{x} = \dfrac{1}{n} \sum\limits_{i=1}^{n} x_i$
Median	Middle value in an ordered array
p^{th} quantile or fractile	Characteristic in an ordered array such that 100 percent of the values are less than it
Mode	Value with greatest frequency of occurrence
Geometric mean	antilog $\left[\dfrac{1}{n} \sum\limits_{i=1}^{n} \log x_i \right] = \sqrt[n]{\prod\limits_{i=1}^{n} x_i}$

Measure of dispersion

Variance	$s^2 = \dfrac{1}{n-1} \sum\limits_{i=1}^{n} (x_i - \bar{x})^2$
Standard deviation	$+ \sqrt{s^2}$
Mean deviation	$\dfrac{1}{n-1} \sum\limits_{i=1}^{n} |x_i - \bar{x}|$
Range	(largest value) $-$ (smallest value)

Measure of skewness

Skewness $\quad Sk = \dfrac{\dfrac{1}{n} \sum\limits_{i=1}^{n} (x_i - \bar{x})^3}{\left[\dfrac{1}{n} \sum\limits_{i=1}^{n} (x_i - \bar{x})^2 \right]^{3/2}}$

Measure of kurtosis (peakedness)

Kurtosis $\quad K = \dfrac{\dfrac{1}{n} \sum\limits_{i=1}^{n} (x_i - \bar{x})^4}{\left[\dfrac{1}{n} \sum\limits_{i=1}^{n} (x_i - \bar{x}) \right]^2}$

In addition to this list, three additional statistics will be useful in evaluating and comparing populations of interest. These are the chi-squared (χ^2) statistic, the Snedcor (F) statistic, and the Student's (t) statistic. These will be described in more detail later.

Finally, it may be observed that statistics are useful in three basic ways: to describe the sample per se, to estimate parameters of the population from which the samples came, and to compare and contrast samples and, thereby, compare and contrast populations from which the samples came.

9.3 POPULATION DISTRIBUTIONS

The measurable characteristic of interest for a given population, say x, typically varies over some range of values. A function $f(x)$ that gives the probability that the random variable x will assume any *particular* value in the range is called the *probability density function* or frequency function. To put it differently, the probability density function $f(x)$ describes how the population is distributed with respect to some scale of qualities or quantities. Probability density functions have the following properties:

1. $f(x) \geq 0$ for all x
2. $\sum\limits_{\text{all } x} f(x) = 1$ for discrete case

3. $\int_{\text{all } x} f(x)dx = 1$ for continuous case

Another useful function, called the *cumulative distribution function*, $F(x)$, gives the probability of obtaining a value of the random variable x that is less than or equal to a particular value in the range. Cumulative distribution functions have the following properties:

1. $F(x)$ is a monotonic nondecreasing function.
2. $F(x) = 0$ for all x less than the smallest value of x in the population.
3. $F(x) = 1$ for all x greater than the largest value of x in the population.

Probability density functions may be either discrete or continuous. An example of a discrete probability density function is given by

$$f(x) = \frac{1}{2} \quad \text{for} \quad x = 0, 1$$

$$f(x) = 0 \quad \text{otherwise} \tag{9-4}$$

This discrete probability density function may be plotted as shown in Figure 9.1a. The plot of the corresponding cumulative distribution function is sketched in Figure 9.1b. Other examples* of discrete probability density functions are the binomial distribution and the Poisson distribution.

A very important example of a continuous probability density function is

*See, for example, ref. 1.

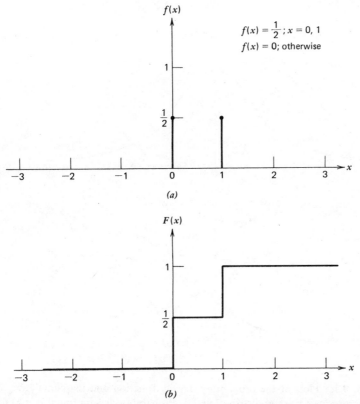

$$f(x) = \frac{1}{2}; x = 0, 1$$
$$f(x) = 0; \text{otherwise}$$

(a)

(b)

FIGURE 9.1. An example of a discrete distribution, showing the probability density function and the cumulative distribution function. (a) Probability density function. (b) Cumulative distribution function.

the normal or Gaussian probability density function, given by

$$f(x) = \frac{1}{\sqrt{2\pi}\,\sigma} e \exp\left[-\frac{1}{2}\left(\frac{x-\mu}{\sigma}\right)^2\right] \quad \text{for} \quad -\infty < x < \infty \quad (9\text{-}5)$$

where μ is the mean and σ is the standard deviation of the population. This normal probability density function is plotted in Figure 9.2a. The plot of the corresponding cumulative distribution function is sketched in Figure 9.2b. It may be noted in (9-5) that the mean μ and standard deviation σ are the only two *parameters* required to give a complete determination of the distribution. Therefore, the normal distribution is termed a two-parameter distribution. The quantity

$$X \equiv \frac{x-\mu}{\sigma} \quad (9\text{-}6)$$

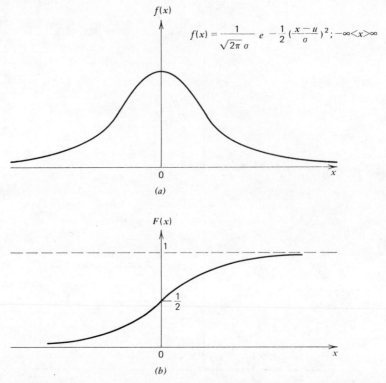

$$f(x) = \frac{1}{\sqrt{2\pi}\,\sigma}\; e - \frac{1}{2}(\frac{x-u}{\sigma})^2; -\infty < x > \infty$$

(a)

(b)

FIGURE 9.2. Plots of the probability density function and the cumulative distribution function for a normal distribution. (*a*) Probability density function. (*b*) Cumulative distribution function.

is defined to be the *standard normal variable* and has a normal distribution with a mean of zero and a standard deviation of unity. Any normal distribution may be transformed into a standard normal distribution if its mean and standard deviation are known. The probability density function for the standard normal distribution is shown in tabular form in Table 9.1. The bell-shape curve of Figure 9.2*a* which represents the normal distribution, is symmetrical about its mean μ, and the inflection points occur at $\mu \pm \sigma$. The tails of the distribution, although they approach the x axis, extend to infinity in both directions. The normal cumulative distribution function is given for any specified value of x by the integrated area under the probability density curve up to the specified value of x. Table 9.2 gives in tabular form values of the cumulative distribution function $F(X)$ for a standard normal distribution.

Other continuous distributions of interest include the chi-squared distribution, Student's t distribution, Snedcor's F distribution, and Weibull distribution.

Table 9.1. Ordinates of the Probability Density Function $f(X)$ for the Standard Normal Distribution, Where $f(X) = \dfrac{1}{\sqrt{2\pi}}\, e^{(-X^2/2)}$

x	.00	.01	.02	.03	.04	.05	.06	.07	.08	.09
.0	.3989	.3989	.3989	.3988	.3986	.3984	.3982	.3980	.3977	.3973
.1	.3970	.3965	.3961	.3956	.3951	.3945	.3939	.3932	.3925	.3918
.2	.3910	.3902	.3894	.3885	.3876	.3867	.3857	.3847	.3836	.3825
.3	.3814	.3802	.3790	.3778	.3765	.3752	.3739	.3725	.3712	.3697
.4	.3683	.3668	.3653	.3637	.3621	.3605	.3589	.3572	.3555	.3538
.5	.3521	.3503	.3485	.3467	.3448	.3429	.3410	.3391	.3372	.3352
.6	.3332	.3312	.3292	.3271	.3251	.3230	.3209	.3187	.3166	.3144
.7	.3123	.3101	.3079	.3056	.3034	.3011	.2989	.2966	.2943	.2920
.8	.2897	.2874	.2850	.2827	.2803	.2780	.2756	.2732	.2709	.2685
.9	.2661	.2637	.2613	.2589	.2565	.2541	.2516	.2492	.2468	.2444
1.0	.2420	.2396	.2371	.2347	.2323	.2299	.2275	.2251	.2227	.2203
1.1	.2179	.2155	.2131	.2107	.2083	.2059	.2036	.2012	.1989	.1965
1.2	.1942	.1919	.1895	.1872	.1849	.1826	.1804	.1781	.1758	.1736
1.3	.1714	.1691	.1669	.1647	.1626	.1604	.1582	.1561	.1539	.1518
1.4	.1497	.1476	.1456	.1435	.1415	.1394	.1374	.1354	.1334	.1315
1.5	.1295	.1276	.1257	.1238	.1219	.1200	.1182	.1163	.1145	.1127
1.6	.1109	.1092	.1074	.1057	.1040	.1023	.1006	.0989	.0973	.0957
1.7	.0940	.0925	.0909	.0893	.0878	.0863	.0848	.0833	.0818	.0804
1.8	.0790	.0775	.0761	.0748	.0734	.0721	.0707	.0694	.0681	.0669
1.9	.0656	.0644	.0632	.0620	.0608	.0596	.0584	.0573	.0562	.0551
2.0	.0540	.0529	.0519	.0508	.0498	.0488	.0478	.0468	.0459	.0449
2.1	.0440	.0431	.0422	.0413	.0404	.0396	.0387	.0379	.0371	.0363
2.2	.0355	.0347	.0339	.0332	.0325	.0317	.0310	.0303	.0297	.0290
2.3	.0283	.0277	.0270	.0264	.0258	.0252	.0246	.0241	.0235	.0229
2.4	.0224	.0219	.0213	.0208	.0203	.0198	.0194	.0189	.0184	.0180
2.5	.0175	.0171	.0167	.0163	.0158	.0154	.0151	.0147	.0143	.0139
2.6	.0136	.0132	.0129	.0126	.0122	.0119	.0116	.0113	.0110	.0107
2.7	.0104	.0101	.0099	.0096	.0093	.0091	.0088	.0086	.0084	.0081
2.8	.0079	.0077	.0075	.0073	.0071	.0069	.0067	.0065	.0063	.0061
2.9	.0060	.0058	.0056	.0055	.0053	.0051	.0050	.0048	.0047	.0046
3.0	.0044	.0043	.0042	.0040	.0039	.0038	.0037	.0036	.0035	.0034
3.1	.0033	.0032	.0031	.0030	.0029	.0028	.0027	.0026	.0025	.0025
3.2	.0024	.0023	.0022	.0022	.0021	.0020	.0020	.0019	.0018	.0018
3.3	.0017	.0017	.0016	.0016	.0015	.0015	.0014	.0014	.0013	.0013
3.4	.0012	.0012	.0012	.0011	.0011	.0010	.0010	.0010	.0009	.0009
3.5	.0009	.0008	.0008	.0008	.0008	.0007	.0007	.0007	.0007	.0006
3.6	.0006	.0006	.0006	.0005	.0005	.0005	.0005	.0005	.0005	.0004
3.7	.0004	.0004	.0004	.0004	.0004	.0004	.0003	.0003	.0003	.0003
3.8	.0003	.0003	.0003	.0003	.0003	.0002	.0002	.0002	.0002	.0002
3.9	.0002	.0002	.0002	.0002	.0002	.0002	.0002	.0002	.0001	.0001

From ref. 1.

Table 9.2. Cumulative Distribution Function $F(X)$ for the Standard Normal Distribution, Where $F(X) = \int_{-\infty}^{X} \frac{1}{\sqrt{2\pi}} e^{(-t^2/2)}dt$.

x	.00	.01	.02	.03	.04	.05	.06	.07	.08	.09
.0	.5000	.5040	.5080	.5120	.5160	.5199	.5239	.5279	.5319	.5359
.1	.5398	.5438	.5478	.5517	.5557	.5596	.5636	.5675	.5714	.5753
.2	.5793	.5832	.5871	.5910	.5948	.5987	.6026	.6064	.6103	.6141
.3	.6179	.6217	.6255	.6293	.6331	.6368	.6406	.6443	.6480	.6517
.4	.6554	.6591	.6628	.6664	.6700	.6736	.6772	.6808	.6844	.6879
.5	.6915	.6950	.6985	.7019	.7054	.7088	.7123	.7157	.7190	.7224
.6	.7257	.7291	.7324	.7357	.7389	.7422	.7454	.7486	.7517	.7549
.7	.7580	.7611	.7642	.7673	.7704	.7734	.7764	.7794	.7823	.7852
.8	.7881	.7910	.7939	.7967	.7995	.8023	.8051	.8078	.8106	.8133
.9	.8159	.8186	.8212	.8238	.8264	.8289	.8315	.8340	.8365	.8389
1.0	.8413	.8438	.8461	.8485	.8508	.8531	.8554	.8577	.8599	.8621
1.1	.8643	.8665	.8686	.8708	.8729	.8749	.8770	.8790	.8810	.8830
1.2	.8849	.8869	.8888	.8907	.8925	.8944	.8962	.8980	.8997	.9015
1.3	.9032	.9049	.9066	.9082	.9099	.9115	.9131	.9147	.9162	.9177
1.4	.9192	.9207	.9222	.9236	.9251	.9265	.9279	.9292	.9306	.9319
1.5	.9332	.9345	.9357	.9370	.9382	.9394	.9406	.9418	.9429	.9441
1.6	.9452	.9463	.9474	.9484	.9495	.9505	.9515	.9525	.9535	.9545
1.7	.9554	.9564	.9573	.9582	.9591	.9599	.9608	.9616	.9625	.9633
1.8	.9641	.9649	.9656	.9664	.9671	.9678	.9686	.9693	.9699	.9706
1.9	.9713	.9719	.9726	.9732	.9738	.9744	.9750	.9756	.9761	.9767
2.0	.9772	.9778	.9783	.9788	.9793	.9798	.9803	.9808	.9812	.9817
2.1	.9821	.9826	.9830	.9834	.9838	.9842	.9846	.9850	.9854	.9857
2.2	.9861	.9864	.9868	.9871	.9875	.9878	.9881	.9884	.9887	.9890
2.3	.9893	.9896	.9898	.9901	.9904	.9906	.9909	.9911	.9913	.9916
2.4	.9918	.9920	.9922	.9925	.9927	.9929	.9931	.9932	.9934	.9936
2.5	.9938	.9940	.9941	.9943	.9945	.9946	.9948	.9949	.9951	.9952
2.6	.9953	.9955	.9956	.9957	.9959	.9960	.9961	.9962	.9963	.9964
2.7	.9965	.9966	.9967	.9968	.9969	.9970	.9971	.9972	.9973	.9974
2.8	.9974	.9975	.9976	.9977	.9977	.9978	.9979	.9979	.9980	.9981
2.9	.9981	.9982	.9982	.9983	.9984	.9984	.9985	.9985	.9986	.9986
3.0	.9987	.9987	.9987	.9988	.9988	.9989	.9989	.9989	.9990	.9990
3.1	.9990	.9991	.9991	.9991	.9992	.9992	.9992	.9992	.9993	.9993
3.2	.9993	.9993	.9994	.9994	.9994	.9994	.9994	.9995	.9995	.9995
3.3	.9995	.9995	.9995	.9996	.9996	.9996	.9996	.9996	.9996	.9997
3.4	.9997	.9997	.9997	.9997	.9997	.9997	.9997	.9997	.9997	.9998

X	1.282	1.645	1.960	2.326	2.576	3.090	3.291	3.891	4.417
$F(X)$	.90	.95	.975	.99	.995	.999	.9995	.99995	.999995
$2[1 - F(X)]$	.20	.10	.05	.02	.01	.002	.001	.0001	.00001

From ref. 1.

9.4 SAMPLING DISTRIBUTIONS

Each statistic of a sample varies from sample to sample and therefore is itself a random variable having its own distribution, which is called a *sampling distribution*. Sampling distributions have certain properties that make them useful because of their relationship to the distribution of the population. As one example, the following theorem may be cited:

Let $\bar{x}$ be the mean of a random sample of size n drawn from an infinite population with mean μ and standard deviation σ. It can be proved then that $\bar{x}$ has a sampling distribution with mean μ and standard deviation $\sigma/\sqrt{n}$. Further, if the population from which the sample was drawn is normal, it can be proved that the sampling distribution of $\bar{x}$ is also normal. Numerous other useful theorems about sampling distributions have been proved. The most powerful and most important theorem in statistics from both the theoretical and applied point of view is the *central limit theorem*. Historically, the central limit theorem required over two centuries to develop, with contributions being made by many eminent mathematicians. The major milestones in development of the central limit theorem include the following:

1713—Bernoulli's theorem proved

1733—DeMoivre's theorem proved

1812—Laplace improved DeMoivre's theorem

1860—Tchebycheff's inequality proved

1901—Liapounoff's theorem proved

1925—Levy's general form of the central limit theorem

1935—Fellar and Lindberg completed the development by giving necessary and sufficient conditions for the central limit theorem

Stated mathematically, the central limit theorem may be written as

$$\lim_{n\to\infty} P\left\{ \frac{\bar{x}-\mu}{\sigma/\sqrt{n}} \le \alpha \right\} = \frac{1}{2\pi} \int_{-\infty}^{\alpha} e^{-x^2/2}\,dx \qquad (9\text{-}7)$$

where the expression in braces is to be read as "the probability that $(\bar{x}-\mu)/(\sigma/\sqrt{n})$ is less than or equal to α."

In words, the central limit theorem may be phrased in the following way: Let $\bar{x}$ be the mean of a random sample of size n drawn from *any* infinite population with mean μ and standard deviation σ. Then for sufficiently large n (practically speaking n of 25 or 30) $\bar{x}$ has an approximately *normal* distribution with mean μ and standard deviation of $\sigma/\sqrt{n}$. Further, the central limit theorem applies to discrete as well as continuous distributions.

Three sampling distributions of great interest are the chi-squared distribution, the t distribution, and the F distribution. The chi-squared distribution may be defined as follows: Let $x_1, x_2, x_3, \ldots, x_\nu$ be ν normal, independent random variables distributed with mean of zero and standard deviation of 1.

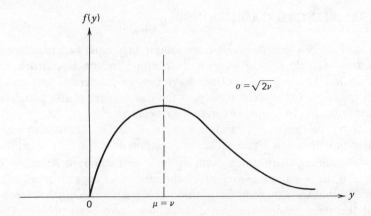

FIGURE 9.3. Sketch of the chi-squared distribution with mean μ and standard deviation σ.

Then the random variable

$$y = \sum_{i=1}^{\nu} x_i^2 \qquad (9\text{-}8)$$

has a chi-squared distribution with ν "degrees of freedom." The number of degrees of freedom is equal to the number of independent random variables that comprise the sum of squares in (9-8).

The chi-squared probability density function $f(y)$ is given by

$$f(y) = \frac{1}{\Gamma\left(\dfrac{\nu}{2}\right)2^{\nu/2}} y^{(\nu/2)-1} e^{-y/2}; \qquad y > 0$$

$$f(y) = 0; \qquad\qquad\qquad \text{otherwise} \qquad (9\text{-}9)$$

The chi-squared probability density function is sketched in Figure 9.3. An important property of chi-squared variables is that the sum of two independent chi-squared variables is also a chi-squared variable. The cumulative distribution function for the chi-squared variable is given in tabular form in Table 9.3. Hypotheses concerning the variance may be tested using the chi-squared distribution as well as goodness-of-fit tests, tests of independence, and maximum likelihood ratio tests.*

The Student's t distribution is a sampling distribution devised by Gossett, who worked in a brewery, an honorable profession, while he was a student of the then professionally questionable activity of mathematical probability. He, therefore, adopted the pen name "Student." Student's t distribution is defined as follows: Let u be a normal random variable with mean of zero and

*See, for example, ref. 1.

Freedom, Where, $F(Y) = \int_0^Y \frac{x^{\cdots}\, e^{\cdots}}{2^{\nu/2}[(\nu-2)/2]!}\, dx$.

ν \ F(Y)	.005	.010	.025	.050	.100	.250	.500	.750	.900	.950	.975	.990	.995
1	.0⁴393	.0³157	.0³982	.0²393	.0158	.102	.455	1.32	2.71	3.84	5.02	6.63	7.88
2	.0100	.0201	.0506	.103	.211	.575	1.39	2.77	4.61	5.99	7.38	9.21	10.6
3	.0717	.115	.216	.352	.584	1.21	2.37	4.11	6.25	7.81	9.35	11.3	12.8
4	.207	.297	.484	.711	1.06	1.92	3.36	5.39	7.78	9.49	11.1	13.3	14.9
5	.412	.554	.831	1.15	1.61	2.67	4.35	6.63	9.24	11.1	12.8	15.1	16.7
6	.676	.872	1.24	1.64	2.20	3.45	5.35	7.84	10.6	12.6	14.4	16.8	18.5
7	.989	1.24	1.69	2.17	2.83	4.25	6.35	9.04	12.0	14.1	16.0	18.5	20.3
8	1.34	1.65	2.18	2.73	3.49	5.07	7.34	10.2	13.4	15.5	17.5	20.1	22.0
9	1.73	2.09	2.70	3.33	4.17	5.90	8.34	11.4	14.7	16.9	19.0	21.7	23.6
10	2.16	2.56	3.25	3.94	4.87	6.74	9.34	12.5	16.0	18.3	20.5	23.2	25.2
11	2.60	3.05	3.82	4.57	5.58	7.58	10.3	13.7	17.3	19.7	21.9	24.7	26.8
12	3.07	3.57	4.40	5.23	6.30	8.44	11.3	14.8	18.5	21.0	23.3	26.2	28.3
13	3.57	4.11	5.01	5.89	7.04	9.30	12.3	16.0	19.8	22.4	24.7	27.7	29.8
14	4.07	4.66	5.63	6.57	7.79	10.2	13.3	17.1	21.1	23.7	26.1	29.1	31.3
15	4.60	5.23	6.26	7.26	8.55	11.0	14.3	18.2	22.3	25.0	27.5	30.6	32.8
16	5.14	5.81	6.91	7.96	9.31	11.9	15.3	19.4	23.5	26.3	28.8	32.0	34.3
17	5.70	6.41	7.56	8.67	10.1	12.8	16.3	20.5	24.8	27.6	30.2	33.4	35.7
18	6.26	7.01	8.23	9.39	10.9	13.7	17.3	21.6	26.0	28.9	31.5	34.8	37.2
19	6.84	7.63	8.91	10.1	11.7	14.6	18.3	22.7	27.2	30.1	32.9	36.2	38.6
20	7.43	8.26	9.59	10.9	12.4	15.5	19.3	23.8	28.4	31.4	34.2	37.6	40.0
21	8.03	8.90	10.3	11.6	13.2	16.3	20.3	24.9	29.6	32.7	35.5	38.9	41.4
22	8.64	9.54	11.0	12.3	14.0	17.2	21.3	26.0	30.8	33.9	36.8	40.3	42.8
23	9.26	10.2	11.7	13.1	14.8	18.1	22.3	27.1	32.0	35.2	38.1	41.6	44.2
24	9.89	10.9	12.4	13.8	15.7	19.0	23.3	28.2	33.2	36.4	39.4	43.0	45.6
25	10.5	11.5	13.1	14.6	16.5	19.9	24.3	29.3	34.4	37.7	40.6	44.3	46.9
26	11.2	12.2	13.8	15.4	17.3	20.8	25.3	30.4	35.6	38.9	41.9	45.6	48.3
27	11.8	12.9	14.6	16.2	18.1	21.7	26.3	31.5	36.7	40.1	43.2	47.0	49.6
28	12.5	13.6	15.3	16.9	18.9	22.7	27.3	32.6	37.9	41.3	44.5	48.3	51.0
29	13.1	14.3	16.0	17.7	19.8	23.6	28.3	33.7	39.1	42.6	45.7	49.6	52.3
30	13.8	15.0	16.8	18.5	20.6	24.5	29.3	34.8	40.3	43.8	47.0	50.9	53.7

(Note that tabled values are Y). From ref. 1.

variance of 1. Let v be an independent chi-squared random variable with k degrees of freedom. The t statistic

$$t = \frac{u}{\sqrt{v/k}}$$

(9-10)

has a Student's t distribution with k degrees of freedom. The probability density function $f(t)$ for the Student's t distribution is given by

$$f(t) = C_t \frac{1}{\left(1 + \dfrac{t^2}{k}\right)^{(k+1)/2}} ; \quad -\infty < t < \infty$$

(9-11)

where

$$C_t = \frac{\Gamma\left(\dfrac{n+1}{2}\right)}{\sqrt{n\pi}\,\Gamma\left(\dfrac{n}{2}\right)}$$

(9-12)

The Student's t distribution is sketched in Figure 9.4. The t distribution is useful in testing hypotheses about mean values and comparisons of means and in determining regression coefficients. The cumulative distribution function for Student's t distribution is shown in tabular form in Table 9.4.

Snedcor's F distribution may be defined as follows: Let u be a chi-square random variable with m degrees of freedom. Let v be an independent chi-squared random variable with n degrees of freedom. Then the F statistic

$$F = \frac{u/m}{v/n} = \frac{nu}{mv}$$

(9-13)

has a Snedcor's F distribution with m and n degrees of freedom. The

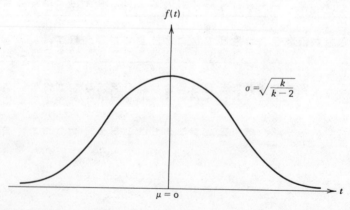

FIGURE 9.4. Sketch of Student's t distribution with mean μ and standard deviation σ.

Table 9.4. Cumulative Distribution Function $F(t)$ for the Student's t Distribution with k Degrees of Freedom, Where $F(t) = \int_{-\infty}^{t} \dfrac{\left(\dfrac{k-1}{2}\right)!}{\left(\dfrac{k-2}{2}\right)! \sqrt{\pi k}\left(1 + \dfrac{x^2}{k}\right)^{(k+1)/2}}\, dx$

k \\ $F(t)$	.75	.90	.95	.975	.99	.995	.9995
1	1.000	3.078	6.314	12.706	31.821	63.657	636.610
2	.816	1.886	2.920	4.303	6.965	9.925	31.598
3	.765	1.638	2.353	3.182	4.541	5.841	12.941
4	.741	1.533	2.132	2.776	3.747	4.604	8.610
5	.727	1.476	2.015	2.571	3.365	4.032	6.859
6	.718	1.440	1.943	2.447	3.143	3.707	5.959
7	.711	1.415	1.895	2.365	2.998	3.499	5.405
8	.706	1.397	1.860	2.306	2.896	3.355	5.041
9	.703	1.383	1.833	2.262	2.821	3.250	4.781
10	.700	1.372	1.812	2.228	2.764	3.169	4.587
11	.697	1.363	1.796	2.201	2.718	3.106	4.437
12	.695	1.356	1.782	2.179	2.681	3.055	4.318
13	.694	1.350	1.771	2.160	2.650	3.012	4.221
14	.692	1.345	1.761	2.145	2.624	2.977	4.140
15	.691	1.341	1.753	2.131	2.602	2.947	4.073
16	.690	1.337	1.746	2.120	2.583	2.921	4.015
17	.689	1.333	1.740	2.110	2.567	2.898	3.965
18	.688	1.330	1.734	2.101	2.552	2.878	3.922
19	.688	1.328	1.729	2.093	2.539	2.861	3.883
20	.687	1.325	1.725	2.086	2.528	2.845	3.850
21	.686	1.323	1.721	2.080	2.518	2.831	3.819
22	.686	1.321	1.717	2.074	2.508	2.819	3.792
23	.685	1.319	1.714	2.069	2.500	2.807	3.767
24	.685	1.318	1.711	2.064	2.492	2.797	3.745
25	.684	1.316	1.708	2.060	2.485	2.787	3.725
26	.684	1.315	1.706	2.056	2.479	2.779	3.707
27	.684	1.314	1.703	2.052	2.473	2.771	3.690
28	.683	1.313	1.701	2.048	2.467	2.763	3.674
29	.683	1.311	1.699	2.045	2.462	2.756	3.659
30	.683	1.310	1.697	2.042	2.457	2.750	3.646
40	.681	1.303	1.684	2.021	2.423	2.704	3.551
60	.679	1.296	1.671	2.000	2.390	2.660	3.460
120	.677	1.289	1.658	1.980	2.358	2.617	3.373
∞	.674	1.282	1.645	1.960	2.326	2.576	3.291

(Note that tabled values are t.) From ref. 1.

probability density function $f(F)$ for the F distribution is given by

$$f(F) = C_f \frac{F^{(m-2)/2}}{\left(1 + \dfrac{mF}{n}\right)^{(m+n)/2}}; \qquad F > 0$$

$$f(F) = 0; \qquad\qquad\qquad \text{otherwise} \qquad (9\text{-}14)$$

where

$$C_f = \frac{\Gamma\left(\dfrac{m+n}{2}\right)}{\Gamma\left(\dfrac{m}{2}\right)\Gamma\left(\dfrac{n}{2}\right)} m^{m/2} n^{n/2} \qquad (9\text{-}15)$$

The Snedcor F distribution is sketched in Figure 9.5. Tabular values for the cumulative distribution function for the F distribution are given in Table 9.5. The F distribution is used to test hypotheses about linear relationships among variances and is the basic statistical tool used in "analysis of variance" techniques.

Finally, the Weibull distribution, a three-parameter distribution, is actually a family of probability density functions. Each of the Weibull probability density functions may be written as

$$f(N) = \frac{b}{N_a - N_0}\left[\frac{N - N_0}{N_a - N_0}\right]^{b-1} e^{-[(N-N_0)/(N_a-N_0)]^b} \qquad (9\text{-}16)$$

where N = specimen life, cycles

$N_0 \geq 0$ = minimum life parameter

N_a = characteristic life parameter occurring at point where 63.2 percent have failed

$b > 0$ = Weibull shape parameter (or Weibull slope)

Several typical Weibull distributions are sketched in Figure 9.6. It may be noted that this probability density function represents a simple exponential

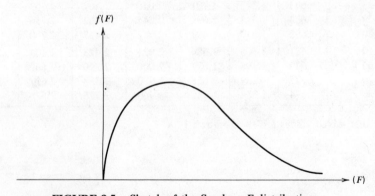

FIGURE 9.5.   Sketch of the Snedcor F distribution.

Table 9.5. Cumulative Distribution Function $G(F)$ for the Snedcor F Distribution with m (numerator) and n (denominator) Degrees of Freedom, Where

$$G(F) = \int_0^F \frac{\left(\frac{m+n-2}{2}\right)! \, m^{m/2} n^{n/2} x^{-[(m-2)/2]}}{\left(\frac{m-2}{2}\right)! \left(\frac{n-2}{2}\right)! (n+mx)^{(m+n)/2}} \, dx$$

$G(F)$	n	m	1	2	3	4	5	6	7	8	9	10	12	15	20	30	60	120	∞
.90		1	39.9	49.5	53.6	55.8	57.2	58.2	58.9	59.4	59.9	60.2	60.7	61.2	61.7	62.3	62.8	63.1	63.3
.95			161	200	216	225	230	234	237	239	241	242	244	246	248	250	252	253	254
.975			648	800	864	900	922	937	948	957	963	969	977	985	993	1000	1010	1010	1020
.99			4,050	5,000	5,400	5,620	5,760	5,860	5,930	5,980	6,020	6,060	6,110	6,160	6,210	6,260	6,310	6,340	6,370
.995			16,200	20,000	21,600	22,500	23,100	23,400	23,700	23,900	24,100	24,200	24,400	24,600	24,800	25,000	25,200	25,400	25,500
.90		2	8.53	9.00	9.16	9.24	9.29	9.33	9.35	9.37	9.38	9.39	9.41	9.42	9.44	9.46	9.47	9.48	9.49
.95			18.5	19.0	19.2	19.2	19.3	19.3	19.4	19.4	19.4	19.4	19.4	19.4	19.5	19.5	19.5	19.5	19.5
.975			38.5	39.0	39.2	39.2	39.3	39.3	39.4	39.4	39.4	39.4	39.4	39.4	39.4	39.5	39.5	39.5	39.5
.99			98.5	99.0	99.2	99.2	99.3	99.3	99.4	99.4	99.4	99.4	99.4	99.4	99.4	99.5	99.5	99.5	99.5
.995			199	199	199	199	199	199	199	199	199	199	199	199	199	199	199	199	199
.90		3	5.54	5.46	5.39	5.34	5.31	5.28	5.27	5.25	5.24	5.23	5.22	5.20	5.18	5.17	5.15	5.14	5.13
.95			10.1	9.55	9.28	9.12	9.01	8.94	8.89	8.85	8.81	8.79	8.74	8.70	8.66	8.62	8.57	8.55	8.53
.975			17.4	16.0	15.4	15.1	14.9	14.7	14.6	14.5	14.5	14.4	14.3	14.3	14.2	14.1	14.0	13.9	13.9
.99			34.1	30.8	29.5	28.7	28.2	27.9	27.7	27.5	27.3	27.2	27.1	26.9	26.7	26.5	26.3	26.2	26.1
.995			55.6	49.8	47.5	46.2	45.4	44.8	44.4	44.1	43.9	43.7	43.4	43.1	42.8	42.5	42.1	42.0	41.8
.90		4	4.54	4.32	4.19	4.11	4.05	4.01	3.98	3.95	3.93	3.92	3.90	3.87	3.84	3.82	3.79	3.78	3.76
.95			7.71	6.94	6.59	6.39	6.26	6.16	6.09	6.04	6.00	5.96	5.91	5.86	5.80	5.75	5.69	5.66	5.63
.975			12.2	10.6	9.98	9.60	9.36	9.20	9.07	8.98	8.90	8.84	8.75	8.66	8.56	8.46	8.36	8.31	8.26
.99			21.2	18.0	16.7	16.0	15.5	15.2	15.0	14.8	14.7	14.5	14.4	14.2	14.0	13.8	13.7	13.6	13.5
.995			31.3	26.3	24.3	23.2	22.5	22.0	21.6	21.4	21.1	21.0	20.7	20.4	20.2	19.9	19.6	19.5	19.3
.90		5	4.06	3.78	3.62	3.52	3.45	3.40	3.37	3.34	3.32	3.30	3.27	3.24	3.21	3.17	3.14	3.12	3.11
.95			6.61	5.79	5.41	5.19	5.05	4.95	4.88	4.82	4.77	4.74	4.68	4.62	4.56	4.50	4.43	4.40	4.37
.975			10.0	8.43	7.76	7.39	7.15	6.98	6.85	6.76	6.68	6.62	6.52	6.43	6.33	6.23	6.12	6.07	6.02
.99			16.3	13.3	12.1	11.4	11.0	10.7	10.5	10.3	10.2	10.1	9.89	9.72	9.55	9.38	9.20	9.11	9.02
.995			22.8	18.3	16.5	15.6	14.9	14.5	14.2	14.0	13.8	13.6	13.4	13.1	12.9	12.7	12.4	12.3	12.1

Table 9.5 (Continued)

G(F)	n	m	1	2	3	4	5	6	7	8	9	10	12	15	20	30	60	120	∞
.90	6		3.78	3.46	3.29	3.18	3.11	3.05	3.01	2.98	2.96	2.94	2.90	2.87	2.84	2.80	2.76	2.74	2.72
.95			5.99	5.14	4.76	4.53	4.39	4.28	4.21	4.15	4.10	4.06	4.00	3.94	3.87	3.81	3.74	3.70	3.67
.975			8.81	7.26	6.60	6.23	5.99	5.82	5.70	5.60	5.52	5.46	5.37	5.27	5.17	5.07	4.96	4.90	4.85
.99			13.7	10.9	9.78	9.15	8.75	8.47	8.26	8.10	7.98	7.87	7.72	7.56	7.40	7.23	7.06	6.97	6.88
.995			18.6	14.5	12.9	12.0	11.5	11.1	10.8	10.6	10.4	10.2	10.0	9.81	9.59	9.36	9.12	9.00	8.88
.90	7		3.59	3.26	3.07	2.96	2.88	2.83	2.78	2.75	2.72	2.70	2.67	2.63	2.59	2.56	2.51	2.49	2.47
.95			5.59	4.74	4.35	4.12	3.97	3.87	3.79	3.73	3.68	3.64	3.57	3.51	3.44	3.38	3.30	3.27	3.23
.975			8.07	6.54	5.89	5.52	5.29	5.12	4.99	4.90	4.82	4.76	4.67	4.57	4.47	4.36	4.25	4.20	4.14
.99			12.2	9.55	8.45	7.85	7.46	7.19	6.99	6.84	6.72	6.62	6.47	6.31	6.16	5.99	5.82	5.74	5.65
.995			16.2	12.4	10.9	10.1	9.52	9.16	8.89	8.68	8.51	8.38	8.18	7.97	7.75	7.53	7.31	7.19	7.08
.90	8		3.46	3.11	2.92	2.81	2.73	2.67	2.62	2.59	2.56	2.54	2.50	2.46	2.42	2.38	2.34	2.31	2.29
.95			5.32	4.46	4.07	3.84	3.69	3.58	3.50	3.44	3.39	3.35	3.28	3.22	3.15	3.08	3.01	2.97	2.93
.975			7.57	6.06	5.42	5.05	4.82	4.65	4.53	4.43	4.36	4.30	4.20	4.10	4.00	3.89	3.78	3.73	3.67
.99			11.3	8.65	7.59	7.01	6.63	6.37	6.18	6.03	5.91	5.81	5.67	5.52	5.36	5.20	5.03	4.95	4.86
.995			14.7	11.0	9.60	8.81	8.30	7.95	7.69	7.50	7.34	7.21	7.01	6.81	6.61	6.40	6.18	6.06	5.95
.90	9		3.36	3.01	2.81	2.69	2.61	2.55	2.51	2.47	2.44	2.42	2.38	2.34	2.30	2.25	2.21	2.18	2.16
.95			5.12	4.26	3.86	3.63	3.48	3.37	3.29	3.23	3.18	3.14	3.07	3.01	2.94	2.86	2.79	2.75	2.71
.975			7.21	5.71	5.08	4.72	4.48	4.32	4.20	4.10	4.03	3.96	3.87	3.77	3.67	3.56	3.45	3.39	3.33
.99			10.6	8.02	6.99	6.42	6.06	5.80	5.61	5.47	5.35	5.26	5.11	4.96	4.81	4.65	4.48	4.40	4.31
.995			13.6	10.1	8.72	7.96	7.47	7.13	6.88	6.69	6.54	6.42	6.23	6.03	5.83	5.62	5.41	5.30	5.19
.90	10		3.29	2.92	2.73	2.61	2.52	2.46	2.41	2.38	2.35	2.32	2.28	2.24	2.20	2.15	2.11	2.08	2.06
.95			4.96	4.10	3.71	3.48	3.33	3.22	3.14	3.07	3.02	2.98	2.91	2.84	2.77	2.70	2.62	2.58	2.54
.975			6.94	5.46	4.83	4.47	4.24	4.07	3.95	3.85	3.78	3.72	3.62	3.52	3.42	3.31	3.20	3.14	3.08
.99			10.0	7.56	6.55	5.99	5.64	5.39	5.20	5.06	4.94	4.85	4.71	4.56	4.41	4.25	4.08	4.00	3.91
.995			12.8	9.43	8.08	7.34	6.87	6.54	6.30	6.12	5.97	5.85	5.66	5.47	5.27	5.07	4.86	4.75	4.64
.90	12		3.18	2.81	2.61	2.48	2.39	2.33	2.28	2.24	2.21	2.19	2.15	2.10	2.06	2.01	1.96	1.93	1.90
.95			4.75	3.89	3.49	3.26	3.11	3.00	2.91	2.85	2.80	2.75	2.69	2.62	2.54	2.47	2.38	2.34	2.30
.975			6.55	5.10	4.47	4.12	3.89	3.73	3.61	3.51	3.44	3.37	3.28	3.18	3.07	2.96	2.85	2.79	2.72
.99			9.33	6.93	5.95	5.41	5.06	4.82	4.64	4.50	4.39	4.30	4.16	4.01	3.86	3.70	3.54	3.45	3.36
.995			11.8	8.51	7.23	6.52	6.07	5.76	5.52	5.35	5.20	5.09	4.91	4.72	4.53	4.33	4.12	4.01	3.90

Table 9.5 (*Continued*)

G(F)	m	n=1	2	3	4	5	6	7	8	9	10	12	15	20	30	60	120	∞
.90	15	3.07	2.70	2.49	2.36	2.27	2.21	2.16	2.12	2.09	2.06	2.02	1.97	1.92	1.87	1.82	1.79	1.76
.95		4.54	3.68	3.29	3.06	2.90	2.79	2.71	2.64	2.59	2.54	2.48	2.40	2.33	2.25	2.16	2.11	2.07
.975		6.20	4.77	4.15	3.80	3.58	3.41	3.29	3.20	3.12	3.06	2.96	2.86	2.76	2.64	2.52	2.46	2.40
.99		8.68	6.36	5.42	4.89	4.56	4.32	4.14	4.00	3.89	3.80	3.67	3.52	3.37	3.21	3.05	2.96	2.87
.995		10.8	7.70	6.48	5.80	5.37	5.07	4.85	4.67	4.54	4.42	4.25	4.07	3.88	3.69	3.48	3.37	3.26
.90	20	2.97	2.59	2.38	2.25	2.16	2.09	2.04	2.00	1.96	1.94	1.89	1.84	1.79	1.74	1.68	1.64	1.61
.95		4.35	3.49	3.10	2.87	2.71	2.60	2.51	2.45	2.39	2.35	2.28	2.20	2.12	2.04	1.95	1.90	1.84
.975		5.87	4.46	3.86	3.51	3.29	3.13	3.01	2.91	2.84	2.77	2.68	2.57	2.46	2.35	2.22	2.16	2.09
.99		8.10	5.85	4.94	4.43	4.10	3.87	3.70	3.56	3.46	3.37	3.23	3.09	2.94	2.78	2.61	2.52	2.42
.995		9.94	6.99	5.82	5.17	4.76	4.47	4.26	4.09	3.96	3.85	3.68	3.50	3.32	3.12	2.92	2.81	2.69
.90	30	2.88	2.49	2.28	2.14	2.05	1.98	1.93	1.88	1.85	1.82	1.77	1.72	1.67	1.61	1.54	1.50	1.46
.95		4.17	3.32	2.92	2.69	2.53	2.42	2.33	2.27	2.21	2.16	2.09	2.01	1.93	1.84	1.74	1.68	1.62
.975		5.57	4.18	3.59	3.25	3.03	2.87	2.75	2.65	2.57	2.51	2.41	2.31	2.20	2.07	1.94	1.87	1.79
.99		7.56	5.39	4.51	4.02	3.70	3.47	3.30	3.17	3.07	2.98	2.84	2.70	2.55	2.39	2.21	2.11	2.01
.995		9.18	6.35	5.24	4.62	4.23	3.95	3.74	3.58	3.45	3.34	3.18	3.01	2.82	2.63	2.42	2.30	2.18
.90	60	2.79	2.39	2.18	2.04	1.95	1.87	1.82	1.77	1.74	1.71	1.66	1.60	1.54	1.48	1.40	1.35	1.29
.95		4.00	3.15	2.76	2.53	2.37	2.25	2.17	2.10	2.04	1.99	1.92	1.84	1.75	1.65	1.53	1.47	1.39
.975		5.29	3.93	3.34	3.01	2.79	2.63	2.51	2.41	2.33	2.27	2.17	2.06	1.94	1.82	1.67	1.58	1.48
.99		7.08	4.98	4.13	3.65	3.34	3.12	2.95	2.82	2.72	2.63	2.50	2.35	2.20	2.03	1.84	1.73	1.60
.995		8.49	5.80	4.73	4.14	3.76	3.49	3.29	3.13	3.01	2.90	2.74	2.57	2.39	2.19	1.96	1.83	1.69
.90	120	2.75	2.35	2.13	1.99	1.90	1.82	1.77	1.72	1.68	1.65	1.60	1.54	1.48	1.41	1.32	1.26	1.19
.95		3.92	3.07	2.68	2.45	2.29	2.18	2.09	2.02	1.96	1.91	1.83	1.75	1.66	1.55	1.43	1.35	1.25
.975		5.15	3.80	3.23	2.89	2.67	2.52	2.39	2.30	2.22	2.16	2.05	1.94	1.82	1.69	1.53	1.43	1.31
.99		6.85	4.79	3.95	3.48	3.17	2.96	2.79	2.66	2.56	2.47	2.34	2.19	2.03	1.86	1.66	1.53	1.38
.995		8.18	5.54	4.50	3.92	3.55	3.28	3.09	2.93	2.81	2.71	2.54	2.37	2.19	1.98	1.75	1.61	1.43
.90	∞	2.71	2.30	2.08	1.94	1.85	1.77	1.72	1.67	1.63	1.60	1.55	1.49	1.42	1.34	1.24	1.17	1.00
.95		3.84	3.00	2.60	2.37	2.21	2.10	2.01	1.94	1.88	1.83	1.75	1.67	1.57	1.46	1.32	1.22	1.00
.975		5.02	3.69	3.12	2.79	2.57	2.41	2.29	2.19	2.11	2.05	1.94	1.83	1.71	1.57	1.39	1.27	1.00
.99		6.63	4.61	3.78	3.32	3.02	2.80	2.64	2.51	2.41	2.32	2.18	2.04	1.88	1.70	1.47	1.32	1.00
.995		7.88	5.30	4.28	3.72	3.35	3.09	2.90	2.74	2.62	2.52	2.36	2.19	2.00	1.79	1.53	1.36	1.00

(Note that tabled values are F.) From ref. 1.

distribution when $b = 1$, a Rayleigh distribution when $b = 2$, and a good approximation to the Gaussian distribution when $b = 3.57$, that is, when mean and median are equal. The Weibull probability density function is usually skewed to the right, going to infinity. The distribution is said to have a nonzero minimum life if its $f(N)$ curve touches the life axis at a value greater than zero. The cumulative distribution function for the Weibull distribution is given by

$$F(N) = 1 - e^{-[(N-N_0)/(N_a-N_0)]^b} \qquad (9\text{-}17)$$

The Weibull distribution is often used for representation of fatigue life data at a constant stress level. Table 9.6 is useful in plotting Weibull cumulative distributions on specially constructed Weibull probability paper, as discussed in Section 9.9.

9.5   STATISTICAL HYPOTHESES

A statistical hypothesis is an assumption or assertion about a population parameter. The assumption made about the population parameter is usually called the *null hypothesis*, H_0. For example, a typical null hypothesis might be written

$$H_0 : \mu \le 50{,}000 \text{ psi} \qquad (9\text{-}18)$$

To *test* the null hypothesis means to establish a procedure for deciding

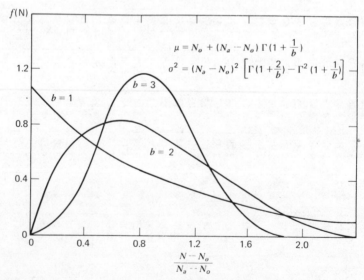

FIGURE 9.6.   Sketch of several typical Weibull distribution curves. (After ref. 2)

Table 9.6. Weibull Ordinate Locations Corresponding
to Percent Failed Values. (5)* (Note: All logs are to the base 10.)

$F(N) \times 100$	$\log \dfrac{1}{1 - F(N)}$	$F(N) \times 100$	$\log \dfrac{1}{1 - F(N)}$
2	0.0088	52	0.3188
4	0.0177	54	0.3372
5	0.0223	55	0.3468
6	0.0269	56	0.3565
8	0.0362	58	0.3768
10	0.0458	60	0.3979
12	0.0555	62	0.4202
14	0.0655	63.2	0.4341
15	0.0706	64	0.4437
16	0.0757	65	0.4559
18	0.0862	66	0.4685
20	0.0969	68	0.4949
22	0.1079	70	0.5229
24	0.1192	72	0.5528
25	0.1249	74	0.5850
26	0.1308	75	0.6021
28	0.1427	76	0.6198
30	0.1549	78	0.6576
32	0.1675	80	0.6990
34	0.1805	82	0.7447
35	0.1871	84	0.7959
36	0.1938	85	0.8239
38	0.2076	86	0.8539
40	0.2218	88	0.9208
42	0.2366	90	1.000
44	0.2518	92	1.097
45	0.2596	94	1.222
46	0.2676	95	1.301
48	0.2840	96	1.398
50	0.3010	98	1.699

*Copyright ASTM, 1916 Race Street, Philadelphia, Pa. 19103. Reprinted with permission.

whether to *accept* or *reject* the hypothesis, usually on the basis of a representative random sample drawn from the population. In testing the null hypothesis, rejection of the null hypothesis H_0 implies acceptance of an alternative hypothesis H_1. For example, if the null hypothesis is as given in (9-18), it implies that the corresponding alternative hypothesis H_1 is given by

$$H_1: \mu > 50,000 \text{ psi} \tag{9-19}$$

In making decisions to accept or reject a hypothesis based on sample data it is possible to make two types of errors:

Type I error: probability of *rejecting* the null hypothesis *when*, in fact, the null hypothesis is *true*. This probability is usually called α, the *significance* of the test.

Type II error: probability of *accepting* the null hypothesis *when*, in fact, it is *false*. This probability is usually called β, and the quantity $(1 - \beta)$ is defined to be the *power* of the test.

The procedure for testing a statistical hypothesis is normally established *before* drawing the sample from the population and typically involves the following steps:

1. Choose the statistic and, hence, the sampling distribution to be used. The sampling distribution must involve the parameter being hypothesized upon.

2. Specify α, the desired significance of the test. That is, specify the desired probability of Type I error.

3. Choose or determine the *critical* region. This is the region in which, if the selected statistic falls, the null hypothesis will be rejected. This involves the specification of α and selection of the sampling distribution.

4. Choose the sample size. This will set bounds to the Type II error probability β.

5. Test the hypothesis and accept or reject H_0 based on the procedure established and the critical region defined in Step 3.

To illustrate this procedure, consider a random variable x with mean μ and variance σ^2 such that

$$y = \frac{x - \mu}{\sigma} \overset{d}{=} N(0, 1) \qquad (9\text{-}20)$$

The statement of (9-20) is to be read "y equals $(x - \mu)/\sigma$ is distributed normally with a mean of 0 and variance of 1."

A large class of statistics may be treated in this way under rather general conditions if sample size n is sufficiently large. For example, x in (9-20) may be the mean of a sample of size n. The population parameters μ and σ will in general be fixed but unknown.

Suppose now that it is desired to test the null hypothesis

$$H_0 : \mu = \mu_0 \qquad (9\text{-}21)$$

That is, it is desired to test the hypothesis that the true value of the population mean μ is μ_0, any real number including zero. Further, it is desired to test this hypothesis on the basis of a random sample of size n drawn from the population. Implied in the statement of the null hypothesis of

(9-21) is the alternative hypothesis

$$H_1 : \mu \neq \mu_0 \tag{9-22}$$

To proceed with the test of null hypothesis H_0 it is first necessary to select an appropriate level of significance α, recalling that

$$\alpha = P\{\text{rejecting } H_0 \,|\, H_0 \text{ is true}\} \tag{9-23}$$

Equation (9-23) is to be read "α is equal to the *probability* of rejecting H_0 *given that H_0 is true*."

After having selected the level of significance α, it is necessary to define the critical (rejection) region for the sample statistic y in (9-20) by finding a value y_0 such that

$$P\{-y_0 \leq y \leq y_0\} = 1 - \alpha \tag{9-24}$$

With the critical region established by (9-24) as depicted in Figure 9.7, the next step is to draw the random sample and calculate

$$(y)_{\text{sample}} = \frac{x - \mu_0}{\sigma} \tag{9-25}$$

either exactly or approximately, depending on whether σ is known or must be estimated. Then, on the basis of (9-25), if y_{sample} as calculated falls into the rejection region as defined by (9-24) and Figure 9.7, the hypothesis H_0 is rejected at the α level of significance. If y_{sample} does not fall into the critical region, the null hypothesis H_0 is not rejected. To be more specific, suppose

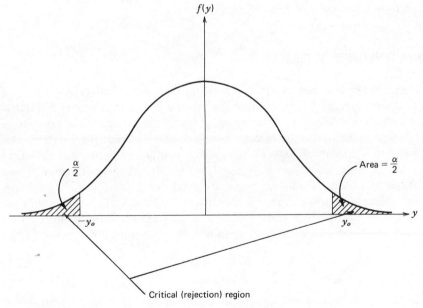

FIGURE 9.7. Graphical representation of the critical (rejection) region for testing $H_o : \mu = \mu_o$ for significance level of α using the sample statistic $y = \dfrac{x - \mu}{\sigma}$.

that one wishes to test the hypothesis

$$H_0 : \mu = 50,000 \text{ psi} \qquad (9\text{-}26)$$

based on the sample mean of 12 specimens drawn from the population. Suppose that independent estimates of the population standard deviation indicate that $\sigma = 2000$ psi. It is desired to test the hypothesis of (9-26) at the 0.05 level of significance. It is established in (9-20) that y has a standard normal distribution, which is shown in Table 9.2. For a significance level of 0.05, with a symmetrical distribution as shown in Figure 9.7, the value of y_0 is the y value corresponding to $(1 - \alpha/2)$ accumulated area under the distribution curve. Table 9.2 shows that y_0 is 1.96. Hence, for the 0.05 significance level the rejection region will include all values of y that are less than -1.96 or greater than 1.96. The next step is to calculate the statistic y based on the sample, namely

$$y_{\text{sample}} = \frac{\bar{x} - \mu_0}{\sigma / \sqrt{n}} \qquad (9\text{-}27)$$

Now suppose that the 12-specimen sample yields a sample mean of 47,500 psi. The statistic of (9-27) would be

$$y_{\text{sample}} = \frac{47,500 - 50,000}{2000 / \sqrt{12}} = -4.2 \qquad (9\text{-}28)$$

Since $y = -4.2$ lies in the rejection region, the hypothesis H_0 that $\mu = 50,000$ psi is rejected at the 0.05 level of significance.

9.6 CONFIDENCE LIMITS

Testing hypotheses such as the one given in (9-26) is intuitively unappealing since the probability of finding that the mean μ is exactly equal to 50,000 psi must be very small indeed. A better approach to the estimation of population parameters is the technique of establishing *confidence limits* on the parameter in question. To establish $100\,(1 - \alpha)$ percent confidence limits, it is necessary only to solve the equations $y = \pm y_0$ from (9-24) for the parameter. These solutions yield the $100\,(1 - \alpha)$ percent confidence limits for this parameter.

For example, suppose that it is desired to find the 95 percent confidence limits on the mean μ of a given population, utilizing the sample mean $\bar{x}$ of a random sample drawn from the population. It is known that

$$y = \frac{\bar{x} - \mu}{\sigma / \sqrt{n}} \overset{d}{=} N(0, 1) \qquad (9\text{-}29)$$

since the sampling distribution for $\bar{x}$ can be shown to have a mean μ and standard deviation $\sigma / \sqrt{n}$ as described in Section 9.4,. in view of the ramifications of the central limit theorem. Specifying a confidence of 95 percent implies a significance level α of 0.05. Consider, then, based on (9-24),

$$P\left\{ -y_0 \leq \frac{\bar{x} - \mu}{\sigma/\sqrt{n}} \leq y_0 \right\} = 1 - \alpha = 0.95 \tag{9-30}$$

From Table 9.2 it may be found that the critical value of y_0 corresponding to a symmetrical two-tailed critical region with $\alpha = 0.05$ is

$$y_0 = 1.96 \tag{9-31}$$

Therefore, (9-30) may be written as

$$P\left\{ -1.96 \leq \frac{\bar{x} - \mu}{\sigma/\sqrt{n}} \leq 1.96 \right\} = 0.95 \tag{9-32}$$

The inequality to the left in (9-32) may be written alone as

$$-1.96 \leq \frac{\bar{x} - \mu}{\sigma/\sqrt{n}} \tag{9-33}$$

which when inverted yields

$$\mu \leq \bar{x} + \frac{1.96\sigma}{\sqrt{n}} \tag{9-34}$$

Similarly, the inequality to the right in (9-32) may be inverted to give

$$\mu \geq \bar{x} - \frac{1.96\sigma}{\sqrt{n}} \tag{9-35}$$

The results of (9-34) and (9-35) may then be incorporated in (9-32) to give

$$P\left\{ \bar{x} - \frac{1.96\sigma}{\sqrt{n}} \leq \mu \leq \bar{x} + \frac{1.96\sigma}{\sqrt{n}} \right\} = 0.95 \tag{9-36}$$

which, when written in corresponding confidence limit notation, becomes

$$C\left\{ \bar{x} - \frac{1.96\sigma}{\sqrt{n}} \leq \mu \leq \bar{x} + \frac{1.96\sigma}{\sqrt{n}} \right\} = 95 \text{ percent} \tag{9-37}$$

To be more specific, suppose that a sample of 25 specimens has been drawn at random from a given population of aluminum bars and it is found that the mean strength calculated for the sample is $\bar{x} = 13,000$ psi. Suppose further that the population standard deviation has somehow been estimated (perhaps from the same sample) to be $\sigma = 2000$ psi. Then the 95 percent confidence limits on the population mean, based on the statistic $\bar{x}$, would be

$$C\left\{ 13,000 - \frac{1.96(2000)}{25} \leq \mu \leq 13,000 + \frac{1.96(2000)}{25} \right\} = 95 \text{ percent}$$

$$\tag{9-38}$$

or

$$C\{12,220 \leq \mu \leq 13,780\} = 95 \text{ percent} \qquad (9\text{-}39)$$

This is to say that with 95 percent confidence it may be predicted that the population mean is in the range from 12,220 psi to 13,780 psi.

It may be noted that the length of the confidence interval $2y_0\sigma/\sqrt{n}$ is a function of the significance level α and also of the number of specimens in the sample. To decrease the length of the confidence interval, and thereby improve the quality of the estimate, one may either reduce the confidence or increase the sample size. Procedures have been established (2) for making point estimates of population parameters and calculating confidence intervals for population parameters of interest in a variety of different types of fatigue tests.

9.7 PROPERTIES OF GOOD ESTIMATORS

Statistics, calculated from random samples, that are used to estimate population parameters are called *estimators*. Good estimators should have, insofar as possible, the following desirable properties: They should be unbiased, consistent, efficient, and sufficient.

An *unbiased estimator* is one that neither consistently underestimates nor overestimates the true value of the parameter. A *consistent estimator* is one that is unbiased and converges more closely to the true value of the population parameter as the sample size is increased. An *efficient estimator* is a consistent estimator whose standard deviation is smaller than the standard deviation of any other estimator for the same population parameter. A *sufficient estimator* is an efficient estimator that utilizes all the information about the parameter that the sample possesses. One always strives to use the best possible estimator within the context of these desirable properties.

9.8 SAMPLE SIZE FOR DESIRED CONFIDENCE

As just observed, the size of the confidence interval is influenced by the sample size. If the data of interest are normally distributed, or can somehow be transformed into a normal distribution, it is possible to predict the width of the confidence interval as a function of sample size. Fatigue data are of this type. To provide information about the number of specimens required to give acceptable prediction accuracy, Table 9.7 lists the minimum number of specimens needed to determine 95 percent confidence intervals of the stated width for a population mean μ, assuming the standard deviation σ is known (2).

Table 9.7. Number of Specimens for 95 percent Confidence Limits on Mean

Width of Interval	95 percent Confidence Limits on Mean μ	Number of Specimens Required, n
$0.2\,\sigma$	$\bar{x} \pm 0.1\,\sigma$	384
$0.4\,\sigma$	$\bar{x} \pm 0.2\,\sigma$	96
$0.6\,\sigma$	$\bar{x} \pm 0.3\,\sigma$	43
$0.8\,\sigma$	$\bar{x} \pm 0.4\,\sigma$	24
$1.0\,\sigma$	$\bar{x} \pm 0.5\,\sigma$	15
$1.2\,\sigma$	$\bar{x} \pm 0.6\,\sigma$	11
$1.4\,\sigma$	$\bar{x} \pm 0.7\,\sigma$	8
$1.6\,\sigma$	$\bar{x} \pm 0.8\,\sigma$	6
$1.8\,\sigma$	$\bar{x} \pm 0.9\,\sigma$	5
$2.0\,\sigma$	$\bar{x} \pm 1.0\,\sigma$	4

The value of n in Table 9.7 is calculated from

$$n = \left(\frac{1.96\sigma}{E} \right)^2 \tag{9-40}$$

where

$$E = \frac{\text{width of interval}}{2} \tag{9-41}$$

The minimum number of specimens needed to determine 95 percent confidence intervals of stated width for a population standard deviation σ, with some estimate of σ available, is given in Table 9.8.

Table 9.8. Number of Specimens of 95 percent Confidence Limits on Standard Deviation

Width of Interval	Number of Specimens, n
$0.14\,\sigma$	385
$0.2\ \sigma$	190
$0.3\ \sigma$	84
$0.4\ \sigma$	47
$0.5\ \sigma$	30
$0.6\ \sigma$	21
$0.7\ \sigma$	16
$0.8\ \sigma$	13
$0.9\ \sigma$	10
$1.0\ \sigma$	8

The number of specimens n in Table 9.8 is calculated from (2)

$$1 + \left(\frac{\text{width of interval}}{2\sigma} \right) = \left(\frac{\chi^2_{0.975}}{n-1} \right)^{1/2} \tag{9-42}$$

9.9 PROBABILITY PAPER

For the normal distribution the cumulation distribution function $F(x)$ plotted versus the random variable x gives an ogee-shaped curve, as shown in Figure 9.2b. It is possible to design special graph paper that has the scale on $F(x)$ distorted in just the right way so that for a normal distribution the plot of $F(x)$ versus x will be a straight line. This type of special graph paper, called *normal probability paper*, is utilized to determine whether data taken from a population of interest is normally distributed. To make such a determination, it is necessary only to plot the data on the normal probability paper in accordance with the plotting technique to be described. If the data plot as a straight line, it may be concluded that the population is normally distributed, and the mean and standard deviation may be read directly from the plot without further calculation. If the data do not plot as a straight line, it must be concluded that the population is not normally distributed.

Probability paper can be constructed for any type of distribution by properly distorting the probability scale of the cumulative distribution plot to yield a straight line when plotted versus the random variable. *Normal* probability paper, as described earlier, is shown in Figure 9.8. Another common type of paper is *logarithmic-normal* probability paper shown in Figure 9.9. Both *normal* probability paper and *log-normal* probability paper are commercially available. Other special paper, such as *Weibull* probability paper, must be either purchased from special suppliers or constructed as described later. An example of Weibull probability paper is shown in Figure 9.11.

To plot sample data in a meaningful way on probability paper, the following procedure may be used:

1. Order the array of data with smallest value of the random variable listed first.
2. Assign a rank q to each point in the ordered array.
3. Determine a plotting position for each data point by dividing its rank q by $n + 1$, where n is the total number of points in the array.
4. Plot the magnitude of the random variable at its proper plotting position on the chosen probability paper.

Having plotted the data, the probability plot is examined for linearity. If the data points fall along a straight line, it may be concluded that the distribution for the data is indeed the distribution for which the probability paper was constructed.

Table 9.9.   Yield Strength Test Data

Test Number	Yield Strength (psi)
1	152,400
2	153,400
3	151,000
4	151,800
5	155,400
6	154,000
7	153,000
8	154,300
9	152,100
10	154,700
11	152,900
12	153,800
13	152,600
14	153,600
15	153,100

For example, suppose it is desired to determine whether the yield strengths of machine parts made from a certain heat of 4340 steel are normally distributed and, if they are, what the mean yield strength and standard deviation will be. To make the determination, a sample of $n = 15$ specimens is tested with the results shown in Table 9.9.

Following the steps outlined earlier, we order the data as shown in Table 9.10, arranged from lowest to highest strength.

Table 9.10.   Ordered Data for Probability Paper Plot

Rank	Test Number	Yield Strength	Plotting Position $100\,q/(n + 1)$ (percent)
1	3	151,000	6.25
2	4	151,800	12.50
3	9	152,100	18.75
4	1	152,400	25.00
5	13	152,600	31.25
6	11	152,900	37.50
7	7	153,000	43.75
8	15	153,100	50.00
9	2	153,400	56.25
10	14	153,600	62.50
11	12	153,800	68.75
12	6	154,000	75.00
13	8	154,300	81.25
14	10	154,700	87.50
15	5	155,400	93.75

Utilizing Table 9.9 we plot the data on normal probability paper as shown in Figure 9.10. The data are close enough to linear on this plot to regard the population as approximately normal. The extreme points on such a plot often deviate from the straight line, and such deviations at the tails are usually disregarded. The mean may be estimated by reading the yield strength at the 50 percent level on the probability scale. Thus the mean yield strength for this population would be estimated to be 152,250 psi. The standard deviation may be estimated by noting from Table 9.2 that for a normal distribution, one standard deviation above the mean occurs at a probability level of 84.13 percent. Thus in Figure 9.10 one standard deviation in strength is defined by the increment between the 50 percent and the 84 percent probability levels. The standard deviation is estimated to be $\sigma = 1250$ psi.

If we desire to plot experimental data on Weibull probability paper, we must first obtain the Weibull paper or construct it by selecting a piece of "square" log-log paper, that is, paper on which the log scales are the same size in both directions. Utilizing the information from Table 9.6, we may construct the probability scale as shown in Figure 9.11.

Plotting position for Weibull paper is established in the same way as for the normal or log-normal paper described earlier. If the data points are reasonably linear on the Weibull paper, a best straight line is drawn through the data and estimates of N_a, N_0, and b are made from the plot. Assuming fatigue life to be the random variable of interest, we estimate the characteristic life N_a by reading the life that corresponds to 63.2 percent on the probability scale. The Weibull shape parameter b is the slope of the straight line on the Weibull plot. Shape parameter b is estimated from tan θ, as indicated in Figure 9.13.

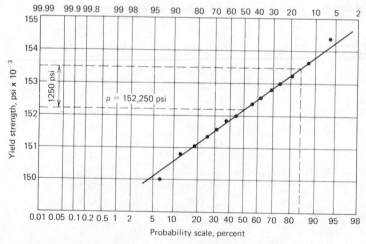

FIGURE 9.10. Normal probability plot of yield strength data from Table 9.9 showing a check for normality and determination of mean μ and standard deviation σ.

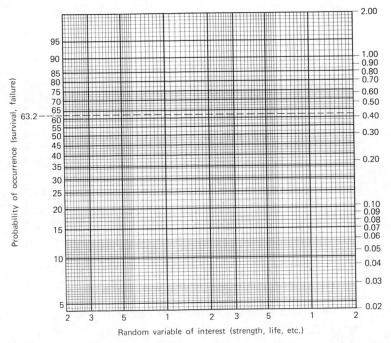

FIGURE 9.11. Example of Weibull probability paper constructed from "square" log-log paper.

The minimum life N_0 is initially assumed zero. If the data do not fit a straight line, one suspects a nonzero minimum life N_0 and appropriate adjustments must be made.

For example, the fatigue data shown in Table 9.11 were all taken at a constant stress level of 50,000 psi. To plot the data on Weibull paper, the plotting position is established as shown in Table 9.11 and the data are

Table 9.11. Fatigue Test Data Taken at a Constant Stress Level of 50,000 psi

Rank	Test Number	Cycles to Failure $\times 10^{-5}$	Plotting Position $100\, q/(n+1)$ (percent)
1	4	4.0	11.1
2	2	5.0	22.2
3	5	6.0	33.3
4	8	7.3	44.4
5	1	8.0	55.6
6	7	9.0	66.7
7	6	10.6	77.8
8	3	13.0	88.9

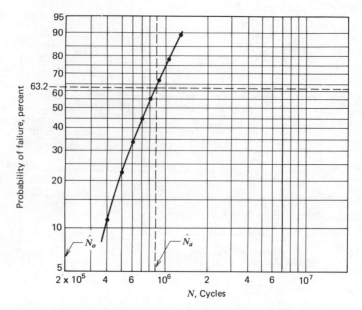

FIGURE 9.12. Weibull plot for fatigue data from Table 9.10. (After ref. 2, copyright ASTM; adapted with permission.)

plotted as indicated in Figure 9.12, assuming the minimum life parameter N_0 to be zero.

The data shown in Figure 9.12 tend to curve downward, suggesting a minimum life parameter greater than zero. To obtain an estimate of minimum life, the life value that the curve approaches asymptotically is selected as the estimate of $\hat{N}_0$. Next, the quantity $N - \hat{N}_0$ is calculated for each data point and plotted on Weibull paper versus the same probability values as before. By trial and error, the process may be repeated to obtain the best estimate of N_0 when the transformed data plot as a straight line, as shown in Figure 9.13.

From Figure 9.13 the characteristic life may be estimated from the life value $N_a - N_0$ corresponding to 63.2 percent. Knowing N_0 used to obtain Figure 9.13 from Figure 9.12, we therefore also know N_a. Of course, if the data plot as a straight line initially, N_0 is zero and N_a is read directly from the Weibull plot at 63.2 percent.

The median life may be estimated by reading the intersection of the straight line with the 50 percent probability coordinate on the Weibull paper. It should be noted that the median and the mean do not in general coincide because the distribution is generally skewed. Knowing the values of N_0, N_a, and b, however, we may calculate the Weibull mean value from the expression for μ given in Figure 9.6.

The slope parameter b may be estimated by measuring the tangent of θ, as shown in Figure 9.13.

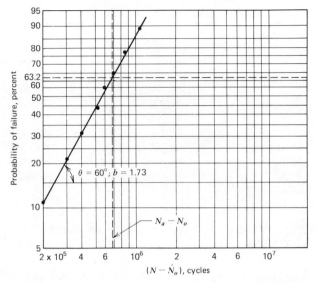

FIGURE 9.13. Transformed Weibull plot based on Figure 9.11 and an estimate of minimum life N_o equal to 2×10^5 cycles. (After ref. 2, copyright ASTM; adapted with permission.)

Finally, it should be noted that if the data include run-out specimens, special plotting techniques are required to determine the Weibull distribution parameters.*

9.10 COMPARISON OF MEANS AND VARIANCES

Frequently it is desired to compare data from two different sources, data taken at two different times, or data taken under two different conditions. In making such comparisons it is necessary to compare the population parameters associated with the two different sets of data. In the case of normally distributed populations, for example, this requires comparisons of the means and variances of the two populations to determine whether they are the same or different. Population means may be compared by using the *t test*, based on the Student's *t* distribution described earlier and tabulated in Table 9.4. Population variances may be compared by using the *F test*, based on the Snedcor *F* distribution described earlier and tabulated in Table 9.5. To illustrate the procedure consider the following example: Suppose that a year ago a sample of eight fatigue specimens from an old heat of material were tested at a constant stress level of 90,000 psi, with data on failure lives as

*See p. 78 of ref. 2.

Table 9.12.   Constant Stress-Level Fatigue Life Data Taken a Year Ago

Specimen Number	Fatigue Life, N cycles	Logarithm of N
1	95,100	4.9777
2	69,000	4.8388
3	94,000	4.9731
4	121,000	5.0828
5	84,100	4.9243
6	108,000	5.0334
7	88,100	4.9445
8	90,000	4.9542

given in Table 9.12. Since experience has indicated that life distribution at a constant stress level is approximately log-normal, the logarithms of the failure lives are also tabulated.

Now suppose that a recent sample of 10 specimens has been tested from a new heat of the same material, again at a constant stress level of 90,000 psi, and it is desired to know whether the two heats of material in fact have the same fatigue properties. The data from the new heat of material are shown in Table 9.13.

An unbiased estimate of the population mean* is given by the sample mean, whence

$$\hat{\mu}_1 = \bar{x}_1 = \frac{1}{n_1} \sum_{i=1}^{n} x_i \qquad (9\text{-}43)$$

Table 9.13.   Constant Stress-Level Fatigue Life Data Taken Recently

Specimen Number	Fatigue Life, N cycles	Logarithm of Life
1	34,000	4.5315
2	44,100	4.6435
3	49,000	4.6902
4	47,000	4.6721
5	42,000	4.6232
6	45,000	4.6532
7	42,000	4.6232
8	49,000	4.6902
9	44,100	4.6435
10	53,000	4.7243

*See, for example, ref. 2.

where $\hat{\mu}_1 =$ estimator of population mean μ, log units
  $\bar{x}_1 =$ sample mean for sample number 1, log units
  $n_1 =$ number of specimens in sample number 1
  $x_i = logarithms$ of lives in sample

Note that the assumption based on experience is that the logarithms of life are normally distributed, not the lives themselves. That is, the distribution of lives is log-normal. To compute the means of log-life, a tabulation of data is made as shown in Tables 9.14 and 9.15.

Utilizing (9-43) and the compilation of Table 9.14, we compute the estimated population mean for the old heat of material to be

$$\hat{\mu}_1 = \frac{39.7288}{8} = 4.9661 \tag{9-44}$$

Table 9.14. Tabulation of Data for Calculation of Mean and Standard Deviation for Sample Taken from Old Heat of Material

n_1	$x_i = \log N_i$	$(x_i - \bar{x}_1)$	$(x_i - \bar{x}_1)^2 \times 10^4$
1	4.8388	− 0.1273	162.1
2	4.9243	− 0.0418	17.5
3	4.9445	− 0.0216	4.7
4	4.9542	− 0.0119	1.4
5	4.9731	0.0070	0.5
6	4.9777	0.0116	1.3
7	5.0334	0.0673	45.3
8	5.0828	0.1167	136.2
	$\sum_{i=1}^{8} x_i = 39.7288$		$\sum_{i=1}^{8} (x_i - \bar{x}_1)^2 = 369.0 \times 10^{-4}$

Table 9.15. Tabulation of Data for Calculation of Mean and Standard Deviation for Sample Taken from New Heat of Material

n_2	$x_i = \log N_i$	$x_i - \bar{x}_2$	$(x_i - \bar{x}_2)^2 \times 10^4$
1	4.5315	− 0.1180	139.2
2	4.6232	− 0.0263	6.9
3	4.6232	− 0.0263	6.9
4	4.6435	− 0.0060	0.4
5	4.6435	− 0.0060	0.4
6	4.6532	0.0037	0.1
7	4.6721	0.0226	5.1
8	4.6902	0.0407	16.6
9	4.6902	0.0407	16.6
10	4.7243	0.0748	56.0
	$\sum_{i=1}^{10} x_i = 46.4949$		$\sum_{i=1}^{10} (x_i - \bar{x}_2)^2 = 248.2 \times 10^{-4}$

Similarly, utilizing the tabulation of Table 9.15, we find that the estimated population mean for the new heat of material is

$$\hat{\mu}_2 = \frac{46.4949}{10} = 4.6495 \qquad (9\text{-}45)$$

An unbiased estimate of the population variance* is given by

$$\hat{\sigma}^2 = \frac{1}{n-1}\Sigma(x_i - \bar{x})^2 \qquad (9\text{-}46)$$

where σ^2 = estimate of population variance, log units
$\bar{x}$ = sample mean, log units
x_i = logarithm of lives in the sample

Utilizing (9-46) and the compilation of Table 9.14, we compute the estimated population variance for the old heat of material to be

$$\hat{\sigma}_1^2 = \frac{1}{8-1}(369.0 \times 10^{-4}) = 0.00527 \qquad (9\text{-}47)$$

Similarly, utilizing the tabulation of Table 9.15, we compute the estimated population variance for the new heat of material to be

$$\hat{\sigma}_2^2 = \frac{1}{10-1}(248.2 \times 10^{-4}) = 0.00276 \qquad (9\text{-}48)$$

With these estimates for means and variances of the two populations, it is desired to compare populations by comparing the means and variances. To compare means, the t test will be used in which

$$t = \frac{|\hat{\mu}_1 - \hat{\mu}_2|}{\sqrt{\dfrac{\hat{\sigma}_1^2}{n_1 - 1} + \dfrac{\hat{\sigma}_2^2}{n_2 - 1}}} \qquad (9\text{-}49)$$

is known to have a Student's t distribution (2) if $\hat{\sigma}_1^2$ and $\hat{\sigma}_2^2$ are not significantly different. If the variances are significantly different, another t statistic must be used (2). To determine whether the variances are significantly different, they will be compared using the F test, in which the statistic

$$F = \frac{\hat{\sigma}_1^2}{\hat{\sigma}_2^2}; \qquad \hat{\sigma}_1^2 > \hat{\sigma}_2^2 \qquad (9\text{-}50)$$

is known to have a Snedcor F distribution (2). Note that this statistic is formed so as to always be a number greater than 1.

To compare the variances it is hypothesized that

$$H_0: \hat{\sigma}_1^2 = \hat{\sigma}_2^2 \qquad (9\text{-}51)$$

A significance level of $\alpha = 0.05$ may be selected to test this null hypothesis,

*See, for example, ref. 2.

which establishes the critical region on the F statistic as

$$P\{F \leq F_0\} = 1 - \frac{\alpha}{2} = 0.975 \qquad (9\text{-}52)$$

It is desired to find the value of F_0 that will make this probability statement correct. Graphically, the rejection region is shown in Figure 9.14. To determine the value F_0 it will be necessary to utilize Table 9.5 for the F distribution. The number of degrees of freedom is one less than the number of specimens tested. Hence, for the *numerator* of (9-50) the number of degrees of freedom will be $n_1 - 1$ or 7 degrees of freedom, and for the denominator of (9-50) the number of degrees of freedom will be $n_2 - 1$ or 9 degrees of freedom. In Table 9.5, then, it may be found that with $m = 7$ and $n = 9$, for a probability level of 0.975, the value of F_0 is 4.20. The value of F_0' is shown in Figure 9.14 is of no consequence in this test because the statistic of (9-50) is constructed so as to always exceed 1, forcing the test to the right tail of the distribution. That is, in Figure 9.14 the rejection region is defined as the region containing values of F greater than 4.20.

From (9-50), then, the statistic F is computed based on sample data given in (9-47) and (9-48) as

$$F = \frac{0.00527}{0.00276} = 1.91 \qquad (9\text{-}53)$$

Thus the statistic does *not* fall into the rejection region, and therefore the hypothesis of (9-51) is not rejected. Therefore, one is justified on a statistically sound basis in assuming the variances to be the same. This permits the use of the t statistic in (9-49) to compare the means. To compare means, a new null

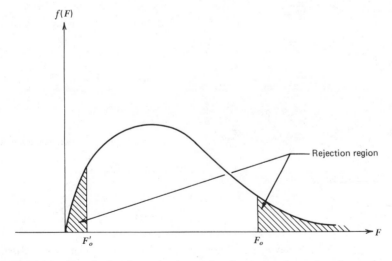

FIGURE 9.14. Sketch showing rejection region for comparison of variances based on the F test.

hypothesis may be stated as

$$H_0: \hat{\mu}_1 = \hat{\mu}_2 \qquad (9\text{-}54)$$

Again, testing the hypothesis at the $\alpha = 0.05$ level of significance, we find the critical region on the t statistic from

$$P\{t \le t_0\} = 1 - \frac{\alpha}{2} = 0.975 \qquad (9\text{-}55)$$

It remains to find the value of t_0 that will make this probability statement correct. Graphically, the rejection region is shown in Figure 9.15. To determine the value t_0 it will be necessary to utilize Table 9.4 for the t distribution. The number of degrees of freedom is the pooled sum for the two samples, or 16 degrees of freedom. In Table 9.4 it may be found that with 16 degrees of freedom at a probability level of 0.975 the value of t_0 is 2.12. That is, in Figure 9.15 the rejection region is defined as the region containing all values of t greater than 2.12 or less than -2.12.

From (9-49), then, the statistic t is computed based on sample data as

$$t = \frac{|4.9661 - 4.6495|}{\sqrt{\dfrac{0.00527}{7} + \dfrac{0.00276}{9}}} = 9.75 \qquad (9\text{-}56)$$

Thus, the statistic t lies well into the rejection region and the hypothesis of (9-54) is therefore rejected. That is, it may be concluded that the means of the two populations are not the same and therefore the populations are not the same. On this basis one would then judge the fatigue properties of the new heat of material to be significantly inferior to the old heat of material.

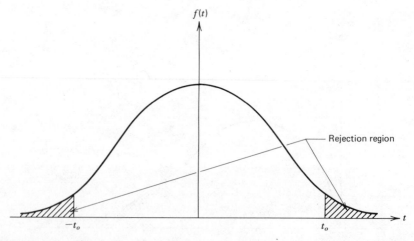

FIGURE 9.15. Sketch showing rejection region for comparison of means based on the t test.

9.11 IN SUMMARY

Although many techniques in statistical analysis of fatigue data have been discussed, many other very useful techniques have also been established.* Some of these other important techniques include:

1. Evaluation of the "scatter" in fatigue data.
2. Establishment of the controlling probability density functions.
3. Establishment of extreme value distributions.
4. Establishment of constant probability of survival curves (S-N-P curves).
5. Determination of whether testing machines, specimen preparation techniques, heat treatment, etc., are giving consistent data. That is, determination of whether or not the data are statistically "in control."
6. Evaluation of the significance of the effects of many intentional or unintentional variables.
7. Evaluation of the applicability of laboratory test data.
8. Determination of the relative accuracy of various theories of failure.
9. Evaluation of given hypotheses about mechanisms of failure.
10. Proof of the existence of significant correlations among experimental variables.
11. Determination of optimum sample size for a given purpose.
12. Establishment of efficient testing procedures and techniques.

It should further be noted that statistical packages have become widely available in many computer laboratories, and well verified programs are commercially available. Such programs are often invaluable to both the analyst and the designer faced with fatigue as a governing mode of failure.

QUESTIONS

1. Carefully define the terms *population parameter* and *sample statistic*, giving examples of each.

2. (a) Clearly define the meaning of *probability density function* (*pdf*) and *cumulative distribution function* (*cdf*) giving the properties of each.
(b) Sketch the *pdf* and *cdf* for a normal distribution, carefully labeling all coordinates.
(c) Is the normal distribution discrete or continuous?

3. (a) Sketch the probability density function and cumulative distribution

*See, for example, ref. 2.

function for

$$f(x) = \tfrac{1}{4} \quad \text{for } x = \pm 1$$
$$f(x) = \tfrac{1}{2} \quad \text{for } x = 0$$
$$f(x) = 0 \quad \text{otherwise}$$

(b) Is this distribution discrete or continuous?

4. List the three more usual distributions encountered in fatigue analysis and design, giving an equation for the probability density function of each one. Carefully define all terms.

5. A fatigue testing program has produced the life data given in the chart for 35 specimens of an aluminum alloy, all tested to failure at a completely reversed cyclic stress amplitude of 26,000 psi. Compute the mean and standard deviation for this sample, assuming it to be normally distributed.

Fatigue Life in Thousands of Cycles (first heat)						
290	490	342	456	517	310	445
540	233	376	410	439	403	315
439	433	473	367	400	358	445
358	351	422	276	560	360	406
400	395	321	356	443	404	362

6. Plot the data of problem 5 on normal probability paper, log-normal paper, and Weibull paper and compare results to find which distribution the data fits more closely. (If necessary, construct your own Weibull paper.)

7. Define and distinguish between "Type I error" and "Type II error," telling how to minimize each one.

8. Interpret in words the following statistical statement:

$$C\{47{,}400 \text{ psi} \leq \mu \leq 52{,}000 \text{ psi}\} = 99.7 \text{ percent}$$

9. (a) From a 16-specimen sample used in a wear-testing experiment, it has been calculated that the mean wear depth of the sample is 3.2 mils and the sample standard deviation is 0.63 mil, based on the assumption that such wear data are normally distributed. Calculate the 95 percent confidence limits on the population mean for wear under such conditions.
(b) If it is desired to maintain the same width of confidence interval, but improve the confidence level to 99 percent, how many additional specimens must be tested?

10. A second sample of 20 specimens is tested from another heat of the same aluminum alloy for which failure life data were shown in problem 5. The failure lives for this second heat, all tested to failure at a completely reversed cyclic stress amplitude of 26,000 psi, are shown in the chart.

Fatigue Life in Thousands of Cycles (second heat)				
318	342	476	468	338
363	379	414	416	374
408	423	370	372	489
452	474	335	312	416

Assuming the distribution to be log-normal, do the following:
(a) Calculate estimates for the population means for both the first and second heats of material (log units).
(b) Calculate estimates for the population variances for both the first and second heats of material (log units).
(c) Compare variances at the 0.05 level of significance.
(d) Compare means at the 0.05 level of significance. Do you conclude that these samples are drawn from the same population, or are they drawn from different populations?

11. If one found, in the fatigue literature, data for the mean and variance for fatigue life of 7075-T6 aluminum alloy and a statement that fatigue life at any given stress level has a log-normal distribution, explain in detail how the family of S-N-P curves of Figure 7.15 could be constructed. As a designer, how would you decide which of the curves to use?

REFERENCES

1. Mood, A. M., *Introduction to the Theory of Statistics*, McGraw-Hill, New York, 1950.
2. *A Guide for Fatigue Testing and the Statistical Analysis of Fatigue Data*, STP-91-A, American Society for Testing and Materials, Philadelphia, 1963.
3. Hardenbergh, D. E. (ed.), *Statistical Methods in Materials Research*, Proceedings of Short Course, Dept. of Engineering Mechanics, Pennsylvania State University, 1956.
4. Finney, D. J., *Probit Analysis*, Cambridge University Press, London, 1952.
5. Prot, E. M., *Fatigue Testing Under Progressive Loading; A New Technique for Testing Materials* (translation by E. J. Ward), WADC TR52-148, September 1952.
6. Dixon, W. J., and Massey, F. J., Jr., *Introduction to Statistical Analysis*, McGraw-Hill, New York, 1957.
7. Gumbel, E. J., "Statistical Theory of Extreme Values and Some Practical Applications," *National Bureau of Standards, Applied Mathematics Series 33*, February 12, 1954.
8. Heller, R. A. (ed.), *Probabilistic Aspects of Fatigue*, STP-511, American Society for Testing and Materials, Philadelphia, 1971.
9. Little, R. E., and Jebe, E. H., *Statistical Design of Fatigue Experiments*, John Wiley & Sons, New York, 1975.

CHAPTER 10

Fatigue Testing Procedures and Statistical Interpretations of Data

10.1 INTRODUCTION

Many different fatigue testing procedures have been devised to provide information suitable for meeting a variety of different objectives. For example, it may be desired to obtain life distribution data at a constant stress level, strength distribution data at a constant life level, significant fatigue design data with the smallest possible sample size or in the shortest possible time, or data to meet other specific needs. Several fatigue testing procedures are outlined in the following paragraphs to meet a variety of objectives. It should be noted that the procedures discussed here are applicable to all types of laboratory and field service fatigue data, no matter what types of testing machines or methods are used. Testing machines and methods have been widely discussed in the literature.*

10.2 STANDARD METHOD

When only a few parts or specimens are available for testing, and an estimate of the whole S-N curve is needed, the testing method often used is the so-called "standard" method. The standard method simply involves running one or two specimens at each of several different stress amplitude levels, recording the stress amplitude and number of cycles to failure. If specimens "run out," they are sometimes run again at a higher stress level, taking due precaution to note that damage or coaxing effects may influence the final failure lives of these specimens. The data are typically plotted on a standard S-N plot as shown in Figure 10.1. Scatter of the data is to be expected, and the question of how to construct a meaningful curve through the data becomes pertinent. Two techniques are used for constructing S-N curves based on such a test. One is to construct a mean curve through the data; the other is to construct a "conservative" curve that lies just below all the data points. The mean curve usually will be a reasonably good estimate of the actual 50 percent probability of survival S-N curve. Based on the mean curve

---

*See, for example, ref. 9.

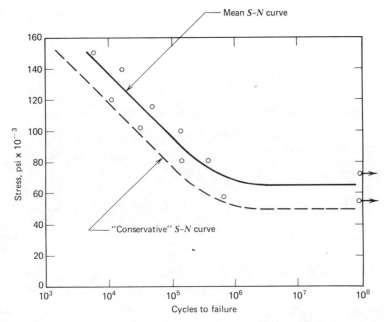

FIGURE 10.1. *S-N* curves resulting from "standard" fatigue testing method.

and some estimate of the standard deviation, reasonable estimates may be made for the *S-N-P* family of curves.

Any type of "conservative" *S-N* curve drawn below the data is rather indefinite and cannot be associated with any specific probability of survival. This suggests that the first technique of constructing a mean curve and associated estimate of the *S-N-P* family is probably superior. However, it is doubtful that much useful statistical information can be developed from the "standard" method, even under the best conditions, because the sample size is so small.

10.3   CONSTANT STRESS LEVEL TESTING

The constant stress level procedure involves testing groups of approximately 15 or more specimens at each of four or more different constant stress levels in the stress range between the fatigue limit and the yield strength of the material, recording failure lives for all tests. Experience has indicated that the distributions of fatigue failure lives at constant stress levels higher than the fatigue limit are log-normal to a good approximation. All the data taken at each constant stress level are therefore plotted on log-normal probability paper to verify the distribution and to determine the mean and variance for log-life at that stress level. Figure 10.2 shows an example of the results of a

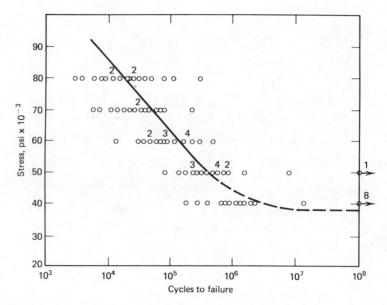

FIGURE 10.2. Fatigue data for constant stress level testing program plotted on standard *S-N* plot. (After ref. 1)

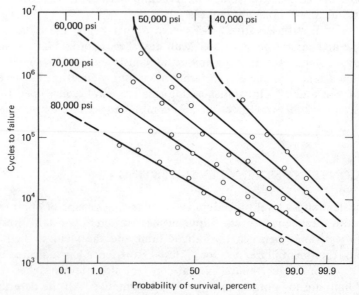

FIGURE 10.3. Fatigue data from constant stress level testing program plotted on log-normal probability paper. (After ref. 1)

constant stress level testing program plotted on a standard S-N plot, and Figure 10.3 shows the same data plotted on log-normal probability paper. It may be noted that the log-normal assumption is accurate at the higher stress levels, but at stress levels near the fatigue limit it is not valid. This is true because near the fatigue limit the data are not homogeneous since they contain some failures and some run-outs; thus the method is not recommended for use at stress levels near the fatigue limit. At the higher stress levels, however, the constant stress level method is efficient and forms a good basis for determining a useful family of S-N-P curves in the finite life range.

10.4 RESPONSE OR SURVIVAL METHOD (PROBIT METHOD)

It has just been noted that the constant stress level method is not an effective means for establishing the fatigue limit on a statistically sound basis. A much more useful method for determining the fatigue limit is the *survival method*, which is also called the *mortality method*, *quantal response method*, or *all-or-nothing method*. This technique may be used to determine the mean and variance of the fatigue limit or the mean and variance of fatigue strength at any specified life.

The survival method involves testing several groups of specimens at closely spaced stress levels spanning the range from about two standard deviations below the fatigue limit to two standard deviations above it. For example, preliminary testing may indicate a fatigue limit around 72,000 psi. On this basis five stress levels might be selected, ranging from 68,000 psi to 76,000 psi at 2000 psi intervals. If 20 specimens were tested at each stress level, the data might plot on S-N coordinates as shown in Figure 10.4. It may be observed that at the lowest selected stress level most specimens are run-outs, whereas at the highest selected stress level only a few specimens are run-outs, a result that might reasonably be expected. It is important to select the stress levels judiciously because if the stress level increments are too large, many groups may have all failures and other groups may have no failures, with little or no intermediate response. Usually, however, if the groups span the stress range of about two standard deviations above and below the mean, the response will be satisfactory.

The results of the survival test shown in Figure 10.4 may next be plotted on normal probability paper, as shown in Figure 10.5 where stress level is the random variable plotted against probability of survival calculated from the data of Figure 10.4. Drawing the best straight line through the probability plot then allows one to read the mean fatigue limit directly, and determine the standard deviation in stress, from the plot.

The mere act of plotting the survival data in this way is *not* the Probit method, as is sometimes erroneously stated. Probit analysis (2) is a rather more complex technique of computing an optimum line to be drawn through

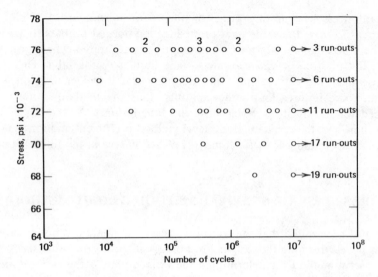

FIGURE 10.4. *S-N* plot showing results of a survival test for determination of fatigue limit.

the survival data points by using a type of least-squares analysis in which each of the data points is weighted in accordance with its distance from the optimum line. For most purposes satisfactory results may be achieved by simply drawing the best straight line through the points by inspection, but with currently available computer codes, least-squares curve fitting has also become a simple matter.

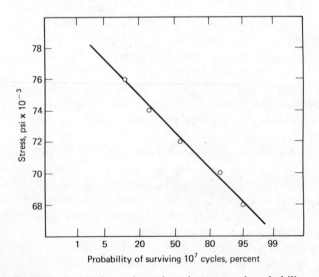

FIGURE 10.5. Survival test data plotted on normal probability paper.

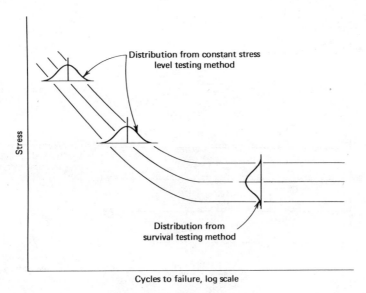

FIGURE 10.6. *S-N-P* curves determined by using constant stress level method in finite life range and survival method in infinite life range.

Utilizing the survival method for determining the fatigue limit and its distribution, and utilizing the constant stress level method for determining the fatigue life and its distribution at each of several stress levels in the finite life range, we may construct a family of *S-N-P* curves. Such a family of curves is shown schematically in Figure 10.6.

10.5 STEP-TEST METHOD

A clear disadvantage of the survival method just described is that a large number of specimens are required, usually about 60 to 100. This makes the survival method costly and time-consuming.

An alternative method, called the step-test method, is a test that forces every specimen to fail. The technique is to subject each specimen to a prescribed block of cycles at each of a series of increasing stress levels, until the specimen finally fails. The risk of such a test lies in the unknown effects of coaxing or latent damage that may be incurred while stepping up through the series of stress levels to reach the failure stress levels.

To start a step test, a stress level is selected at about 70 percent of the estimated mean fatigue limit. The specimen is then tested at that stress level until either failure occurs or the run-out life is achieved, say at 10^7 cycles. If failure occurs, the stress level and failure life are recorded. If run-out occurs, the stress level is increased approximately 0.7 of the estimated stresswise

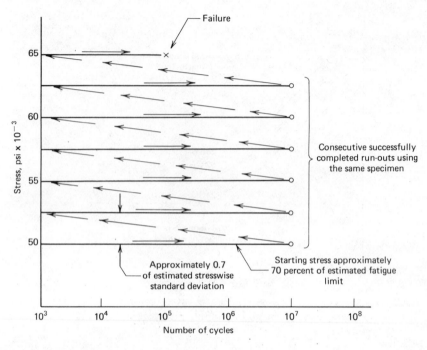

FIGURE 10.7. Plot showing the progressive testing associated with each specimen using the step-test method. (After ref. 1)

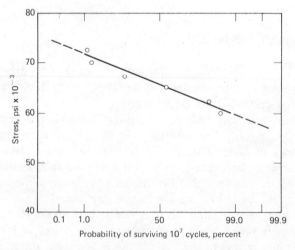

FIGURE 10.8. Step-test results plotted on normal probability paper. (After ref. 1)

standard deviation and the same specimen is run again at the new stress level. Again, if the specimen fails, the data are recorded; and if run-out occurs, the stress level is again increased for a new run using the same specimen. This procedure is continued until the specimen does fail. For a typical specimen, the method is illustrated in Figure 10.7. A minimum of 10 to 15 specimens is required for a meaningful test.

When the step-test data have been acquired, they may be plotted on normal probability paper, as shown in Figure 10.8. The plot is made by calculating the percent of the total sample that has survived and plotting this probability versus the stress level, as shown in Figure 10.8. From the probability plot the mean fatigue limit and stresswise standard deviation may be determined in the usual way, as discussed in Section 9.9.

10.6 PROT METHOD

In 1948 a unique rapid method for determining the fatigue limit was proposed by Marcel Prot (3). This testing technique, now called the Prot method, involves the application of a steadily increasing stress level with applied cycles until the specimen fails. The failure stress is then related to the fatigue limit through the rate of stress increase and two material constants. Clearly the effects of coaxing must be known, or known to be small, if this method is to be used. The Prot method has been used successfully for many steel alloys, titanium alloys, and even aluminum alloys. The details of the testing and analysis associated with the Prot method are as follows: A specimen is placed in a testing machine with the initial stress level well below the estimated fatigue limit, usually in the range between zero and 70 percent of the fatigue limit. When the test is started and cycles begin to accumulate, the stress level is increased systematically with increasing cycles so that on the average the increase in stress is linear with applied cycles. The stress level may be increased either in small steps or continuously, and the test is always carried forth until failure of the specimen occurs. The change in stress level might be accomplished by the metered flow of water or steel shot into a loading container, a gearbox coupled to the drive spindle that in turn drives a lead screw that moves a dead weight along a calibrated loading beam, or some other method of progressively increasing the load. There are no run-outs in a Prot test. A group of 15 or 20 specimens would typically be tested using the same rate of stress increase. This rate of stress increase is called the Prot rate α and has dimensions of psi per cycle. A second group of specimens is then run at a different Prot rate, and even a third or fourth group may be run at different Prot rates. The results of such a series of tests might be as shown in Figure 10.9, where the results are shown for three groups of specimens all started at an initial stress level of 52,000 psi but tested at three different Prot rates. The dashed lines in Figure 10.9 indicate the mean path of specimen

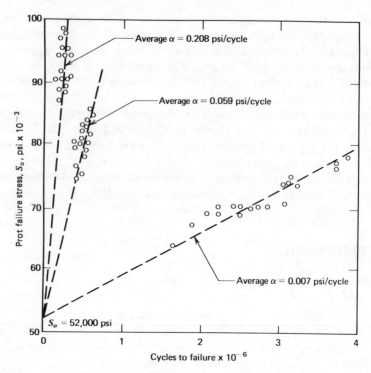

FIGURE 10.9.   Prot test results showing failure stress versus cycles to failure for three different Prot rates. (After ref. 1)

stress history for each of the three Prot rates used. The data to be recorded for each test include the starting stress level S_o, Prot rate α, number of cycles N_α, and Prot failure stress S_α. It was proposed by Prot that the fatigue limit E may then be calculated from

$$S_\alpha = E + K\alpha^n \tag{10-1}$$

where K and n are materials constants. The Prot method is to plot S_α versus α^n, as shown in Figure 10.10, with the objective of finding a value of n that results in a linear relationship between S_α and α^n. For most ferrous alloys this value of n lies between 0.45 and 0.50, and the assumption that $\alpha^n = \sqrt{\alpha}$ is widely used for steel alloys. For certain nonferrous alloys the magnitude of n may be as low as 0.15. With the curve of Figure 10.10 available, the constant K in (10-1) may be evaluated by selecting two different Prot rates. Once having determined K and n for the material, we may calculate the fatigue limit from 10-1 as

$$E = S_\alpha - K\alpha^n \tag{10-2}$$

where all information on the right side of the equation is provided directly

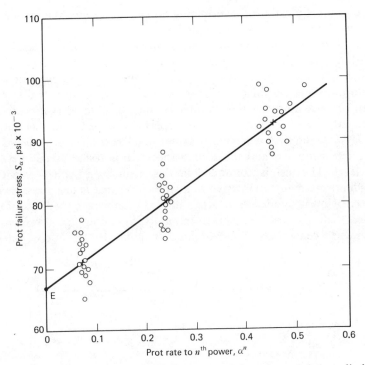

FIGURE 10.10. Prot test results showing the determination of fatigue limit E by plotting failure stress S_α versus α^n. (After ref. 1)

from the Prot test data. The solution of (10-2) is the equivalent of extrapolating the straight line of Figure 10.10 back to the zero Prot rate ordinate where the intercept is interpreted to be the fatigue limit E.

An important concept of the Prot test is that once having established the constants K and n for a material, it is possible to determine a fatigue limit value for each individual specimen tested. The mean fatigue limit and stresswise standard deviation may then be found directly in terms of E. This technique has been widely used in aircraft quality control testing since it is simple and fast, and it gives a definite value for fatigue limit for each specimen tested. The method has been questioned for use in research work, and many have declared it inefficient compared to other methods when new materials are being evaluated.

10.7   STAIRCASE OR UP-AND-DOWN METHOD

A very useful testing method for determining the mean and variance of fatigue strength at any specified life is the up-and-down testing method. The method is equally valid for determining fatigue limit, since the fatigue limit is simply the fatigue strength at infinite life.

To perform an up-and-down test a group of at least 15 specimens is selected to evaluate the fatigue strength at the life of interest. The first specimen is tested at a stress level a little higher than the estimated fatigue strength until it either fails or runs out at the life of interest. If the specimen fails before reaching the life of interest, the stress level is *decreased* by a preselected increment and the second specimen is tested at this new lower stress level. If the first specimen runs out, the stress level is *increased* by the preselected increment and the second specimen is tested at this new higher stress level. The test is continued in this manner in sequence, with each succeeding specimen being tested at a stress level that is one stress increment above or below its predecessor, depending on whether the predecessor ran out or failed. Thus the test is sequential in nature, has an up-and-down character, and tends to center on the mean value of failure for the life of interest. A

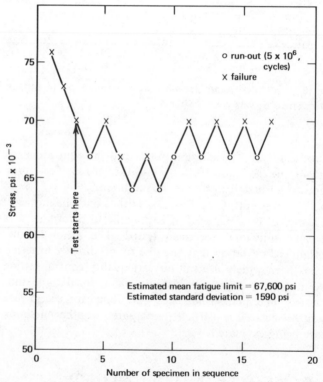

FIGURE 10.11. Up-and-down fatigue test used to determine median (mean) fatigue strength at 5×10^6 cycles for a 4340 steel alloy.

typical up-and-down test is illustrated in Figure 10.11. It may be noted that the up-and-down test is not considered to have started until the first reversal occurs, as indicated in Figure 10.11. The starting stress level therefore is arbitrary and does not influence the result. However, it is usually wise to try to start at a stress level higher than the estimated mean to economize on time by ensuring failure of the first specimen.

Up-and-down test data may be analyzed statistically by observing certain characteristics of the normal distribution and the binomial distribution, then approximating one by the other. Although the details of the development are beyond the scope of this discussion, the results are simple and easy to use. The procedure for analyzing up-and-down data is as follows:

1. Estimate the mean fatigue strength S_m corresponding to the life level of interest.

2. Estimate the standard deviation σ for the material, based on experience.

3. Test the first specimen at a stress level $S_m + d$, where d is the step size, usually taken to be approximately one standard deviation. Continue the test at this stress level until the specimen either fails or runs out at the prescribed life of interest.

4. If the first specimen fails, test the second specimen at a stress level one increment d *lower* than the previous stress level. If the first specimen runs out, test the second specimen at a stress level one increment d *higher* than the previous stress level.

5. Continue the test sequentially until at least 15 to 30 specimens have been tested.

6. When the test has been completed, determine whether failures or run-outs are the less frequent event. Only the less frequent event is used in the analysis.

7. Tabulate the data as follows in a five-column table: In column I list all stress levels experienced by the less frequent event. In column II the number 0 is assigned to the lowest stress level, 1 is assigned to the next higher stress level, etc., until all stress levels are numbered in order of increasing magnitude. In column III list the number of times the event occurred at each stress level. In column IV tabulate the product of column II times column III. In column V tabulate the product of the square of column II times column III.

8. From the tabulation of Step 7, sum column IV and set the sum equal to A. Also from the tabulation, sum column V and set the sum equal to B.

9. Calculate the statistical estimate of mean fatigue strength at the prescribed life from

$$\hat{S}_m = S_0 + d\left[\frac{A}{N} \pm \frac{1}{2}\right]$$

(10-3)

where $\hat{S}_m$ = statistical estimate of mean fatigue strength at prescribed life, psi

S_0 = lowest stress level at which the less frequent event occurred, psi

d = step size, psi

N = total number of less frequent events

A = sum defined in Step 8

The plus sign ($+$) is used if the less frequent event is runout, and the minus sign ($-$) is used if the less frequent event is failure.

10. Calculate the statistical estimate of the population standard deviation from

$$\hat{\sigma} = 1.62\, d\left[\frac{NB - A^2}{N^2} + 0.029\right] \quad \text{if} \quad \frac{NB - A^2}{N^2} \geq 0.3$$

$$\hat{\sigma} = 0.53\, d \qquad\qquad\qquad \text{if} \quad \frac{NB - A^2}{N^2} < 0.3$$

(10-4)

where $\hat{\sigma}$ = statistical estimate of standard deviation, psi

d = step size, psi

N = total number of less frequent event

A, B = sums defined in Step 8

11. To obtain confidence limits on the estimated mean fatigue limit it is necessary to find the standard deviation σ_m of the estimate of the mean fatigue limit $\hat{S}_m$. This is given by

$$\sigma_m = \frac{G}{\sqrt{N}}\,\sigma \tag{10-5}$$

The factor G is a nonlinear function of d/σ and must be obtained graphically from Figure 10.12. Since the population standard deviation σ is not known, the estimator $\hat{\sigma}$ of (10-4) must be used to obtain σ. Confidence limits are then obtained for the mean fatigue limit as

$$C\{\hat{S}_m - y_0\sigma_m \leq \mu \leq \hat{S}_m + y_0\sigma_m\} = 100(1 - \alpha) \tag{10-6}$$

where μ = true mean fatigue strength at prescribed life

α = significance level of test

y_0 = ordinate defining standard normal rejection region based on selection of α

$\hat{S}_m$ = estimate of mean fatigue limit from (10-3)

σ_m = standard deviation of estimated mean fatigue limit from (10-5)

C = confidence in percent

To illustrate the use of the up-and-down analysis consider the test data plotted in Figure 10.11.

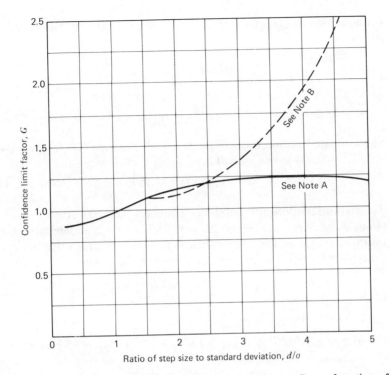

FIGURE 10.12. Graphical display of confidence limit factor G as a function of d/σ for use in up-and-down method. Note A: Solid branch to be used if mean falls on a testing level. Note B: Dashed branch to be used if mean falls midway between two testing levels. Note C: Interpolate between solid and dashed branches for intermediate levels of the mean. (After ref. 1)

1. In the data of Figure 10.11 eight failures and seven run-outs are observed. Therefore *run-outs* are the less frequent event.
2. Table 10-1 may be constructed.

Table 10.1. Tabular Information Required for Computing Mean and Standard Deviation From Up-and-Down Test Data

I	II	III	IV	V
64,000	0	2	0	0
67,000	1	5	5	5
70,000	2	0	0	0

3. Summing column III yields $N = 7$.
   Summing column IV yields $A = 5$.
   Summing column V yields $B = 5$.

4. Estimated mean fatigue strength at 5×10^6 cycles is calculated from (10-3) as

$$\hat{S}_m = 64{,}000 + 3{,}000 \left[\frac{5}{7} + \frac{1}{2} \right] \qquad (10\text{-}7)$$

or

$$S_m = 67{,}000 \text{ psi} \qquad (10\text{-}8)$$

5. The estimated standard deviation is calculated from (10-4), since $(NB - A^2)/N^2 = 0.204$ is less than 0.3, as

$$\hat{\sigma} = 0.53(3000) = 1590 \text{ psi} \qquad (10\text{-}9)$$

6. To establish 95 percent confidence limits on the mean, the value of y_o in (10-6) is read from Table 9.2 for $\alpha = 0.05$ with a symmetrical two-tailed rejection region as 1.96. The factor G is then read at a value of $(d/\sigma) = (3{,}000/1{,}590) = 1.9$ to be $G = 1.12$, whence (10-5) gives

$$\sigma_m = \frac{1.12}{\sqrt{7}}(1590) = 675 \text{ psi} \qquad (10\text{-}10)$$

and (10-6) finally becomes

$$C\{67{,}600 - 1.96(675) \le \mu \le 67{,}600 + 1.96(675)\} = 95 \text{ percent} \qquad (10\text{-}11)$$

or

$$C\{66{,}780 \le \mu \le 68{,}920\} = 95 \text{ percent} \qquad (10\text{-}12)$$

Thus, with 95 percent confidence it would be predicted that the mean fatigue strength at 5×10^6 cycles lies in the range from 66,280 to 68,920 psi. Similar techniques are available for establishing confidence limits on the standard deviation,* but since the primary purpose of the up-and-down test is to establish the mean, the details of establishing confidence limits on standard deviation are omitted here.

It should be observed that the up-and-down method is efficient in establishing the mean value but is not very useful in defining the distribution. Other methods, such as the survival method, are more useful for defining the distribution.

10.8   EXTREME VALUE METHOD

Finally, a method sometimes used to establish extreme value probability S-N curves is the so-called *extreme value method* or *least-of-n* method. For this

*See p. 325 of ref. 4.

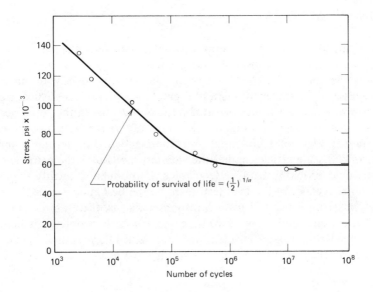

FIGURE 10.13. Extreme value probability *S-N* plot based on least-of-*n* testing. (After ref. 1)

type of test the technique is to select a group of *n* specimens to be run simultaneously in *n* different identical fatigue testing machines, all at the same stress level. When the first specimen of the group fails, the stress level and number of cycles to failure are recorded; then all other machines are stopped and the specimens are discarded.

Next, a second group of *n* specimens is run at a new stress level, again recording data for the first failure and discarding all the other specimens. The procedure is repeated for several different stress levels at and above the fatigue limit.

Finally the first-of-*n* failure data are plotted on an *S-N* plot and a best-fit curve drawn through the data, as shown in Figure 10.13. The curve drawn in this way may be shown (5) to correspond to a probability of survival given by

$$P\{\text{survival}\} = \left(\frac{1}{2}\right)^{1/n} \qquad (10\text{-}13)$$

where *n* is the number of specimens in each test group. This technique is especially appropriate for providing design information since designers are always concerned with extreme value probabilities of survival to ensure design success. This testing technique provides such information without specific knowledge of the form of the distribution.

10.9   IN SUMMARY

Although the preceding discussions are not exhaustive, they indicate that careful planning and execution of a fatigue testing program is necessary if the collection and analysis of data are to meet the test objectives with maximum effectiveness, minimum investment, and greatest achievable statistical significance. For producing designer-oriented basic S-N-P curves, the combined use of constant stress level testing in the finite life range with up-and-down testing in the very long life range is recommended. For testing a limited number of full-scale articles, either the step-test method or the extreme value method would probably be a better choice. If production quality control is the test objective, use of the Prot test might be the best choice since the test is relatively fast and every test gives a response; that is, there are no run-outs. Careful attention should always be given to the desired objectives before a testing procedure is finally selected, and statistical interpretation of the data should be carefully planned before the test is conducted.

QUESTIONS

1.   Of the various fatigue testing methods discussed, which methods would you suggest using to obtain an accurate representation of the S-N-P curves for an alloy steel over the full life range from about 10^4 cycles to infinite life? Explain each of these methods in a short discussion.

2.   For inexpensive quality control of production items, and a short turn-around time on test results, which fatigue testing methods would you suggest? Explain in a brief discussion.

3.   A constant stress level testing program has been conducted using SAE 1045 steel specimens, all from a single heat of material. Three completely reversed stress levels were investigated in the "short life" range. The test results from 20-specimen samples for each of these three constant stress level tests are shown in the chart, in which the life data have been ordered from shortest to longest lifetime.

(a)   Using log-normal probability paper, plot the fatigue life data for these three stress levels.

(b)   From the plots determine the log mean life and the log standard deviation associated with each stress level.

(c)   For each stress level, convert the data of part (b) to mean life in cycles and high-side and low-side standard deviation in cycles.

(d)   Explain how the information of part (c) could be used to construct the S-N-P curves in the "short life" range for this material.

4.   Using the same material as for the tests of problem 3, an up-and-down testing program was run in the "long life" range, using 5×10^7 cycles as the run-out criterion. Fifty-four specimens were tested sequentially with the results shown in Figure Q10.4.

Specimen Number	Fatigue Life (thousands of cycles)*		
	$\sigma_a = 55{,}000$ psi	$\sigma_a = 61{,}000$ psi	$\sigma_a = 67{,}000$ psi
1	174	78	30
2	236	82	31
3	242	85	31
4	257	86	32
5	271	95	33
6	295	98	33
7	319	101	38
8	352	102	39
9	357	105	41
10	377	105	44
11	415	123	45
12	438	126	45
13	458	126	45
14	493	127	46
15	552	137	46
16	578	147	46
17	685	148	50
18	696	158	54
19	1390	162	64
20	1676	175	75

*Data extracted from p. 9 of ref. 10

(a) Estimate the mean fatigue strength for this alloy at infinite life (5×10^7 cycles data assumed to give infinite life data for ferrous alloys) from the up-and-down test results.

(b) Estimate the stresswise standard deviation at infinte life from the up-and-down test results.

(c) Calculate the 95 percent confidence limits on the mean fatigue limit for this material.

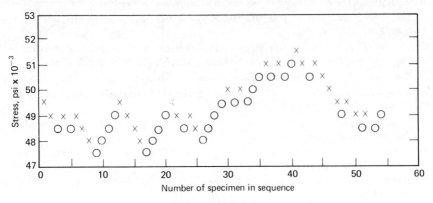

FIGURE Q10.4.

(d)   Explain how the information of parts (a) and (b) could be used to construct the S-N-P curves in the "long life" range for this material.

5.   Using pertinent results from problems 3 and 4, construct a family of S-N-P curves for SAE 1045 steel covering the life range from 10^4 cycles to infinity. Show the curves for probability of survival equal to 0.01, 0.05, 0.10, 0.50, 0.90, 0.95, and 0.99.

6.   (a)   In Prot tests of a steel alloy, it has been found that if a Prot rate of 0.04 psi per cycle is used, the mean Prot failure stress based on a sample of 15 specimens is 78,500 psi. If a Prot rate of 0.25 psi per cycle is used, the mean Prot failure stress based on a sample of 16 is 95,500 psi. Accepting that $n = 0.5$ for steel, estimate the mean fatigue limit for this steel alloy.

(b)   Prot testing is being used as a quality control method, and a Prot rate of 0.1 psi per cycle is being used to test periodic samples of the steel alloy described in part (a). If the mean fatigue limit is not permitted to vary more than ± 10 percent from the mean value estimated in part (a) and your quality control test produced a Prot failure stress of 82,000 psi, would you stop production and look for a problem or would you allow production to continue?

7.   Three identical components are tested in three identical testing machines, all at a stress amplitude of 80,000 psi. All of the machines are started at the same time. The first component fails after 80,000 cycles. Based on this information, estimate how many components would be expected to survive at least 80,000 cycles if a different group of 13 similar components were to be tested.

REFERENCES

1.   Hardenbergh, D. E. (ed.), *Statistical Methods in Materials Research*, Proceedings of Short Course, Department of Engineering Mechanics, Pennsylvania State University, 1956.

2.   Finney, D. J., *Probit Analysis*, Cambridge University Press, London, 1952.

3.   Prot, E. M., *Fatigue Testing Under Progressive Loading; A New Technique for Testing Materials* (translation by E. J. Ward), WADC TR52-148, September 1952.

4.   Dixon, W. J., and Massey, F. J., Jr., *Introduction to Statistical Analysis*, McGraw Hill, New York, 1957.

5.   Gumbel, E. J., "Statistical Theory of Extreme Values and Some Practical Applications," *National Bureau of Standards Applied Mathematics Series*, No. 33, February 12, 1954.

6.   Mood, A. M., *Introduction to the Theory of Statistics*, McGraw Hill, New York, 1950.

7.   *A Guide for Fatigue Testing and the Statistical Analysis of Fatigue Data*, STP-91-A, American Society for Testing and Materials, Philadelphia, 1963.

8.   Heller, R. A. (ed.), *Probabilistic Aspects of Fatigue STP-511*, American Society for Testing and Materials, Philadelphia, 1972.

9.   Swanson, S. R. (ed.), *Handbook of Fatigue Testing*, STP-655, American Society for Testing and Materials, Philadelphia, 1974.

10.   *Symposium on Statistical Aspects of Fatigue*, STP-121, American Society for Testing and Materials, Philadelphia, 1952.

CHAPTER 11

Low-Cycle Fatigue

11.1 INTRODUCTION

Two domains of cyclic loading were identified in Section 7.1. One domain is that for which the cyclic loads are relatively low, strain cycles are confined largely to the elastic range, and long lives or high numbers of cycles to failure are exhibited. This behavior, which has been extensively discussed in the preceding chapters, has traditionally been called high-cycle fatigue. The other domain is that for which the cyclic loads are relatively high, significant amounts of plastic strain are induced during each cycle, and short lives or low numbers of cycles to failure are exhibited if these relatively high loads are repeatedly applied. This type of behavior has been commonly called *low-cycle fatigue* or, more recently, cyclic strain-controlled fatigue. The transition from low-cycle fatigue behavior to high-cycle fatigue behavior generally occurs in the range from about 10^4 to 10^5 cycles, and many investigators now define the low-cycle fatigue range to be failure in 50,000 cycles or less (1). Although the usual objective of an engineering designer is to provide long life, there are several circumstances in which the low-cycle fatigue or strain-controlled life response are of great importance. For example, in the design of high-performance devices such as missiles and rockets, the total design lifetime may be only a few hundred or a few thousand cycles from launch to delivery, and low-cycle fatigue analysis and design methods are of direct interest. In the design of other high-performance devices, such as aircraft gas turbine blades and wheels, nuclear pressure vessels and fuel elements, or steam turbine rotors and shells, the occurrence of occasional large mechanical or thermal transients during operation may give rise to significant damage accumulation due to a few hundred or a few thousand of these large cycles over the design lifetime, so that low-cycle fatigue design methods are of great importance. Even if the loads on a machine or structure are nominally low, the material at the root of any critical notch will experience local plasticity that is cyclically strain controlled because of the constraints imposed by the surrounding bulk of elastic material, and the methods of low-cycle or strain-controlled fatigue will again be important in life prediction of such components.

Many of the basic concepts of low-cycle fatigue have already been discussed in Section 8.5, where the local stress-strain approach to prediction of crack initiation was shown to embody the important ideas of low-cycle fatigue. Although low-cycle fatigue has been the subject of intensive study during the past two decades, and great progress has been made in understanding the phenomenon, much remains to be done before the designer has reliable life prediction tools that properly account for effects of nonzero mean stress and strain, multiaxial states of stress, and cumulative damage, especially under conditions of elevated temperature.

11.2   THE STRAIN CYCLING CONCEPT

If a typical S-N curve, such as the one illustrated in Figure 11.1, is examined in the low-cycle fatigue region, it will be found that over the range from a quarter cycle up to around 10^3 cycles the fatigue strength is nearly constant and close to the ultimate strength of the material. That is, the S-N curve remains relatively flat throughout this region in which the material cyclically experiences general yielding and gross plastic deformation. In this region of macroscopic plastic behavior the fatigue life is much more accurately described as a function of the cyclic *strain amplitude* than the cyclic stress amplitude. Stress-strain behavior under these circumstances is characterized by a stress-strain hysteresis loop, as shown in Figure 11.2, with evidence of a measurable plastic strain in the specimen or machine part. This plastic behavior is typically nonlinear and history dependent, and it has been observed that the stress-strain response of most materials changes significantly with cyclic straining into the plastic range. Some materials exhibit

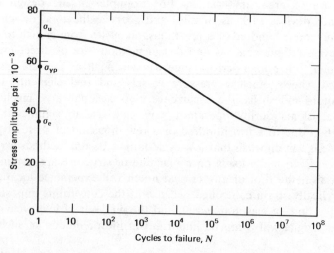

FIGURE 11.1.   Typical S-N curve showing characteristic flattening in the low-cycle fatigue range.

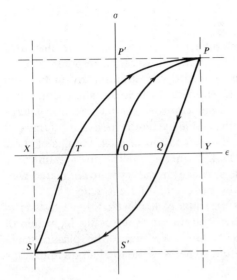

FIGURE 11.2. Hysteresis loop associated with cyclic loading that produces low-cycle fatigue damage.

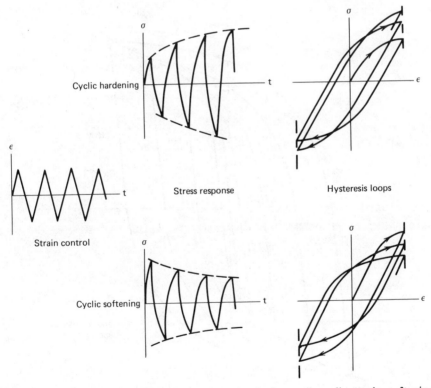

FIGURE 11.3. Illustration of cyclic strain hardening and cyclic strain softening phenomena under strain control. (From ref. 2, copyright ASTM; reprinted with permission.)

cyclic strain hardening and others exhibit cyclic strain softening, as illustrated in Figure 11.3. As discussed in Section 8.5, the stress-strain response of most materials changes significantly with applied cyclic strains early in life, but typically the hysteresis loops tend to stabilize so that the stress amplitude remains reasonably constant under strain control over the remaining large portion of the fatigue life. Based on the stable hysteresis loops for a family of different constant strain amplitudes, a curve passed through the tips of these hysteresis loops, as shown in Figure 11.4, defines a "cyclic" stress-strain curve for the material. In Figure 8.17 cyclic stress-strain curves were compared with static or monotonic stress-strain curves for several different materials.

The usual method of displaying the results of low-cycle fatigue tests is to plot the logarithm of strain amplitude or strain range versus the logarithm of number of cycles (or reversals) to failure. Sometimes the plastic strain amplitude or strain range is plotted, and sometimes the total strain amplitude

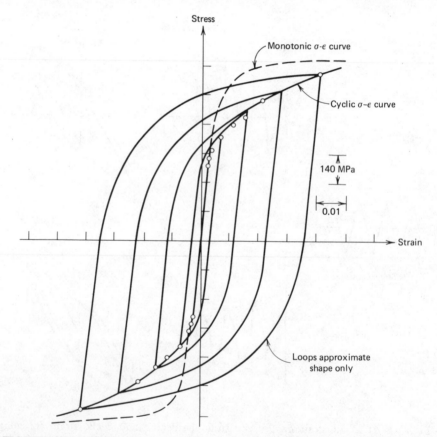

FIGURE 11.4. Cyclic stress-strain curve compared to monotonic stress-strain curve for SAE 4340 steel. (From ref. 2, copyright ASTM; reprinted with permission.)

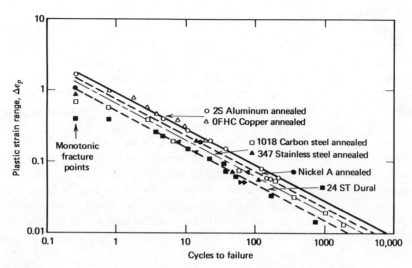

FIGURE 11.5. Low-cycle fatigue plot for several different materials. (From ref. 3, copyright American Society for Metals, 1959; reprinted with permission.)

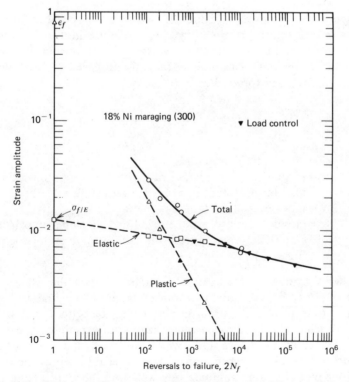

FIGURE 11.6. Strain amplitude versus life for 18 percent Ni maraging steel, separately showing elastic, plastic, and total components of the strain (From ref. 2, copyright ASTM; reprinted with permission.)

or strain range is plotted as the ordinate. Early experimental investigations had indicated that if the plastic strain amplitude were plotted versus cycles to failure on a log-log plot, the data would approximate a straight line with a slope of about -0.5. Subsequent investigations have indicated that the slope ranges from about -0.5 to -0.7. Such plots seem to be remarkably similar for a wide range of materials (3), as indicated in Figure 11.5. Experimental evidence accumulated by various investigators in recent years seems to indicate that the cyclic life is better related to total strain than to plastic strain, especially at the longer life end of the low-cycle range. An example of a plot of strain amplitude versus life is shown in Figure 11.6, separately showing the plastic strain amplitude and the total strain amplitude for a nickel-steel alloy. The data of Figure 11.6 may be interpreted in light of the earlier discussion of Figure 8.23 and the discussion of Section 11.3 that follows.

11.3   THE STRAIN-LIFE CURVE AND LOW-CYCLE FATIGUE RELATIONSHIPS

Data such as those shown in Figure 11.5 led to the proposal of an empirical equation relating plastic strain range $\Delta\varepsilon_p$ to failure life N_f for completely reversed strain cycling under uniaxial stress conditions in the low-cycle fatigue regime. The relationship, independently proposed by Manson (4) and Coffin (5, 6), may be expressed as

$$\frac{\Delta\varepsilon_p}{2} = \varepsilon_f'(2N_f)^c \qquad (11\text{-}1)$$

where $\Delta\varepsilon_p/2$ = plastic strain amplitude

ε_f' = fatigue ductility coefficient, defined as strain intercept at one load reversal, that is, at $2N_f = 1$ (see Figure 11.7)

$2N_f$ = total reversals to failure

c = fatigue ductility exponent defined as slope of plastic strain amplitude versus reversals to failure curve on log-log plot (see Figure 11.7)

Later work by many investigators, capitalizing upon the Manson-Coffin equation of (11-1), has indicated that total strain amplitude, the sum of elastic strain amplitude, plus plastic strain amplitude may be better correlated to life. Figure 11.7 schematically shows a generalization of experimental evidence of the type presented in Figure 11.6, which indicates that the elastic strain amplitude and the plastic strain amplitude may each be represented as linear on a log-log plot of strain amplitude versus life. As shown in Figure 11.7, the total strain amplitude is the sum of elastic plus plastic components. This has been modeled mathematically by Morrow, et al. (7), and has already been

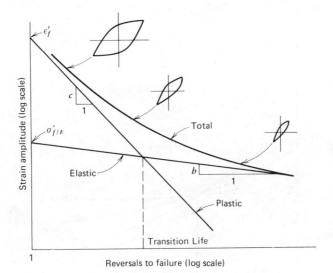

FIGURE 11.7. Schematic representation of elastic, plastic, and total strain amplitude versus fatigue life. (From ref. 2, copyright ASTM; reprinted with permission.)

presented in (8-116) as

$$\frac{\Delta\varepsilon}{2} = \frac{\sigma_f'}{E}(2N_f)^b + \varepsilon_f'(2N_f)^c \tag{11-2}$$

where constants b and σ_f'/E are the slope and one-reversal intercept of the elastic curve in Figure 11.7, and constants c and ε_f' are the slope and one-reversal intercept of the plastic curve in Figure 11.7. Typically (2) b ranges from about -0.05 to -0.15 and c ranges from about -0.5 to -0.8. The quantities σ_f' and ε_f' may be approximated as discussed in Section 8.5. From an energy-based argument, the exponents b and c may be approximated as a function of the cyclic strain-hardening exponent n' (See Figure 8.17) as (16)

$$b = \frac{-n'}{1 + 5n'} \tag{11-3}$$

and

$$c = \frac{-1}{1 + 5n'} \tag{11-4}$$

From the schematic representation of (11-2) in Figure 11.7, it may be noted that at short lives the plastic strain amplitude component dominates, whereas at longer lives the elastic strain amplitude component dominates. The point at which the elastic and plastic curves intersect has been called the "transition life." A thoughtful consideration of (11-2) and Figure 11.7 leads to the observation that for design lives less than the transition life, materials with

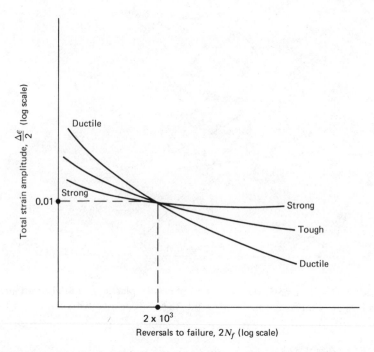

FIGURE 11.8. Idealized representation of cyclic strain failure resistance of various types of materials. (After ref. 2, copyright ASTM; reprinted with permission.)

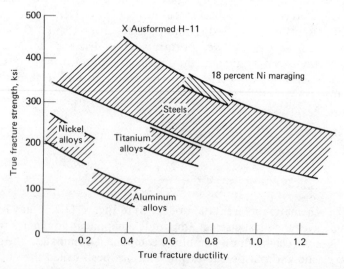

FIGURE 11.9. Monotonic fracture strength-ductility combinations presently attainable for various alloy classes. (From ref. 2, copyright ASTM; reprinted with permission.)

high fatigue ductility (high fracture ductility) are superior, whereas design lives greater than the transition life demand materials with high values of true fracture strength. This contradictory set of requirements, namely high strength with high ductility, requires careful consideration on the part of the designer to match the appropriate material to the application, with careful attention to the magnitude of operating strain amplitude. This point is emphasized in Figure 11.8 where three idealized materials are shown, one very high strength material (strong), one very ductile material (ductile), and one whose properties lie between the two extremes (tough). These curves cross at about 10^3 cycles (2×10^3 reversals) at a total cyclic strain amplitude of about 0.01. Thus, one would select the "strong" material for design life requirements greater than about 10^3 cycles, pick the "ductile" material for design life requirements shorter than about 10^3 cycles, and "optimize" with the "tough" material for spectrum loading of a more complicated nature. It is interesting to note that all types of materials seem to have about the same fatigue resistance for total strain amplitude around 0.01, corresponding to a failure life of about 10^3 cycles. Presently attainable combinations of true fracture strength and true fracture ductility are illustrated in Figure 11.9.

11.4 THE INFLUENCE OF NONZERO MEAN STRAIN AND NONZERO MEAN STRESS

The effects of nonzero mean *strain* under low-cycle fatigue conditions have been studied by relatively few investigators, principally for the case of tensile mean strain. The experimental results of these few investigations (9, 10, 11) indicate that the effect of a compressive mean strain on low-cycle fatigue life is essentially the same as the effect of a tensile mean strain if their magnitudes are the same. These results also indicate that mean strain effects are of primary importance only in the operating range where the *plastic* strain component dominates, that is, at design lives less than the transition life for the material.

The effects of nonzero mean *stress* are of primary importance only in the operating range where the *elastic* strain component dominates, that is, at design lives greater than the transition life of the material. The topic of nonzero mean stress influence has already been discussed extensively in Section 7.9, addressed again in (8-114) and (8-115), and will not be discussed further. The effects of nonzero mean strain, however, should be considered further.

The stress-strain diagram and corresponding strain-time plot of Figure 11.10 show a typical uniaxial loading path that leads to a nonzero mean strain. It may be observed that during the first loading cycle the strain increases from zero to a maximum value at B and subsequently cycles from B through C to D and back again to B repeatedly, assuming that the stress-strain

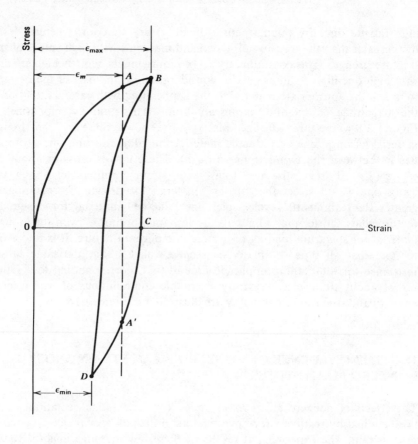

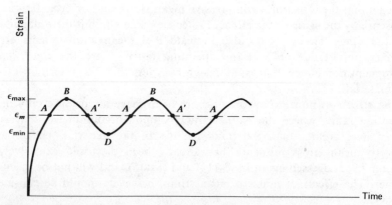

FIGURE 11.10. Stress-strain diagram and corresponding strain-time plot for uniaxial loading that leads to a nonzero mean strain. (After ref. 8)

properties do not change with additional cycles. In Figure 11.10, point B corresponds to the maximum strain in the cycle ε_{max}, and point D corresponds to minimum strain in the cycle ε_{min}, after the initial load application. The mean strain ε_m is the arithmetic mean of ε_{max} and ε_{min}. To account for the nonzero mean strain induced by this type of loading, one investigator (8) suggests the following empirical relationship between total strain range $\Delta\varepsilon$

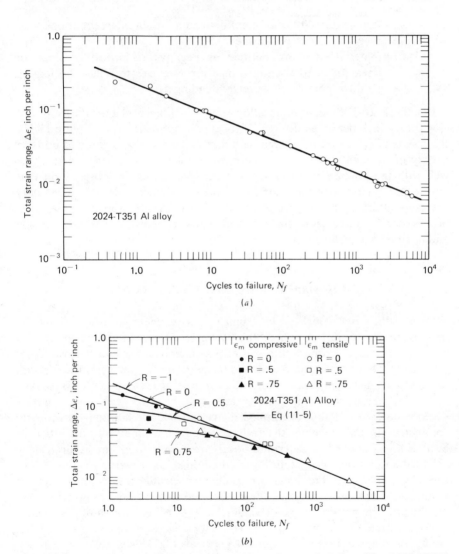

FIGURE 11.11. Effects of nonzero mean strain on low-cycle fatigue life of 2024-T351 aluminum alloy. (*a*) Completely reversed strain cycling data. (*b*) Strain cycling data for tensile and compressive mean strains compared to (11-5). (From ref. 8)

and life N_f:

$$\Delta\varepsilon = \frac{2(1 - R)\varepsilon'_f}{\left[(4N_f - 1)(1 - R)^a + (2)^a\right]^{1/a}} \tag{11-5}$$

where $R = \varepsilon_{min}/\varepsilon_{max}$ if $|\varepsilon_{max}| \geq |\varepsilon_{min}|$
$R = \varepsilon_{max}/\varepsilon_{min}$ if $|\varepsilon_{max}| < |\varepsilon_{min}|$
$\varepsilon'_f =$ fatigue ductility coefficient defined as strain intercept at one load reversal, or one half cycle
$a =$ material constant, defined as negative reciprocal of slope of linear fit of total strain range versus cycles to failure on log-log plot, for completely reversed loading

For the 2024-T351 aluminum alloy shown in Figure 11.11a, the values of $\varepsilon'_f = 0.1934$ inch per inch and $a = 2.51$ were determined; and in Figure 11.11b the results of (11-5) are compared with data for the same alloy tested under a variety of different tensile and compressive mean strains. These data agree well with the predictions of (11-5) and support the earlier observation that tensile and compressive mean strains of the same magnitude have essentially the same effect on life. Application of (11-5) to A-302 pressure vessel steel produced good agreement with data and similar conclusions regarding tensile and compressive mean strain.

11.5   CUMULATIVE DAMAGE IN LOW-CYCLE FATIGUE

Just as in the case of high-cycle fatigue, the assessment of low-cycle fatigue damage under conditions where the cyclic strain amplitude ranges over a spectrum of values requires the use of a cumulative damage theory. Many investigators have studied cumulative damage effects in low-cycle fatigue with the general conclusion that a linear damage rule of the Palmgren-Miner type yields acceptable results if local stress-strain behavior can be accurately determined as a function of applied loads and if cycle counting is properly conducted. If, for example, the fatigue modified Neuber rule described in Section 8.5 is utilized to determine local stress and strain amplitude history and if the rain-flow cycle counting method, also described in Section 8.5, is properly applied to the local strain-time spectrum, results of using the Palmgren-Miner linear damage rule of (8-4) have been found to give acceptable life predictions. The range of values for $\Sigma(n/N)$ corresponding to failure under a variety of spectrum loading conditions has been reported to be from about 0.6 to about 1.6 for a variety of materials (8). This is a narrower range for low-cycle fatigue than has been observed for high-cycle fatigue. It should be noted, however, that for multiaxial states of stress, the linear damage rule may be much less reliable (12).

11.6   INFLUENCE OF MULTIAXIAL STATES OF STRESS

A detailed treatment of the effects of multiaxial states of stress on low-cycle fatigue behavior is beyond the scope of this discussion, but proposed techniques have been presented for estimating low-cycle fatigue life under these conditions.* The proposed technique involves the definition of an *equivalent stress* and an *equivalent total strain range*, both calculable on the basis of the discussion of Section 5.4 from the multiaxial states of stress and strain. Life estimates based on the equivalent total strain range under conditions of a multiaxial state of stress can then be made from uniaxial state-of-stress low-cycle fatigue data expressed as total strain range versus cycles to failure. Many questions still remain to be answered regarding the validity of this technique; however, it seems to be the best approach available at the present time. If fatigue properties are anisotropic, or become anisotropic because of cyclic plasticity effects on material properties, then the concept of a single equivalent stress or strain is no longer valid since the direction of loading relative to the directional fatigue properties may become a factor of importance. Much remains to be done before the designer has reliable tools for handling the influence of multiaxial states of stress on prediction of low-cycle fatigue life. All the foregoing problems are further complicated if the low-cycle fatigue process is caused by or accompanied by elevated temperatures. Some of these problems are discussed in Chapter 13.

11.7   RELATIONSHIP OF THERMAL FATIGUE TO LOW-CYCLE FATIGUE

Low-cycle fatigue has been described as a progressive failure phenomenon brought about by the cyclic application of strains that extend into the plastic range to produce failure in about 10^5 cycles or less. Although the strains are often produced mechanically, it is perhaps even more usual to find the cyclic strains produced by a cyclic thermal field. If a machine part undergoes cyclic temperature changes and if the natural thermal expansions and contractions are either wholly or partially constrained, cyclic strains and stresses result. These cyclic strains produce fatigue failure just as if the strains were produced by external mechanical loading. Thus, all the results described earlier for low-cycle fatigue (and high-cycle fatigue if strains are all elastic) are at least qualitatively applicable to thermal fatigue as well. It must be recognized, however, that thermal fatigue problems involve not only all the complexities of mechanical loading but, in addition, all the temperature-induced problems as well.

*See refs. 8, 13, and p. 165ff of 14.

A series of results (14) in which low-cycle fatigue tests involving mechanical strain cycling at constant elevated temperature are compared with thermally cycled constrained specimens, in which the strain cycle was induced by a cyclic temperature field, are depicted in Figure 11.12. These results include mechanically induced strain cycling at fixed temperatures of 350°C to 500°C, as well as thermally cycled tests with fully constrained specimens cycled from 200°C to 500°C at a mean cycle temperature of 350°C.

It may be observed from the results shown in Figure 11.12 that for equal values of plastic strain range the number of cycles to failure was much less for the thermally cycled specimens than for the mechanically cycled specimens, even though in one case the mechanically cycled specimens were tested at 100°C above the maximum temperature of 500°C used with the thermally cycled specimens. To bring the thermal fatigue curve in Figure 11.12 into coincidence with the isothermal 350°C mechanically strain cycled test requires multiplication of the strain at any N_f value by a factor of approximately 2.5. Although similar results for several other materials have been noted, testing of Inconel (17) indicates a good direct correlation between the results of thermally cycled and mechanically cycled specimens. Thus, it is clear that although the thermal low-cycle fatigue phenomenon and the mechanical low-cycle fatigue phenomenon are very similar, and are mathematically expressible by the same types of equations, the use of mechanical low-cycle fatigue results to predict thermal low-cycle fatigue performance must be undertaken with care.

Some of the differences between thermally and mechanically induced low-cycle fatigue that give rise to the apparent discrepancy in low-cycle

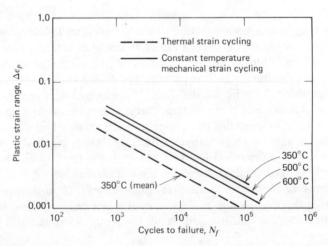

FIGURE 11.12. Comparison of mechanical strain cycling at elevated temperatures with thermal strain cycling in the low-cycle fatigue region for AISI type 347 stainless steel. (After ref. 14, p. 256)

fatigue data for these two cases are*

1. In thermal fatigue the plastic strain tends to become concentrated in the hottest regions of the body, since the yield point is locally reduced at these hottest regions.

2. In thermal fatigue there is often a localized region of strain developed by virtue of plastic flow during the compressive branch of the strain cycle to produce a bulging at the hottest region. This is followed by a necking tendency adjacent to the bulge caused by plastic flow during the tensile branch of the strain cycle upon cooling.

3. Cyclic variations in temperature may, in and of themselves, have important effects upon the material properties and ability to resist low-cycle fatigue failure.

4. There may be interaction effects caused by superposition of simultaneous variations in temperatures and strains.

5. Rates at which the strain cycling is induced may have an important effect, since the testing speeds in thermal fatigue tests are often greatly different from the rates used in mechanical low-cycle fatigue tests.

For these reasons caution is necessary in the prediction of thermal low-cycle fatigue behavior from mechanical low-cycle fatigue results or vice versa.

11.8   IN SUMMARY

It may be concluded from the discussions of this chapter that the designer's tools for predicting failure and designing equipment to operate in a low-cycle fatigue environment are empirical and sketchy. Intensive research efforts in the low-cycle fatigue failure domain are being pursued in an effort to improve the designer's ability to successfully avert failure by the low-cycle fatigue failure mode, especially under conditions of nonzero mean strain, nonzero mean stress, multiaxial states of stress and strain, spectrum loading with a mixture of low-cycle and high-cycle fatigue loading, and for materials that have a significant degree of anisotropy.

11.9   USING THE IDEAS

Suppose that it is desired to design a high-performance, short, structural load link for use in an air-to-air missile where light weight is of the utmost importance. The link may be regarded as a two-force member under direct axial loading P that fluctuates in accordance with the estimate plotted in Figure 11.13 for a "typical" 5 seconds of loading. This "typical" loading

*See pp. 270–272 of ref. 14.

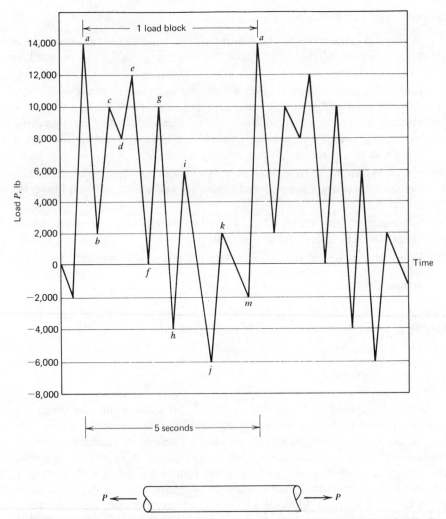

FIGURE 11.13. Loading spectrum for short-life, high-performance load link for missile application.

block is assumed to repeat time after time. If the material is SAE 4340 steel heat treated to 350 BHN (see Figure 8.17) and true fracture ductility has been determined to be $\varepsilon_f = 0.48$, what size should the cross-sectional area be made to support the load for a design life of 6 minutes if a safety factor of 1.2 is desired?

An assessment of this design task indicates that the problem is relatively challenging, even under the specified uniaxial loading conditions and even if the further decision is made to avoid any complications associated with stress or strain concentrations. The task involves spectrum loading, cycle counting, nonzero mean strain, and cumulative damage concepts in low-cycle fatigue. To obtain an estimate of proper link size, the rain flow cycle counting method of Section 8.5 may be employed to determine the maximum and minimum value of each load reversal in the "typical" 5-second block of loading. The stress and strain are related to the load through the area, as yet unknown. Therefore, if a value for area is assumed, a maximum stress and minimum stress value may be computed for the maximum and minimum load of each reversal in the 5-second block. From these stress peaks, corresponding values of maximum and minimum strain may be found from the cyclic stress-strain curve for SAE 4340 steel in Figure 8.17. Equation (11-5) may then be utilized to determine the theoretical failure life in cycles associated with each level within the 5-second loading block.

Finally, the Palmgren-Miner linear damage rule may be used to estimate the damage associated with each 5-second loading block and subsequently with the 72 repeated blocks covering the 6-minute design life. The area must then be adjusted and the process repeated until the Palmgren-Miner rule sums to unity for the 6-minute history. Only then can the safety factor be imposed.

To proceed with the details of the solution outlined, consider one load block from Figure 11.13 and divide each force peak magnitude by a trial area of $A = 0.10$ inch^2 to determine peak stresses. These values of stress are plotted versus time in Figure 11.14. Figure 8.17 is then used to find corresponding values of strain. Table 11.1 summarizes the computational results in going from Figure 11.13 to Figure 11.14 and shows the corresponding strain values obtained from Figure 8.17.

The final column in Table 11.1 is computed from (11-5), in which the value of $a = -c$ is determined from the cyclic strain-hardening exponent $n' = 0.14$ taken from Figure 8.17 and inserted into (11-4) to give

$$c = \frac{-1}{1 + 5(0.14)} = -0.59 \tag{11-6}$$

Table 11.1.   Determination of strain-time history for trial area of 0.10 inch^2

Reversal Number	σ_{max}	σ_{min}	ε_{max}	ε_{min}	$\Delta\varepsilon$	R	N_f
1, 10	140,000	− 60,000	0.0200	− 0.0020	0.0220	− 0.10	87
2, 5	120,000	20,000	0.0075	0.0007	0.0068	0.09	1,100
3, 4	100,000	80,000	0.0045	0.0027	0.0018	0.60	10,516
6, 7	100,000	0	0.0045	0	0.0045	0	2,215
8, 9	60,000	− 40,000	0.0020	− 0.0013	0.0007	− 0.65	51,770
11, 12	20,000	− 20,000	0.0007	− 0.0007	0.0014	− 1.0	15,975

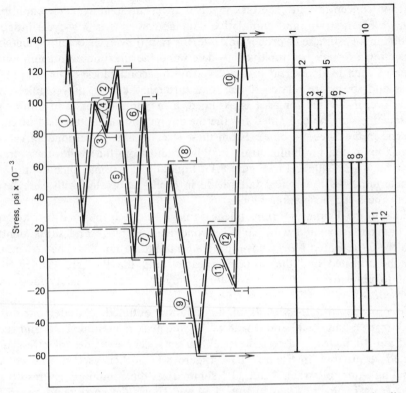

FIGURE 11.14. Rain flow cycle count of stress cycles imposed by the loading of Figure 11.13 on a link with cross-sectional area of 0.10 inch².

so that

$$a = -\frac{1}{c} = 1.70 \tag{11-7}$$

Thus, (11-5) becomes

$$\Delta\varepsilon = \frac{2(1 - R)(0.48)}{\left[(4N_f - 1)(1 - R)^{1.70} + (2)^{1.70}\right]^{1/1.70}} \tag{11-8}$$

or

$$\Delta\varepsilon = \frac{0.96(1 - R)}{\left[(4N_f - 1)(1 - R)^{1.70} + 3.25\right]^{0.59}} \tag{11-9}$$

For reversal numbers 1 and 10 in Table 11.1, this equation becomes, with $\Delta\varepsilon = 0.0220$ and $R = -0.10$,

$$0.0220 = \frac{0.96(1 + 0.10)}{\left[(4N_f - 1)(1 + 0.1)^{1.70} + 3.25\right]^{0.59}} \tag{11-10}$$

from which $N_f = 87$ cycles. The remaining values of N_f are computed in the same way. Next, utilizing the Palmgren-Miner linear damage rule, we find that the damage associated with one 5-second loading block is

$$D_b = \frac{1}{87} + \frac{1}{1100} + \frac{1}{10,516} + \frac{1}{2215} + \frac{1}{51,770} + \frac{1}{15,975} \quad (11\text{-}11)$$

or

$$(D)_{5\,\text{sec}} = 0.013 \quad (11\text{-}12)$$

Finally, the total damage for the 6-minute design life is estimated to be

$$(D)_{6\,\text{min}} = \frac{6(60)}{5}(0.013) = 0.94 \quad (11\text{-}13)$$

Thus, the trial area was close but slightly larger than necessary. Another iteration might be used to bring the magnitude of $(D)_{6\,\text{min}}$ closer to unity. Assuming the solution to be close enough for a first estimate, the safety factor of 1.2 would be applied to the area to obtain a first design estimate of $A = 0.12$ inch^2 for this application. An alternative solution to this problem might be to neglect the influence of nonzero mean strain and use (11-2) for determination of N_f, followed by the same type of cumulative damage assessment. For very short design lives it is important to consider the influence of nonzero mean strains, but for longer lives they may usually be neglected with little error. However, at the longer lives, nonzero mean stresses must be considered, as discussed in Chapter 7.

QUESTIONS

1. Distinguish between low-cycle fatigue and high-cycle fatigue, giving the characteristics of each.

2. Distinguish between "load-controlled" fatigue tests and "strain-controlled" fatigue tests, using sketches of hysteresis loops to clearly identify the primary differences.

3. Determine the "transition life" for the 18 percent Ni maraging steel shown in Figure 11.6 and the cyclic strain amplitude associated with this life.

4. Certain cyclic properties for two steel materials have been found as shown in the chart.*

Material	σ'_{yp} (psi)	n'	ε'_f	σ'_f (psi)
SAE 1015 (BHN 80)	35,000	0.22	0.95	120,000
SAE 4340 (BHN 409)	120,000	0.15	0.48	290,000

*See Chap. 3 of ref. 18.

(a)   For each of these two materials calculate the number of cycles to failure for total strain amplitude values of 0.05, 0.01, and 0.005. Which of these materials would be the best choice for each of the three specified total strain amplitudes?

(b)   How does the 18 percent Ni maraging steel of Figure 11.6 compare with the two materials shown in the chart? Does your conclusion agree with the results shown in Figure 11.8?

(c)   What would your estimate be for the transition life of the 4340 steel alloy described?

5.   The axial loading spectrum shown in Figure Q11.5 is to be applied in repeated block fashion to a solid bar of square cross section made of the 4340 steel alloy shown in Figure 8.17. The square bar may be regarded as a two-force member under direct axial loading P that fluctuates in accordance with the repeated 10-second loading block shown. What size should the square bar be to support the load for a design life of 4 minutes if a safety factor of 1.15 is desired?

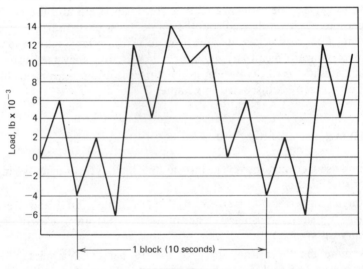

FIGURE Q11.5.

6.   A new low-cycle thermal fatigue testing machine is to be constructed. The specimen is to be heated by using it as a direct resistance across the output of a welding transformer. It is desired to be able to produce either mechanical low-cycle fatigue or thermal low-cycle fatigue in this device and to study their similarities, differences, and interactions. Discuss the problems that you would foresee in the design, fabrication, and use of such a device; suggest solutions if you can.

REFERENCES

1. *Manual on Low Cycle Fatigue Testing*, STP-465, American Society for Testing and Materials, Philadelphia, 1969.

2. Landgraf, R. W., "The Resistance of Metal to Cyclic Deformation," *Achievement of High Fatigue Resistance in Metal and Alloys*, STP-467, American Society for Testing and Materials, Philadelphia, 1970.

3. Tavernelli, J. F., and Coffin, L. F., Jr., "A Compilation and Interpretation of Cyclic Strain Fatigue Tests on Metals," *ASM Transactions of American Society for Metals*, **51** (1959): 438–453.

4. Manson, S. S., "Behavior of Materials Under Conditions of Thermal Stress," NACA TN-2933, National Advisory Committee for Aeronautics, Cleveland 1954.

5. Coffin, L. F., Jr., "A Study of the Effects of Cyclic Thermal Stresses in a Ductile Metal," *ASME Transactions*, **16** (1954): 931–950.

6. Coffin, L. F., Jr., "Design Aspects of High Temperature Fatigue with Particular Reference to Thermal Stresses," *ASME Transactions*, **78** (1955): 527–532.

7. Morrow, J., Martin, J. F., and Dowling, N. E., "Local Stress-Strain Approach to Cumulative Fatigue Damage Analysis," Final Report, T & A. M. Report No. 379, Department of Theoretical and Applied Mechanics, University of Illinois, Urbana, Ill., January 1974.

8. Ohji, K., Miller, W. R., and Marin, J., "Cumulative Damage and Effect of Mean Strain in Low Cycle Fatigue of a 2024-T351 Aluminum Alloy," *Journal of Basic Engineering*, ASME Transactions, **88**, Series D (1966): 801.

9. Sachs, G., Gerberich, W. W., Weiss, V., and Latorre, J. V., "Low Cycle Fatigue of Pressure Vessel Materials," *ASTM Proceedings*, **60** (1961): 512–529.

10. Sessler, J. G., and Weiss, V., "Low Cycle Fatigue Damage in Pressure Vessel Materials," *Journal of Basic Engineering*, ASME Transactions, **85**, Series D (1963): 539–547.

11. Manson, S. S., "Thermal Stresses in Design, Part 21—Effect of Mean Stress and Strain on Cyclic Life," *Machine Design*, **32**, No. 16 (August 4, 1969): 129–135.

12. Blatherwick, A. A., and Viste, N. D., "Cumulative Damage Under Biaxial Fatigue Stress," *Materials Research and Standards*, Vol. 7, American Society for Testing and Materials, Philadelphia, 1967, pp. 331–336.

13. Krempl, E., *The Influence of State of Stress on Low-Cycle Fatigue of Structural Materials: A Literature Survey and Interpretation Report*, STP-549, American Society for Testing and Materials, Philadelphia, 1974.

14. Manson, S. S., *Thermal Stress and Low Cycle Fatigue*, McGraw-Hill, New York, 1966.

15. Baldwin, E. E., Sokol, G. J., and Coffin, L. F., Jr., "Cyclic Strain Fatigue Studies on AISI Type 347 Stainless Steel," *ASTM Proceedings*, **57** (1957): 567–586.

16. Morrow, J., *Internal Friction, Damping and Cyclic Plasticity*, STP-378, American Society for Testing and Materials, Philadelphia, 1965, p. 45.

17. Swindeman, R. W. and Douglas, D. A., "The Failure of Structural Metals Subjected to Strain-Cycling Conditions," *Journal of Basic Engineering*, ASME Transactions, **81**, Series D (1959): 203–212.

CHAPTER 12

Stress Concentration

12.1 INTRODUCTION

Failures in machines and structures virtually always initiate at sites of local *stress concentration* caused by geometrical or microstructural discontinuities. These stress concentrations, or *stress raisers*, often lead to local stresses many times higher than the nominal net section stress that would be calculated without considering stress concentration effects. An intuitive appreciation of the stress concentration associated with a geometrical discontinuity may be developed by thinking in terms of "force flow" through a member as it is subjected to external loads. The sketches of Figure 12.1 illustrate the concept. In Figure 12.1*a* the rectangular flat plate of width *w* and thickness *t* is fixed at the lower edge and subjected to a total force *F* uniformly distributed along the upper edge. The dashed lines each represent a fixed quantum of force, and the local spacing between lines is therefore an indication of the local force intensity, or stress. In Figure 12.1*a* the lines are uniformly spaced throughout the plate, and the stress σ is uniform and calculable as

$$\sigma = \frac{F}{wt} \tag{12-1}$$

In the sketch of Figure 12.1*b* a flat rectangular plate of the same thickness has been subjected to the same total force *F*, but the plate has been made wider and notched to provide the same net section width *w* at the site of the notch. The lines of force flow may be visualized in very much the same way that streamlines would be visualized in the steady flow of a fluid through a channel with the same shape as the plate cross section. It may be noted in Figure 12.1*b* that no force can be supported across the notch, and therefore the lines of force flow must pass around the root of the notch. In so doing, force flow lines crowd together locally near the root of the notch, producing a higher force intensity, or stress, at the notch root. Thus, the local stress is raised or concentrated near the notch root, and even though the net section nominal stress is still properly calculated by (12-1), the actual local stress at the root of the notch may be many times higher than the calculated nominal stress. Many common examples of stress concentration may be cited, some of which are illustrated in Figure 12.2. Discontinuities at the roots of gear teeth, at the corners of keyways in shafting, at the roots of screw threads, at the fillets of shaft shoulders, around rivet holes and bolt holes, and in the

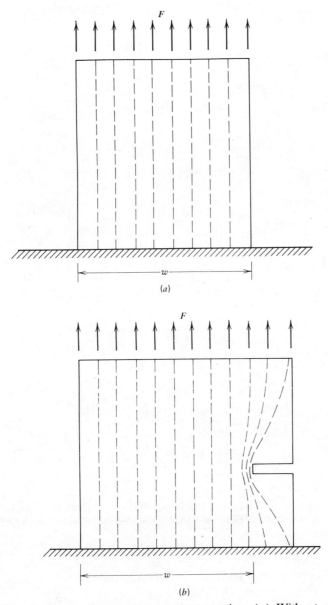

FIGURE 12.1. Intuitive concept of stress concentration. (*a*) Without stress concentration. (*b*) With stress concentration.

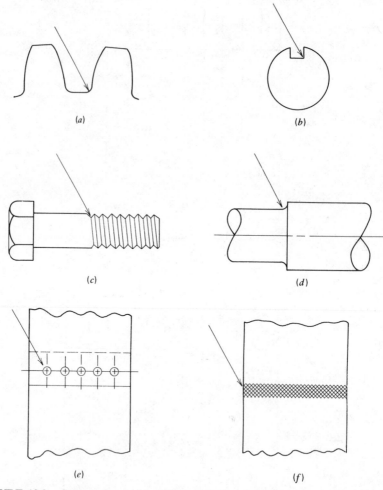

FIGURE 12.2. Some common examples of stress concentration. (*a*) Gear teeth. (*b*) Shaft keyway. (*c*) Bolt threads. (*d*) Shaft shoulder. (*e*) Riveted or bolted joint. (*f*) Welded joint.

neighborhood of welded joints all constitute stress raisers that usually must be considered by a designer. The seriousness of the stress concentration depends on the type of loading, the type of material, and the size and shape of the discontinuity.

12.2   STRESS CONCENTRATION EFFECTS

For the purposes of studying stress concentration effects, stress raisers may be classified as being either *highly local* or *widely distributed*. Highly local stress raisers are those for which the volume of material containing the concentration of stress in negligibly small compared to the overall volume of the stressed member. Widely distributed stress raisers are those for which the volume of material containing the concentration of stress is a significant portion of the overall volume of the stressed member. Thus, for the case of a highly local stress concentration the overall size and shape of the stressed part would not be significantly changed by yielding in the region of the stress concentration, whereas for a widely distributed stress concentration the overall size and shape of the stressed part would be subject to significant changes by virtue of yielding in the region of stress concentration. For example, small holes and fillets would usually be regarded as highly local stress concentrations; curved hooks or eye-and-clevis joints would be categorized as widely distributed stress concentration.

With the foregoing definitions made, stress concentration effects may be classified as shown in Table 12.1. From the table it may be noted that the effects of stress concentration must be considered for all combinations of geometry, loading, and material except one, namely, the case of a highly local stress concentration in a ductile material subjected to static loading. In this case the local yielding is usually negligible and a stress concentration factor of unity may be used. All other cases must be analyzed for potential failure because of the effects of stress concentration.

The final column in the table lists K_t or K_f as stress concentration factors. The factor K_t is the theoretical elastic stress concentration factor, which is defined to be the ratio of the actual maximum local stress in the region of the discontinuity to the nominal net section stress calculated by simple theory as if the discontinuity exerted no stress concentration effect; that is,

$$K_t = \frac{\text{actual maximum stress}}{\text{nominal stress}} \qquad (12\text{-}2)$$

Table 12.1.   Effects of Stress Concentration

Type of Stress Concentration	Type of Loading	Type of Material	Type of Failure	Stress Concentration Factor
Widely distributed	Static	Ductile	Widely distributed yielding	K_t (modified)
Widely distributed	Static	Brittle	Brittle fracture	K_t
Widely distributed	Cyclic	Any	Fatigue failure	K_f
Highly local	Static	Ductile	No failure (redistribution occurs)	1
Highly local	Static	Brittle	Brittle fracture	K_t
Highly local	Cyclic	Any	Fatigue failure	K_f

It should be noted that the value of K_t is valid only for stress levels within the elastic range, and it must be suitably modified if stresses are in the plastic range. Such a modification is described in Section 12.4.

The factor K_f is the fatigue stress concentration factor, which is defined to be the ratio of the effective fatigue stress that actually exists at the root of the discontinuity to the nominal fatigue stress calculated as if the notch has no stress concentrating effect. This definition is made for the high-cycle fatigue range and must be suitably modified for the low-cycle fatigue range, as will be noted in Sections 12.4 and 12.6. Thus, the fatigue stress concentration factor may be defined as

$$K_f = \frac{\text{effective fatigue stress}}{\text{nominal fatigue stress}} \tag{12-3}$$

12.3   STRESS CONCENTRATION FACTORS FOR THE ELASTIC RANGE

Stress concentration factors are determined in a variety of different ways, including direct measurement of strain, utilization of photoelastic techniques, application of the principles of the theory of elasticity, and finite element analysis. Photoelastic stress analysis was, until recently, the most widely used technique for determining the stress distributions and stress concentration factors associated with various geometrical discontinuities. The technique is based on the birefringent or doubly refracting behavior of many transparent solids upon the application of strain-producing external loads. Interpretation of the interference patterns produced by passing polarized light through strained models made of the birefringent material yields a quantitative measure of local stress distributions throughout the body, from which stress concentration factors can be calculated. In recent years the use of finite element modeling has largely supplanted photoelastic analysis in determination of stress concentration factors. Numerical values for a wide variety of geometries and types of loading are presented in Reference 4. A few of the more common cases drawn from Reference 4 are reproduced in Figures 12.3 through 12.8.

The determination of theoretical stress concentration factors through application of the principles of theory of elasticity is a difficult problem. Such solutions have been obtained for only a few geometrical shapes, including certain two-dimensional problems, such as a circular or elliptical hole or groove in an infinite sheet, and axially symmetric three-dimensional problems, such as a deep hyperbolic groove in a solid bar.

A solution for the stress distribution around a small elliptical hole in a large plate subjected to tensile loading was developed by C. E. Inglis (5) in 1913. This case is illustrated in Figure 12.9. The resulting equations for the stress

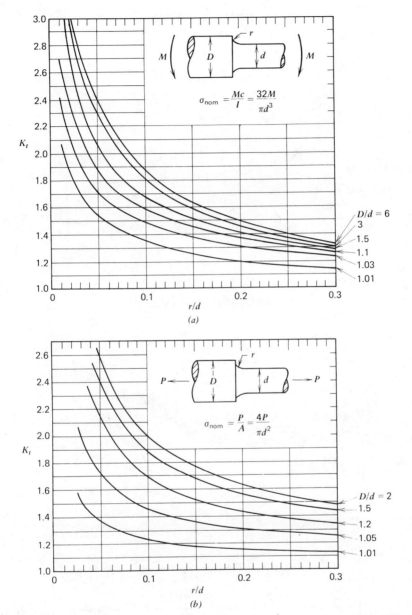

FIGURE 12.3. Stress concentration factors for a shaft with a fillet subjected to (*a*) bending, (*b*) axial load, or (*c*) torsion. (From ref. 4; adapted with permission from John Wiley & Sons, Inc.)

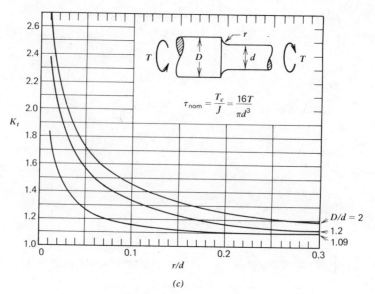

$$\tau_{nom} = \frac{T_c}{J} = \frac{16T}{\pi d^3}$$

(c)

FIGURE 12.3. (*Continued*)

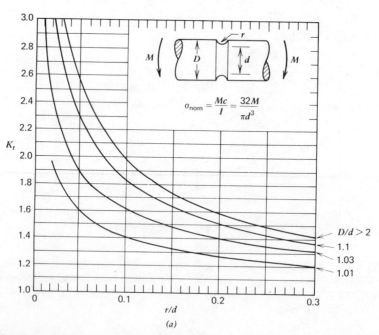

$$\sigma_{nom} = \frac{Mc}{I} = \frac{32M}{\pi d^3}$$

(a)

FIGURE 12.4. Stress concentration factors for a shaft with a groove subjected to (*a*) bending, (*b*) axial load, or (*c*) torsion. (From ref. 4; adapted with permission from John Wiley & Sons, Inc.)

406

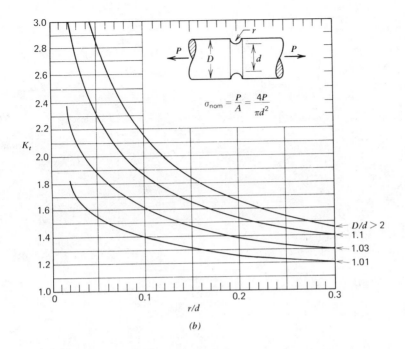

(b)

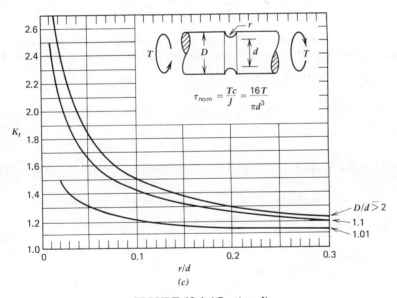

(c)

FIGURE 12.4 (*Continued*)

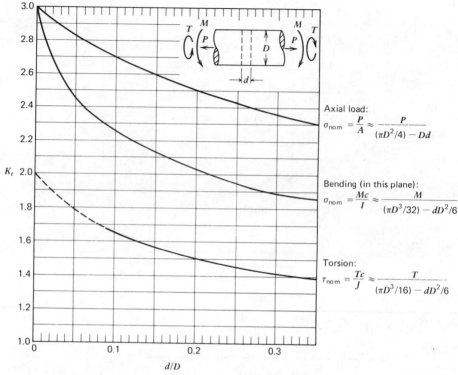

FIGURE 12.5. Stress concentration factors for a shaft with a radial hole subjected to axial load, bending, or torsion. (From ref. 4; adapted with permission from John Wiley & Sons, Inc.)

field are complex. However, for an elliptical hole whose major axis is along the x axis of a thin sheet loaded in the y direction, the stresses along the x axis are

$$\sigma_x = \sigma_n \left[-A + Ax(x^2 - a^2 + a\rho)^{-1/2} - Ba^2x(x^2 - a^2 + a\rho)^{-3/2} \right] \quad (12\text{-}4)$$

$$\sigma_y = \sigma_n \left[1 - C + Cx(x^2 - a^2 + a\rho)^{-1/2} + Ba^2x(x^2 - a^2 + a\rho)^{-3/2} \right] \quad (12\text{-}5)$$

$$\tau_{xy} = 0 \quad (12\text{-}6)$$

where $A = \left(1 - \sqrt{\rho/a} \right)^{-2}$ (12-7)

$$B = \left(1 - \sqrt{\rho/a} \right)^{-1} \quad (12\text{-}8)$$

$$C = \left(1 - \sqrt{\rho/a} \right)^{-2} \left(1 - \left(2\sqrt{\rho/a} \right) \right) \quad (12\text{-}9)$$

$a = $ semimajor axis
$b = $ semiminor axis
$\rho = $ minimum radius at end of the major axis
$\sigma_n = $ nominal net section stress

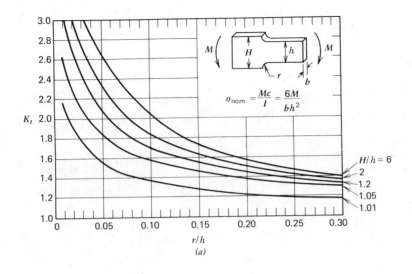

(a)

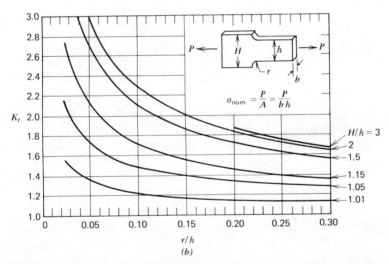

(b)

FIGURE 12.6. Stress concentration factors for a flat bar with a shoulder fillet subjected to (*a*) bending or (*b*) axial load. (From ref. 4; adapted with permission from John Wiley & Sons, Inc.)

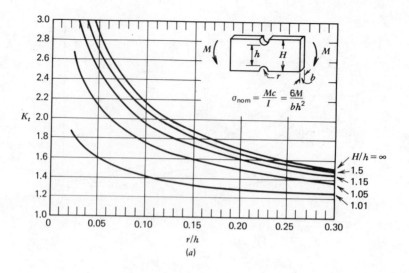

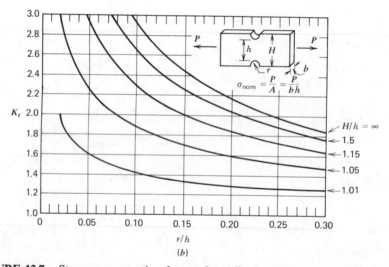

FIGURE 12.7. Stress concentration factors for a flat bar with a notch subjected to (a) bending or (b) axial load. (From ref. 4; adapted with permission from John Wiley & Sons, Inc.)

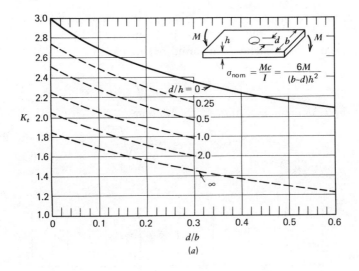

(a)

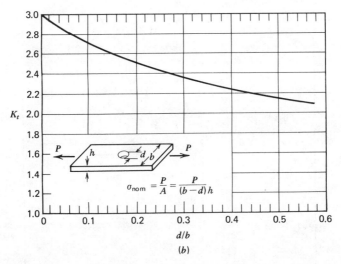

(b)

FIGURE 12.8. Stress concentration factors for a flat plate with a central hole subjected to (*a*) bending or (*b*) axial load. (From ref. 4; adapted with permission from John Wiley & Sons, Inc.)

The critical point of stress concentration occurs at the ends of the major axis of the elliptical hole, as shown in Figure 12.9, where from (12-5) the stress component σ_y reaches a maximum value of

$$\sigma_{y\,max} = \sigma_n\left(1 + 2\sqrt{\frac{a}{\rho}}\,\right) \tag{12-10}$$

so that the stress concentration factor becomes

$$K_t = \frac{\sigma_{y\,max}}{\sigma_n} = 1 + 2\sqrt{\frac{a}{\rho}} \tag{12-11}$$

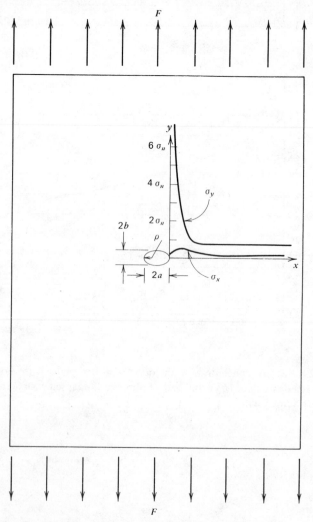

FIGURE 12.9. Stress concentration at edge of an elliptical hole in a thin sheet under tension.

It may be noted that for long, narrow elliptical holes the value of K_t becomes very high. Equation (12-11) is also approximately correct for a shallow circumferential groove in a circular shaft subjected to tension or bending if a is interpreted to be the depth of the groove and ρ the radius of curvature at the bottom of the groove.

For a shallow circumferential groove in a circular shaft subjected to torsion, the stress concentration factor for a groove depth a and bottom radius of curvature ρ is given by

$$(K_t)_{\text{torsion}} = 1 + \sqrt{\frac{a}{\rho}} \qquad (12\text{-}12)$$

A wide range of notch geometries was investigated by Neuber (6), who in 1946 presented a mathematical solution by which K_t can be satisfactorily estimated for most grooves and notches of interest. To do this he made use of the solutions for a shallow elliptical notch and a deep hyperbolic notch in an infinite plate. For the intermediate region between shallow and deep notches, Neuber devised a quadratic relationship that satisfied the end conditions and gave approximate values of K_t in the intermediate region. The results of Neuber's work are included in ref. 4.

As suggested at the outset, however, many geometries and types of loading produce stress concentration effects that do not yield to mathematical analysis. For these cases experimental techniques or finite element analyses are utilized to determine the stress concentration factors. Finite element stress analysis has already been cited as the most widely used technique for this purpose. Other techniques sometimes used include short-gage-length mechanical, optical, or electrical extensometers, electric resistance strain gages, brittle lacquer techniques, x-ray diffraction methods, and photostress techniques (7).

12.4 STRESS CONCENTRATION FACTORS AND STRAIN CONCENTRATION FACTORS FOR THE PLASTIC RANGE

With a sharply notched specimen or machine part, it is clear that even a moderate load may produce actual stresses at the root of the notch that exceed the yield point of the material locally. The local yielding causes a redistribution of stresses, and the theoretical elastic stress concentration factor K_t no longer describes the ratio of actual to nominal stresses accurately, since the actual maximum stress is relatively lower compared to the nominal stress than it would be if the material remained elastic. That is, the stress concentration factor is diminished in magnitude by local plastic flow, whereas the local strain is made larger than would be predicted by elastic theory.

Mathematical solutions of elastic-plastic stress and strain distributions around notches are relatively difficult to obtain, even using numerical solutions and digital computer techniques. One of the more successful approximations for stress concentration due to a circular hole in a very wide plate under tension has been given as (8)

$$K = 1 + 2\frac{E_s}{E} \tag{12-13}$$

where E = Young's modulus
E_s = secant modulus
K = stress concentration factor

Figure 12.10 illustrates values of stress concentration factor and strain concentration factor computed for a specific case by (12-13) and compared with values measured using very small strain gages. The agreement between calculated and measured values is good.

It has been further suggested (9) that a generalization of (12-13) be made to define a *plastic stress concentration factor* K_p for any type of geometrical discontinuity as

$$K_p = 1 + (K_t - 1)\frac{E_s}{E} \tag{12-14}$$

where K_t is the theoretical elastic stress concentration factor for whatever type of discontinuity is involved, E_s is the secant modulus, and E is Young's modulus. This expression gives acceptably accurate estimates for most cases.

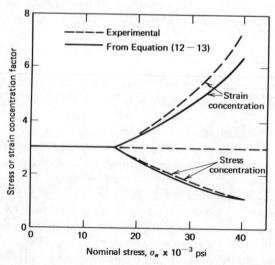

FIGURE 12.10.   Effect of plasticity on stress and strain concentration. (After ref. 2)

12.5   STRESS CONCENTRATION FACTORS FOR MULTIPLE NOTCHES

Sometimes it will be found that one stress raiser is superimposed upon another, such as a notch within a notch or a notch in a fillet. Although accurate calculation of the overall stress concentration factor is difficult for such combinations, reasonable estimates can be made (7). Figure 12.11a illustrates a large notch with a smaller notch at its root. To estimate the combined influence of these notches, the stress concentration factor K_{t1} for the large notch is determined as if the small notch did not exist. This permits an estimate of the stress σ_n' at the root of the large notch by multiplying the nominal stress σ_n by K_{t1} to give

$$\sigma_n' = K_{t1}\sigma_n \qquad (12\text{-}15)$$

Now, assuming that this stress σ_n' occurs throughout the zone within the dashed line near the notch root of Figure 12.11a, σ_n' becomes the nominal stress as far as the small notch is concerned. This is true because the entire small notch lies within the σ_n' field. The next step is to determine the stress concentration factor K_{t2} for the small notch acting alone and multiply it times σ_n' to obtain the actual stress at the notch root. Thus

$$\sigma_{\text{actual}} = K_{t2}\sigma_n' \qquad (12\text{-}16)$$

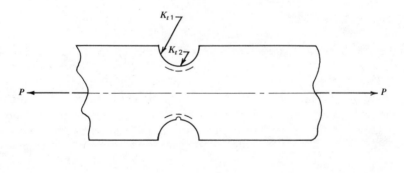

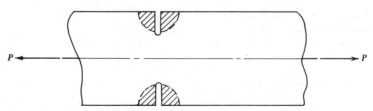

FIGURE 12.11.   Stress concentration effects due to superimposed multiple notches. (After ref. 7); reprinted with permission from McGraw Hill Book Company.)

or, utilizing (12-15),

$$\sigma_{actual} = K_{t1}K_{t2}\sigma_n \tag{12-17}$$

Thus, the combined theoretical stress concentration factor K_{tc} for the multiple notch is the *product* of the stress concentration factors for the two notches considered individually, whence

$$K_{tc} = K_{t1}K_{t2} \tag{12-18}$$

This has been verified photoelastically (7). The fatigue stress concentration factor for this case would depend on q (see Section 12.6) and could be calculated from (12-22) by substituting K_{tc} for K_t in that equation. If the small notch depth extends beyond the dashed line in Figure 12.11a, the value of K_{tc} will be less than the product $K_{t1}K_{t2}$ and would have to be estimated by using some technique such as photoelastic stress analysis or finite element modeling.

A technique for estimating a conservative limiting value for K_{tc} has been sketched in Figure 12.11b. The technique is to assume the notch of 12.11a to be filled in as shown by the crosshatched area of Figure 12.11b to leave a single deep, narrow notch. The theoretical stress concentration factor for this assumed single deep, narrow notch will always be greater than the stress concentration factor for the multiple notch.

12.6   FATIGUE STRESS CONCENTRATION FACTORS AND NOTCH SENSITIVITY INDEX

Unlike the theoretical stress concentration factor K_t, the fatigue stress concentration factor K_f is a *function of the material* as well as geometry and type of loading. To account for the influence of material characteristics, a *notch sensitivity index* q has been defined to relate the actual effect of a notch on fatigue strength of a material to the effect that might be predicted solely on the bases of elastic theory. The definition of notch sensitivity index q is given by

$$q = \frac{K_f - 1}{K_t - 1} \tag{12-19}$$

where K_f = fatigue stress concentration factor
$\quad\quad K_t$ = theoretical stress concentration factor
$\quad\quad q$ = notch sensitivity index valid for high cycle fatigue range.

The reason for subtracting unity from the numerator and denominator in this definition is to provide a scale for q that ranges from zero for no notch effect to unity for full notch effect. That is, for full notch effect K_f is equal to K_t. The notch sensitivity index has been found to be a function of both material and notch radius.

Although it was asserted for many years that K_f is almost always smaller than K_t, more recent experimental results have indicated that there are many cases, especially for the finer grained materials, such as quenched and tempered steels, where q is close to unity. Also, for coarser grained materials, such as annealed or normalized aluminum alloys, q approaches unity for stress raisers with notch radii that exceed about one-quarter inch. In view of these results, it is tempting to recommend the use of $K_f = K_t$ throughout all fatigue design since the error in doing so would be on the safe side. However, such an oversimplification would ignore several important notch sensitivity effects that should be appreciated. These effects include the following:

1. In some cases, under fatigue loading an alloy steel with superior *static* properties will be found *not* to have superior fatigue properties when compared to a plain carbon steel, because of the difference in notch sensitivity.

2. There is a tendency to improperly assess the effects of tiny scratches and cavities unless notch sensitivity effects are recognized.

3. Serious errors in applying the results from models to large structures may be made if notch sensitivity effects are not recognized.

4. In critical design situations, inefficiencies may accrue if notch sensitivity effects are not considered.

Scatter in the experimental data is a serious problem in evaluating notch sensitivity index, as may be seen in the experimental results shown in Figure 12.12. The notch sensitivity index for a range of steels and an aluminum alloy are shown in Figure 12.13 for axial, bending, and torsional loading. These curves provide sufficient accuracy for most design applications and clearly demonstrate that the notch sensitivity index is a function of both the material and the notch root radius.

Many design relationships have been proposed for relating notch sensitivity index to notch radius. One of the more useful of these relationships is based on the work of Neuber (6), which expresses the fatigue stress concentration factor as

$$K_f = 1 + \frac{K_t - 1}{1 + \sqrt{\rho'/r}} \qquad (12\text{-}20)$$

where r is the contour radius of the notch root and ρ' is a material constant related to the grain size of the material. An experimentally determined plot (10) of $\sqrt{\rho'}$ versus tensile strength for steel alloys is shown in Figure 12.14. Substituting the Neuber result of (12-20) back into the definition of notch sensitivity index given in (12-19) yields

$$q = \frac{1}{1 + \sqrt{\rho'/r}} \qquad (12\text{-}21)$$

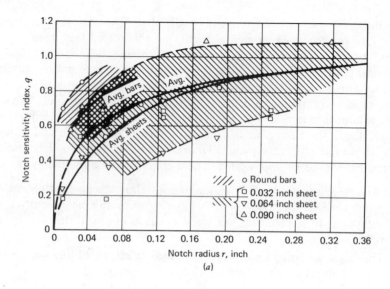

(a)

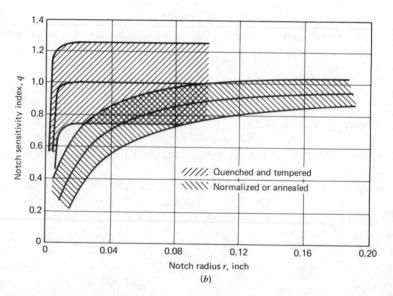

(b)

FIGURE 12.12. An indication of scatter in the experimental determination of notch sensitivity index for alloys of aluminum and steel. (*a*) 24ST aluminum under completely reversed axial loading. (*b*) Steel alloys under alternating bending. (After ref. 1; reprinted with permission from McGraw Hill Book Company.)

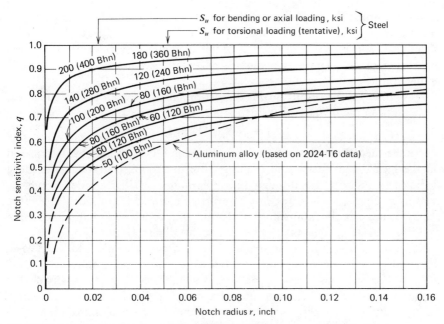

FIGURE 12.13. Curves of notch sensitivity index versus notch radius for a range of steels and an aluminum alloy subjected to axial, bending, and torsional loading. (After ref. 7; reprinted with permission from McGraw Hill Book Company.)

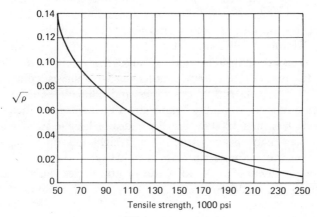

FIGURE 12.14. A plot showing $\sqrt{\rho'}$ versus tensile strength for steel alloys. (From ref. 10.)

It may be noted that this expression for q is independent of the value of K_t. Other investigators have sometimes represented the value of q as a function of K_t and also sometimes as a function of the local stress gradient. However, the expression of (12-21) yields results of acceptable accuracy for most design applications.

To summarize the results of this discussion, an expression for fatigue stress concentration factor may be written from (12-19) as

$$K_f = q(K_t - 1) + 1 \qquad (12\text{-}22)$$

where the theoretical elastic stress concentration factor K_t may be determined on the basis of geometry and loading from handbook charts such as those depicted in Figures 12.3 through 12.8. The notch sensitivity index q may be read from charts, such as the one shown in Figure 12.13, or calculated from (12-21) using experimental values of the material constant ρ' determined from experimental data, such as are represented by Figure 12.14.

For uniaxial states of cyclic stress it is sometimes convenient to use K_f as a "strength reduction factor" rather than as a "stress concentration factor." That is, for uniaxial stressing only, a designer may choose to divide the fatigue limit by K_f rather than multiplying the applied nominal cyclic stress by K_f. Although conceptually it is clearly more correct to think of K_f as a stress concentration factor, computationally it is equivalent, and often simpler, to use K_f as a strength reduction factor. For multiaxial states of stress, however, K_f must be used as a stress concentration factor, since an appropriate value of K_f used as a strength reduction factor would be undefined.

The fatigue stress concentration factor (or strength reduction factor) determined from (12-22) is strictly applicable only in the high-cycle fatigue range, that is, for cycle lives of $10^5 - 10^6$ cycles and greater. It has earlier been noted in Section 12.2 that for ductile materials and static loads, effects of stress concentration may usually be neglected. Thus, in the intermediate and low-cycle life range from a quarter cycle (static load) up to about $10^5 - 10^6$ cycles, the stress concentration factor changes from unity to K_f. As shown in Figure 12.15, the notched and unnotched S-N curves converge as they approach the low-cycle end of the range and coincide at the quarter cycle point A. Many materials exhibit fatigue stress concentration factors very near unity for lives less than 1000 cycles. Estimates of fatigue stress concentration factor are often made by constructing a straight line on a semilogarithmic S-N plot from the ultimate strength at a life of 1 cycle to the unnotched fatigue strength divided by K_f at a life of 10^6 cycles. Such a straight line construction is shown in Figure 12.15. The ratio of unnotched to notched fatigue strength values read at a specified life then becomes the estimate of fatigue stress concentration factor to be used for that life.

Figure 12.16 indicates how the fatigue stress concentration factor varies with cycle life for materials that exhibit the fatigue characteristics depicted in Figure 12.15.

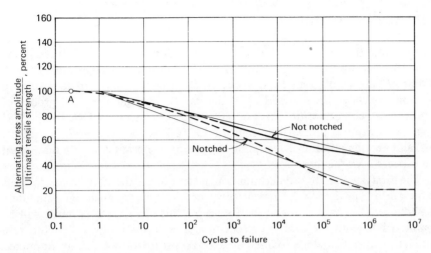

FIGURE 12.15. *S-N* curves for notched and unnotched specimens subjected to completely reversed axial loading. (After ref. 11, *Fatigue and Fracture of Metals*, by W. M. Murray, by permission of The MIT Press, Cambridge, Massachusetts, Copyright 1952.)

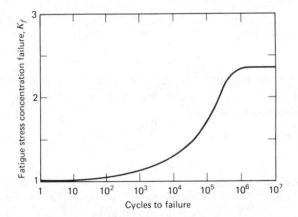

FIGURE 12.16. Variation in fatigue stress concentration factor with cycles of life for material of Figure 12.15 (From ref. 11, *Fatigue and Fracture of Metals*, by W. M. Murray, by permission of The MIT Press, Cambridge, Massachusetts, Copyright 1952.)

Finally, it should be noted that experimental investigations have indicated that for fatigue of *ductile materials* the fatigue stress concentration factor should be applied *only* to the *alternating component* of stress and not to the steady component of stress that exists in any nonzero mean cyclic stress. For fatigue loading of *brittle* materials, the stress concentration factor should be applied to the steady component as well.

12.7 USING THE IDEAS

To illustrate the use of fatigue stress concentration factors (strength reduction factors) for a uniaxial cyclic state of stress, consider a steel bar of 0.5-inch diameter subjected to an axial cyclic force ranging from zero to 10,000 lb tension. As shown in Figure 12.17a, the bar has a circumferential groove of semicircular contour with radius of 0.05 inch. The material of the bar is AISI 4340 steel with an ultimate strength of 150,000 psi, a yield point of 120,000 psi, and an elongation in 2 inches of 15 percent. Determine the life that might be expected for this bar.

The first step in this determination is to obtain the S-N curve for this material or, if data are unavailable, the S-N curve must be estimated. Although data could probably be found for this alloy, an S-N curve will be estimated here to indicate the procedure. For ferrous alloys with ultimate strength values below approximately 160,000 psi to 180,000 psi, an estimated S-N curve may be constructed on a semilog plot by connecting a straight line from the point σ_u at 1 cycle to the point $\sigma_u/2$ at 10^6 cycles, with a second straight line extending horizontally to the right from the 10^6 cycle point, as illustrated by the S-N curve BD in Figure 12.17d. A similar construction may be made to obtain the "notched" S-N curve, except the point at 10^6 cycles is obtained by dividing the unnotched fatigue limit σ_e by K_f, using K_f in this case as a strength reduction factor. Utilizing (12-22), then,

$$K_f = q(K_t - 1) + 1$$

Both K_t and q are required for this calculation. The value of K_t may be found from Figure 12.4b. To utilize this chart, we first tabulate from the bar and groove dimensions that

$$r = 0.05 \text{ inch}$$
$$D = 0.50 \text{ inch} \tag{12-23}$$
$$d = 0.40 \text{ inch}$$

from which we calculate that

$$\frac{r}{d} = \frac{0.05}{0.40} = 0.125 \tag{12-24}$$

and

$$\frac{D}{d} = \frac{0.50}{0.40} = 1.25 \tag{12-25}$$

With these values, the chart of Figure 12.4b yields

$$K_t = 1.9 \tag{12-26}$$

A value for q may be obtained from (12-21) utilizing the graph of Figure 12.14 to obtain a value of $\sqrt{\rho'}$ for steel with 150,000 psi ultimate strength. From Figure 12.14, then,

$$\sqrt{\rho'} = 0.035 \tag{12-27}$$

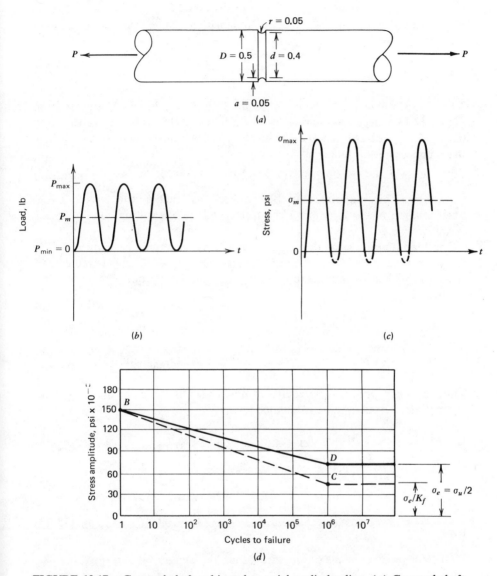

FIGURE 12.17. Grooved shaft subjected to axial cyclic loading. (*a*) Grooved shaft geometry. (*b*) Cyclic load-time plot. (*c*) Cyclic effective stress-time plot. (*d*) Estimated *S-N* curves for 4340 steel shaft.

With this value established, the notch sensitivity index may be calculated from (12-21) as

$$q = \frac{1}{1 + 0.035/\sqrt{0.05}} = 0.87 \tag{12-28}$$

As a check on this calculation, the value of q may be read directly from Figure 12.13, where good agreement with the value 0.87 is noted. Substituting the results of (12-26) and (12-28) into (12-22), we may calculate the fatigue stress concentration factor as

$$K_f = 0.87(1.9 - 1.0) + 1 = 1.78 \tag{12-29}$$

Now, since this material may be regarded as ductile, the fatigue stress concentration factor K_f is properly applied to the alternating component of cyclic stress *only*.

The load-time plot for this axially loaded bar is shown in Figure 12.17b. Since the load cycles from a minimum load of $P_{min} = 0$ to a maximum load of $P_{max} = 10,000$ lbs, the mean load P_m and alternating load P_a may be calculated as

$$P_m = \frac{P_{max} + P_{min}}{2} = \frac{10,000 + 0}{2} = 5000 \text{ lbs} \tag{12-30}$$

$$P_a = \frac{P_{max} - P_{min}}{2} = \frac{10,000 - 0}{2} = 5000 \text{ lbs} \tag{12-31}$$

Since the net section diameter at the notch root is 0.4 inch, the mean and alternating stresses may be calculated as

$$\sigma_m = \frac{P_m}{A} = \frac{5000[4]}{\pi[0.4]^2} = 39,800 \text{ psi} \tag{12-32}$$

and

$$\sigma_a = K_f\left[\frac{P_a}{A}\right] = 1.78\left[\frac{5000[4]}{\pi[0.4]^2}\right] = 70,600 \text{ psi} \tag{12-33}$$

The effective value of maximum stress then is given as

$$\sigma_{max} = \sigma_a + \sigma_m = 70,600 + 39,800 \tag{12-34}$$

$$\sigma_{max} = 110,400 \text{ psi} \tag{12-35}$$

This effective cyclic stress is sketched in Figure 12.17c.

The fatigue limit for this material in unnotched condition may be obtained from curve *BD* of Figure 12.17d as

$$\sigma_e = 75,000 \text{ psi} \tag{12-36}$$

To proceed with the analysis, reference is made to (7-15) through (7-18), which are summarized in Table 7.2. Noting from (12-32) that $\sigma_m = 39,800$ psi we may determine that (7-17) governs since

$$0 \le \sigma_m \le \left[\frac{\sigma_{yp} - \sigma_e}{1 - (\sigma_e/\sigma_u)} \right] \qquad (12\text{-}37)$$

or

$$0 \le \sigma_m \le \left[\frac{120{,}000 - 75{,}000}{1 - 0.50} \right] \qquad (12\text{-}38)$$

or

$$0 \le \sigma_m \le 90{,}000 \qquad (12\text{-}39)$$

Thus, (7-17), which governs, may be written as

$$\sigma_{max} - \sigma_m \left[1 - \frac{\sigma_e}{\sigma_u} \right] = \sigma_e \qquad (12\text{-}40)$$

or

$$\sigma_{max-f} = \sigma_e + \sigma_m \left[1 - \frac{\sigma_e}{\sigma_u} \right] \qquad (12\text{-}41)$$

where σ_{max-f} is the maximum stress in the cycle that can just be tolerated and still obtain infinite life. Thus,

$$\sigma_{max-f} = 75{,}000 + 39{,}800(1 - 0.50) \qquad (12\text{-}42)$$

or

$$\sigma_{max-f} = 94{,}900 \text{ psi} \qquad (12\text{-}43)$$

Now, from (12-35) the actual effective maximum stress in the cycle is 110,400 psi. Therefore, the safety factor on infinite life, n_∞, is

$$n_\infty \frac{\sigma_{max-f}}{\sigma_{max}} = \frac{94{,}900}{110{,}400} = 0.86 \qquad (12\text{-}44)$$

This safety factor is less than unity so failure would be expected in some finite number of cycles. An estimate of the failure life may be made by plotting on Figure 12.17d a curve based on the high cycle fatigue stress concentration factor $K_f = 1.78$ as computed in (12-29). This value is plotted at C as shown in Figure 12.17d. The "notched" curve is then constructed by connecting point B to point C.

A new value of stress concentration factor then may be estimated at any desired intermediate life. For example, at a life of 10^5 cycles the unnotched curve has an ordinate of 87,000 psi and the notched curve a value of 62,000 psi. The ratio of these two is the fatigue stress concentration factor at 10^5 cycles. Thus,

$$(K_f)_{10^5} = \frac{87{,}000}{62{,}000} = 1.40 \qquad (12\text{-}45)$$

Next, the alternating stress value computed in (12-33) must be modified to

$$\sigma_a = 1.45 \left[\frac{5000[4]}{\pi[0.4]^2} \right] = 55,700 \text{ psi} \qquad (12\text{-}46)$$

whence

$$\sigma_{max} = \sigma_a + \sigma_m = 55,700 + 39,800 \qquad (12\text{-}47)$$

$$\sigma_{max} = 95,500 \text{ psi} \qquad (12\text{-}48)$$

For this finite life case, the fatigue strength at 10^5 cycles may be obtained from Figure 12.17d as

$$\sigma_N = 87,000 \text{ psi} \qquad (12\text{-}49)$$

Then equation (12-37) becomes

$$0 \le \sigma_m \le \left[\frac{120,000 - 87,000}{1 - 0.60} \right] \qquad (12\text{-}50)$$

or

$$0 \le \sigma_m \le 82,500 \qquad (12\text{-}51)$$

Since σ_m still lies in this range, (12-41) still governs and becomes

$$\sigma_{max-f} = 87,000 + 39,800(1 - 0.60) \qquad (12\text{-}52)$$

or

$$\sigma_{max-f} = 102,920 \text{ psi} \qquad (12\text{-}53)$$

The safety factor for 10^5 cycles then becomes

$$n_{10^5} = \frac{102,920}{95,500} = 1.08 \qquad (12\text{-}54)$$

This indicates failure in approximately 10^5 cycles since the safety factor is approximately 1.

Suppose now that the same grooved shaft is subjected to a smaller released tensile loading of 5,000 lb, but also to a simultaneous in-phase released cyclic torsional moment of 400 in-lb as well. What would be the safety factor on infinite life design under this more complicated loading condition?

To make the safety factor estimate it is necessary to incorporate the stress concentration concepts just discussed into the multiaxial cyclic failure theory discussion of Sections 7.11, 7.12, and 7.13. Analyzing the grooved shaft of Figure 12.18, we may note that the cyclic tensile force produces critical points uniformly all around the root of the groove, and the cyclic torsional moment likewise produces critical points all around the root of the groove. Selecting a "typical" critical point at the root of the groove, we find that an elemental volume at that point will include tensile stress σ_x and torsional stress τ_{xy}, as shown in Figure 12.18, each of which must include an appropriate stress concentration factor.

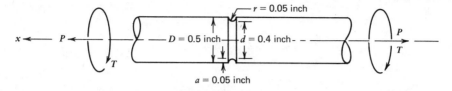

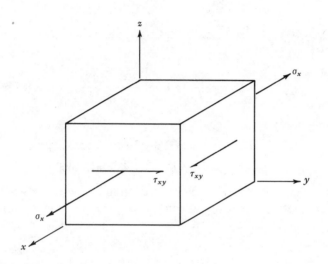

FIGURE 12.18. Grooved shaft subjected to simultaneous cyclic axial tension and in-phase cyclic torsion.

The fatigue stress concentration factor for the released tensile load has already been calculated in (12-29) as

$$(K_f)_{ten} = 1.78 \tag{12-55}$$

For the torsional loading case, using (12-24) and (12-25) with Figure 12.4c, it may be determined that $(K_t)_{tor} = 1.4$, and from (12-22) then

$$(K_f)_{tor} = 0.87(1.4 - 1) + 1 = 1.35 \tag{12-56}$$

Noting then that

$$P_m = \frac{P_{max} + P_{min}}{2} = \frac{5000 + 0}{2} = 2500 \text{ lb} \tag{12-57}$$

$$P_a = \frac{P_{max} - P_{min}}{2} = \frac{5000 - 0}{2} = 2500 \text{ lb} \tag{12-58}$$

$$T_m = \frac{T_{max} + T_{min}}{2} = \frac{400 + 0}{2} = 200 \text{ in-lb} \tag{12-59}$$

$$T_a = \frac{T_{max} - T_{min}}{2} = \frac{400 - 0}{2} = 200 \text{ in-lb} \tag{12-60}$$

It may be calculated that

$$\sigma_{x-m} = \frac{P_m}{A} = \frac{2500(4)}{\pi(0.4)^2} = 19,900 \text{ psi} \tag{12-61}$$

$$\sigma_{x-a} = (K_f)_{\text{ten}} \frac{P_a}{A} = 1.78 \left[\frac{2500(4)}{\pi(0.4)^2} \right] = 35,400 \text{ psi} \tag{12-62}$$

$$\tau_{xy-m} = \frac{T_m r}{J} = \frac{200(0.2)32}{\pi(0.4)^4} = 15,915 \text{ psi} \tag{12-63}$$

$$\tau_{xy-a} = (K_f)_{\text{tor}} \frac{T_a r}{J} = 1.35 \left[\frac{200(0.2)32}{\pi(0.4)^4} \right] = 21,485 \text{ psi} \tag{12-64}$$

From this the effective actual values of σ_x and τ_{xy} are

$$\sigma_{x-\max} = 19,900 + 35,400 = 55,300 \text{ psi} \tag{12-65}$$

$$\tau_{xy-\max} = 15,915 + 21,485 = 37,400 \text{ psi} \tag{12-66}$$

Note that the effects of stress concentration are now properly incorporated in these components of stress. Utilizing the general stress cubic equation (4-23), then, with this state of stress it becomes

$$\sigma^3 - \sigma^2 \sigma_x + \sigma\left(-\tau_{xy}^2\right) = 0 \tag{12-67}$$

which has solutions

$$\sigma_1 = \frac{\sigma_x}{2} + \sqrt{\left(\frac{\sigma_x}{2}\right)^2 + \tau_{xy}^2} \tag{12-68}$$

$$\sigma_2 = 0$$

$$\sigma_3 = \frac{\sigma_x}{2} - \sqrt{\left(\frac{\sigma_x}{2}\right)^2 + \tau_{xy}^2} \tag{12-69}$$

From these solutions, with reference to the discussions of Section 7.13, and results of (12-65) and (12-66), it may be deduced that

$$\sigma_{1\max} = \frac{55,300}{2} + \sqrt{\left(\frac{55,300}{2}\right)^2 + (37,400)^2} = 74,160 \text{ psi} \tag{12-70}$$

$$\sigma_{1\min} = 0 \tag{12-71}$$

$$\sigma_{1m} = \frac{74,160 + 0}{2} = 37,080 \text{ psi} \tag{12-72}$$

$$\sigma_{2\max} = 0 \tag{12-73}$$

$$\sigma_{2\min} = 0 \tag{12-74}$$

$$\sigma_{2m} = 0 \tag{12-75}$$

$$\sigma_{3\max} = 0 \tag{12-76}$$

$$\sigma_{3\,min} = \frac{55,300}{2} - \sqrt{\left(\frac{55,300}{2}\right)^2 + (37,400)^2} = -18,860 \text{ psi} \quad (12\text{-}77)$$

$$\sigma_{3m} = -9430 \text{ psi} \quad (12\text{-}78)$$

Since the material is ductile, the distortion energy multiaxial fatigue theory of Section 7.11 is appropriate. For this purpose, then, from (7-54) and (7-55)

$$U_{d\,max} = \left[\frac{1+\nu}{3E}\right]\left[\frac{(74,160-0)^2}{2} + \frac{(0-0)^2}{2} + \frac{(0-74,160)^2}{2}\right]$$

$$(12\text{-}79)$$

or

$$U_{d\,max} = \left[\frac{1+\nu}{3E}\right][5.50 \times 10^9] \quad (12\text{-}80)$$

and

$$U_{dm} = \left[\frac{1+\nu}{3E}\right]\left[\frac{(37,080-0)^2}{2} + \frac{(0+9430)^2}{2} + \frac{(-9430-37,080)^2}{2}\right]$$

$$(12\text{-}81)$$

or

$$U_{dm} = \left[\frac{1+\nu}{3E}\right][1.81 \times 10^9] \quad (12\text{-}82)$$

Next, utilizing the results of (12-80) and (12-82), together with (7-64) through (7-67), and setting $\sigma_N = \sigma_e$ for infinite life,
failure is predicted to occur if

$$\left[\frac{3E}{1+\nu}\right]^{1/2}\left[\frac{1+\nu}{3E}\right]^{1/2}\left[(5.5\times10^9)^{1/2} - 2(1.81\times10^9)^{1/2}\right] \geq \sigma_{yp}$$

$$(12\text{-}83)$$

or if

$$\left[\frac{3E}{1+\nu}\right]^{1/2}\left[\frac{1+\nu}{3E}\right]^{1/2}\left[(5.5\times10^9)^{1/2} - (1.81\times10^9)^{1/2}\right] \geq \sigma_e \quad (12\text{-}84)$$

or if

$$\left[\frac{3E}{1+\nu}\right]^{1/2}\left[\frac{1+\nu}{3E}\right]^{1/2}\left[(5.5\times10^9)^{1/2} - (1.81\times10^9)^{1/2}(1-0.5)\right] \geq \sigma_e$$

$$(12\text{-}85)$$

or if

$$\left[\frac{3E}{1+\nu}\right]^{1/2}\left[\frac{1+\nu}{3E}\right]^{1/2}\left[(5.5\times10^9)^{1/2}\right] \geq \sigma_{yp} \quad (12\text{-}86)$$

Equation (12-83) reduces to
failure is predicted to occur if

$$- 10{,}930 \geq \sigma_{yp} \qquad (12\text{-}87)$$

which is meaningless since it will never predict failure. The remaining three failure prediction equations then become
failure is predicted to occur if

$$31{,}620 \geq \sigma_e \qquad (12\text{-}88)$$

or if

$$52{,}890 \geq \sigma_e \qquad (12\text{-}89)$$

or if

$$74{,}160 \geq \sigma_{yp} \qquad (12\text{-}90)$$

Since $\sigma_e = 75{,}000$ psi and $\sigma_{yp} = 120{,}000$ psi, failure is not predicted and infinite life may be expected from this design. To determine the safety factor on infinite life, the fatigue and static properties in (12-88) through (12-90) are divided by safety factor n and solved for n. Thus,

$$n_1 = \frac{\sigma_e}{31{,}620} = \frac{75{,}000}{31{,}620} = 2.37 \qquad (12\text{-}91)$$

$$n_2 = \frac{\sigma_e}{52{,}890} = \frac{75{,}000}{52{,}890} = 1.42 \qquad (12\text{-}92)$$

$$n_3 = \frac{\sigma_{yp}}{74{,}160} = \frac{120{,}000}{74{,}160} = 1.62 \qquad (12\text{-}93)$$

The governing safety factor is the smallest value, so the safety factor on infinite life for the given grooved shaft under specified cyclic torsional and tensile loads is estimated to be $n_\infty = 1.4$.

QUESTIONS

1. (a) The theoretical stress concentration factor K_t is a function of what variables or parameters?
(b) The fatigue stress concentration factor K_f is a function of what variables or parameters?
(c) What generalizations may be made, if any, regarding the magnitude of K_f relative to K_t?
(d) The notch sensitivity index q ranges in value from 0 to 1. What is the significance of each of these limiting values?
2. Using the "force flow" concept, describe how one would assess the relative severity of various types of geometrical discontinuities in a machine part subject to a given set of external loads. Use a series of clearly drawn sketches to augment your explanation.

3.   Using appropriate sketches, explain how one may estimate the magnitude of the stress concentration factor that should be used at a specified life in the finite life region of the S-N curve.

4.   A 2.0-inch-wide rectangular bar of annealed 1040 steel is shown in Figure Q12.4, with a 0.25-inch-diameter hole drilled through, as shown. The properties of the 1040 steel are $S_u = 54{,}000$ psi, $S_{yp} = 48{,}000$ psi, $e = 50$ percent in 2 inches, and $S_e = 27{,}000$ psi. The bar is to be subjected to a completely reversed alternating direct force of 8000 lb. A safety factor of 1.5 is to be used in the design. Find the required thickness t, based on infinite life design.

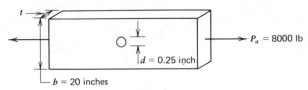

t —
$d = 0.25$ inch
$P_a = 8000$ lb
$b = 20$ inches

FIGURE Q12.4.

5.   A shouldered shaft of solid circular cross section, having the dimensions shown in Figure Q12.5, is subjected to a duty cycle in which 5000 cycles of completely reversed torsional moment of 22,000 in-lb amplitude are followed by 8000 cycles of "released" torsion (from 0 up to a maximum torque in one direction and back to 0 constitutes a cycle) for which the maximum torsional moment is 25,000 in-lb. A total of 50 duty cycles of this kind must be sustained by the part with 99.7 percent probability of success. The material is 4340 steel heat treated to an ultimate strength of 150,000 psi, a yield strength of 120,000 psi, and the mean S-N properties of curve BD in Figure 12.17d. The stresswise standard deviation for this material in fatigue may be taken as 3000 psi.

(a)   List in detail the step-by-step procedure required to find the shaft diameter d that will provide 50 duty cycles with 99.7 percent probability of success without omitting any important considerations or concepts. Be very clear and specific in your comments.

(b)   Execute the numerical solution.

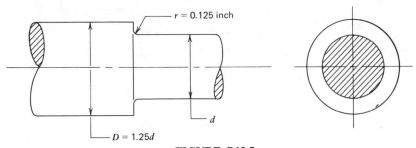

$r = 0.125$ inch
d
$D = 1.25d$

FIGURE Q12.5.

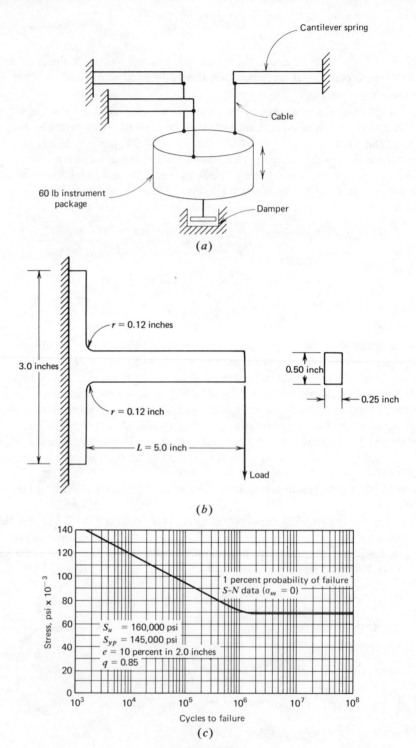

Cantilever spring

Cable

60 lb instrument
package

Damper

(a)

r = 0.12 inches

3.0 inches

r = 0.12 inch

0.50 inch

0.25 inch

L = 5.0 inch

Load

(b)

140

120

100

Stress, psi x 10⁻³

80

60

40

20

0

1 percent probability of failure
S-N data ($\sigma_m = 0$)

S_u = 160,000 psi
S_{yp} = 145,000 psi
e = 10 percent in 2.0 inches
q = 0.85

10^3 10^4 10^5 10^6 10^7 10^8

Cycles to failure

(c)

FIGURE Q12.6.

6.   A special instrument package in an experimental aircraft weighs 60 lbs. It is supported by three small cables, each cable attached to the end of a cantilever beam spring as shown schematically in Figure Q12.6*a*.

The beams are symmetrically placed and equally loaded by the 60-lb instrument package. The system is damped so that it does not vibrate, but during various maneuvers the instrument package is subjected to vertical acceleration ranging from −1g to +9g's. That is, the total effective tension force on the three cables ranges from 0 to 600 lbs during the mission. The mission *loading* spectrum for one "typical" mission is found to be

0 to 600 lbs for 100 cycles
60 to 400 lbs for 10,000 cycles
200 to 300 lbs for 50,000 cycles
0 to 200 lbs for 100,000 cycles

The cantilever beam has the dimensions given in Figure Q12.6*b*, and the material properties are shown on the *S-N* curve of Figure Q12.6*c*.
(a)   Calculate the fatigue stress concentration factor K_f at the critical point.
(b)   Determine the *nominal stress spectrum* at the critical point for the "typical" mission.
(c)   Determine the *actual stress spectrum* at the critical point for the "typical" mission.
(d)   Determine the *equivalent completely reversed* stress spectrum at the critical point for the "typical" mission.
(e)   Determine the total number of missions that can be flown before failure of a cantilever spring would be expected (at the 1 percent probability of failure level).

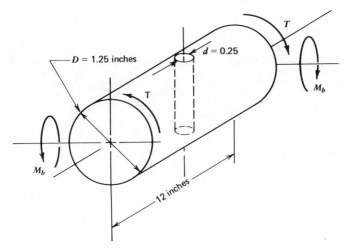

FIGURE Q12.7.

7.   Refer to Figure Q12.7 in which an oscillating shaft of 1.25 inch diameter has a 0.25-inch-diameter hole all the way through it. By the way the shaft is loaded, it is subjected to a completely reversed torsional moment of 8300 in-lb and and in-phase completely reversed cyclic bending moment in the plane of the through-hole axis of 3700 in-lb. If the shaft is made of 4340 steel with the properties shown in curve *BD* of Figure 12.17, how many shaft oscillations would you expect to be completed before failure of the shaft takes place? (Assume that the critical point for bending and torsion coincide.)

REFERENCES

1.   Sines, G., and Waisman, J.L. (ed.), *Metal Fatigue*, McGraw-Hill, New York, 1959.

2.   Grover, H.J., *Fatigue of Aircraft Structures*, NAVAIR 01-1A-13, Washington, D.C., 1966.

3.   Forrest, P.G., *Fatigue of Metals*, Pergamon Press, New York, 1962.

4.   Peterson, R.E., *Stress Concentration Factors*, John Wiley & Sons, New York, 1974.

5.   Inglis, C.E., "Stresses in a Plate Due to the Presence of Cracks and Sharp Corners," *Transactions of the Institute of Naval Architects*, **55**, Pt. 1 (1913).

6.   Neuber, H., *Theory of Notch Stresses*, J.W. Edwards, Ann Arbor, Mich., 1946.

7.   Juvinall, R.C., *Stress, Strain, and Strength*, McGraw-Hill, New York, 1967.

8.   Stowell, E.Z., "Stress and Strain Concentration at a Circular Hole in an Infinite Plate," NACA TN2073, April 1950.

9.   Hardrath, H.F., and Ohman, L., "A Study of Elastic and Plastic Stress Concentration Factors Due to Notches and Fillets in Flat Plates," NACA TN2566, December 1951.

10.   Kuhn, P., and Hardrath, H.F., "An Engineering Method for Estimating Notch-size Effect in Fatigue Tests of Steel," NACA TN2805, 1952.

11.   Murray, W.M. (ed.), *Fatigue and Fracture of Metals*, John Wiley & Sons, New York, 1952.

12.   Paul, F.W., and Faucett, T.R., "The Superposition of Stress Concentration Factors," *Journal of Engineering for Industry*, February 1962.

CHAPTER 13

Creep, Stress Rupture, and Fatigue

13.1 INTRODUCTION

Creep in its simplest form is the progressive accumulation of plastic strain in a specimen or machine part under stress at elevated temperature over a period of time. Creep failure occurs when the accumulated creep strain results in a deformation of the machine part that exceeds the design limits. *Creep rupture* is an extension of the creep process to the limiting condition where the stressed member actually separates into two parts. *Stress rupture* is a term used interchangeably by many with creep rupture; however, others reserve the term stress rupture for the rupture termination of a creep process in which steady-state creep is never reached and use the term creep rupture for the rupture termination of a creep process in which a period of steady-state creep has persisted. Figure 13.1 illustrates these differences. The interaction of creep and stress rupture with cyclic stressing and the fatigue process has not yet been clearly understood but is of great importance in many modern high-performance engineering systems.

Creep strains of engineering significance are not usually encountered until the operating temperatures reach a range of approximately 35 to 70 percent of the melting point on a scale of absolute temperature. The approximate melting temperature for several substances is shown in Table 13.1.

Early creep studies were reported by a French engineer who noted time-dependent cable elongations that exceeded elastic predictions in connection with wire rope for bridge suspension. Not until after World War I, however, did creep failure become important as a failure mode. Since that time many applications have developed in which creep failure may govern the design. Load-carrying members operating in the temperature range from 1000°F to 1600°F are found in power plants, refineries, and chemical processing plants. Furnace parts are routinely exposed to temperatures of 1600°F to 2200°F. Gas turbine rotor blades are subjected to temperatures of 1200°F to 2200°F, together with high centrifugal stresses. Rocket nozzles and spacecraft nose cones are subjected to even higher temperatures for brief periods of time. Skin temperatures of Mach 7 aircraft have been estimated at about 5000°F,

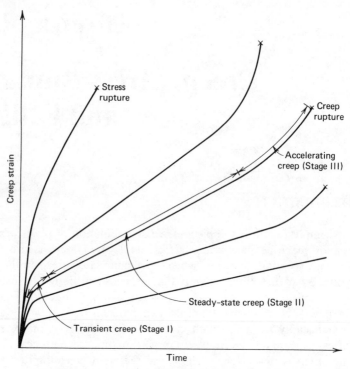

FIGURE 13.1. Illustration of creep and stress rupture.

Table 13.1. Melting Temperatures

Material	°F	°C
Hafnium carbide	7030	3887
Graphite (sublimes)	6330	3500
Tungsten	6100	3370
Tungsten carbide	5190	2867
Magnesia	5070	2800
Molybdenum	4740	2620
Boron	4170	2300
Titanium	3260	1795
Platinum	3180	1750
Silica	3140	1728
Chromium	3000	1650
Iron	2800	1540
Stainless steels	2640	1450
Steel	2550	1400
Aluminum alloys	1220	660
Magnesium alloys	1200	650
Lead alloys	605	320

From ref. 1.

436

with aerodynamic and structural consequences of creep deformation, creep buckling, and stress rupture becoming critical design considerations.

Not only is excessive deformation due to creep an important consideration, but other consequences of the creep process may also be important. These might include creep rupture, thermal relaxation, dynamic creep under cyclic loads or cyclic temperatures, creep and rupture under multiaxial states of stress, cumulative creep effects, and effects of combined creep and fatigue.

Creep deformation and rupture are initiated in the grain boundaries and proceed by sliding and separation. Thus, creep rupture failures are intercrystalline, in contrast, for example, to the transcrystalline failure surface exhibited by room temperature fatigue failures. Although creep is a plastic flow phenomenon, the intercrystalline failure path gives a rupture surface that has the appearance of brittle fracture. Creep rupture typically occurs without necking and without warning. Current state-of-the-art knowledge does not permit a reliable prediction of creep or stress rupture properties on a theoretical basis. Further, there seems to be little or no correlation between the creep properties of a material and its room temperature mechanical properties. Therefore, test data and empirical methods of extending these data are relied on heavily for prediction of creep behavior under anticipated service conditions.

Metallurgical stability under long-time exposure to elevated temperatures is mandatory for good creep resistant alloys. Prolonged time at elevated temperatures acts as a tempering process, and any improvement in properties originally gained by quenching may be lost. Resistance to oxidation and other corrosive media are also usually important attributes for a good creep resistant alloy. Larger grain size may also be advantageous since this reduces the length of grain boundary, where much of the creep process resides.

13.2 PREDICTION OF LONG-TERM CREEP BEHAVIOR

Much time and effort has been expended in attempting to devise good short-time creep tests for accurate and reliable prediction of long-term creep and stress rupture behavior. It appears, however, that really reliable creep data can be obtained only by conducting long-term creep tests that duplicate actual service loading and temperature conditions as nearly as possible. Unfortunately, designers are unable to wait for years to obtain design data needed in creep failure analysis. Therefore, certain useful techniques have been developed for approximating long-term creep behavior based on a series of short-term tests.

Data from creep testing may be cross plotted in a variety of different ways. The basic variables involved are stress, strain, time, temperature, and perhaps strain rate. Any two of these basic variables may be selected as plotting coordinates, with the remaining variables to be treated as parametric constants for a given curve. Three commonly used methods for extrapolating short-time creep data to long-term applications are the abridged method, the mechanical acceleration method, and the thermal acceleration method. These three methods are discussed in the paragraphs that follow. In any testing method, however, it should be noted that creep testing guide lines usually dictate that test periods of less than 1 percent of the expected life are not deemed to give significant results. Tests extending to at least 10 percent of the expected life are preferred where feasible.

Abridged Method

In the abridged method of creep testing the tests are conducted at several different stress levels and at the contemplated operating temperatures. As shown in Figure 13.2, the data are plotted as creep strain versus time for a family of stress levels, all run at constant temperature. The curves are plotted out to the laboratory test duration and then extrapolated to the required design life. The design specifications will dictate a limiting design strain that will intersect the design life, as shown in the figure, to yield a value for design stress. It is important to recognize that such extrapolations are not able to predict the potential of failure by creep rupture prior to reaching the creep design life.

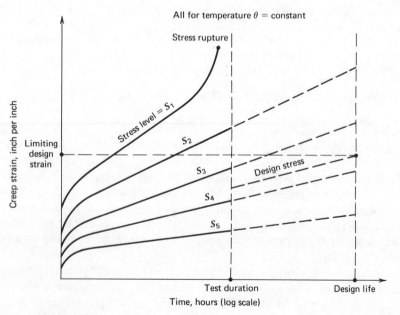

FIGURE 13.2.   Illustration of abridged method of creep testing.

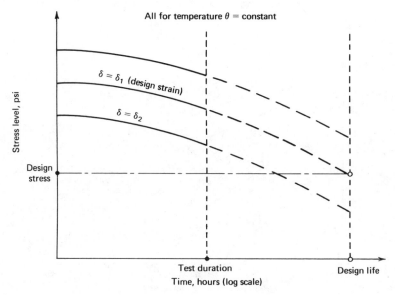

FIGURE 13.3.   Illustration of the mechanical acceleration method of creep testing.

Mechanical Acceleration Method

In the mechanical acceleration method of creep testing, the stress levels used in the laboratory tests are significantly higher than the contemplated design stress levels, so the limiting design strains are reached in a much shorter time than in actual service. The data taken in the mechanical acceleration method are plotted as stress level versus time, as shown in Figure 13.3, for a family of constant strain curves all run at a constant temperature. As shown in the figure, the stress rupture curve may also be plotted by this method. The constant strain curves are plotted out to the laboratory test duration and then extrapolated to the design life. The point at which the limiting design strain curve intersects the design life dictates the design stress to be used, as illustrated in Figure 13.3.

Thermal Acceleration Method

The thermal acceleration method involves laboratory testing at temperatures much higher than the actual service temperature expected. As shown in Figure 13.4, the data are plotted as stress versus time for a family of constant temperatures where the creep strain produced is constant for the whole plot. It may be noted that stress rupture data may also be plotted in this way. The data are plotted out to the laboratory test duration and then extrapolated to the design life. The point at which the design temperature curve intersects the design life dictates the design stress level, as illustrated in the figure.

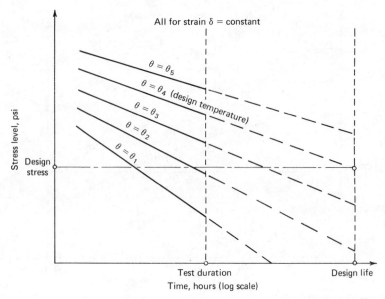

FIGURE 13.4.   Illustration of the thermal acceleration method of creep testing.

13.3   THEORIES FOR PREDICTING CREEP BEHAVIOR

Several different theories have been proposed in recent years to correlate the results of short-time elevated temperature tests with long-term service performance at more moderate temperatures. The more accurate and useful of these proposals to date are the Larson-Miller theory and the Manson-Haferd theory.

Larson-Miller Parameter

The Larson-Miller theory (2) postulates that for each combination of material and stress level there exists a unique value of a parameter P that is related to temperature and time by the equation

$$P = (\theta + 460)(C + \log_{10} t) \qquad (13\text{-}1)$$

where P = Larson-Miller parameter, constant for a given material and stress level

θ = temperature, °F

C = constant, usually assumed to be 20

t = time in hours to rupture or to reach a specified value of creep strain

This equation was investigated both for creep and rupture for some 28 different materials by Larson and Miller with good success. By using (13-1) it

Table 13.2. Equivalent Conditions Based on Larson-Miller Parameter

Operating Condition	Equivalent Test Condition
10,000 hours at 1000°F	13 hours at 1200°F
1,000 hours at 1200°F	12 hours at 1350°F
1,000 hours at 1350°F	12 hours at 1500°F
1,000 hours at 300°F	2.2 hours at 400°F

is a simple matter to find a short-term combination of temperature and time that is equivalent to any desired long-term service requirement. For example, for any given material at a specified stress level the test conditions listed in Table 13.2 should be equivalent to the operating conditions.

Good agreement between theory and experiment has been observed by using the Larson-Miller parameter for a wide variety of materials, including several plastics, in predicting long-term creep behavior and stress rupture performance.

Manson-Haferd Parameter

The Manson-Haferd theory (3) postulate that for a given material and stress level there exists a unique value of a parameter P' that is related to temperature and time by the equation

$$P' = \frac{\theta - \theta_a}{\log_{10} t - \log_{10} t_a} \tag{13-2}$$

where P' = Manson-Haferd parameter, constant for a given material and stress level

θ = temperature, °F

t = time is hours to rupture or to reach a specified value of creep strain

θ_a, t_a = material constants

Table 13.3. Constants for Manson-Haferd Equation (3)

Material	Creep or Rupture	θ_a	$\log_{10} t_a$
25–20 stainless steel	Rupture	100	14
18–8 stainless steel	Rupture	100	15
S-590 alloy	Rupture	0	21
DM steel	Rupture	100	22
Inconel X	Rupture	100	24
Nimonic 80	Rupture	100	17
Nimonic 80	0.2 percent plastic strain	100	17
Nimonic 80	0.1 percent plastic strain	100	17

In the Manson-Haferd equation values of the constants for several materials are shown in Table 13.3. The Manson-Haferd theory has been found to give good agreement with experimental results once the materials constants have been evaluated.

13.4   CREEP UNDER UNIAXIAL STATE OF STRESS

Uniaxial creep and stress rupture tests of 100 hours (4 days), 1000 hours (42 days), and 10,000 hours (420 days) duration are common, with some longer duration testing of 100,000 hours (11.5 years) being performed in some cases. Certain recent high-performance applications have given rise to short-term creep testing that measures duration in minutes rather than hours or years. For example, in some cases creep test durations of 1000 minutes, 100 minutes, 10 minutes, and 1 minute have been used. An example of such data (4) for several different materials tested is shown in Figure 13.5. For short-time tests, however, at temperatures below about 300°F for aluminum alloys and below about 700°F for steel alloys, creep may be neglected.

It is of interest to note that with an increase in temperature the static ultimate strength, yield strength, and modulus of elasticity all tend to decrease, whereas the elongation and reduction in area tend to increase. The stress concentration effect due to a geometrical notch is also reduced at elevated temperatures.

Many relationships have been proposed to relate stress, strain, time, and temperature in the creep process. If one investigates experimental creep strain versus time data, it will be observed that the data are close to linear for a wide variety of materials when plotted on log strain versus log time coordinates. Such a plot is shown, for example, in Figure 13.6 for three different materials. An equation describing this type of behavior is

$$\delta = At^a \tag{13-3}$$

where δ = true creep strain
t = time
A, a = empirical constants

Differentiating (13-3) with respect to time gives

$$\dot{\delta} = aAt^{(a-1)} \tag{13-4}$$

or, setting $aA = b$ and $(1 - a) = n$,

$$\dot{\delta} = bt^{-n} \tag{13-5}$$

This equation represents a variety of different types of creep strain versus time curves, depending on the magnitude of the exponent n. If n is zero, the behavior is termed *constant creep rate*, and the creep strain is given as

$$\delta = b_1 t + C_1 \tag{13-6}$$

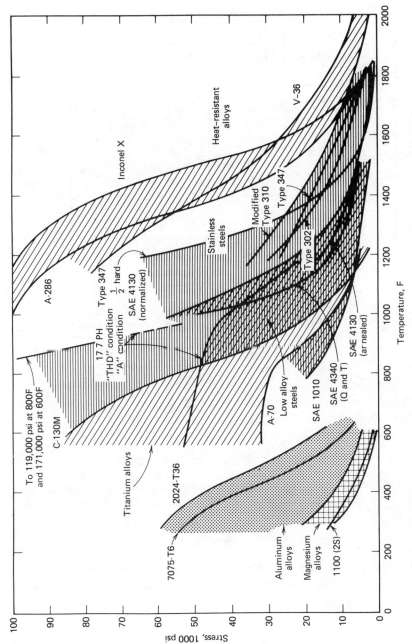

FIGURE 13.5. Short-time creep tests for several materials showing curves of stress versus temperature for 3 percent total deformation in 10 minutes. (From ref. 4), copyright American Society for Metals, 1958; reprinted with permission.)

443

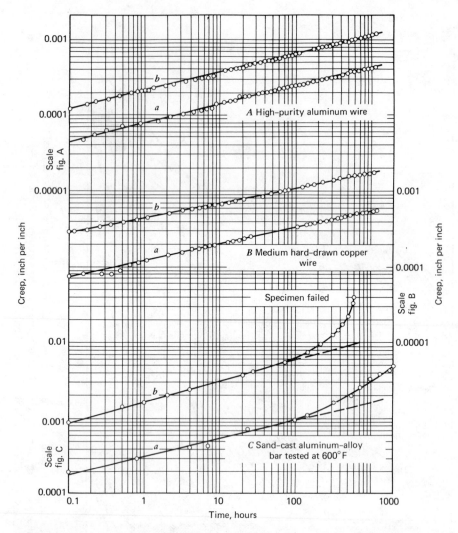

FIGURE 13.6. Creep curves for three materials plotted on log-log coordinates (From ref. 5.)

This type of creep behavior is most commonly found at high temperatures. If the exponent n is unity, the behavior is termed *logarithmic creep*, and the creep strain is given by

$$\delta = b_2 \ln t + C_2 \tag{13-7}$$

This type of creep behavior is displayed by rubber, glass, and certain types of concrete, as well as by metals at lower temperatures.

If the exponent n lies between zero and 1, the behavior is termed *parabolic creep*, and the creep strain is given by

$$\delta = b_3 t^m + C_3 \tag{13-8}$$

This type of creep behavior occurs at intermediate and high temperatures. The coefficient b_3 increases exponentially with stress and temperature, and the exponent m decreases with stress and increases with temperature. The influence of stress level σ on creep rate can often be represented by the empirical expression

$$\dot{\delta} = B\sigma^N \tag{13-9}$$

Assuming the stress σ to be independent of time, we may integrate (13-9) to yield the creep strain

$$\delta = Bt\sigma^N + C' \tag{13-10}$$

If the constant C' is small compared with $Bt\sigma^N$, as it often is, the result is called the *log-log stress-time* creep law, given as

$$\delta = Bt\sigma^N \tag{13-11}$$

As long as the instantaneous deformation on load application and the Stage I transient creep are small compared to Stage II steady-state creep, (13-11) is useful as a design tool. With this expression one may calculate the stress required at a specified temperature to hold creep deformation within specified limits. In Table 13.4 the constants B and N are evaluated for three materials and temperatures, where time t is in days.

Suppose, for example, that it is desired to design a solid circular tension member 5 feet long, made of 1030 steel, to support a 10,000 lb load for 10 years at 750°F without exceeding 0.1 inch creep deformation. It is also desired to incorporate a safety factor of 1.25 in the design. To solve this design problem, (13-11) may be solved for the failure stress σ_f as

$$\sigma_f = \left[\frac{\delta}{Bt} \right]^{1/N} \tag{13-12}$$

or

$$\sigma_f = \left[\frac{0.1/(5 \times 12)}{(48 \times 10^{-38})(10 \times 365)} \right]^{1/6.9} \tag{13-13}$$

Table 13.4.   Constants for Log-Log Stress-Time Creep Law

Material	Temperature	B	N
1030 steel	750°F	48×10^{-38}	6.9
1040 steel	750°F	16×10^{-46}	8.6
Ni-Cr-Mo steel	850°F	10×10^{-20}	3.0

or

$$\sigma_f = 23,200 \text{ psi} \tag{13-14}$$

where the constants B and N were selected from Table 13.4. The design stress σ_d then is

$$\sigma_d = \frac{\sigma_f}{n} = \frac{23,200}{1.25} = 18,500 \text{ psi} \tag{13-15}$$

using a safety factor n of 1.25. The diameter d may then be calculated from

$$\sigma_d = \frac{F}{A} = \frac{4F}{\pi d^2} \tag{13-16}$$

which may be solved for d as

$$d = \sqrt{\frac{4F}{\pi \sigma_d}} = \sqrt{\frac{4(10,000)}{\pi(18,500)}} \tag{13-17}$$

or

$$d = 0.84 \text{ inch} \tag{13-18}$$

If it is necessary to consider all stages of the creep process, the creep strain expression becomes much more complex. The most general expression for the creep process is*

$$\delta = \frac{\sigma}{E} + k_1 \sigma^m + k_2(1 - e^{-qt})\sigma^n + k_3 t \sigma^p \tag{13-19}$$

where
$$\delta = \text{total creep strain}$$
$$\frac{\sigma}{E} = \text{initial elastic strain}$$
$$k_1 \sigma^m = \text{initial plastic strain}$$
$$k_2(1 - e^{-qt})\sigma^n = \text{anelastic strain}$$
$$k_3 t \sigma^p = \text{viscous strain}$$
$$\sigma = \text{stress}$$
$$E = \text{modulus of elasticity}$$
$$m = \text{reciprocal of strain-hardening exponent}$$
$$k_1 = \text{reciprocal of strength coefficient}$$
$$q = \text{reciprocal of Kelvin retardation time}$$
$$k_2 = \text{anelastic coefficient}$$
$$n = \text{empirical exponent}$$
$$k_3 = \text{viscous coefficient}$$
$$p = \text{empirical exponent}$$
$$t = \text{time}$$

To utilize this empirical nonlinear expression in a design environment requires specific knowledge of the constants and exponents that characterize

*See p. 438 of ref. 6.

the material and temperature of the application. In all cases it must be recognized that stress rupture may intervene to terminate the creep process, and the prediction of this occurrence is difficult.

13.5 CREEP UNDER MULTIAXIAL STATE OF STRESS

Many service applications, such as pressure vessels, piping, turbine rotors, and others, may involve creep conditions under a multiaxial state of stress. To determine creep strain and deformation under a multiaxial state of stress, it will be convenient to utilize the plastic deformation equations developed earlier as equations (5-66), (5-67), and (5-68). These equations are

$$\delta_1 = \frac{1}{D}\left[\sigma_1' - \frac{1}{2}(\sigma_2' + \sigma_3')\right] \tag{13-20}$$

$$\delta_2 = \frac{1}{D}\left[\sigma_2' - \frac{1}{2}(\sigma_1' + \sigma_3')\right] \tag{13-21}$$

$$\delta_3 = \frac{1}{D}\left[\sigma_3' - \frac{1}{2}(\sigma_1' + \sigma_2')\right] \tag{13-22}$$

where $\delta_1, \delta_2, \delta_3$ = principal true strains
$\sigma_1', \sigma_2', \sigma_3'$ = principal true stresses
D = modulus of plasticity, a function of the amount of prior plastic strain

Dividing these equations by time t, they become creep rate equations of the form

$$\frac{\delta_1}{t} = \dot{\delta}_1 = \frac{1}{Dt}\left[\sigma_1' - \frac{1}{2}(\sigma_2' + \sigma_3')\right] \tag{13-23}$$

$$\frac{\delta_2}{t} = \dot{\delta}_2 = \frac{1}{Dt}\left[\sigma_2' - \frac{1}{2}(\sigma_1' + \sigma_3')\right] \tag{13-24}$$

$$\frac{\delta_3}{t} = \dot{\delta}_3 = \frac{1}{Dt}\left[\sigma_3' - \frac{1}{2}(\sigma_1' + \sigma_2')\right] \tag{13-25}$$

These creep rate equations may be adapted to creep under a uniaxial tensile stress by setting $\sigma_2' = \sigma_3' = 0$ to yield

$$\dot{\delta}_1 = \frac{\sigma_1'}{Dt} \tag{13-26}$$

Now, utilizing the distortion energy theory developed in (6-42), we may write an *equivalent true stress*, σ_e', for the multiaxial state of stress as

$$\sigma_e' = \frac{\sqrt{2}}{2}\left[(\sigma_1' - \sigma_2')^2 + (\sigma_2' - \sigma_3')^2 + (\sigma_3' - \sigma_1')^2\right]^{1/2} \tag{13-27}$$

Adapting (13-27) to the uniaxial stress case by setting $\sigma_2' = \sigma_3' = 0$, it becomes

$$\sigma_e' = \sigma_1' \tag{13-28}$$

Combining the results of (13-26) and (13-28), then,

$$\sigma_e' = \sigma_1' = \dot{\delta}_1 Dt \tag{13-29}$$

However, from (13-9) the creep rate $\dot{\delta}$ may be written as

$$\dot{\delta}_1 = B(\sigma_1')^N \tag{13-30}$$

and (13-29) therefore becomes

$$\sigma_1' = B(\sigma_1')^N Dt \tag{13-31}$$

Solving this expression for $(1/Dt)$ then yields

$$\frac{1}{Dt} = B(\sigma_1')^{N-1} = B(\sigma_e')^{N-1} \tag{13-32}$$

Substituting the result of (13-32) back into (13-23), (13-24), and (13-25) then gives

$$\dot{\delta}_1 = B(\sigma_e')^{N-1}\left[\sigma_1' - \frac{1}{2}(\sigma_2' + \sigma_3')\right] \tag{13-33}$$

$$\dot{\delta}_2 = B(\sigma_e')^{N-1}\left[\sigma_2' - \frac{1}{2}(\sigma_3' + \sigma_1')\right] \tag{13-34}$$

$$\dot{\delta}_3 = B(\sigma_e')^{N-1}\left[\sigma_3' - \frac{1}{2}(\sigma_1' + \sigma_2')\right] \tag{13-35}$$

To simplify these relationships the following ratios are defined:

$$\alpha = \frac{\sigma_2'}{\sigma_1'} \tag{13-36}$$

$$\beta = \frac{\sigma_3'}{\sigma_1'} \tag{13-37}$$

Utilizing these definitions and (13-27), we may write the expressions of (13-33), (13-34), and (13-35) in a useful form as

$$\delta_1 = Bt(\sigma_1')^N\left[\alpha^2 + \beta^2 - \alpha\beta - \alpha - \beta + 1\right]^{(N-1)/2}\left[1 - \frac{\alpha}{2} - \frac{\beta}{2}\right] \tag{13-38}$$

$$\delta_2 = Bt(\sigma_1')^N\left[\alpha^2 + \beta^2 - \alpha\beta - \alpha - \beta + 1\right]^{(N-1)/2}\left[\alpha - \frac{\beta}{2} - \frac{1}{2}\right] \tag{13-39}$$

$$\delta_3 = Bt(\sigma_1')^N\left[\alpha^2 + \beta^2 - \alpha\beta - \alpha - \beta + 1\right]^{(N-1)/2}\left[\beta - \frac{\alpha}{2} - \frac{1}{2}\right] \tag{13-40}$$

These three equations completely define the principal creep strains in terms of the principal creep stresses and the experimentally determined uniaxial tensile creep parameters B and N. Predictions of creep behavior in any multiaxial state of stress can be made by these equations based only on the results of a simple uniaxial creep test.

13.6 CUMULATIVE CREEP CONCEPTS

There is at the present time no universally accepted method for estimating the creep strain accumulated as a result of exposure for various periods of time at different temperatures and stress levels. However, several different techniques for making such estimates have been proposed. The simplest of these is a linear hypothesis suggested by E. L. Robinson (7). A generalized version of the Robinson hypothesis may be written as follows: If a design limit of creep strain δ_D is specified, it is predicted that the creep strain δ_D will be reached when

$$\sum_{i=1}^{k} \frac{t_i}{L_i} = 1 \qquad (13\text{-}41)$$

where t_i = time of exposure at the ith combination of stress level and temperature

L_i = time required to produce creep strain δ_D if entire exposure were held constant at the ith combination of stress level and temperature

Stress rupture may also be predicted by (13-41) if the L_i values correspond to stress rupture. This prediction technique gives relatively accurate results if the creep deformation is dominated by Stage II steady-state creep behavior. Under other circumstances the method may yield predictions that are seriously in error.

Other cumulative creep prediction techniques that have been proposed include the time-hardening rule, the strain-hardening rule, and the life-fraction rule. The *time-hardening rule* is based on the assumption that the major factor governing the creep rate is the length of exposure at a given temperature and stress level, no matter what the past history of exposure has been. This concept is illustrated in Figure 13.7, where the total accumulated creep strain δ_T would be obtained by adding the individual δ_i values to give

$$\delta_T = \delta_1 + \delta_2 + \cdots + \delta_n \qquad (13\text{-}42)$$

where each δ_i is the strain associated with operation at the ith combination of stress and temperature. Note that the operational path moves vertically from curve to curve along lines of constant time.

The *strain-hardening rule* is based on the assumption that the major factor governing the creep rate is the amount of prior strain, no matter what the past history of exposure has been. This proposal is illustrated in Figure 13.8, where again the total creep strain would be calculated by adding the δ_i increments shown. Note that under this postulate the operational path moves horizontally from curve to curve along lines of constant strain.

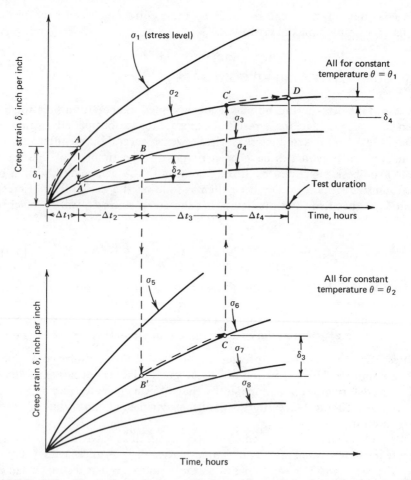

FIGURE 13.7. Illustration of the prediction of cumulative creep strain by the time-hardening rule.

The *life-fraction rule* is a compromise between the time-hardening rule and the strain-hardening rule. As illustrated in Figure 13.9, instead of moving from curve to curve along lines of constant time or lines of constant strain, the operational path from curve to curve is an intermediate path calculated to arrive on the new curve at that point that represents the same ratio of exposure time to total life as had been experienced at the terminus of operation along the old curve. For example, in Figure 13.9 the total lifetime required to reach the design limit on creep strain at stress level σ_1 is designated as L_1. Likewise, the total lifetime required to reach the design limit on creep strain at stress level σ_2 is designated as L_2. In moving from the σ_1

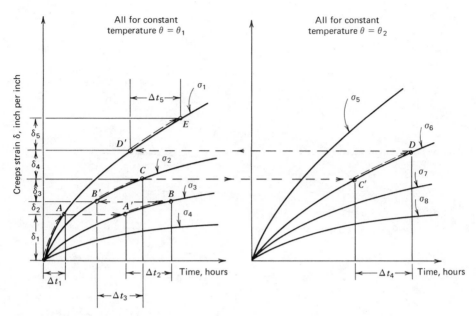

FIGURE 13.8. Illustration of the prediction of cumulative creep strain by the strain-hardening rule.

curve to the initial point on the σ_2 curve, it is postulated that

$$\frac{l_{2i}}{L_2} = \frac{l_{1f}}{L_1} \tag{13-43}$$

whence

$$l_{2i} = \frac{L_2}{L_1} l_{1f} \tag{13-44}$$

where l_{2i} is the initial time coordinate for operation along the σ_2 curve and l_{1f} is the final time coordinate for operation along the σ_1 curve. As in the other methods, the total creep strain is found by adding all the δ_i increments, as shown in Figure 13.9. Of all the techniques postulated, the life-fraction rule is probably the most accurate.

13.7 COMBINED CREEP AND FATIGUE

Neither the fatigue process nor the creep process is well understood, so it comes as no suprise to find that when creep and fatigue phenomena occur simultaneously, the combined process is not well understood, and reliable failure prediction under these circumstances is difficult. Yet there are several

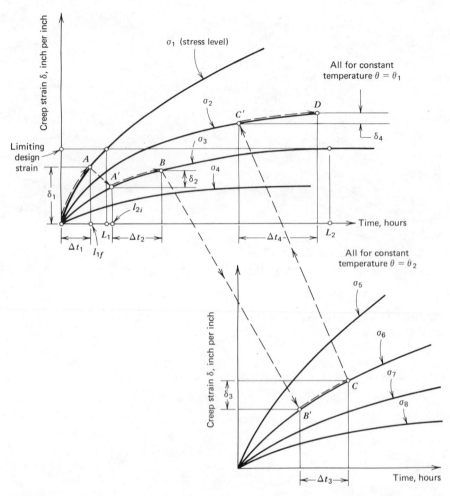

FIGURE 13.9. Illustration of the prediction of cumulative creep strain by the life-fraction rule.

important high-performance applications of current interest in which conditions persist that lead to combined creep and fatigue. For example, aircraft gas turbines and nuclear power reactors are subjected to this combination of failure modes. To make matters worse, the duty cycle in these applications might include a sequence of events including fluctuating stress levels at constant temperature, fluctuating temperature levels at constant stress, and periods during which both stress and temperature are simultaneously fluctuating. Further, there is evidence to indicate that the fatigue and creep processes interact to produce a synergistic response.

To illustrate one problem, the effect of changing the operating stress for a short period of time midway through a creep test on a lead alloy produced the results* shown in Figure 13.10. If the stress is temporarily increased during period AB, the creep rate increases to produce a superposed transient creep behavior. If the stress is temporarily decreased during period AB, the creep rate is decreased. Returning to the original stress value following period AB, the creep curves generally tend to converge toward the curve for uninterrupted creep. However, detailed examination of the data indicates that the ultimate creep strain is not a simple function of the stress change. For example, small negative increments in stress can actually result in a larger ultimate creep strain than would occur otherwise. This is illustrated in the inset of Figure 13.10a and more clearly in Figure 13.10b.

The single interruption experiment, however, does not necessarily indicate how the creep strain would accumulate over a large number of stress cycles. Diverse results have been reported for creep tests under cyclic stress conditions, depending on the alloys and test conditions used. Tests run on SAE 4130 steel at 430°C subjected to stress cycles consisting of 1 hour of a high stress level where creep was important followed by 1 hour at a low stress level where creep was negligible (9) led to the conclusion that the cyclic stress interruption had no significant effect on the creep strain accumulated at the higher temperature. That is, N hours at the higher stress level under the cycling test condition and N hours continuously at the higher stress level produced about the same creep strain. On the other hand, for an aluminum alloy subjected to the same type of comparative test, the average creep rate under cyclic testing was 0.3 inch per inch per hour loaded, compared to an average creep rate of 0.11 inch per inch per hour loaded for the continuously loaded case. Thus, the creep strain accumulation is absolutely greater under conditions where the load is periodically removed or reduced than if the load is applied continuously for the same period of time. This behavior is illustrated in Figure 13.11.

In testing titanium alloy RC-130A, the stress rupture life as a function of stress was compared for cyclic versus continuously applied loads (9). As shown in Figure 13.12, under some conditions the total elapsed time to failure is increased by a factor of 5 if the load is removed 1 hour in every 2. At a stress level of 40,000 psi the life under constant loading is about 80 hours, whereas the life under interrupted loading is about 400 hours. This means that the actual time under load required to produce stress rupture has been increased by a factor of about 2.5. A magnesium alloy tested in a similar way also exhibited a marked increase in rupture life under conditions of interrupted stressing. An aluminum alloy similarly tested, however, ruptured in about the same total time whether the stress was interrupted or not. Thus, it

*See p. 402 of ref. 8.

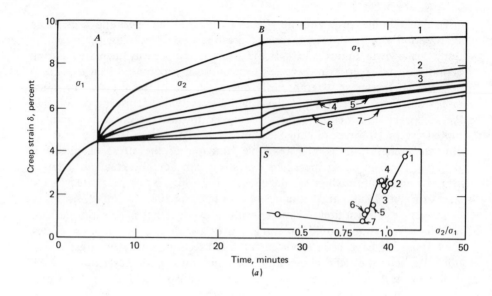

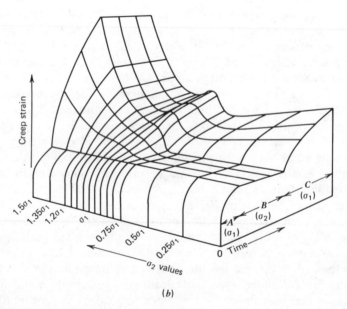

FIGURE 13.10. Effect of changing stress level during a creep test. (From ref. 8) (*a*) Effect of changing stress level from σ_1 to σ_2 and back to σ_1 during creep test on lead at 22°C. Note that S is creep strain after 50 minutes. (*b*) Three-dimensional plot showing influences of interposed stress σ_2 during period *B* of the creep test.

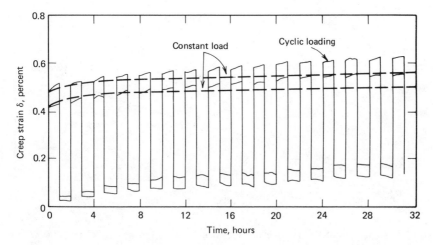

FIGURE 13.11. Creep curves for an aluminum alloy sheet material at 150°C comparing the creep strain as a function of time for constant and cyclic loading condition. (From ref. 8.)

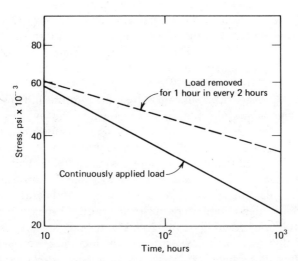

FIGURE 13.12. Time to rupture as a function of applied stress for RC-130-A titanium sheet showing the effect of interrupted stressing. (From ref. 8.)

may be observed that interrupted stressing may accelerate, retard, or leave unaffected the time under stress required to produce stress rupture. The same observation can also be made with respect to creep rate.

Temperature cycling at constant stress level may also produce a variety of responses, depending on material properties and the details of the temperature cycle. For example, creep tests on S-816 cobalt-base alloy at 816°C are

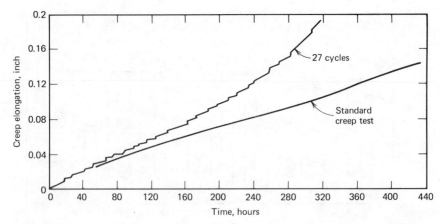

FIGURE 13.13. Creep curve for cobalt-base alloy S-816 at 816°C and 18,000 psi compared with similar test in which periodic 2-minute overheat cycles to 900°C are superposed. (From ref. 8.)

shown in Figure 13.13 for a test at 18,000 psi compared with a similar test in which a 2-minute overheat to 900°C was imposed once every 12 hours (11). It may be observed that under this cyclic temperature pattern the accumulated creep elongation was increased and the stress rupture life was decreased compared to the constant temperature tests. In contrast, tests on M-252 nickel-chromium alloy tested at 816°C with zero-stress 2-minute overheat cycles to 1093°C every 12 hours increased the rupture life by a factor of nearly 3, as shown in Figure 13.14*a*. Even when the overheat cycles are applied under stress, the rupture life is still significantly increased, as indicated in Figure 13.14*b*. Thus, the nickel-chromium alloy is seriously affected by short-duration high-temperature pulses, whereas the cobalt-base alloy is not seriously affected. In fact, if the additional creep strain can be accommodated in the design, the rupture life may be significantly improved by imposing periodic high-temperature pulses at a reduced stress level.

Assessing the brief results just presented, and other similar results, it may be concluded that no general law exists by which cumulative creep and stress rupture response under temperature cycling at constant stress or stress cycling at constant temperature in the creep range can be accurately predicted. However, some recent progress has been made in developing life prediction techniques for combined creep and fatigue. For example, a procedure sometimes used to predict failure under combined creep and fatigue conditions for isothermal cyclic stressing is to assume that the creep behavior is controlled by the mean stress σ_m and that the fatigue behavior is controlled by the stress amplitude σ_a, with the two processes combining linearly to produce failure. This approach is similar to the development of the Goodman diagram described in Chapter 7 except that instead of an intercept of σ_u on the σ_m axis,

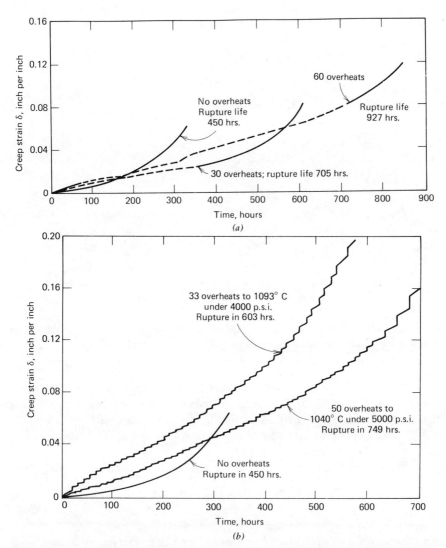

FIGURE 13.14. Creep test results for Nickel-Chromium alloy M-252 showing influence of periodic temperature overheat cycles. Creep tests were run at 816°C and 24,000 psi. (From ref. 8) (*a*) Two-minute overheat cycles applied under zero stress every 12 hours. (*b*) Two-minute overheat cycles applied at reduced stress every 12 hours.

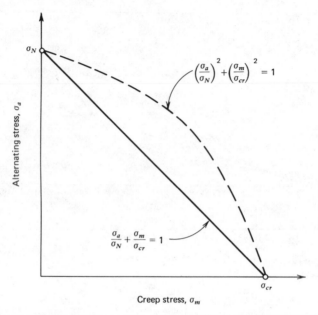

FIGURE 13.15. Failure prediction diagram for combined creep and fatigue under constant temperature conditions.

as shown in Figure 7.59, the intercept used is the *creep-limited static stress* σ_{cr}, as shown in Figure 13.15. The creep-limited static stress corresponds either to the design limit on creep strain at the design life or to creep rupture at the design life, depending on which failure mode governs. The linear failure prediction rule then may be stated as

   Failure is predicted to occur under combined isothermal creep and fatigue if

$$\frac{\sigma_a}{\sigma_N} + \frac{\sigma_m}{\sigma_{cr}} \geq 1 \qquad (13\text{-}45)$$

An elliptic relationship is also shown in Figure 13.15, which may be written as

   Failure is predicted to occur under combined isothermal creep and fatigue if

$$\left(\frac{\sigma_a}{\sigma_N}\right)^2 + \left(\frac{\sigma_m}{\sigma_{cr}}\right)^2 \geq 1 \qquad (13\text{-}46)$$

The linear rule is usually (but not always) conservative. In the higher temperature portion of the creep range the elliptic relationship usually gives better agreement with data. For example, in Figure 13.16*a* actual data for combined isothermal creep and fatigue tests are shown for several different temperatures using a cobalt-base S-816 alloy. The elliptic approximation is clearly better at higher temperatures for this alloy. Similar data are shown in

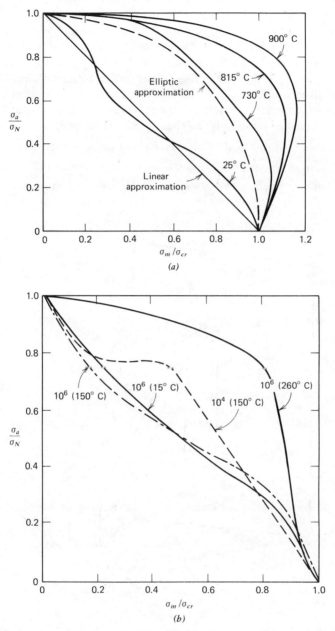

FIGURE 13.16. Combined isothermal creep and fatigue data plotted on coordinates suggested in Figure 13.15. (*a*) Data for S-816 alloy for 100 hour life, where σ_N is fatigue strength for 100 hour life and σ_{cr} is creep rupture stress for 100 hour life. (From refs. 8 and 12). (*b*) Data for 2024 aluminum alloy, where σ_N is fatigue strength for life indicated on respective curves and σ_{cr} is creep stress for corresponding time to rupture. (From refs. 8 and 13.)

Figure 13.16*b* for 2024 aluminum alloy. Detailed studies of the relationships among creep strain, strain at rupture, mean stress, and alternating stress amplitude over a range of stresses and constant temperatures involve extensive, complex testing programs. The results of one study of this type (13) are shown in Figure 13.17 for S-816 alloy at two different temperatures.

Several other empirical methods have recently been proposed for the purpose of making life predictions under more general conditions of combined creep and low-cycle fatigue. These methods include:

1. Frequency modified stress and strain-range method (14).
2. Total time to fracture versus time-of-one-cycle method (15).
3. Total time to fracture versus number of cycles to fracture method (16).

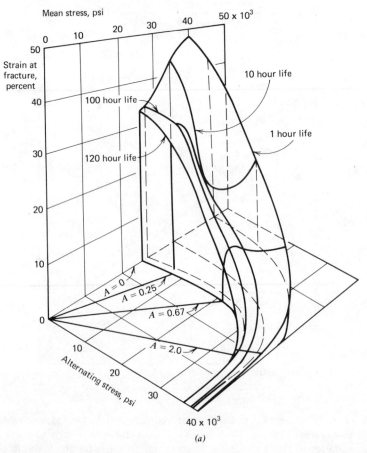

FIGURE 13.17.   Strain at fracture for various combinations of mean and alternating stresses in unnotched specimens of S-816 alloy. (*a*) Data taken at 816°C. (*b*) Data taken at 900°C. (From refs. 8 and 12.)

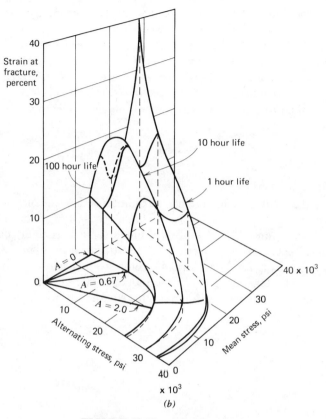

FIGURE 13.17. (*Continued*)

4. Summation of damage fractions using interspersed fatigue with creep method (17).
5. Strain-range partitioning method (18).

Although all these proposed methods have not yet been investigated thoroughly, it is worthwhile to briefly examine each concept. The frequency modified strain-range approach of Coffin was developed by including frequency dependent terms in the basic Manson-Coffin equation, cited earlier as (8-116) and (11-2). The resulting equation can be expressed as

$$\Delta \varepsilon = A N_f^a \nu^b + B N_f^c \nu^d \tag{13-47}$$

where the first term on the right of the equation represents the elastic component of strain range, and the second term represents the plastic component. The constants A and B are the intercepts, respectively, of the elastic and plastic strain components at $N_f = 1$ cycle and $\nu = 1$ cycle per minute. The exponents a, b, c, and d are constants for a particular material at a given temperature. When the constants are experimentally evaluated, this expression provides a relationship between total strain range $\Delta\varepsilon$ and cycles to failure N_f. For example, Figure 13.18 displays the results of this type of analysis for 1Cr-1Mo-1/4V rotor steel at 1000°F. The empirically determined constants to be used in (13-47) for this case are $A = 0.0097$, $B = 2.80$, $a = -0.095$, $b = 0.080$, $c = -0.831$, and $d = 0.162$.

The total time to fracture versus time-of-one-cycle method is based on the expression

$$t_f = \frac{N_f}{\nu} = Ct_c^k \tag{13-48}$$

where t_f is the total time to fracture in minutes, ν is frequency expressed in cycles per minute, N_f is total cycles to failure, $t_c = 1/\nu$ is the time for one cycle in minutes, and C and k are constants for a particular material at a particular temperature for a particular total strain range. Figure 13.19 shows a plot of t_f versus t_c for 1Cr-Mo-1/4V rotor steel. It may be noted that for strain ranges of 0.006 and greater, the plots are straight lines and (13-48) is valid. However, at smaller strain ranges this equation is not acceptable.

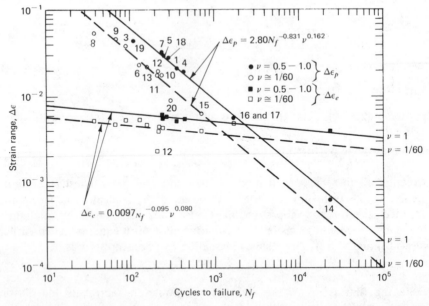

FIGURE 13.18. Elastic and plastic strain ranges versus cycles to failure for 1Cr-1Mo -1/4V rotor steel at 1000°F in air, and best-fit curves using the frequency-modified strain range method of Coffin. (After ref. 20, copyright Society for Experimental Stress Analysis, 1973; reprinted with permission.)

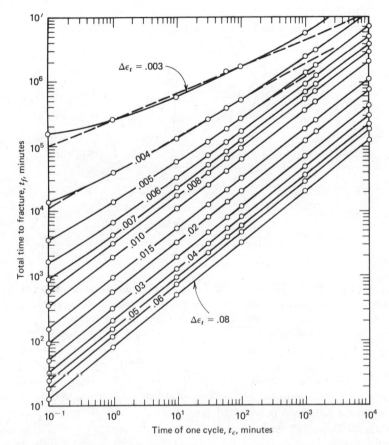

FIGURE 13.19. Variation of total time to fracture versus time of one cycle for several different total strain ranges for 1Cr-1Mo-1/4V rotor steel at 1000°F in air. (After ref. 20, copyright Society for Experimental Stress Analysis, 1973; reprinted with permission.)

The total time to fracture versus number-of-cycles method characterizes the fatigue-creep interaction as

$$t_f = DN_f^{-m} \tag{13-49}$$

which is identical to (13-48) if $D = C^{1/(1-k)}$ and $m = k/(1-k)$. However, it has been postulated that there are three different sets of constants D and m: one set for continuous cycling at varying strain rates, a second set for cyclic relaxation, and a third set for cyclic creep. This concept is illustrated in Figure 13.20. A plot of t_f versus N_f is shown in Figure 13.21 for 1Cr-1Mo-1/4V rotor steel.

The interspersed fatigue and creep analysis proposed by the Metal Properties Council involves the use of a specified combined test cycle on unnotched

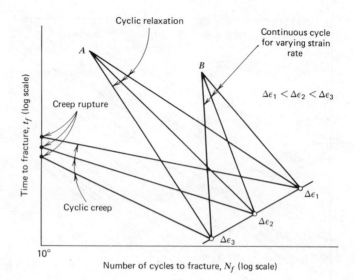

FIGURE 13.20. Postulated curves of t_f versus N_f showing three different sets of constants: one for varying strain rate, one for cyclic relaxation, and one for cyclic creep. (After ref. 20, copyright Society for Experimental Stress Analysis, 1973; reprinted with permission.)

bars. The test cycle consists of a specified period at constant tensile load followed by various numbers of fully reversed strain-controlled fatigue cycles. The specified test cycle is repeated until failure occurs. For example, in one investigation the specified combined test cycle consisted of 23 hours at constant tensile load followed by either 1.5, 2.5, 5.5, or 22.5 fully reversed strain-controlled fatigue cycles. The hysteresis loop for this type of combined cycle is illustrated in Figure 13.22. The failure data then are plotted as fatigue damage fraction versus creep damage fraction, as illustrated in Figure 13.23.

The fatigue damage fraction is the ratio of total number of fatigue cycles N_f' included in the combined test cycle divided by the number of fatigue cycles N_f to cause failure if no creep time were interspersed. The creep damage fraction is the ratio of total creep time t_{cr} included in the combined test cycle divided by the total creep life to failure t_f if no fatigue cycles were interspersed. A "best-fit" curve through the data provides the basis for making a graphical estimate of life under combined creep and fatigue conditions, as shown in Figure 13.23.

The strain-range partitioning method is based on the concept that any cycle of completely reversed inelastic strain may be partitioned into the following strain range components: completely reversed plasticity, $\Delta\varepsilon_{pp}$; tensile plasticity reversed by compressive creep, $\Delta\varepsilon_{pc}$; tensile creep reversed by compressive plasticity, $\Delta\varepsilon_{cp}$; and completely reversed creep, $\Delta\varepsilon_{cc}$. The first letter of each subscript in the notation, c for creep and p for plastic deformation, refers to

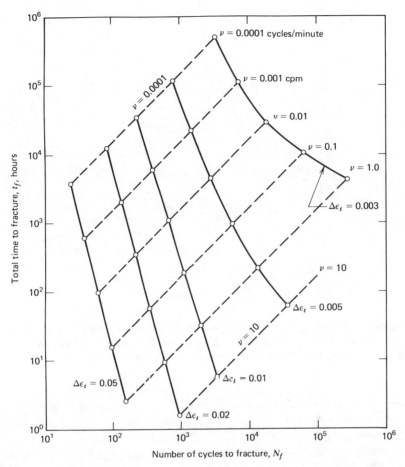

FIGURE 13.21. Plot of t_f versus N_f for 1Cr-1Mo-$\frac{1}{4}$V rotor steel at 1000°F in air. (After ref. 20, copyright Society for Experimental Stress Analysis, 1973; reprinted with permission.)

the type of strain imposed during the tensile portion of the cycle, and the second letter refers to the type of strain imposed during the compressive portion of the cycle. The term *plastic deformation* or *plastic flow* in this context refers to *time-independent* plastic strain that occurs by crystallographic slip within the crystal grains. The term *creep* refers to *time-dependent* plastic deformation that occurs by a combination of diffusion within the grains together with grain boundary sliding between the grains. The concept is illustrated in Figure 13.24.

It may be noted in Figure 13.24 that tensile inelastic strain, represented as $\overline{AD}$ is the sum of plastic strain $\overline{AC}$ plus creep strain $\overline{CD}$. Also, compressive inelastic strain $\overline{DA}$ is the sum of plastic strain $\overline{DB}$ plus creep strain $\overline{BA}$. In

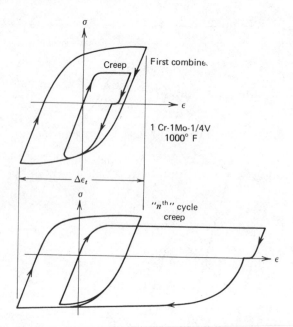

FIGURE 13.22. Typical hysteresis loops for interspersed creep-fatigue tests on 1Cr-1Mo-$\frac{1}{4}$V rotor steel at 1000°F using Metal Properties Council method. (After ref. 20, copyright Society for Experimental Stress Analysis, 1973; reprinted with permission.)

general, $\overline{AC}$ will not be equal to $\overline{DB}$, nor will $\overline{CD}$ be equal to $\overline{BA}$. However, since we are dealing with a closed hysteresis loop, $\overline{AD}$ does equal $\overline{DA}$. The partitioned strain ranges are obtained in the following manner (19): The completely reversed portion of the plastic strain range, $\Delta\varepsilon_{pp}$, is the smaller of the two plastic flow components, which in Figure 13.24 is equal to $\overline{DB}$. Likewise, the completely reversed portion of the creep strain range, $\Delta\varepsilon_{cc}$, is the smaller of the two creep components, which in Figure 13.24 is equal to $\overline{CD}$. As can be seen graphically, the difference between the two plastic components must be equal to the difference between the two creep components, or $\overline{AC}$-$\overline{DB}$ must equal $\overline{BA}$-$\overline{CD}$. This difference then is either $\Delta\varepsilon_{pc}$ or $\Delta\varepsilon_{cp}$, in accordance with the notation just defined. For the case illustrated in Figure 13.24, the difference is $\Delta\varepsilon_{pc}$, since the tensile plastic strain component is greater than the compressive plastic strain component. It follows from this discussion that the sum of the partitioned strain ranges will necessarily be equal to the total inelastic strain range, or the width of the hysteresis loop.

It is next assumed that a unique relationship exists between cyclic life to failure and each of the four strain range components listed. Available data indicate that these relationships are of the form of the basic Manson-Coffin expression (11-2), as indicated, for example, in Figure 13.25 for a Type 316 stainless steel alloy at 1300°F. The governing life prediction equation, or

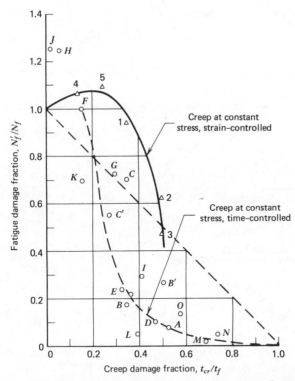

FIGURE 13.23. Plot of fatigue damage fraction versus creep damage fraction for 1Cr-1Mo-$\frac{1}{4}$V rotor steel at 1000°F in air, using the method of the Metal Properties Council. (After ref. 20, copyright Society for Experimental Stress Analysis, 1973; reprinted with permission.)

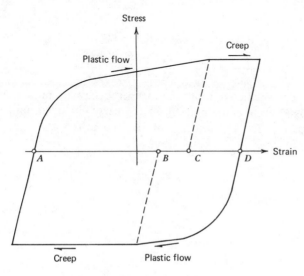

FIGURE 13.24. Typical hysteresis loop.

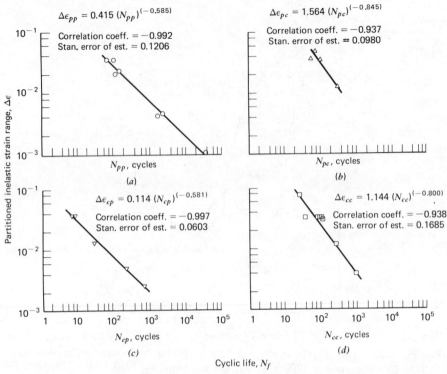

FIGURE 13.25.

FIGURE 13.25. Summary of partitioned strain-life relations for Type 316 stainless steel at 1300°F. (After ref. 21) (*a*) *pp*-type strain range. (*b*) *pc*-type strain range. (*c*) *cp*-type strain range. (*d*) *cc*-type strain range.

"interaction damage rule," is then postulated to be

$$\frac{1}{N_{\text{pred}}} = \frac{F_{pp}}{N_{pp}} + \frac{F_{pc}}{N_{pc}} + \frac{F_{cp}}{N_{cp}} + \frac{F_{cc}}{N_{cc}} \qquad (13\text{-}50)$$

where N_{pred} is the predicted total number of cycles to failure under the combined *straining* cycle containing all of the pertinent strain range components. The terms F_{pp}, F_{pc}, F_{cp}, and F_{cc} are defined as

$$F_{pp} = \frac{\Delta\varepsilon_{pp}}{\Delta\varepsilon_p}$$

$$F_{pc} = \frac{\Delta\varepsilon_{pc}}{\Delta\varepsilon_p}$$

$$(13\text{-}51)$$

$$F_{cp} = \frac{\Delta\varepsilon_{cp}}{\Delta\varepsilon_p}$$

$$F_{cc} = \frac{\Delta\varepsilon_{cc}}{\Delta\varepsilon_p}$$

for any selected inelastic strain range $\Delta\varepsilon_p$, using information from a plot of experimental data such as that shown in Figure 13.25. The partitioned failure lives N_{pp}, N_{pc}, N_{cp}, and N_{cc} are also obtained from Figure 13.25. The use of (13-50) has, in several investigations (21–26), shown the predicted lives to be acceptably accurate, with most experimental results falling within a scatter band of $\pm 2N_f$ of the predicted value. Results of tests on Type 316 stainless steel at 1300°F in air are plotted in Figure 13.26 versus the strain-range partitioning prediction based on (13-50).

More recent investigations (27) have indicated that improvements in predictions by the strain-range partitioning method may be achieved by using the "creep" ductility and "plastic" ductility of a material determined in the actual service environment, to "normalize" the strain versus life equations prior to using (13-50). Procedures for using the strain-range partitioning

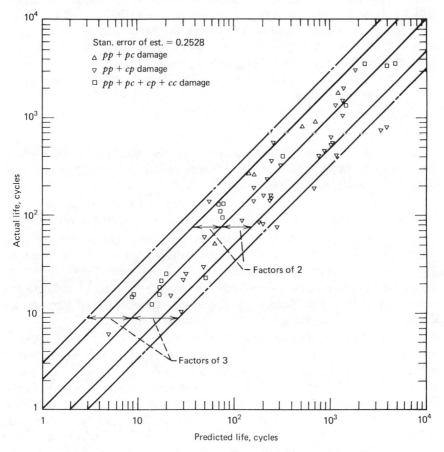

FIGURE 13.26. Actual life versus life predicted by strain-range partitioning method. See (13-50). (From ref. 21.)

method under conditions of multiaxial loading have also been proposed (25) but remain to be verified more fully. The strain-range partitioning method has also recently been applied to the "interspersed fatigue and creep" data produced by the Metal Properties Council, discussed earlier, with good success (24).

Although both the frequency modified strain-range method and the strain-range partitioning method are regarded as reasonably promising prediction techniques, the strain-range partitioning technique seems to have the unique capability of providing a tool for analyzing the more complex loading histories to which many actual structures or components may be subjected. In complex cases, where the strain range cannot be partitioned accurately, the assumption can be made that the most damaging strain-range component dominates, and the lower bound on life may be predicted based on this conservative assumption.

Although much remains to be learned about the creep-fatigue interaction, initial experience with the strain-range partitioning analysis seems promising (26). The method has potential for guiding the generation of creep-fatigue data and interpreting the effects of frequency, hold time, temperature, and environment. However, even though good progress has recently been made, there is still much to be understood about the creep-fatigue interaction.

QUESTIONS

1. Carefully define the terms *creep*, *creep rupture*, and *stress rupture*, citing the similarities that relate these three failure modes and the differences that distinguish them from one another.

2. Construct a table of temperature ranges for which creep may become a significant problem, using the materials of Table 13.1 as entries in your table.

3. List and describe several methods that have been used for extrapolating short-time creep data to long-term applications. What are the potential pitfalls in using these methods?

4. List and describe several methods that have been used to estimate the cumulative creep under conditions of varying stress level and temperature during exposure of a machine part to its operating environment.

5. A new high-temperature alloy is to be used for a 0.125-inch-diameter tensile support member for an impact-sensitive instrument that weighs 200 lb and costs $300,000. The instrument and its support are to be enclosed in a test vessel for 3000 hours at 1600°F. A laboratory test on the new alloy used a 0.125-inch-diameter specimen loaded with a 200-lb mass and was found to fail by stress rupture after 100 hours at 1800°F. Based on the test results, determine whether the tensile support is adequate for this application.

6. A sand-cast aluminum alloy creep tested at 600°F results in the data shown in Figure 13.6C, curve *b*, when subjected to a stress level of 10,000 psi.

A solid square support bracket of this material is 1 inch × 1 inch × 4 inches long and is subjected to a direct tensile load of 10,000 lb. The ambient temperature cycle is 380°F for 1200 hours, then 600°F for 2 hours, and then 810°F for 15 seconds. How many cycles would you predict could be survived by the bracket if the design criterion is that the bracket must not elongate more than 0.10 inch?

7. A sand-cast aluminum alloy that has been creep tested at 600°F at a stress level of 10,000 psi results in the data shown as curve *b* in Figure 13.6C. A hollow cylindrical support bracket of this material has a 2.0-inch outside diameter, a 0.175-inch wall thickness, and is 5.0 inches long. The bracket is subjected to a direct tensile load of 5 tons. Each ambient temperature cycle is 400°F for 1000 hrs., then 600°F for 3 hours, and then 800°F for 0.5 minute.
(a) How many cycles would you predict could be survived by the bracket if the design criterion is that the bracket must not elongate more than 0.025 inch?
(b) If the bracket were internally pressurized at 1750 psi using a clever end-sealing arrangement so that the pressure produced no axial component of stress in the bracket wall, would you predict that the bracket would achieve a longer or shorter life before failure, according to the criterion of part (a)?
(c) In the case of part (b), what other design criteria, if any, might require investigation?

8. From the data plotted in the curves shown in Figure Q13.8, evaluate the constants *B* and *N* of (13-9) for the material shown.

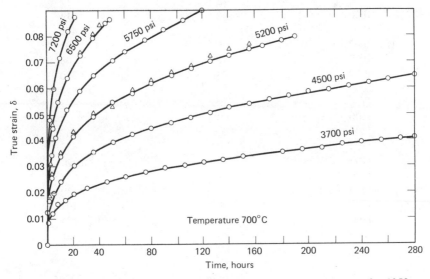

FIGURE Q13.8. (From ref. 29, copyright American Society for metals, 1958; reprinted with permission.)

9. A cylindrical pressure vessel of Ni-Cr-Mo steel has closed ends and a diameter of 10 inches. The tank is to operate continuously for 5 years at a temperature of 850°F under an internal pressure of 10,000 psi. If a safety factor of 1.20 is desired, and permanent deformation must not exceed 3 percent, what should the minimum thickness of the cylindrical pressure vessel wall be made?

10. The creep rupture data from a series of tests on Cr-Mo-V steel are shown in the chart.

(a) Using these data, plot the creep rupture curves for the various temperatures on a single plot of stress versus log-time.

(b) Using the data for a stress level of 70,000 psi at a temperature of 900°F, predict the time to rupture at the same stress level if the temperature is 1100°F; use the Larson Miller Theory.

(c) Compare the prediction of part (b) with the data of part (a), calculating the percent error in the predicted creep rupture life.

Creep Rupture Data for Cr-Mo-V Steel(28)

Test Temperature (°F)	Stress (psi $\times 10^{-3}$)	Rupture Time (hours)
900	90	37
900	82	975
900	78	3,581
900	70	9,878
1000	80	7
1000	75	17
1000	68	213
1000	60	1,493
1000	56	2,491
1000	49	5,108
1000	43	7,390
1000	38	10,447
1100	70	1
1100	60.5	18
1100	50	167
1100	40	615
1100	29	2,220
1100	22	6,637
1200	40	19
1200	30	102
1200	25	125
1200	20	331
1350	20	3.7
1350	15	8.9
1350	10	31.8

11. The creep curves for alloy "x" at 1200°F are shown in Figure Q13.11. The design limit on creep strain for this application is 0.01 inch per inch. A tension member at 1200°F is subjected to the following stress history, applied in sequence:

1. 1000 psi for 10 hrs
2. 2000 psi for 20 hrs
3. 3000 psi for 20 hrs
4. 4000 psi for 10 hrs

Based on this stress history and the data shown in the curves of Figure Q13.11, estimate the total creep strain at the end of the 60-hour design life by using the (a) time-hardening rule, (b) strain-hardening rule, and (c) life-fraction rule.

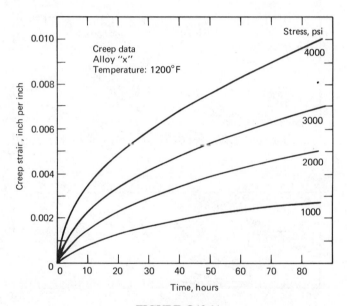

FIGURE Q13.11.

12. An aircraft engine disk made of "Alloy Z" is subjected to a stress-time-temperature history during a "typical" 2-hour flight, as shown in the chart. The design limit on creep strain is 0.1 inch per inch. Using the creep curves shown in Figure Q13.12 for this "Alloy Z" material, estimate the total creep strain per 2-hour flight by using the (a) time-hardening rule, (b) strain-hardening rule, and (c) life-fraction rule.

	Stress (psi $\times 10^{-3}$)	Temperature (°F)	Time (hour)
1.	110	1200	0.05
2.	150	1000	0.10
3.	110	1200	0.01
4.	60	1400	0.01
5.	80	1400	0.01
6.	110	1000	1.32
7.	150	1000	0.10
8.	160	1000	0.10
9.	150	1000	0.10
10.	110	1200	0.10
11.	150	1000	0.10

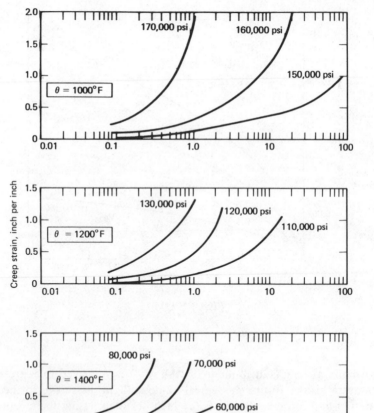

FIGURE Q13.12.

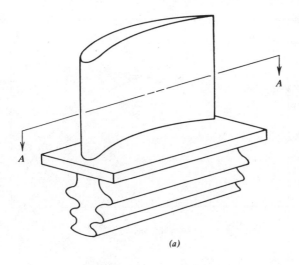

(a)

Typical Mission Profile

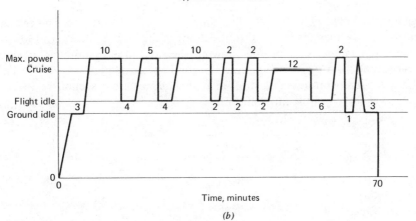

Time, minutes

(b)

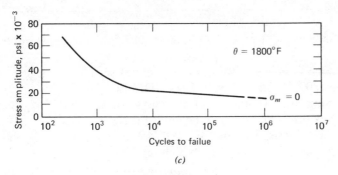

(c)

FIGURE Q13.13.

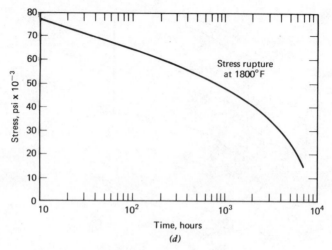

FIGURE Q13.13.   (*Continued*)

13.   A turbine blade in the high-pressure stage of an aircraft gas turbine engine is subjected to both cyclic loads and elevated temperature. The critical area A-A of the blade is in the airfoil section, as shown in Figure Q13.13*a*.

The estimated mission profile showing the hold-time in minutes at each power setting for a typical 70-minute mission is given in Figure Q13.13*b*. The chart gives temperatures and stresses at the critical cross section related to each power setting in the mission profile.

Power Setting	Metal Temperature (°F)	Stress (psi × 10⁻³)
Ground idle	1600	30
Flight idle	1620	36
Cruise	1680	50
Maximum Power	1750	56

Using the elevated temperature fatigue data and the stress rupture data given,

(a)   determine fatigue life (missions to failure) separately, neglecting creep effects (use 1800°F data shown in Figure Q13.13*c*), and

(b)   determine stress rupture life (missions to failure) separately, neglecting fatigue effects (use 1800°F data) shown in Figure Q13.13*d*.

(c)   Using the elliptic relationship of (13-46), determine the life under combined creep and fatigue (missions to failure).

(d)   Does fatigue or stress rupture appear to be the dominant influence in this design environment?

14. In combined creep and fatigue failure analysis the "strain-range partitioning" method seems to hold good promise for predicting failure. Describe the basic ideas behind this approach, define the four basic "partitioned" strain-range components, and indicate qualitatively how failure may be predicted.

REFERENCES

1. Juvinall, R.C., *Engineering Considerations of Stress, Strain, and Strength*, McGraw-Hill, New York, 1967.

2. Larson, F.R., and Miller, J., "Time-Temperature Relationships for Rupture and Creep Stresses," *ASME Transactions*, **74** (1952): 765.

3. Manson, S.S., and Haferd, A.M., "A Linear Time-Temperature Relation for Extrapolation of Creep and Stress Rupture Data," NACA Technical Note 2890, March 1953.

4. Van Echo, J.A., "Short-Time Creep of Structural Sheet Metals," *Short-Time High-Temperature Testing*, American Society for Metals, 1958.

5. Sturm, R.G., Dumont, C., and Howell, F.M., "A Method of Analyzing Creep Data," *ASME Transactions*, **58** (1936): A62.

6. Polakowski, N.H., and Ripling, E.J., *Strength and Structure of Engineering Materials*, Prentice-Hall, Englewood Cliffs, N.J., 1966.

7. Robinson, E.L., "Effect of Temperature Variation on the Long-Time Rupture Strength of Steels" *ASME Transactions*, **74** (1952): 777–781.

8, Kennedy, A.J., *Processes of Creep and Fatigue in Metals*, John Wiley & Sons, New York, 1963.

9. Simmons, W.F., and Cross, H.C., *Symposium on Effect of Cyclic Heating and Stressing on Metals at Elevated Temperatures*, STP-165, American Society for Testing and Materials, Philadelphia, 1954.

10. Guarnieri, G.J., *Symposium on Effect of Cyclic Heating and Stressing on Metals at Elevated Temperatures*, STP-165, American Society of Testing and Materials, Philadelphia, 1954.

11. Rowe, J.P., and Freeman, J.W., NACA Tests, Note 4224, 1958.

12. Demoney, F.W., and Lazan, B.J., WADC Tech Report 53-510, 1954.

13. Vitovec, F., and Lazan, B.J., *Symposium on Metallic Materials for Service at Temperatures Above 1600 F.*, STP-174, American Society for Testing and Materials, Philadelphia, 1956.

14. Coffin, L.F., Jr., "The Effect of Frequency on the Cyclic Strain and Low Cycle Fatigue Behavior of Cast Udimet 500 at Elevated Temperature," *Metallurgical Transactions*, **12** (November 1971): 3105–3113.

15. Conway, J.B., and Berling, J.T., "A New Correlation of Low-Cycle Fatigue Data Involving Hold Periods," *Metallurgical Transactions*, **1** No. 1 (January 1970): 324–325.

16. Ellis, J.R., and Esztergar, E.P., "Considerations of Creep-Fatigue Interaction in Design Analysis," *Symposium on Design for Elevated Temperature Environment*, ASME, (May 1971): 29–33.

17. Curran, R.M., and Wundt, B.M., "A Program to Study Low-Cycle Fatigue and Creep Interaction in Steels at Elevated Temperatures," *Current Evaluation of 2-1/4 Chrome 1 Molybdenum Steel in Pressure Vessels and Piping*, ASME (1972): 49–82.

18. Manson, S.S., Halford, G.R., and Hirschberg, M.H., "Creep-Fatigue Analysis by Strain-Range Partitioning," *Symposium on Design for Elevated Temperature Environment*, ASME, (May 1971): 12–24.

19. Manson, S.S., Halford, G.R., and Hirschberg, M.H., NASA Technical Memo TMX-67838, Lewis Research Center, Cleveland, May 1971.

20. Leven, M.M., "The Interaction of Creep and Fatigue for a Rotor Steel," *Experimental Mechanics*, **13**, No. 9 (September 1973): 353–372.

21. Saltsman, J.F., and Halford, G.R., "Application of Strain-Range Partitioning to the Prediction of Creep-Fatigue Lives of AISI Types 304 and 316 Stainless Steel," NASA Technical Memo TMX-71898, Lewis Research Center, Cleveland, September 1976.

22. Annis, C.G., Van Wanderham, M.C., and Wallace, R.M., "Strain-Range Partitioning Behavior of an Automotive Turbine Alloy," Final Report NASA TR 134974, February 1976.

23. Hirschberg, M.H., and Halford, G.R., "Use of Strain-Range Partitioning to Predict High-Temperature Low Cycle Fatigue Life," NASA TN D-8072, January 1976.

24. Saltsman, J.F., and Halford G.R., "Application of Strain-Range Partitioning to the Prediction of MPC Creep-Fatigue Data for 2-1/4 Cr-1 Mo Steel," NASA TMX-73474, December 1976.

25. Manson, S.S., and Halford, G.R., "Treatment of Multiaxial Creep-Fatigue by Strain-Range Partitioning," NASA TMX-73488, December 1976.

26. *Characterization of Low Cycle High Temperature Fatigue by the Strain-Range Partitioning Method*, AGARD Conference Proceedings No. 243, distributed by NASA, Langley Field, Va. April 1978.

27. Halford G.R., Saltsman, J.F., and Hirshberg, M.H., "Ductility Normalized-Strainrange Partitioning Life Relations for Creep-Fatigue Life Predictions," NASA TM-73737, October 1977.

28. Goldhoff, R.M., and Gill, R.F., "Discussion to Manson, Succop and Brown," *Transactions, American Society for Metals*, (1959): 911.

29. Parker, E.R., "Modern Concepts of Flow and Fracture," *Transactions, American Society for Metals*, **50** (1958): 52.

CHAPTER 14

Fretting, Fretting Fatigue, and Fretting Wear

14.1 INTRODUCTION

Service failure of mechanical components due to fretting fatigue has gradually come to be recognized as a failure mode of major importance, both in terms of frequency of occurrence and seriousness of the failure consequences. Fretting wear has also presented major problems in certain applications. Both fretting fatigue and fretting wear, as well as fretting corrosion, are directly attributable to *fretting action*. Basically, fretting action has, for many years, been defined as a combined mechanical and chemical action in which the contacting surfaces of two solid bodies are pressed together by a normal force and are caused to execute oscillatory sliding relative motion, wherein the magnitude of normal force is great enough and the amplitude of the oscillatory sliding motion is small enough to significantly restrict the flow of fretting debris away from the originating site (1). More recent definitions of fretting action have been broadened to include cases in which contacting surfaces periodically separate and then reengage, as well as cases in which the fluctuating friction-induced surface tractions produce stress fields that may ultimately result in failure. The complexities of fretting action have been discussed by numerous investigators, who have postulated the combination of many mechanical, chemical, thermal, and other phenomena that interact to produce fretting. Among the postulated phenomena are plastic deformation caused by surface asperities plowing through each other, welding and tearing of contacting asperities, shear and rupture of asperities, friction generated subsurface shearing stresses, dislodging of particles and corrosion products at the surfaces, chemical reactions, debris accumulation and entrapment, abrasive action, microcrack initiation, surface delamination, and others (2–12, 24–28).

Damage to machine parts due to fretting action may be manifested as corrosive surface damage due to fretting corrosion, loss of proper fit or change in dimensions due to fretting wear, or accelerated fatigue failure due to fretting fatigue. Typical sites of fretting damage include interference fits; bolted, keyed, splined, and riveted joints; points of contact between wires in wire ropes and flexible shafts; friction clamps; small amplitude-of-oscillation bearings of all kinds; contacting surfaces between the leaves of leaf springs; and all other places where the conditions of fretting persist. Thus, the efficiency and reliability of the design and operation of a wide range of mechanical systems are related to the fretting phenomenon. For example, systems in which fretting is a potential failure mode would include military and civilian ground transport vehicles, weapons systems, ships and submarines, helicopters, aircraft power plants, turbines and compressors, air frames, missile components, nuclear power plants, precision instrumentation, controls and control systems, transmissions, universal joints, and a broad spectrum of other machines and components.

Although fretting fatigue, fretting wear, and fretting corrosion phenomena are potential failure modes in a wide variety of mechanical systems, and much research effort has been devoted to the understanding of the fretting process, there are very few quantitative design data available, and no generally applicable design procedure has been established for predicting failure under fretting conditions. However, even though the fretting phenomenon is not fully understood, and a good general model for prediction of fretting fatigue or fretting wear has not yet been developed, significant progress has been made in establishing an understanding of fretting and the variables of importance in the fretting process.

14.2 VARIABLES OF IMPORTANCE IN THE FRETTING PROCESS

It has been suggested that there may be more than 50 variables that play some role in the fretting process (13). Of these, however, there are probably only eight that are of major importance; they are:

1. The magnitude of relative motion between the fretting surfaces
2. The magnitude and distribution of pressure between the surfaces at the fretting interface
3. The state of stress, including magnitude, direction, and variation with respect to time in the region of the fretting surfaces
4. The number of fretting cycles accumulated
5. The material from which each of the fretting members is fabricated, including surface condition
6. Cyclic frequency of relative motion between the two members being fretted
7. Temperature in the region of the two surfaces being fretted
8. Atmospheric environment surrounding the surfaces being fretted

These variables interact so that a quantitative prediction of the influence of any given variable is very dependent upon all the other variables in any specific application or test. Also, the combinations of variables that produce a very serious consequence in terms of fretting fatigue damage may be quite different from the combinations of variables that produce serious fretting wear damage. No general techniques yet exist for quantitatively predicting the influence of the important variables of fretting fatigue and fretting wear damage, although many special cases have been investigated. However, it has been observed that certain trends usually exist when the variables just listed are changed. For example, fretting damage tends to increase with increasing contact pressure until a nominal pressure of a few thousand pounds per square inch is reached, and further increases in pressure seem to have relatively little direct effect. The state of stress is important, especially in fretting fatigue, and is discussed further in Section 14.3. Fretting damage accumulates with increasing numbers of cycles at widely different rates, depending on specific operating conditions. Fretting damage is strongly influenced by the material properties of the fretting pair, surface hardness, roughness, and finish. No clear trends have been established regarding frequency effects on fretting damage, and although both temperature and atmospheric environment are important influencing factors, their influences have not been clearly established. A clear presentation of the current state of knowledge relative to these various parameters is given, however, in Reference 25.

14.3   FRETTING FATIGUE

Fretting fatigue is fatigue damage directly attributable to fretting action. It has been suggested that premature fatigue nuclei may be generated by fretting through either abrasive pit-digging action, asperity-contact microcrack initiation (15), friction-generated cyclic stresses that lead to the formation of microcracks (14), or subsurface cyclic shear stresses that lead to surface delamination in the fretting zone (28). Under the abrasive pit-digging hypothesis, it is conjectured that tiny grooves or elongated pits are produced at the fretting interface by the asperities and abrasive debris particles moving under the influence of oscillatory relative motion. A pattern of tiny grooves would be produced in the fretted region with their longitudinal axes all approximately parallel and in the direction of fretting motion, as shown schematically in Figure 14.1.

The asperity-contact microcrack initiation mechanism is postulated to proceed by virtue of the contact force between the tip of an asperity on one surface and another asperity on the mating surface as they move back and forth. If the initial contact does not shear one or the other asperity from its base, the repeated contacts at the tips of the asperities give rise to cyclic or fatigue stresses in the region at the base of each asperity. It has been estimated (10) that under such conditions the region at the base of each asperity is subjected to large local stresses that probably lead to the nucleation of fatigue microcracks at these sites. As shown schematically in Figure

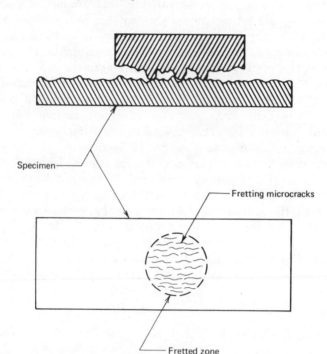

FIGURE 14.1. Idealized schematic illustration of the stress concentrations produced by the abrasive pit-digging mechanism.

14.2, it would be expected that the asperity contact mechanism would produce an array of microcracks whose longitudinal axes would be generally perpendicular to the direction of fretting motion.

The friction-generated cyclic stress fretting hypothesis (12) is based on the observation that when one member is pressed against the other and caused to undergo fretting motion, the tractive friction force induces a compressive tangential stress component in a volume of material that lies ahead of the fretting motion, and a tensile tangential stress component in a volume of material that lies behind the fretting motion, as shown in Figure 14.3a. When the fretting direction is reversed, the tensile and compressive regions change places. Thus, the volume of material adjacent to the contact zone is subjected to a cyclic stress that is postulated to generate a field of microcracks at these sites. Further, the geometrical stress concentration associated with the clamped joint may contribute to microcrack generation at these sites (24). As shown in Figure 14.3c, it would be expected that the friction-generated microcrack mechanism would produce an array of microcracks whose longitudinal axes

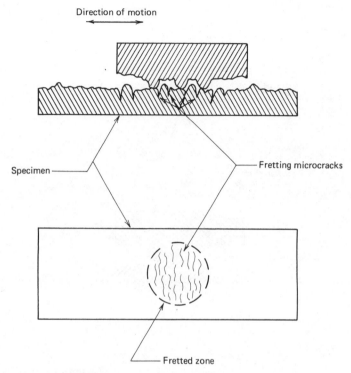

Direction of motion

Specimen

Fretting microcracks

Fretted zone

FIGURE 14.2. Idealized schematic illustration of the stress concentrations produced by the asperity-contact microcrack initiation mechanism.

would be generally perpendicular to the direction of fretting motion. These cracks would lie in a region adjacent to the fretting contact zone.

In the delamination theory of fretting (28), it is hypothesized that the combination of normal and tangential tractive forces transmitted through the asperity contact sites at the fretting interface produce a complex multiaxial state of stress, accompanied by a cycling deformation field, which produces subsurface peak shearing stress and subsurface crack nucleation sites. With further cycling, the cracks propagate approximately parallel to the surface, as in the case of the surface fatigue phenomenon, finally propagating to the surface to produce a thin wear sheet, which "delaminates" to become a particle of debris.

Supporting evidence has been generated to indicate that under various circumstances each of the four mechanisms is active and significant in producing fretting damage. For example, the data shown in Figure 14.4 indicate that the direction of fretting motion greatly influences subsequent fatigue strengths. If uniaxial fatigue stressing is applied in the same direction

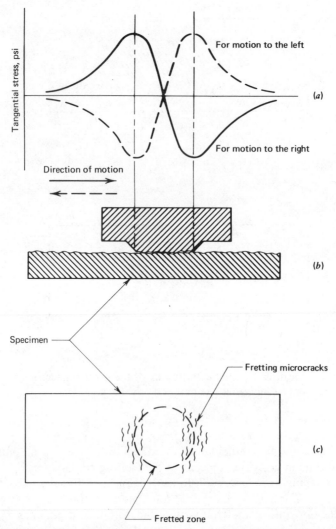

FIGURE 14.3. Idealized schematic illustration of the tangential stress components and microcracks produced by the friction-generated microcrack initiation mechanism.

as the fretting motion, a much greater reduction in fatigue strength is noted than for specimens in which fatigue stressing is applied perpendicular to the direction of fretting motion. Both fretting conditions produce a marked decrease in fatigue properties when compared to nonfretted virgin fatigue data for the same material. These data indicate that the asperity-contact microcrack initiation mechanism probably dominates, but the pit-digging mechanism is probably also active (15). The delamination phenomenon may well be active in both cases. Since fatigue failures in these tests initiated in the

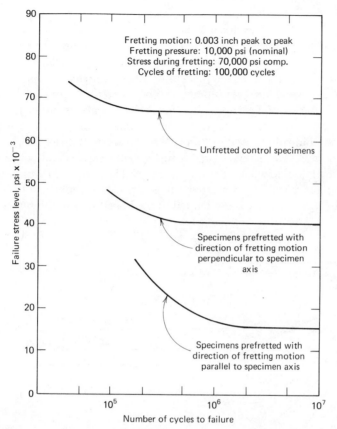

FIGURE 14.4. *S-N* data for specimens fretted in different directions and subsequently evaluated in a completely reversed fatigue test. Specimens and fretting shoes were all 4340 steel heat treated to Rockwell C-35. Tests were run at 1500 cpm in 75°F atmospheric air.

fretted zone rather than adjacent to the contact region, it is probable that the friction-induced microcrack mechanism was overshadowed by other fretting mechanisms for these particular tests.

The influence of the state of stress in the member during the fretting is shown for several different cases (16) in Figure 14.5, including static tensile and compressive mean stresses during fretting. An interesting observation in Figure 14.5 is that fretting under conditions of compressive mean stress, either static or cyclic, produces a drastic reduction in fatigue properties. This, at first, does not seem to be in keeping with the concept that compressive stresses are beneficial in fatigue loading. However, it was deduced (17), as shown in Figure 14.6, that the compressive stresses during fretting shown in Figure 14.5 actually resulted in local residual compressive stresses in the fretted region. Likewise, the tensile stresses during fretting shown in Figure 14.5 actually resulted in local residual compressive stresses in the fretted region. The conclusion, therefore, is that local compressive stresses are beneficial in minimizing fretting fatigue damage.

Further evidence of the beneficial effects of compressive residual stresses in minimizing fretting fatigue damage is illustrated in Figure 14.7, where the results of a series of Prot tests (see Section 10.6) are reported for steel and titanium specimens subjected to various combinations of shot-peening and fretting or cold-rolling and fretting. It is clear from these results that the residual compressive stresses produced by shot-peening and cold-rolling are effective in minimizing the fretting damage. The reduction in scatter of the fretted fatigue properties for titanium is especially important to a designer because design stress is closely related to the lower limit of the scatter band.

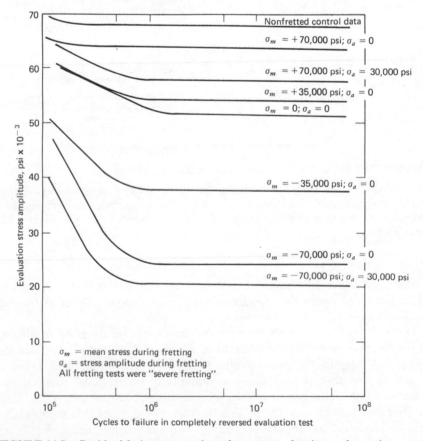

FIGURE 14.5.   Residual fatigue properties subsequent to fretting under various states of stress.

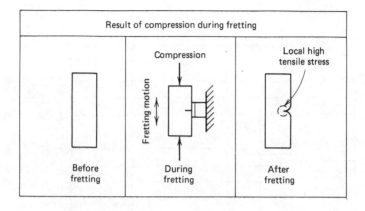

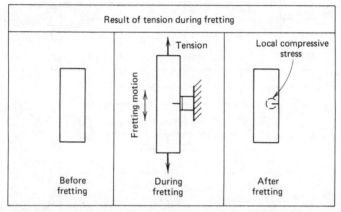

FIGURE 14.6. Sketches illustrating an explanation of why specimens fretted under compressive mean stresses as shown in Figure 14.5 exhibit more fretting damage than specimen fretted under tensile mean stresses.

The concept of using a fretting fatigue strength reduction factor or damage factor has been proposed (1) as a means for modifying the fatigue strength of a material to account for fretting conditions that may persist in a given design. If E_v is defined as the virgin fatigue strength at a specified life and E_f is the fretted fatigue strength at the same life, it should be possible to define a fretting fatigue damage factor D_f such that

$$E_f = (1 - D_f)E_v \qquad (14\text{-}1)$$

where the range on the damage factor D_f is from zero to unity. When D_f is zero, there is no fretting fatigue damage; and when D_f is unity, the fatigue strength is reduced to zero by the fretting action. Unfortunately, the damage factor D_f is a function of all the fretting variables described earlier, and the functional relationship has not been defined. A graphical display of D_f as a

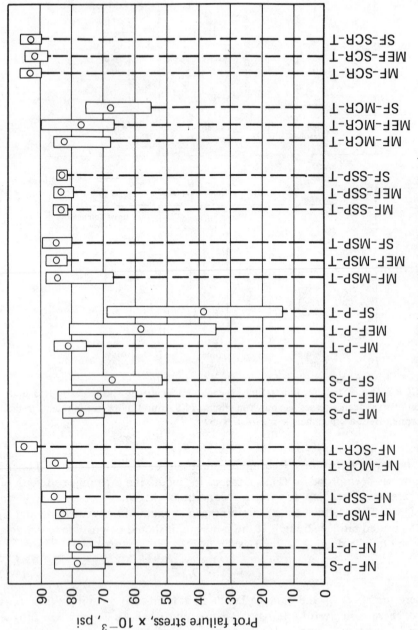

Test conditions used (see table for key symbols)

Tabular summary for test results of Figure 14.7

Test Condition Used	Code Designation	Sample Size	Mean Prot Failure Stress, psi	Unbiased Standard Deviation, psi
Nonfretted, polished, SAE 4340 steel	NF-P-S	15	78,200	5,456
Nonfretted, polished, Ti-140-A titanium	NF-P-T	15	77,800	2,454
Nonfretted, mildly shot-peened, Ti-140-A titanium	NF-MSP-T	15	83,100	1,637
Nonfretted, severely shot-peened, Ti-140-A titanium	NF-SSP-T	15	85,700	2,398
Nonfretted, mildly cold-rolled, Ti-140-A titanium	NF-MCR-T	15	85,430	1,924
Nonfretted, severely cold-rolled, Ti-140-A titanium	NF-SCR-T	15	95,400	2,120
Mildly fretted, polished, SAE 4340 steel	MF-P-S	15	77,280	4,155
Medium fretted, polished, SAE 4340 steel	MeF-P-S	15	71,850	5,492
Severely fretted, polished, SAE 4340 steel	SF-P-S	15	67,700	6,532
Mildly fretted, polished, Ti-140-A titanium	MF-P-T	15	81,050	3,733
Medium fretted, polished, Ti-140-A titanium	MeF-P-T	15	58,140	15,715
Severely fretted, polished, Ti-140-A titanium	SF-P-T	15	38,660	19,342
Mildly fretted, mildly shot-peened, Ti-140-A titanium	MF-MSP-T	15	84,520	5,239
Medium fretted, mildly shot-peened, Ti-140-A titanium	MeF-MSP-T	15	84,930	2,446
Severely fretted, mildly shot-peened, Ti-140-A titanium	SF-MSP-T	15	84,870	2,647
Mildly fretted, severely shot-peened, Ti-140-A titanium	MF-SSP-T	15	83,600	1,474
Medium fretted, severely shot-peened, Ti-140-A titanium	MeF-SSP-T	15	83,240	1,332
Severely fretted, severely shot-peened, Ti-140-A titanium	SF-SSP-T	15	83,110	1,280
Mildly fretted, mildly cold-rolled, Ti-140-A titanium	MF-MCR-T	15	82,050	4,313
Medium fretted, mildly cold-rolled, Ti-140-A titanium	MeF-MCR-T	15	76,930	8,305
Severely fretted, mildly cold-rolled, Ti-140-A titanium	SF-MCR-T	15	67,960	5,682
Mildly fretted, severely cold-rolled, Ti-140-A titanium	MF-SCR-T	15	93,690	1,858
Medium fretted, severely cold-rolled, Ti-140-A titanium	MeF-SCR-T	15	91,950	2,098
Severely fretted. severely cold-rolled. Ti-140-A titanium	SF-SCR-T	15	93,150	1,365

FIGURE 14.7. Fatigue properties of fretted steel and titanium specimens with various degrees of shot-peening and cold-rolling. (See ref. 11)

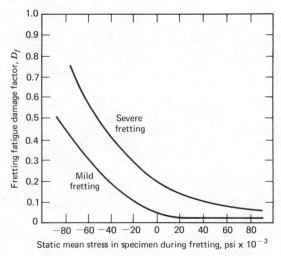

FIGURE 14.8. Fretting-fatigue damage factor as a function of static mean stress during fretting.

function of state-of-stress during fretting is shown in Figure 14.8 for a very limited range of conditions. Although the concept is workable, it would be a monumental task to develop such data to define the damage factor for all fretting conditions of interest.

Recent efforts to apply the tools of fracture mechanics to the problem of life prediction under fretting fatigue conditions have produced encouraging preliminary results that may ultimately provide designers with a viable quantitative approach for the first time (31). These studies emphasize that the principal effect of fretting in the fatigue failure process is to accelerate crack initiation and the early stages of crack growth, and they suggest that when cracks have reached a sufficient length, the fretting no longer has a significant influence on crack propagation. At this point the fracture mechanics description of crack propagation described in Section 8.6 becomes valid. It is further suggested that in many cases the fretting fatigue lives are primarily determined by crack propagation, with crack initiation occuring so early in life as to be a factor of second order importance. These observations seem to be supported by experimental evidence obtained for 7075-T7351 aluminum alloy (31). Based on these hypotheses, the "equivalent initial quality method" is suggested for determining an equivalent initial flaw size for fretting (32). The equivalent "initial flaw size" is the size of a hypothetical flaw that, if it existed at the beginning of the fatigue loading history, would result in the actual observed lifetime under fretting conditions, as calculated by the laws of fracture mechanics. In the work reported (31), using 7075-T7351 aluminum, the calculated equivalent initial flaw sizes averaged about 0.36 mm for the fretting fatigue conditions imposed. It was further observed that the equivalent initial flaw sizes did not appear to be significantly dependent upon stress ratio R, maximum stress σ_{max}, or the presence of interspersed tensile or compressive "overload" cycles. Based on these preliminary results, it would appear that fretting fatigue design procedures based on the fracture mechanics approach to crack growth may become viable tools in the future.

In the final analysis, it is necessary to evaluate the seriousness of fretting fatigue damage in any specific design by running simulated service tests on specimens or components. Within the current state-of-the-art knowledge in the area of fretting fatigue, there is no other safe course of action open to the designer.

14.4 FRETTING WEAR

Fretting wear is a change in dimensions through wear directly attributable to the fretting process between two mating surfaces. It is thought that the abrasive pit-digging mechanism, the asperity-contact microcrack initiation mechanism, and the wear-sheet delamination mechanism may all be important in most fretting wear failures. As in the case of fretting fatigue, there has been no good model developed to describe the fretting wear phenomenon in a way useful for design. An expression for weight loss due to fretting has be proposed (7) as

$$W_{total} = \left(k_0 L^{1/2} - k_1 L\right)\frac{C}{F} + k_2 SLC \qquad (14\text{-}2)$$

where W_{total} = total specimen weight loss
$\qquad L$ = normal contact load
$\qquad C$ = number of fretting cycles
$\qquad F$ = frequency of fretting
$\qquad S$ = peak-to-peak slip between fretting surfaces
$\quad k_0, k_1, k_2$ = constants to be empirically determined

This equation has been shown to give relatively good agreement with experimental data over a range of fretting condition using mild steel specimens (7). However, weight loss is not of direct use to a designer. Wear depth is of more interest. Prediction of wear depth in an actual design application must in general be based on simulated service testing. To illustrate how wear depth might be predicted, let us consider the problem of estimating the fretting wear depth at the contact sites between the supporting grids and fuel pin cladding in the core of a nuclear power reactor. To make such an estimate, information is required on the spectrum of amplitudes and frequencies generated in the operation of the reactor, as well as laboratory test data giving the depth of wear as a function of fretting amplitude and total number of cycles of fretting.

Figure 14.9 shows a plot of the response of a carefully instrumented reactor core element under simulated operating conditions. It may be noted that for this particular case there is only one important characteristic frequency at approximately 78 cycles per second (Hz). Figure 14.10 shows a plot of the cumulative distribution function for fretting amplitude under simulated operating conditions developed from carefully instrumented experiments. Knowing the characteristic frequency from Figure 14.9 and the probability of having a fretting amplitude in a given range from Figure 14.10, one may estimate the number of cycles of any given fretting amplitude range per hour or year of reactor operation. It then remains to relate this information to fretting wear depth.

Simple constant-amplitude fretting wear experiments under simulated reactor conditions may be run in the laboratory for a range of amplitudes and at the normal force of interest. The results of such a test series are shown in Figure 14.11 by plotting wear depth versus cycles of operation for several different fretting amplitudes.

Finally, to predict the total fretting wear depth after one year of actual operation, the cycles-of-amplitude estimates based on Figures 14.9 and 14.10 may be incorporated consecutively in Figure 14.11. This is done graphically by following along a given amplitude curve in Figure 14.11 for the estimated

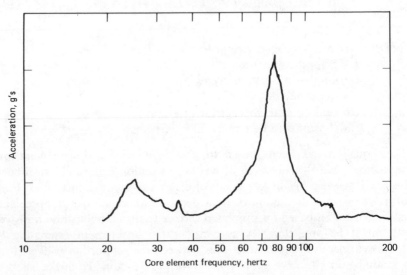

FIGURE 14.9. Response of core element under simulated reactor operating conditions.

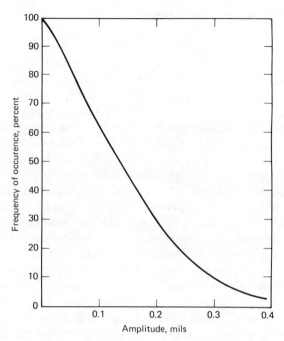

FIGURE 14.10. Cumulative distribution data showing probability of occurrence of fretting amplitude under simulated reactor operating conditions.

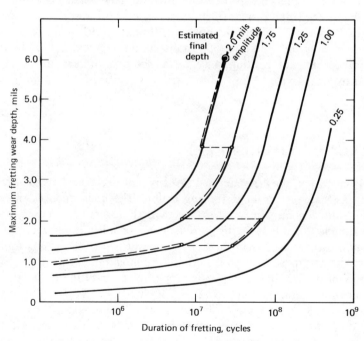

FIGURE 14.11. Wear depth as a function of number of cycles of fretting for several fretting amplitudes.

number of cycles at that amplitude during one year, then moving horizontally along a constant wear depth curve to the second amplitude curve. Next, the second amplitude curve is followed for the predicted number of cycles per year for that amplitude. The procedure is repeated, moving from curve to curve along lines of constant wear depth, as shown in Figure 14.11 by the dashed path line. When all amplitudes have been accounted for in this way, the total estimated wear depth may be read from Figure 14.11 at the terminus of the dashed curve. This general procedure may be used for virtually any fretting wear design prediction. It does require simulated service tests in the laboratory to develop basic data for the specific design application.

Some investigators have suggested that estimates of fretting wear depth may be based on the classical adhesive or abrasive wear equations, in which wear depth is proportional to load and total distance slid, where the total distance slid is calculated by multiplying relative motion per cycle times number of cycles. Although there are some supporting data for such a procedure (29), more investigation is required before it could be recommended as an acceptable approach for general application.

If fretting wear at a support interface, such as between tubes and support plates of a steam generator or heat exchanger or between fuel pins and support grids of a reactor core, produces loss of fit at a support site, impact fretting may occur. Impact fretting is fretting action induced by the small lateral relative displacements between two surfaces when they impact together, where the small displacements are caused by Poisson strains or small tangential "glancing" velocity components. Impact fretting has only recently been addressed in the literature (30), but it should be noted that under certain circumstances impact fretting may be a potential failure mode of great importance.

14.5   FRETTING CORROSION

Fretting corrosion may be defined as any corrosive surface involvement resulting as a direct result of fretting action. The consequences of fretting corrosion are generally much less severe than for either fretting wear or fretting fatigue. Note that the term *fretting corrosion* is not being used here as a synonym for fretting, as in much of the early literature on this topic, but rather as defined in Section 14.1. Perhaps the most important single parameter in minimizing fretting corrosion is proper selection of the material pair for the application. Table 14.1 lists a variety of material pairs grouped according to their resistance to fretting corrosion (18). Cross comparisons from one investigator's results to another must be made with care because testing conditions varied widely.

Table 14.1. Fretting Corrosion Resistance of Various Material Pairs (18)

Material Pairs Having Good Fretting Corrosion Resistance

Sakmann and Rightmire	Lead	on	Steel
	Silver plate	on	Steel
	Silver plate	on	Silver plate
	'Parco-lubrized' steel	on	Steel
Gray and Jenny	Grit blasted steel plus lead plate	on	Steel (very good)
	1/16 in. nylon insert	on	Steel (very good)
	Zinc and iron phosphated (Bonderizing) steel	on	Steel (good with thick coat)
McDowell	Laminated plastic	on	Gold plate
	Hard tool steel	on	Tool steel
	Cold-rolled steel	on	Cold-rolled steel
	Cast iron	on	Cast iron with phosphate coating
	Cast iron	on	Cast iron with rubber cement
	Cast iron	on	Cast iron with tungsten sulphide coating
	Cast iron	on	Cast iron with rubber insert
	Cast iron	on	Cast iron with Molykote lubricant
	Cast iron	on	Stainless steel with Molykote lubricant

Material Pairs Having Intermediate Fretting Corrosion Resistance

Sakmann and Rightmire	Cadmium	on	Steel
	Zinc	on	Steel
	Copper alloy	on	Steel
	Zinc	on	Aluminium
	Copper plate	on	Aluminium
	Nickel plate	on	Aluminium
	Silver plate	on	Aluminium
	Iron plate	on	Aluminium
Gray and Jenny	Sulphide coated bronze	on	Steel
	Cast bronze	on	"Parco-lubrized" steel
	Magnesium	on	"Parco-lubrized" steel
	Grit-blasted steel	on	Steel
McDowell	Cast iron	on	Cast iron (rough or smooth surface)
	Copper	on	Cast iron
	Brass	on	Cast iron
	Zinc	on	Cast iron
	Cast iron	on	Silver plate
	Cast iron	on	Copper plate
	Magnesium	on	Copper plate
	Zirconium	on	Zirconium

Table 14.1. (*Continued*)

Material Pairs Having Poor Fretting Corrosion Resistance

Sakmann and	Steel	on	Steel
Rightmire	Nickel	on	Steel
	Aluminium	on	Steel
	Al-Si alloy	on	Steel
	Antimony plate	on	Steel
	Tin	on	Steel
	Aluminium	on	Aluminium
	Zinc plate	on	Aluminium
Gray and Jenny	Grit blast plus silver plate	on	Steel*
	Steel	on	Steel
	Grit blast plus copper plate	on	Steel
	Grit blast plus tin plate	on	Steel
	Grit blast and aluminium foil	on	Steel
	Be-Cu insert	on	Steel
	Magnesium	on	Steel
	Nitrided steel	on	Chromium plated steel[†]
McDowell	Aluminium	on	Cast iron
	Aluminium	on	Stainless steel
	Magnesium	on	Cast iron
	Cast iron	on	Chromium plate
	Laminated plastic	on	Cast iron
	Bakelite	on	Cast iron
	Hard tool steel	on	Stainless steel
	Chromium plate	on	Chromium plate
	Cast iron	on	Tin plate
	Gold plate	on	Gold plate

*Possibly effective with light loads and thick (0.005 inch) silver plate.
[†]Some improvement by heating chromium plated steel to 538°C for 1 hour.

14.6 MINIMIZING OR PREVENTING FRETTING DAMAGE

The minimization or prevention of fretting damage must be carefully considered as a separate problem in each individual design application because a palliative in one application may significantly accelerate fretting damage in a different application. For example, in a joint that is designed to have no relative motion, it is sometimes possible to reduce or prevent fretting by increasing the normal pressure until all relative motion is arrested. However, if the increase in normal pressure does not *completely* arrest the relative motion, the result may be significantly increasing fretting damage instead of preventing it.

Nevertheless, there are several basic principles that are generally effective in minimizing or preventing fretting. These include:

1. Complete separation of the contacting surfaces.
2. Elimination of all relative motion between the contacting surfaces.
3. If relative motion cannot be eliminated, it is sometimes effective to superpose a large unidirectional relative motion that allows effective lubrication. For example, the practice of driving the inner or outer race of an oscillatory pivot bearing may be effective in eliminating fretting.
4. Providing compressive residual stresses at the fretting surface; this may be accomplished by shot-peening, cold-rolling, or interference fit techniques.
5. Judicious selection of material pairs.
6. Use of interposed low shear modulus shim material or plating, such as lead, rubber, or silver.
7. Use of surface treatments or coatings as solid lubricants.
8. Use of surface grooving or roughening to provide debris escape routes and differential strain matching through elastic action.

Of all these techniques, only the first two are completely effective in preventing fretting. The remaining concepts, however, may often be used to minimize fretting damage and yield an acceptable design.

QUESTIONS

1. Give a definition for *fretting* and distinguish among the failure phenomena of fretting fatigue, fretting wear, and fretting corrosion.

2. List the variables thought to be of primary importance in fretting-related failure phenomena.

3. Several hypotheses have been advanced to explain fretting action. These include pit-digging action, asperity-contact microcrack initiation mechanism, friction-generated cyclic stress fretting hypothesis, and the delamination theory of fretting. Briefly describe the physical basis for each of these theories, using appropriate sketches.

4. In a series of fretting fatigue tests, the conditions of 0.003-inch peak-to-peak relative motion and 10,000 psi nominal contact pressure are defined to be "severe" fretting for SAE 4340 steel on 4340 steel. This material has the fatigue properties shown in Figure 12.17*d*. If the data of Figure 14.8 are valid for the 4340 steel used, a design life 10^5 cycles is required, and a safety factor of 1.2 is desired, compute the required diameter for a support member of 4340 steel if the load is to fluctuate cyclically from 6000 lb compression to 2000 lb tension and if a contacting collar results in severe fretting of the support member during operation.

5.   In the analysis of flow-induced vibration data of heat exchanger tubes, the fretting amplitude behavior can be approximated by the accompanying chart. The characteristic frequency of the tube in the support plate is about 50 Hz. If Figure 14.11 represents pertinent constant-amplitude fretting wear data under a simulated heat exchanger environment for a similar material combination, estimate the fretting wear depth per year of operation, assuming operation is continuous.

Fretting Amplitude (mils)	Frequency of Occurrence (percent)
> 0.5	1.0
0.4–0.5	5.0
0.3–0.4	10.0
0.2–0.3	30.0
0.1–0.2	45.0
< 0.2	9.0

6.   Fretting corrosion has proved to be a problem in aircraft splines of steel on steel. Suggest one or more measures that might be taken to improve the resistance of the splined joint to fretting corrosion.

7.   List several basic principles that are generally effective in minimizing or preventing fretting.

REFERENCES

1.   Collins, J.A., "Fretting-Fatigue Damage-Factor Determination," *Journal of Engineering for Industry*, ASME Transactions 87, No. 8, Series B (August 1965): 298–302.

2.   Godfrey, D., "Investigation of Fretting by Microscopic Observation," NACA Report 1009, 1951 (formerly TN-2039, February 1950).

3.   Bowden, F.P., and Tabor, D., *The Friction and Lubrication of Solids*, Oxford University Press, Amen House, London, 1950.

4.   Godfrey, D., and Bailey, J.M., "Coefficient of Friction and Damage to Contact Area During the Early Stages of Fretting; I—Glass, Copper, or Steel Against Copper," NACA TN-3011, September 1953.

5.   Merchant, M.E., "The Mechanism of Static Friction," *Journal of Applied Physics*, 11, No. 3 (1940): 232.

6.   Bisson, E.E., Johnson, R.L., Swikert, M.A., and Godfrey, D., "Friction, Wear, and Surface Damage of Metals as Affected by Solid Surface Films," NACA TN-3444, May 1955.

7.   Uhlig, H.H., "Mechanism of Fretting Corrosion," *Journal of Applied Mechanics*, ASME Transactions, 76 (1954): 401–407.

8.   Feng, I.M., and Rightmire, B.G., "The Mechanism of Fretting," *Lubrication Engineering*, 9 (June 1953): 134ff.

9.   Feng, I.M., "Fundamental Study of the Mechansim of Fretting," Final Report, Lubrication Laboratory, Massachusetts Institute of Technology, Cambridge, 1955.

10.   Corten, H.T., "Factors Influencing Fretting Fatigue Strength," T. & A.M. Report No. 88, Department of Theoretical and Applied Mechanics, University of Illinois, Urbana, June 1955.

11. Starkey, W.L., Marco, S.M., and Collins, J.A., "Effects of Fretting on Fatigue Characteristics of Titanium-Steel and Steel-Steel Joints," ASME Paper 57-A-113, New York, 1957.

12. Milestone, W.D., "Fretting and Fretting-Fatigue in Metal-to-Metal Contacts," ASME Paper 71-DE-38, New York, 1971.

13. Collins, J.A., "A Study of the Phenomenon of Fretting-Fatigue with Emphasis on Stress-Field Effects," Dissertation, Ohio State University, Columbus, 1963.

14. Milestone, W.D., "An Investigation of the Basic Mechanism of Mechanical Fretting and Fretting-Fatigue at Metal-to-Metal Joints, with Emphasis on the Effects of Friction and Friction-Induced Stresses," Dissertation, Ohio State University, Columbus, 1966.

15. Collins, J.A., and Tovey, F.M., "Fretting Fatigue Mechanisms and the Effect of Direction of Fretting Motion on Fatigue Strength," *Journal of Materials*, American Society for Testing and Materials, 7, No. 4 (December 1972).

16. Collins, J.A., Smith, R.L., and Stormont, C.W., "Static and Cyclic Non-Zero Mean Stress Effects During Fretting on Subsequent Fatigue Strength" (to be published).

17. Collins, J.A., and Marco, S.M., "The Effect of Stress Direction During Fretting on Subsequent Fatigue Life," *ASTM Proceedings*, **64** (1964): 547.

18. Heywood, R.B., *Designing Against Fatigue of Metals*, Reinhold, New York, 1962.

19. Sakmann, B.W., and Rightmire, B.G., "An Investigation of Fretting Corrosion Under Several Conditions of Oxidation," NACA TN-1942, June 1948.

20. McDowell, J.R., "Fretting Corrosion Tendencies of Several Combinations of Metals," *ASTM Symposium on Fretting Corrosion*, STP-144, America Society for Testing and Materials, Philadelphia, 1952.

21. Gray, H.C., and Jenny, R.W., "An Investigation of Chafing on Aircraft Engine Parts," *SAE Journal*, 52 (November 1944).

22. Stowers, C.F., and Rabinowicz, E., "The Mechanism of Fretting Wear," Paper No. 72-Lub-20, ASME, 1972.

23. Harris, W.J., "The Influence of Fretting on Fatigue," AGARD Advisory Report No. 45, Harford House, London, June 1972.

24. Wright, G.P., and O'Connor, J.J., "The Influence of Fretting and Geometric Stress Concentrations on the Fatigue Strength of Clamped Joints," *Proceedings, Institution of Mechanical Engineers*, **186** (1972).

25. Waterhouse, R.B., *Fretting Corrosion*, Pergamon Press, New York, 1972.

26. "Fretting in Aircraft Systems," AGARD Conference Proceedings CP161, distributed through NASA, Langley Field, Va., 1974.

27. "Control of Fretting Fatigue," Report No. NMAB-333, National Academy of Science, National Materials Advisory Board, Washington, D.C., 1977.

28. Suh, N.P., Jahanmir, S., Fleming, J., and Abrahamson, E.P., "The Delamination Theory of Wear—II," Progress Report, Materials Processing Lab, Mechanical Engineering Dept., MIT Press, Cambridge, September 1975.

29. Lyons, H., "An Investigation of the Phenomenon of Fretting-Wear and Attendant Parametric Effects Towards Development of Failure Prediction Criteria," Ph.D. Dissertation, Ohio State University, Columbus, 1978.

30. Ko, P.L., "Experimental Studies of Tube Fretting in Steam Generators and Heat Exchangers," ASME/CSME Pressure Vessels and Piping Conference, Nuclear and Materials Division, Montreal, Canada, June 1978.

31. Alic, J.A., "Fretting Fatigue in the Presence of Periodic High Tensile or Compressive Loads," Final Scientific Report, Grant No. AFOSR77-3422, Wright Patterson AFB, Ohio, April 1979.

32. Rudd, J. L., and Gray, T.D., "Equivalent Initial Quality Method, "Air Force Flight Dynamics Lab Technical Memo No. AFFDL-TM-76-83-FBE, 1976.

CHAPTER 15

Shock and Impact

15.1 INTRODUCTION

When forces or displacements are rapidly applied to a machine member, it is often found that the stress levels and deformations induced are very much larger than would be generated by the same forces or displacements applied very gradually. Such rapidly applied loads or displacements are usually called *shock* or *impact loads*. Shock or impact loading may be generated in a machine or structure either by the collision of moving bodies or simply by the sudden application of a force or motion to the structure. Whether the loading on a structure should be considered as *quasi-static* or impact loading is often judged by comparing the time of application of the load, or *rise time*, with the longest natural period of the structure. If the rise time is less than about one-half of the longest natural period, it is necessary to consider the loading as impact or shock. If the rise time is more than about three times the longest natural period, the loading may be considered to be quasi-static. For the case of shock loading, a designer must be concerned with not only the magnitude of the load, but also the time during which it rises from its initial to its final value and the *impulse*, which is the area under the force versus time curve. In contrast, for quasi-static loading the maximum value of the load is usually the only parameter of interest to the designer.

Shock or impact loads are generated in machines and structures in a variety of ways. For example, a *rapily moving load*, such as a train moving across a bridge, generates shock loads. *Direct impact loads*, such as the drop of a forging hammer, or *suddenly applied loads*, such as produced during combustion in the power stroke of an internal combustion engine, produce shock loads. *Inertial loads* produced by large accelerations, such as during the crashing of an aircraft or automobile, in many instances generate conditions of shock or impact. Although shock or impact loading is usually considered as a one-time-of-application event, there are some cases, such as in railroad car couplers or forging hammer components, in which the impact loads are repeated many times to produce failure by *impact fatigue*. If the repeated impacts between two surfaces produce local elastic cyclic deformations, and if the impact-induced cyclic deformations result in a field of subsurface micro-cracks that grow and coalesce in accordance with the surface fatigue behavior described in Chapter 3, the phenomenon is called *impact wear*.

500

When impact takes place at the interface of a constraining structure, such as between a support plate and the tubes of a heat exchanger, the impact forces may induce fretting due to the small, lateral relative displacements between the two surfaces when they impact together. These small displacements, caused by Poisson strains or small tangential "glancing" velocity components, give rise to the possibility of failure by *impact fretting*.

It should be observed not only that shock or impact loads result in states of stress that are significantly more serious than for quasi-static loading, but also that the material properties may be significantly influenced by rapidly applied loads or displacements. As will be discussed in Section 15.11, ultimate strength, yield strength, and ductility may all be significantly different under impact loading conditions than for static or quasi-static loading.

15.2 ENERGY METHOD OF APPROXIMATING STRESS AND DEFLECTION UNDER IMPACT LOADING CONDITIONS

For simple machine members the maximum stress or deflection under impact conditions may be approximated by utilizing the concept of conservation of energy, wherein the external work done on a structure must be equal to the potential energy of strain stored in the structure, if losses are assumed to be negligible. To utilize this energy method, one proceeds by equating the external work to the stored strain energy, putting the energy expressions in terms of stress or deflection, and solving for the stress or deflection.

Consider, for example, the simple tension member shown in Figure 15.1. The mass of weight W is allowed to fall through height h before it contacts the pan at the end of the tension bar. The resistance offered by the bar brings the weight W to a halt, and in the process it stretches the bar a distance y as shown, thereby storing strain energy in the bar. In using the energy method to estimate the maximum stress in the bar due to this impact loading, it will be assumed that:

1. The inertial resistance of tension bar and weight pan is negligible; that is, the mass of the bar and pan is much smaller than the striking mass.
2. The deflection of the bar is directly proportional to the applied force and is not a function of time.
3. The material obeys Hooke's law, that is, remains in the linear elastic range.
4. No energy is lost in the impact.

Under these assumptions the maximum stress may be estimated by the energy method described earlier. The external energy EE may be calculated as the change in potential energy of the striking mass during its fall, whence

$$EE = W(h + y_{\max}) \qquad (15\text{-}1)$$

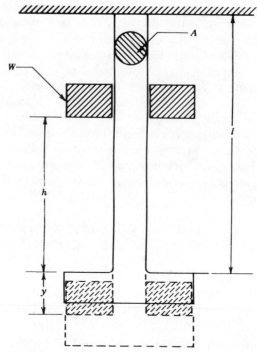

FIGURE 15.1. Simple tension member subjected to impact loading due to a falling weight.

Since the bar is assumed to remain in the linear elastic range, Hooke's law states that

$$\sigma_{max} = E\varepsilon = \frac{Ey_{max}}{l} \qquad (15\text{-}2)$$

from which the end deflection y may be expressed in terms of stress σ, length l, and modulus of elasticity E as

$$y_{max} = \frac{\sigma_{max} l}{E} \qquad (15\text{-}3)$$

Utilizing this expression for the deflection, we may write (15-1) as

$$EE = W\left(h + \frac{\sigma_{max} l}{E}\right) \qquad (15\text{-}4)$$

The strain energy SE stored in the bar at the time of maximum deflection y may be expressed as the product of the average force applied times the deflection, or

$$SE = F_{ave} y_{max} = \left(\frac{0 + F_{max}}{2}\right) y_{max} \qquad (15\text{-}5)$$

However, for this simple tension bar of cross-sectional area A

$$F_{max} = \sigma_{max} A \tag{15-6}$$

which may be substituted into (15-5), together with (15-3), to give

$$SE = \left(\frac{\sigma_{max} A}{2}\right)\left(\frac{\sigma_{max} l}{E}\right) = \frac{\sigma_{max}^2}{2E}(Al) \tag{15-7}$$

Equating the external energy expression of (15-4) to the strain energy expression of (15-7) then gives

$$W\left(h + \frac{\sigma_{max} l}{E}\right) = \frac{\sigma_{max}^2}{2E}(Al) \tag{15-8}$$

or

$$\sigma_{max}^2\left(\frac{Al}{2E}\right) - \sigma_{max}\left(\frac{Wl}{E}\right) - Wh = 0 \tag{15-9}$$

Dividing equation (15-9) by $(Al/2E)$ gives

$$\sigma_{max}^2 - \left(\frac{2W}{A}\right)\sigma_{max} - \frac{2WhE}{Al} = 0 \tag{15-10}$$

which may be solved by the quadratic formula to give a maximum value of

$$\sigma_{max} = \frac{W}{A}\left[1 + \sqrt{1 + \frac{2hEA}{Wl}}\right] \tag{15-11}$$

This is an energy-method estimate of the maximum stress that will be developed in the tension bar due to impact of weight W falling from rest through height h. A similar expression for the maximum end deflection y_{max} may be written by combining (15-11) with (15-3) to give

$$y_{max} = \frac{Wl}{AE}\left[1 + \sqrt{1 + \frac{2hEA}{Wl}}\right] \tag{15-12}$$

It is interesting to note in (15-11) that under impact loading the maximum stress may be reduced not only by increasing the cross-sectional area A but also by decreasing the modulus of elasticity E or increasing the length l of the bar. Thus, it is apparent that the impact situation is quite different from the case of static loading in which the stress in the bar is independent of the modulus of elasticity and the length of the bar. The bracketed expression of (15-11) or (15-12) is called the *impact factor*.

It is also interesting to consider the limiting case of (15-11) in which the drop height h is zero. This case, in which the striking mass is held just in contact with the weight pan and then released from zero height, is called a *suddenly applied load*. From (15-11), with $h = 0$, the stress developed in a bar subjected to a suddenly applied load is

$$\left(\sigma_{max}\right)_{\substack{\text{suddenly}\\\text{applied}}} = 2\frac{W}{A} = 2\left(\sigma_{max}\right)_{\text{static}} \tag{15-13}$$

Thus, the maximum stress developed by suddenly applying load W to the end of the bar is twice as large as the maximum stress developed in the bar by slowly applying the same load W. Similarly, from (15-12), the maximum deflection generated by a suddenly applied load is twice the deflection produced under static loading, whence

$$(y_{max})_{\substack{suddenly \\ applied}} = 2(y_{max})_{static} \qquad (15\text{-}14)$$

If the weight of the bar shown in Figure 15.1 is not negligibly small, as was assumed, the expressions of (15-11) and (15-12) are somewhat modified to account for the fact that some of the energy of the falling weight is used to accelerate the mass of the bar. If the bar is assumed to have a weight of q per unit length of the bar, then (15-11) and (15-12) are modified to

$$\sigma_{max} = \frac{W}{A}\left[1 + \sqrt{1 + \frac{2hEA}{Wl}\left(\frac{1}{1 + \dfrac{ql}{3W}}\right)}\,\right] \qquad (15\text{-}15)$$

and

$$y_{max} = \frac{Wl}{AE}\left[1 + \sqrt{1 + \frac{2hEA}{Wl}\left(\frac{1}{1 + \dfrac{ql}{3W}}\right)}\,\right] \qquad (15\text{-}16)$$

If the stress level generated by the impact exceeds the yield point of the material, Hookes law no longer holds and the equations just developed are not valid. However, it is still possible to estimate stress and deflection under such conditions if a stress versus strain diagram is available for the material. For a bar of cross-sectional area A and length l the stress-strain diagram may be converted to a curve of force P versus deflection y, as shown in Figure 15.2a. Assuming that the material properties remain constant, the force-deflection diagram of Figure 15.2a may be used to determine the elongation of the bar even though the impact stress exceeds the elastic limit. To do this it may be observed that for any arbitrarily chosen elongation y_i an area $OADF$ under the force-deflection curve is defined. This area is the strain energy $(SE)_{y_i}$ required to produce the elongation y_i. Further, this amount of strain energy must be equal to the external energy of the falling mass, or

$$(SE)_{y_i} = W(h_i + y_i) \qquad (15\text{-}17)$$

The area $OADF$ may be determined from the force-deflection diagram for any assumed y_i so that the only unknown in (15-17) for a specified weight W_1 is h_i. Solving for h_i, it may be plotted versus y_i, as shown in Figure 15.2b, and by assuming a range of values for y_i, a whole curve may be plotted for weight W_1. Other similar curves may be obtained for other weights, as shown. Care must be taken to note that when the external energy $W(h + y)$ exceeds the

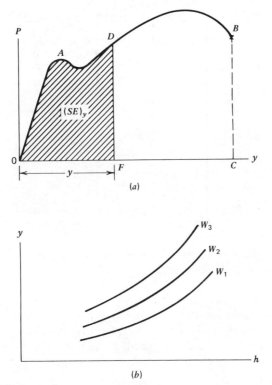

FIGURE 15.2. Illustration of method for estimating deflection under impact loading when the impact stresses exceed the yield point of the material. (*a*) Force-deflection curve. (*b*) Deflection as a function of drop height for a family of weights.

area $OABC$ in Figure 15.2a, the falling weight will fracture the bar. With the result of Figure 15.2b, it is then a simple matter to read directly the deflection y produced by any weight W falling through a specified height h.

It follows from the preceding discussion that any change in the form, shape, or properties of the bar that results in a reduction of the total area $OABC$ under the force-deflection diagram will adversely affect the ability of the bar to withstand shock or impact loads. To further illustrate this observation, the two specimens shown in Figure 15.3 may be compared under static and impact loading. Note that both specimens have the same length, have the same minimum cross-sectional area, and are made of the same material. The effects of stress concentration will be neglected for this comparison.

Under conditions of static loading, it may be observed that the load P_{fy} required to produce first yielding is for *both* bars in Figure 15.3

$$P_{fy} = A_1\sigma_{yp} \tag{15-18}$$

where σ_{yp} is the yield point strength for the material.

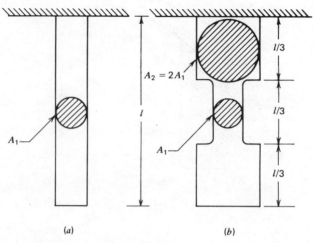

FIGURE 15.3. Configuration of two specimens used in comparing the effects of geometry under static and impact loading.

Next, the total strain energy stored in each of the bars may be calculated at the time of first yielding. This total strain energy stored is, of course, also equal to the external energy required to initiate yielding in each case. Utilizing (15-7), we may write the total strain energy U_a stored in the bar of Figure 15.3a as

$$U_a = \frac{\sigma_{yp}^2}{2E}(A_1 l) \qquad (15\text{-}19)$$

In calculating the total strain energy U_b stored in the bar of Figure 15.3b, it must be recognized that the stress level in the two end segments will be lower than the stress level in the central segment by the ratio A_1/A_2. Thus, again utilizing (15-7) in piecewise fashion,

$$U_b = \frac{\sigma_{yp}^2}{2E}A_1\left(\frac{l}{3}\right) + 2\left\{\frac{\left[\sigma_{yp}\left(\frac{A_1}{A_2}\right)\right]^2}{2E}A_2\left(\frac{l}{3}\right)\right\} \qquad (15\text{-}20)$$

Since the area A_2 is given in Figure 15.3 to be twice A_1, (15-20) becomes

$$U_b = \frac{\sigma_{yp}^2}{2E}\left(\frac{A_1 l}{3}\right) + \frac{\sigma_{yp}^2}{8E}\left(\frac{4A_1 l}{3}\right) \qquad (15\text{-}21)$$

or

$$U_b = \frac{2}{3}\left[\frac{\sigma_{yp}^2}{2E}(A_1 l)\right] \qquad (15\text{-}22)$$

or

$$U_b = \frac{2}{3} U_a \qquad (15\text{-}23)$$

From this result it is evident that the energy stored in the specimen of Figure 15.3*b* when yielding begins is only two-thirds of the energy stored in the specimen of Figure 15.3*a* when yielding begins, in spite of the larger total volume in the specimen of Figure 15.3*b*. In other words, the impact resistance of design *a*, with its smaller total volume, is significantly superior to the impact resistance of design *b* in Figure 15.3. Furthermore, if stress concentration effects had been considered, design *b* would have been even worse by comparison.

The key to success in designing members for maximum resistance to impact loading is to assure that the maximum stress is as uniformly distributed as possible throughout the largest possible volume of material. Grooved and notched members are very poor in impact resistance because for these configurations there always exist small, highly stressed volumes of material. A relatively minor impact load may in such cases produce fracture. Thus, members that contain holes, fillets, notches, or grooves may be subject to abrupt fracture under impact or shock loading.

15.3 STRESS WAVE PROPAGATION UNDER IMPACT LOADING CONDITIONS

Although the energy method just described provides a means for estimating the maximum stress and deflection under impact loading conditions, it does not provide a good physical model of the actual behavior of the member. A more exact consideration of the stresses and strains generated within a simple member under impact loading may be made by considering the uniform elastic bar in Figure 15.4*a* subjected to a suddenly applied axial force *P* distributed uniformly over the free end of the bar. At the instant of application of force *P* at the free end of the bar, a very thin segment of the bar directly under the load *P* is set into motion; the remainder of the bar, remote from the loading, remains undisturbed for some finite length of time. As time passes, the deformation produced directly under *P* propagates along the bar in the axial direction in the form of an elastic deformation wave. Behind the wave front the bar is deformed and the particles are in motion. Ahead of the wave front the bar remains undisturbed and at rest. If the length of the bar is relatively large, the time required for the elastic deformation wave front to propagate its full length becomes significant and must be considered.

The governing equation for one-dimensional wave propagation in an elastic medium may be developed by considering a small slice of the struck bar as shown in Figure 15.4*b*. If the displacement in the *x* direction is denoted by *u*, the axial strain ε in the elemental slice may be computed by dividing the

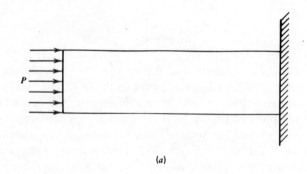

(a)

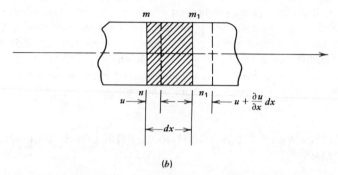

(b)

FIGURE 15.4.   Uniform elastic bar subjected to an impact load P at the free end. (a) Bar subjected to impact loading. (b) Element of bar subjected to impact loading.

change in length of the element by its original length dx. Thus,

$$\varepsilon = \frac{u + \dfrac{\partial u}{\partial x}dx - u}{dx} = \frac{\partial u}{\partial x} \tag{15-24}$$

Utilizing Hooke's law and the expression for direct stress at plane m-n of Figure 15.4b, we find that

$$\frac{P_{mn}}{A} = \sigma = E\varepsilon \tag{15-25}$$

whence

$$P_{mn} = EA\varepsilon \tag{15-26}$$

This expression may be combined with (15-24) to yield

$$P_{mn} = EA\frac{\partial u}{\partial x} \tag{15-27}$$

where it has been tacitly assumed that there is a simple direct stress in the x

direction accompanied by a corresponding lateral strain of $v(\partial u/\partial x)$, and the inertia effects of particle motion in the lateral direction have been neglected. This assumption is valid as long as the wavelength of the axial wave is large compared to the cross section of the bar. The magnitudes of the force $P_{m_1 n_1}$ at plane $m_1 - n_1$ of the bar may be written as

$$P_{m_1 n_1} = P_{mn} + \frac{\partial}{\partial x}(P_{mn})dx \qquad (15\text{-}28)$$

or

$$P_{m_1 n_1} = AE\frac{\partial u}{\partial x} + AE\frac{\partial}{\partial x}\left(\frac{\partial u}{\partial x}\right)dx \qquad (15\text{-}29)$$

The resultant force acting on the slice $m - m_1 - n_1 - n$ therefore is

$$F = P_{m_1 n_1} - P_{mn} \qquad (15\text{-}30)$$

or

$$F = AE\left[\frac{\partial u}{\partial x} + \frac{\partial^2 u}{\partial x^2}dx\right] - AE\frac{\partial u}{\partial x} \qquad (15\text{-}31)$$

or

$$F = AE\frac{\partial^2 u}{\partial x^2}dx \qquad (15\text{-}32)$$

The equation of motion for the small element may then be written from

$$F = ma \qquad (15\text{-}33)$$

as

$$AE\frac{\partial^2 u}{\partial x^2}dx = (\rho A dx)\frac{\partial^2 u}{\partial t^2} \qquad (15\text{-}34)$$

which may be reduced to

$$\frac{\partial^2 u}{\partial t^2} = \frac{E}{\rho}\frac{\partial^2 u}{\partial x^2} \qquad (15\text{-}35)$$

where u is displacement in the axial direction, t is time, E is modulus of elasticity, and ρ is the density of the material in the bar. This is the one-dimensional wave equation for propagation of a plane wave in an elastic medium.

It can be shown by substitution that any function $f(x + \sqrt{E/\rho}\ t)$ is a solution to (15-35), as is any function $f_1(x - \sqrt{E/\rho}\ t)$. Therefore, the general solution for the differential equation of motion given in (15-35) may be written as

$$u(x,t) = f\left(x + \sqrt{E/\rho}\ t\right) + f_1\left(x - \sqrt{E/\rho}\ t\right) \qquad (15\text{-}36)$$

This general solution may be physically interpreted in the following way: It may be noted that for any specified time $t = t_0$ the function f_1 in (15-36) is a

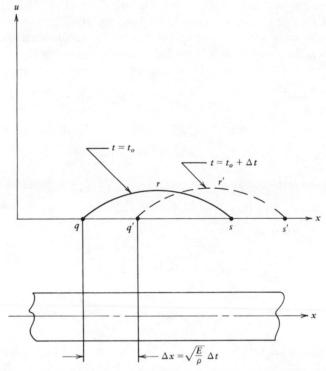

FIGURE 15.5. Graphical interpretation of the general solution of the one-dimensional wave equation.

function of x only and may be represented by some curve, such as qrs in Figure 15.5, the shape of which depends on f_1. After a time interval Δt has passed, the argument of f_1 becomes $x - \sqrt{E/\rho}(t_0 + \Delta t)$. If the abscissa is simultaneously shifted by a distance $\Delta x = \sqrt{E/\rho}\,\Delta t$, the argument and the function remain unchanged. This is indicated in Figure 15.5 where the unchanged wave shape has been translated a distance Δx along the x axis to the location $q'r's'$, with no change in shape. It may be surmised therefore that the f_1 term of (15-36) represents a displacement wave traveling in the x direction with a constant velocity c of

$$c = \frac{\Delta x}{\Delta t} = \sqrt{\frac{E}{\rho}} \tag{15-37}$$

Similarly, the f term in (15-36) represents a displacement wave traveling in the opposite direction with the same propagation velocity c. Therefore, the general solution given in (15-36) represents two displacement waves traveling along the bar in opposite directions, each with a constant wave propagation velocity of $c = \sqrt{E/\rho}$.

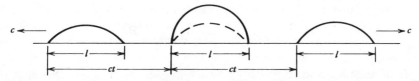

FIGURE 15.6. Splitting of displacement wave into two propagating waves for the case of zero initial velocity.

From (15-37) it may be noted that the wave propagation velocity c is a function of material only. For example, for steel the wave propagation velocity would be calculated as

$$c = \sqrt{\frac{30 \times 10^6 \text{lb/in.}^2 \times 144\text{in.}^2/\text{ft}^2}{490\text{lb/ft}^3 \times \dfrac{1}{32.2}\text{sec}^2/\text{ft}}} \tag{15-38}$$

or

$$c = 16{,}850 \text{ ft/sec} \tag{15-39}$$

The functions f and f_1 are determined in each particular case by the initial conditions at time $t = 0$. The initial conditions may be stated as

$$u_{t=0} = f(x) + f_1(x) \tag{15-40}$$

$$\left(\frac{\partial u}{\partial t}\right)_{t=0} = c\left[\,f'(x) - f_1'(x)\right] \tag{15-41}$$

For the particular case of a bar in which the initial velocity is everywhere zero and the initial displacement is given by $u = F(x)$, for example, the initial condition becomes

$$u_{t=0} = F(x) \tag{15-42}$$

and

$$\left(\frac{\partial u}{\partial t}\right)_{t=0} = 0 \tag{15-43}$$

For this case the initial conditions are satisfied by taking

$$f(x) = f_1(x) = 1/2 F(x) \tag{15-44}$$

The physical interpretation of this case is that the initial displacement splits into two equal parts, the two parts being propagated as two waves traveling in opposite directions along the bar, as shown in Figure 15.6.

15.4 PARTICLE VELOCITY AND WAVE PROPAGATION VELOCITY

Note again that the bar of Figure 15.4a is fixed at one end and subjected to an axial compressive load P applied suddenly to the free end. The load P is uniformly distributed over the cross-sectional area A of the bar. When the load P is first applied, it produces a compressive stress of $\sigma = P/A$ in a very thin layer at the struck end of the bar; the remainder of the bar remains unstressed. The compressive stress is then transmitted to the next adjacent layer of material, and so on, down the length of the bar. That is, a *wave* of compression travels along the bar. Behind the *wave front* the bar is stressed to $\sigma = P/A$, whereas ahead of the wave front the stress remains at zero. The velocity with which the wave front propagates along the bar is called the *wave propagation velocity*, c. Thus, as shown in Figure 15.7, after some finite time t a portion of the bar of length $l_c = ct$ will be compressed while the remainder of the bar remains unstressed.

It should also be noted that when the load P is suddenly applied to the end of the bar, the particles in the thin affected layer will be set into motion with a *particle velocity*, v, in the direction of the applied force. It is important to recognize that the particle velocity v is different and distinct from the wave propagation velocity c. To illustrate, it may be observed that instead of the suddenly applied compressive load shown in Figure 15.4a one could have applied a sudden tensile load. In so doing the propagation of a tensile wave front would be initiated and the tensile wave front would propagate along the bar toward the fixed end with velocity c, just as the compressive wave does. However, the particle velocity in the stressed zone for the tensile case is directed away from the fixed end instead of toward the fixed end as in the compressive case. Thus, one may observe that for the case of a compressive stress wave the particle velocity v and wave propagation velocity c are in the same direction, whereas for a tensile stress wave the particle velocity is in the opposite direction to the wave propagation velocity.

An expression may be obtained for the particle velocity v in the compressed zone as follows: Utilizing Hooke's law,

$$\frac{\sigma}{E} = \varepsilon = \frac{\Delta l_c}{l_c} \tag{15-45}$$

where l_c is the length of the compressed zone and Δl_c is the change in length of the compressed zone due to the applied compressive stress σ. Since l_c has

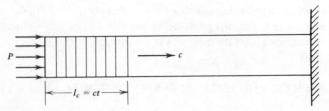

FIGURE 15.7. Bar subjected to impact loading, showing stressed zone and unstressed zone at time t.

already been defined to be equal to ct, (15-45) may be rewritten as

$$\Delta l_c = \left(\frac{\sigma}{E}\right)l_c = \frac{\sigma}{E}(ct) \tag{15-46}$$

The velocity of the struck end of the bar, which is equal to the velocity of the particles in the compressed zone, is therefore

$$v = \frac{\Delta l_c}{t} = \frac{1}{t}\left(\frac{\sigma}{E}\right)ct \tag{15-47}$$

whence

$$v = \frac{c\sigma}{E} \tag{15-48}$$

If the problem is now viewed in a different way, the earlier expression for wave propagation velocity c may be confirmed through a consideration of impulse and momentum. To do so, it may be observed that at the start all particles in the bar, including those in the compressed length l_c, were at rest. Following the impact, the particles in the compressed zone have acquired the velocity v. Thus, the change in momentum may be calculated as

$$\Delta \mathfrak{M} = Mv \tag{15-49}$$

or

$$\Delta \mathfrak{M} = (\rho Act)v \tag{15-50}$$

The impulse of the applied compressive force may be written as

$$\mathfrak{I} = Pt = \sigma At \tag{15-51}$$

Equating the impulse to the change in momentum gives

$$\sigma At = \rho Actv \tag{15-52}$$

or

$$v = \frac{\sigma}{\rho c} \tag{15-53}$$

If (15-53) is then equated to (15-48), the result is

$$\frac{c\sigma}{E} = \frac{\sigma}{\rho c} \tag{15-54}$$

or

$$c = \sqrt{\frac{E}{\rho}} \tag{15-55}$$

which confirms the earlier result obtained in (15-37).

Further, the expression for particle velocity in (15-53) may also be rewritten, by utilizing (15-55), as

$$v = \frac{\sigma}{\sqrt{E\rho}} \tag{15-56}$$

It may be observed from these expressions that the wave front velocity c is independent of the applied compressive stress, whereas the particle velocity v is proportional to the compressive stress σ. To view it another way, the expressions (15-48) and (15-56) may be rewritten to give

$$\sigma = E\left(\frac{v}{c}\right) \tag{15-57}$$

or

$$\sigma = v\sqrt{E\rho} \tag{15-58}$$

It may be observed, then, that the magnitude of the compressive stress in the wave is determined by the modulus of elasticity of the material and the ratio of particle velocity to wave propagation velocity. If an absolutely rigid body moving with velocity v squarely strikes the end of a bar in the axial direction, the compressive stress at the surface of contact would be given by (15-57) or (15-58). If the velocity of the striking mass were above a certain limit determined by the yield strength, modulus of elasticity, and density of the bar, a local plastic deformation would take place. This plastic deformation would be induced even if the mass of the striking body were very small.

The energy contained in the compressive wave as it propagates consists of two parts: potential energy of strain and kinetic energy of the moving particles.

For the linear elastic range an expression for strain energy SE may be written as

$$SE = F_{ave}\Delta l_c = \frac{F_{max}\Delta l_c}{2} \tag{15-59}$$

which may be rewritten using (15-46) as

$$SE = \left[\frac{\sigma A}{2}\right]\left[\left(\frac{\sigma}{E}\right)ct\right] \tag{15-60}$$

or

$$SE = \frac{Act\sigma^2}{2E} \tag{15-61}$$

An expression may also be written for the kinetic energy KE in the wave as

$$KE = \frac{1}{2}Mv^2 \tag{15-62}$$

or

$$KE = \frac{1}{2}[\rho Act]\left[\frac{\sigma}{\sqrt{E\rho}}\right]^2 \tag{15-63}$$

which becomes

$$KE = \frac{Act\sigma^2}{2E} \tag{15-64}$$

Thus, the total energy TE associated with the wave is

$$TE = SE + KE = \frac{1}{2}\left(\frac{Act\sigma^2}{E}\right) + \frac{1}{2}\left(\frac{Act\sigma^2}{E}\right) \qquad (15\text{-}65)$$

and the energy in the propagating wave is one-half potential energy of strain and one-half kinetic energy of the moving mass. If it is assumed that no energy losses of any kind occur in the impact or the propagation, the total energy TE in the wave should be equal to the external energy EE imparted by the suddenly applied force; that is,

$$EE = TE \qquad (15\text{-}66)$$

To verify that this is true, the external energy, or work done by the compressive force $A\sigma$ as it moves through the distance Δl_c, may be written as

$$EE = (A\sigma)\left(\frac{\sigma}{E}\right)ct = \frac{Act\sigma^2}{E} \qquad (15\text{-}67)$$

Comparing (15-67) with (15-65) we note that (15-66) is indeed verified.

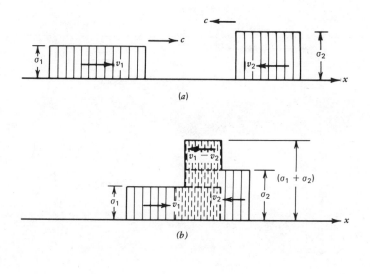

(a)

(b)

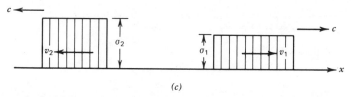

(c)

FIGURE 15.8. Interactions between two compressive waves propagating along a bar in opposite directions.

15.5   STRESS WAVE BEHAVIOR AT FREE AND FIXED ENDS

Certain important features of the governing differential equation (15-35) may be noted. This differential equation is linear, which means that if two solutions are found, their sum will also be a solution. That is, the principle of superposition holds. With this concept in mind, a few simple cases of wave propagation along a bar may be examined.

Consider first the case of two compressive waves traveling along a bar toward each other with wave front propagation velocity c, as illustrated in Figure 15.8. If both waves are compressive, the resultant compressive stress in the overlap zone is obtained directly by adding the component magnitude σ_1 and σ_2. The resultant particle velocity in the overlap zone is obtained by vector addition of the component particle velocities v_1 and v_2. Thus, the magnitude of the particle velocity in the overlap zone is $|v_1 - v_2|$ and the

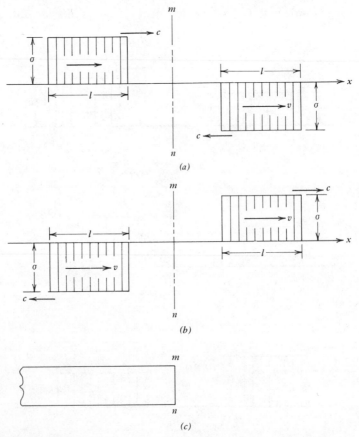

FIGURE 15.9.   Interaction between a compression and tension wave of same magnitude and length propagating along a bar in opposite directions.

direction is that of v_2 since σ_2 is larger than σ_1 and the particle velocity is proportional to stress. Figure 15.8a shows the two compressive waves as they approach each other, Figure 15.8b shows the overlapping superposition of the waves as they pass, and Figure 15.8c shows the waves proceeding undisturbed after they have completely passed each other.

Next, assume that a compression wave and a tension wave of the same magnitude and length are moving toward each other along a bar, as shown in Figure 15.9a. When the two waves interact at plane $m - n$, the magnitude of the stress is zero and the particle velocity is $2v$. After the waves pass, they return to their original configuration, as shown in 15.9b. It may be noted that at the plane $m - n$ before, during, and after the interaction, the stress is always zero. This is characteristic of the boundary condition at the free end

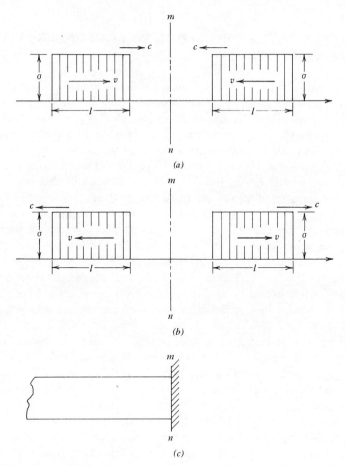

FIGURE 15.10. Interaction between two compression waves of same magnitude and length propagating along a bar in opposite directions.

of a bar, as shown in Figure 15.9c. It may be deduced from Figure 15.9, then, that at the free end of a bar a compressive wave is reflected as a similar tension wave, and a tension wave is reflected as a similar compressive wave.

Likewise, two identical compressive waves moving toward each other along a bar may be considered, as shown in Figure 15.10a. When these two waves interact at plane $m - n$, the stress at $m - n$ is doubled and the particle velocity is zero. After the waves pass, they maintain their original configuration, as shown in Figure 15.10b. It may be noted that at the plane $m - n$ before, during, and after the interaction, the particle velocity is always zero. This is characteristic of the boundary condition at the fixed end of a bar, as shown in Figure 15.10c. It may be deduced from Figure 15.10, then, that at the fixed end of a bar a compression wave is reflected unchanged as a compression wave, and a tension wave is reflected unchanged as a tension wave.

15.6 STRESS WAVE PROPAGATION IN A BAR UNDER SUDDENLY APPLIED AXIAL FORCE

The bar shown in Figure 15.4a, which is fixed at one end and subjected to a suddenly applied axial force at the free end, may be analyzed in terms of stress wave propagation, as shown in Figure 15.11. When the force is first applied, it initiates a compressive stress wave of magnitude $-\sigma$, which propagates along the bar with wave front propagation velocity c toward the fixed end, as shown in Figure 15.11a. The particle velocity in the compressed zone is v toward the fixed end. The event continues in this way until the compressive wave front strikes the fixed end at time $t = l/c$ when it reflects from the fixed end as a similar compressive wave of magnitude $-\sigma$ traveling back toward the free end, as shown in Figure 15.11c. Thus, in the zone behind the reflected wave front the stress level is -2σ and the particle velocity is zero. When the reflected wave front reaches the free end at time $t = 2l/c$, it is reflected from the free end as a tension wave of magnitude σ leaving a net compressive stress of $-\sigma$ in its wake as it travels again toward the fixed end, as shown in Figure 15.11e. Behind the wave front the particle velocity is v away from the fixed end. When the tension wave front reaches the fixed end at $t = 3l/c$, it is reflected as a similar tension wave, as shown in Figure 15.11g. At this time the stress magnitude and particle velocity behind the wave front are both zero. When the wave front again reaches the free end at $t = 4l/c$, the whole bar is unstressed and at rest, just as it was prior to the impact. Since the tensile wave front next reflects as a compression wave of magnitude $-\sigma$ propagating toward the fixed end, the whole cycle is ready to repeat, which it does indefinitely since we have assumed no internal friction damping or energy dissipation in the material. This assumption is, of course, incorrect for any real material.

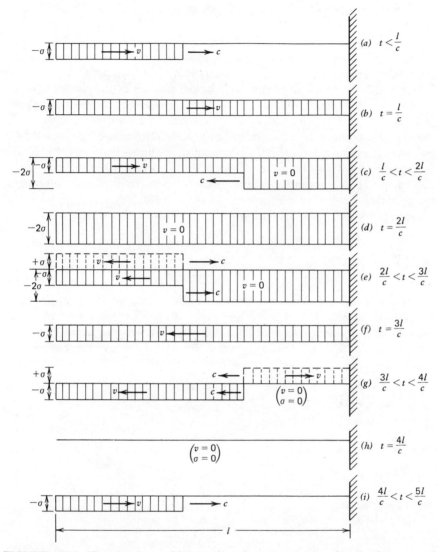

FIGURE 15.11. Propagation and interaction of stress waves in a bar fixed at one end and subjected to a suddenly applied axial load at the free end.

15.7   STRESS WAVE ATTENUATION DUE TO HYSTERETIC DAMPING

In any real material there will be a phase difference between the stress and the strain because of hysteresis or internal friction damping. As shown in Figure 15.12, this phase difference may be represented by the phase angle ϕ.

In view of the phase difference between the $\sigma - t$ and $\varepsilon - t$ plots, as shown in Figure 15.12, consideration of the stress-strain ($\sigma - \varepsilon$) curve under cyclic loading reveals that the curve does not exactly retrace itself as the stress is cycled, but rather a *hysteresis loop* develops, as shown in Figure 15.13. The area enclosed within this hysteresis loop for one cycle represents the energy dissipated per unit volume per cycle because of internal friction damping or hysteresis.

An expression for the area within the loop may be developed as

$$\Delta E = \int_{\text{one cycle}} \sigma(t) d\varepsilon(t) = \int_{t=0}^{t=2\pi/\omega} \sigma_0 \cos \omega t \left[-\varepsilon_0 \omega \sin (\omega t - \phi) \right] dt$$

$$(15\text{-}68)$$

where the stress and strain are assumed to be cosine waves of amplitude σ_0 and ε_0 respectively with a phase difference of ϕ. The integration limits are established for one cycle based on the well known frequency expression $f = \omega/2\pi$ and the fact that the period for one cycle is the reciprocal of frequency. When integrated, expression (15-68) yields

$$\Delta E = \pi \varepsilon_0 \sigma_0 \sin \phi \qquad (15\text{-}69)$$

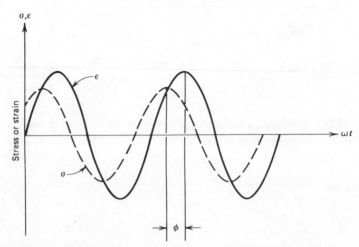

FIGURE 15.12.   Representation of the phase difference between stress and strain due to hysteretic or internal friction damping.

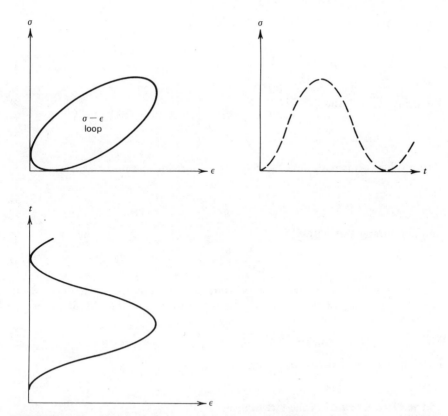

FIGURE 15.13. Hysteresis loop on the stress-strain diagram due to internal friction damping.

From (15-61) we observed that the strain energy stored in the stressed zone is

$$SE = \frac{Act\sigma^2}{2E} \tag{15-70}$$

and if it is noted that the volume of the stressed zone is Act, the strain energy stored per unit volume in the stressed zone is

$$u = W = \frac{SE}{vol} = \frac{\sigma^2}{2E} \tag{15-71}$$

For convenience, this expression may be rewritten, since $c = \sqrt{E/\rho}$, as

$$u = W = \frac{\sigma^2}{2E}\left(\frac{\rho}{\rho}\right) = \frac{\sigma^2}{2c^2\rho} \tag{15-72}$$

Now, from the basic concept that for a linear elastic material the stored strain energy is equal to the work done on the member, and the work done is the average force times the distance, for small amounts of damping it may be written that

$$W \doteq \left| \frac{\left(\frac{\sigma A}{2} \right) (\varepsilon l)}{Al} \right|_{\text{max}} = \frac{1}{2} |\sigma \varepsilon|_{\text{max}} = \frac{\sigma_0 \varepsilon_0}{2} \qquad (15\text{-}73)$$

Equating (15-72) and (15-73), then,

$$\frac{\sigma_0 \varepsilon_0}{2} = \frac{\sigma^2}{2c^2 \rho} \qquad (15\text{-}74)$$

Putting this result back into (15-69) and expressing the result as ΔW

$$\Delta W = \frac{\pi \sigma^2}{\rho c^2} \sin \phi \qquad (15\text{-}75)$$

and combining this with (15-72)

$$\frac{\Delta W}{W} = 2\pi \sin \phi \qquad (15\text{-}76)$$

For small values of ϕ, which is a good assumption for most engineering materials, the sine of the angle may be replaced by the angle itself to give

$$\frac{\Delta W}{W} \doteq 2\pi \phi \qquad (15\text{-}77)$$

Now the length of the compressed zone is

$$l_c = ct \qquad (15\text{-}78)$$

and the frequency of cyclic stressing is

$$f = \frac{\omega}{2\pi} \qquad (15\text{-}79)$$

Thus the period T for one cycle is

$$T = \frac{1}{f} = \frac{2\pi}{\omega} \qquad (15\text{-}80)$$

whereupon the wave length is

$$\lambda = (l_c)_{t=T} = cT = \frac{2\pi c}{\omega} \qquad (15\text{-}81)$$

Thus the damping energy ΔW is dissipated during one cycle over a wave length of $2\pi c/\omega$.

Referring now to the sketch of Figure 15.14, we may reason that the energy dW dissipated in small element dx within the stressed zone is proportional to the energy ΔW dissipated in the whole wave length of the stressed zone. Thus,

$$dW = \Delta W \left[\frac{dx}{\left(\frac{2\pi c}{\omega} \right)} \right] \qquad (15\text{-}82)$$

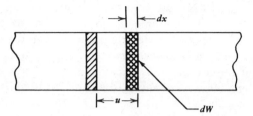

FIGURE 15.14.   Energy loss along the length of a bar due to hysteretic damping.

or

$$\frac{dW}{dx} = -\frac{\Delta W \omega}{2\pi c} \qquad (15\text{-}83)$$

where the minus sign is introduced to account for the fact that the energy dissipation causes a decrease in energy level along the bar with increasing x.

Now, combining (15-77) with (15-33),

$$\frac{dW}{dx} = -\frac{W(2\pi\phi)\omega}{2\pi c} \qquad (15\text{-}84)$$

or

$$\frac{dW}{W} = -\frac{\phi\omega}{c} dx \qquad (15\text{-}85)$$

Integration of (15-85) yields

$$W = W_0 e^{-\frac{\phi\omega x}{c}} \qquad (15\text{-}86)$$

This expression may be written in terms of stress by using the result of (15-72), as

$$\frac{\sigma^2}{2c^2\rho} = \frac{\sigma_0^2}{2c^2\rho} e^{-\frac{\phi\omega x}{2c}} \qquad (15\text{-}87)$$

or

$$\sigma = \sigma_0 e^{-\frac{\phi\omega x}{2c}} \qquad (15\text{-}88)$$

If one defines a new constant, α, called the stress attenuation factor, as

$$\alpha \equiv \frac{\phi\omega}{2c} \qquad (15\text{-}89)$$

the expression of (15-88) may be rewritten as

$$\sigma = \sigma_0 e^{-\alpha x} \qquad (15\text{-}90)$$

Note that the implication of this result is that at $x = 0$, which is the site of the impact, the stress is σ_0. However, at points farther away from the impact site, that is, for x greater than zero, the stress level dies off exponentially with distance away from the impact site. The rate of decay is controlled by the stress attenuation factor α.

By inspecting α in (15-89), we note that for greater damping (larger ϕ), higher frequency or faster rise time (larger ω), or a material with lower modulus of elasticity E or higher density ρ (smaller c), the rate of decay is greater and the attenuation of impact stress level more pronounced. It may be observed here that the use of the impact rise time as one-quarter period of oscillation is acceptable in estimating the circular frequency ω for these calculations.

15.8 STRESS WAVE PROPAGATION IN A BAR STRUCK ON THE END BY A MOVING MASS

If, instead of a suddenly applied force on the free end of a bar, the free end of a bar is struck by a moving mass, the resulting stress wave behavior is a little different because the striking mass is decelerated and the force on the end of the bar is diminished with time. To investigate the behavior of such a system it will be assumed that there is no hysteretic damping loss and that stress levels remain in the elastic range. The configuration of such a system is sketched in Figure 15.15a.

Noting that the striking body has total mass M and the bar has a cross-sectional area A, it will be convenient to define the mass of the striking body per unit cross-sectional area of the bar as

$$m = \frac{M}{A} \tag{15-91}$$

Also, if the force developed by the mass acting on the end of the bar is P, the compressive stress at the struck end of the bar is given by

$$\sigma = \frac{P}{A} \tag{15-92}$$

Further, if the velocity of the striking mass at the time of initial impact is v_0, then the particle velocity v at the struck end of the bar at the time of impact will be

$$(v)_{t=0} = v_0 \tag{15-93}$$

and, using (15-58), the stress σ_0 at the struck end of the bar at the time of impact will be

$$\sigma_0 = v_0 \sqrt{E\rho} \tag{15-94}$$

Since the bar begins to offer resistance to the moving mass at the time of impact, the velocity of the moving body is gradually decreased, and therefore the pressure on the end of the bar will be gradually decreased. Consequently, the compressive stress is gradually decreased, as shown in 15.15b, until finally the velocity of the moving body is brought to zero and the compressive stress of the original impact wave also vanishes.

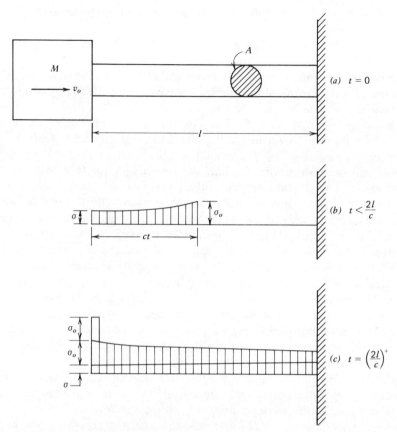

FIGURE 15.15. Stress wave propagation in a bar fixed at one end and struck on the other end by a rigid mass moving in the axial direction.

If we define the variable compressive stress at the struck end of the bar to be σ, and the variable velocity of the rigid striking body (which is also the particle velocity at the end of the bar) to be v, we may write the equation of motion for the end of the bar as

$$M\frac{dv}{dt} + \sigma A = 0 \qquad (15\text{-}95)$$

or, utilizing (15-91),

$$m\frac{dv}{dt} + \sigma = 0 \qquad (15\text{-}96)$$

Now, if the expression for velocity v from (15-56) is substituted into (15-96), we have

$$\frac{m}{\sqrt{E\rho}}\frac{d\sigma}{dt} + \sigma = 0 \qquad (15\text{-}97)$$

which may be solved explicitly for stress σ at the end of the bar as

$$\sigma = \sigma_0 e^{-\frac{\sqrt{E\rho}}{m}t} \text{ for } t < \frac{2l}{c} \tag{15-98}$$

This expression for the stress at the struck end of the bar is valid for any time t less than the time required for the stress wave to make the first round trip to the fixed end and back again, namely, any time t less than $2l/c$.

When the time t is equal to $2l/c$, the wave front of magnitude σ_0 has traveled all the way down the bar to the fixed end and back again to the struck end. Presuming that the impact is still in progress, and the striking mass is still in contact with the bar, the returning wave front will see the struck end as a fixed end since the striking body cannot suddenly change its velocity. Thus the compressive wave front σ_0 is again reflected as a compressive wave at the struck end and the stress level there takes a sudden jump of $2\sigma_0$, as shown in Figure 15.15c. This sudden increase of $2\sigma_0$ in stress at the struck end of the bar will occur throughout the duration of the impact at the end of every time interval $T = 2l/c$, and a separate expression for σ must therefore be obtained for each of these intervals.

For the first interval, where $0 < t < T$, (15-98) gives the total compressive stress σ at the struck end of the bar. For the second interval, where $T < t < 2T$, the condition depicted in Figure 15.15c is obtained and the total stress σ at the struck end is the sum of two waves moving away from the struck end and one wave moving toward the struck end. If these observations are generalized, it may be noted that the resultant stress σ at the struck end may be obtained by summing at that site all the waves moving away from the struck end plus all the waves moving toward the struck end.

Let $s_1(t), s_2(t), s_3(t), \ldots, s_n(t)$ designate the total compressive stress at the struck end due to all waves moving *away* from this end after the intervals of time $T, 2T, 3T, \ldots, nT$, respectively. Any given wave coming back toward the struck end is simply the same wave sent away from the struck end during the preceding interval delayed by time T owing to its travel across and back the length of the bar. Thus the total compressive stress produced by these returning waves at the struck end is obtained in any given interval by substituting $(t - T)$ for t in the stress expression for the compressive stress produced by waves sent away from the struck end during the preceding interval.

The general expression then for the total compressive stress at the struck end during any time interval $nT < t < (n + 1)T$ is obtained by summing the compressive stress due to all waves traveling away from the struck end during that interval plus the compressive stress due to all waves traveling toward the struck end during that interval, whence

$$\sigma = s_n(t) + s_{n-1}(t - T) \tag{15-99}$$

Further, by a similar argument, it may be reasoned that the particle velocity at the struck end during a given interval is the difference between the

linkage becomes unstable and collapses or buckles. At the point where the maximum available resisting moment exactly equals the upsetting moment, the system is on the verge of buckling and is said to be *critical*. The axial load that produces this condition is called the *critical buckling load*. Thus, the critical buckling load is the value of P_a that satisfies the condition

$$M_r = M_u \tag{16-1}$$

The upsetting moment M_u for the linkage of Figure 16.1c is

$$M_u = \left(\frac{2\delta P_a}{L\cos\alpha}\right)\frac{L}{2}\cos\alpha + P_a\delta = 2P_a\delta \tag{16-2}$$

and the resisting moment M_r is

$$M_r = (k\delta)\frac{L}{2}\cos\alpha \tag{16-3}$$

where k is the spring rate of the lateral spring system.

Thus, the critical value of axial load P_a is given from (16-1) as

$$\frac{k\delta L\cos\alpha}{2} = 2\delta(P_a)_{cr} \tag{16-4}$$

or

$$P_{cr} = \frac{kL}{4}\cos\alpha \doteq \frac{kL}{4} \tag{16-5}$$

Any axial load that exceeds $kL/4$ in magnitude will cause the mechanism to collapse since the spring is not stiff enough to provide a resisting moment large enough to balance the upsetting moment caused by applied axial force P_a.

The behavior of perfectly straight, ideal elastic columns is quite analogous to the model of Figure 16.1 except that in the case of a column the resisting moment must be provided by the beam itself. Consequently, in the case of column buckling the bending spring rate of the column will become important and the length, cross-sectional dimensions, and modulus of elasticity will all influence the buckling resistance.

16.3 BUCKLING OF A PINNED-END COLUMN

Very analogous to the four-bar linkage with auxiliary spring support loaded axially as shown in Figure 16.1a is the axisymmetric ideal pinned-end column loaded axially, as shown in Figure 16.2a. As long as the axial load P is less than the critical buckling load, the column is stable and any small lateral disturbing force, say at the midspan of the column, will cause a small deflection at midspan that will disappear upon removal of the disturbing force. However, if the axial load P exceeds the critical buckling load, the application of a small lateral disturbing force at midspan leads to large

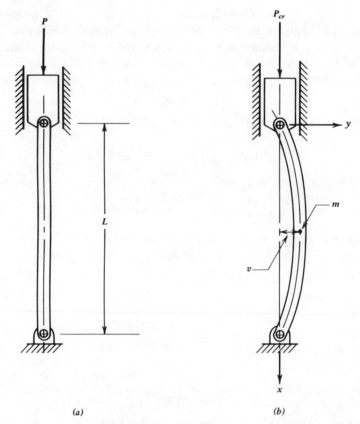

(a)   (b)

FIGURE 16.2.   Ideal axisymmetric pinned-end column subjected to axial buckling load.

deflections and buckling of the column because the maximum *resisting* moment available because of column stiffness is not large enough to balance the *upsetting* moment generated by axial load P acting at a moment arm equal to the midspan column bending deflection. As can be seen in Figure 16.2b, if moments are taken about point m, the upsetting moment M_u is

$$M_u = Pv \tag{16-6}$$

where v is the lateral midspan deflection of the column. Thus, the force P produces an upsetting moment that tends to bend the column even further, which in turn produces a larger eccentricity, v, and therefore a larger upsetting moment. The elastic forces produced in the column by the bending action tend to resist further bending. When the maximum available resisting moment exactly equals the upsetting moment, the column is at the point of incipient buckling and the axial load at this time is called the *critical buckling load*, P_{cr}, for the column.

The critical buckling load for a column can be calculated by using the basic differential equation* for the deflection curve of a beam subjected to a bending moment M, which is

$$EI\frac{d^2v}{dx^2} = -M \tag{16-7}$$

where $v(x)$ is the lateral column deflection at any location along the axis, I is the moment of inertia of the cross section about the axis around which bending takes place, and E is the modulus of elasticity for the material. Referring to Figure 16.2*b* and (16-6), at the time when the critical buckling load is applied, we calculate the upsetting moment as

$$(M_u)_{cr} = P_{cr}v \tag{16-8}$$

whereupon the governing differential equation of (16-7) becomes

$$EI\frac{d^2v}{dx^2} = -P_{cr}v \tag{16-9}$$

or, defining k^2 as

$$k^2 = \frac{P_{cr}}{EI,} \tag{16-10}$$

$$\frac{d^2v}{dx^2} + k^2v = 0 \tag{16-11}$$

The general solution of (16-11) is

$$v = A\cos kx + B\sin kx \tag{16-12}$$

where A and B are constants of integration that may be determined from the boundary conditions, which are

$$v = 0 \ at \ x = 0$$
$$v = 0 \ at \ x = L \tag{16-13}$$

referring again to Figure 16.2*b*. Thus, (16-12) may be evaluated at $x = 0$ to yield the result

$$A = 0 \tag{16-14}$$

and evaluating (16-12) at $x = L$ yields

$$0 = B\sin kL \tag{16-15}$$

This expression may be satisfied either by

$$B = 0 \tag{16-16}$$

which is the trivial case of a straight column, or by

$$\sin kL = 0 \tag{16-17}$$

*See, for example, ref. 2, 3, or 4.

The smallest nonzero value of k that satisfies (16-17) is that which makes the argument equal to π; that is,

$$kL = \pi \qquad (16\text{-}18)$$

or, utilizing (16-10),

$$\sqrt{\frac{P_{cr}}{EI}}\, L = \pi \qquad (16\text{-}19)$$

and, solving for the critical buckling load P_{cr},

$$P_{cr} = \frac{\pi^2 EI}{L^2} \qquad (16\text{-}20)$$

This expression for the smallest critical load that will produce buckling in a pinned-end column is called Euler's equation for buckling of a pinned-end column, and P_{cr} is called *Euler's critical load* for a pinned-end column.

A series of larger values of P_{cr}, corresponding to $n\pi$ in (16-19), with $n = 1$, 2, 3,..., also, produces valid solutions that correspond to higher modes and corresponding deflection curves, as illustrated in Figure 16.3. The higher

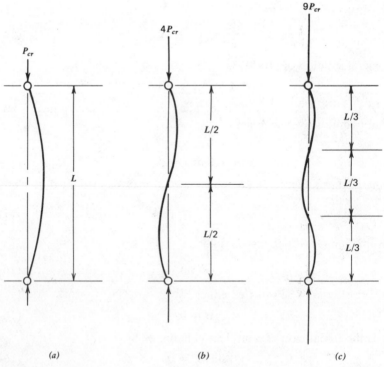

(a)   (b)   (c)

FIGURE 16.3. Buckling modes for a pinned-end column. (*a*) $n = 1$. (*b*) $n = 2$. (*c*) $n = 3$.

modes of buckling shown in Figures 16.3*b* and 16.3*c* can be developed only by supplying external lateral constraints at the inflection points, or nodes, and supplying buckling loads four times and nine times, respectively, as much as the Euler critical load required for the fundamental buckling mode shown in Figure 16.3*a*. These higher modes are not of practical importance to the designer because failure by the fundamental mode occurs long before the higher modes can develop, unless lateral bracing members are used.

The shape of the deflection curve at the point of incipient buckling is given by combining (16-12), (16-14), and (16-18) for the fundamental buckling mode as

$$v(x) = B \sin \frac{\pi x}{L} \tag{16-21}$$

where the constant B represents the undetermined midspan amplitude of the deflection. The deflection equation for higher modes would become

$$v(x) = B \sin \frac{n \pi x}{L} \tag{16-22}$$

where B would be the undetermined amplitude of the deflection curve. These deflection equations are valid as long as the deflections remain relatively small, and within this deflection range the amplitude is indeterminate. This indefinite value for amplitude of the buckled column arises from the use of an approximate expression, $d^2 v / dx^2$, for the curvature of the beam in development of (16-7). If the exact expression for curvature is used, the value of deflection is determinate. The development of the shape of the elastic curve from the exact differential equation, called the *elastica*, is beyond the scope of this presentation.*

16.4   THE INFLUENCE OF END SUPPORT ON COLUMN BUCKLING

Pinned ends were assumed in the column buckling development for Figure 16.3, but not all columns have pinned ends. Figure 16.4 illustrates several other types of end constraint frequently encountered in column design. One could write the governing differential equation and boundary conditions for each of these various types of end constraint to obtain expressions for the critical elastic buckling loads. The results of such analyses may be conveniently summarized by introducing the concept of *effective column length* into the Euler buckling equation (16-20). The *effective length* L$_e$ *of any column is defined as the length of a pinned-pinned column that would buckle at the same critical load as the actual column*. The direction of the applied column load must remain parallel to the *original* column axis. Using the concept of equivalent length, we can write the critical buckling load expression for any type of column end constraint by putting the proper value of equivalent

*For development of the *elastica*, see, p. 76 of ref. 4.

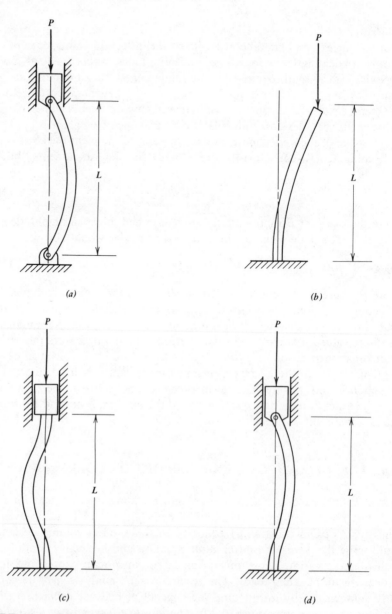

FIGURE 16.4. Frequently encountered end constraints for columns. (*a*) Pinned-pinned ($L_e = L$). (*b*) Fixed-free ($L_e = 2L$). (*c*) Fixed-fixed ($L_e = L/2$). (*d*) Fixed-pinned ($L_e \doteq 0.7L$).

Table 16.1.   Effective Lengths for Several Types of
Column End Constraint

End Constraint	Effective Lenth L_e for Actual Column Length L
Both ends pinned	$L_e = L$
One end pinned, one end fixed	$L_e \doteq 0.7L$
One end fixed, one end free	$L_e = 2L$
Both ends fixed	$L_e = 0.5L$

length into the Euler equation (16-20). Thus, for any elastic column, the
critical buckling load is given by

$$P_{cr} = \frac{\pi^2 EI}{L_e^2} \tag{16-23}$$

where the values of L_e for several types of end constraint are given in Table
16.1.

For real columns it will often be found that the ends may be partially but
not completely fixed, in which case the equivalent length will lie between the
pinned-pinned case of $L_e = L$ and the fixed-fixed case of $L_e = 0.5L$, depending on the effective rotational spring rate at the end points. For example, if
one considers a column with equal rotational spring rates at both ends, it is
possible to calculate the effective length ratio L_e/L as a function of the end
constraint to column stiffness ratio $K/(EI/L)$, where K is the rotational
spring rate at the ends in in-lb/radian, E is Young's modulus of elasticity in
psi, I is the minimum moment of inertia for the cross section in inches[4], L is
the actual column length, and L_e is the effective length of the column to be
used in (16-23) for determining the critical buckling load.* The effective
length ratio for this configuration is plotted in Figure 16.5 as a function of the
end constraint to column stiffness ratio. It may be noted that it takes only a
relatively small amount of end constraint to significantly reduce the effective
length, which corresponds to a significant increase in the critical buckling
load. On the other hand, it is virtually impossible to achieve full end fixity in
an actual column structure.

16.5   INELASTIC BEHAVIOR IN COLUMN BUCKLING

In the failure analysis of any machine or structure, it is essential to consider
all potential failure modes to discover which failure mode governs in each
range of operation. The design and analysis of columns is no exception. The
Euler equation for critical buckling load is restricted to the elastic range.

*See p. 256 of ref. 9.

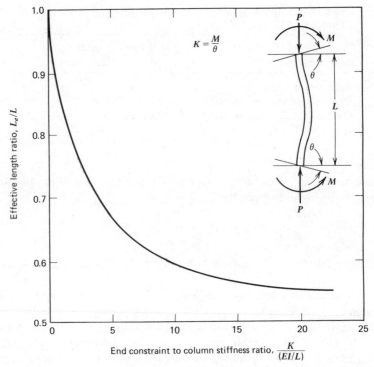

FIGURE 16.5. Plot of effective length ratio as a function of end constraint to column stiffness ratio. (From ref. 9; with permission from McGraw Hill Book Company.)

Thus, if the column is relatively short and stiff, the critical buckling load may be higher than the load at which compressive yielding takes place; therefore, the yielding failure mode governs and the Euler elastic buckling equation is not applicable.

To account for the inelastic behavior of a column when the induced stress levels exceed the elastic range, Engesser suggested in 1889 that the Euler expression for critical buckling load be modified by using the *tangent modulus* E_t, rather than Young's modulus E. The tangent modulus is defined to be the local slope of the engineering stress-strain curve for the material, or

$$E_t = \frac{d\sigma}{d\epsilon} \qquad (16\text{-}24)$$

The tangent modulus may be conveniently determined graphically from a scale plot of the engineering stress-strain diagram. The critical buckling load equation, often called the *tangent-modulus equation* or the *Euler-Engesser equation*, may be written from (16-23) as

$$P_{cr} = \frac{\pi^2 E_t I}{L_e^2} \tag{16-25}$$

or, alternatively,

$$P_{cr} = \frac{\pi^2 E_t A}{(L_e/k)^2} \tag{16-26}$$

where k is the minimum radius of gyration for the cross section of the column, A is the cross-sectional area, and the ratio (L_e/k) is often called the *effective slenderness ratio* for the column. Clearly, the critical buckling load is greatly enhanced by designing a column with a low slenderness ratio.

The Euler-Engesser equation (16-26), a relatively simple expression for determination of the critical load for a column, gives good agreement with experimental results in both the elastic and inelastic range.*

Aside from the Euler-Engesser relationship, a number of empirical column design relationships have been developed to account for the effects of inelastic behavior. Of these, the secant formula is of special interest since it allows direct consideration of initial eccentricity or column crookedness. The secant formula may be expressed as

$$\frac{P_{cr}}{A} = \frac{\sigma_{yp}}{1 + \frac{ec}{k^2} \sec\left[\frac{L_e}{2k}\sqrt{\frac{P_{cr}}{AE_t}}\right]} \tag{16-27}$$

where σ_{yp} is the yield strength of the material, e is the eccentricity of the axial load with respect to the centroidal axis of the column cross section, c is the distance from the centroidal axis to the outer fiber, k is the appropriate radius of gyration, L_e is the appropriate equivalent column length, E_t is the tangent modulus, A is the cross-sectional area of the column, and the ratio P_{cr}/A is defined as the *critical unit load* for the column. Although the critical unit load has the dimensions of stress, *it should not be treated as a stress* because of its nonlinear character. The secant formula is not very convenient for calculation purposes since the critical unit load cannot be explicitly isolated, but with the aid of a computer, or by appropriate graphical techniques, it can be employed satisfactorily. It should be observed also that (16-27) is undefined for zero eccentricity but for very small eccentricity approaches the Euler buckling curve as a limit for long columns and approaches the simple compressive yielding curve as a limit for short columns. This is shown in Figure 16.6.

It is important to recognize that selection of the appropriate radius of gyration and appropriate column length depends on which plane of buckling is under consideration. Column-end constraints are sometimes different in the

*See p. 585 of ref. 1.

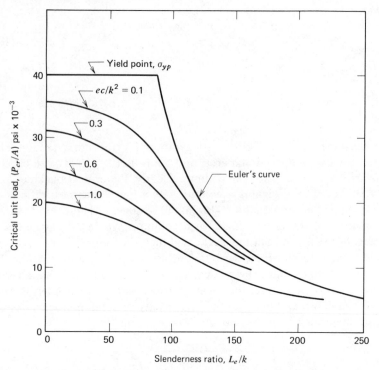

FIGURE 16.6. Secant buckling equation plotted for various eccentricities. Euler's curve and the compressive yielding curve, as shown, are asymptotes as the column eccentricity approaches zero. Modulus of elasticity $E = 30 \times 10^6$ psi.

two planes, as for example in a connecting rod that has essentially pinned ends in one plane and fixed ends in the other plane. Two separate column analyses might be required in such a case to define the more critical buckling condition.

Finally, it should be noted that all the foregoing developments are for *primary buckling* of the column as a whole, where the shape of the cross section does not change significantly. In certain cases, usually involving thin-walled sections such as tubes or rolled shapes, *local buckling* may occur, in which a significant local change in the cross section takes place. Local buckling must be considered separately, and the final design of the column must be based on its ability to resist both primary buckling and local buckling under the applied loads.

16.6   USING THE IDEAS

An aluminum alloy hollow tube of 3.0 inches outside diameter and wall thickness of 0.030 inch is 36 inches in length. The tube is pinned at each end

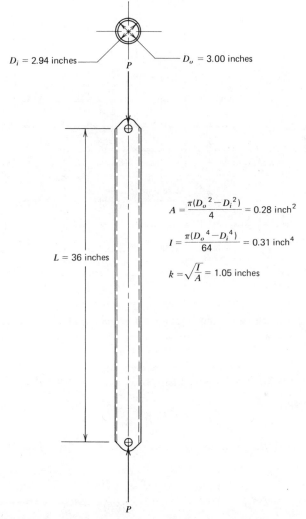

The following appear as labels within the figure:

$D_i = 2.94$ inches

$D_o = 3.00$ inches

P

$L = 36$ inches

$$A = \frac{\pi(D_o^{\,2} - D_i^{\,2})}{4} = 0.28 \text{ inch}^2$$

$$I = \frac{\pi(D_o^{\,4} - D_i^{\,4})}{64} = 0.31 \text{ inch}^4$$

$$k = \sqrt{\frac{I}{A}} = 1.05 \text{ inches}$$

P

FIGURE 16.7.   Illustration of an aluminum tube used as a column.

and must support an axial load of 18,000 lb, as shown in Figure 16.7. The engineering stress-strain curve for the material is given in Figure 16.8. It is desired to estimate the factor of safety on buckling of this column based on Euler's equation; the Euler-Engesser equation; the secant formula, assuming zero eccentricity; and the secant formula, assuming a load eccentricity of 0.15 inch off the axial center line of the column.

The estimated factor of safety in all cases is based on the ratio of critical *buckling load* to applied *column load*. It is not acceptable to use a ratio of material strength to applied stress because of the highly nonlinear behavior of

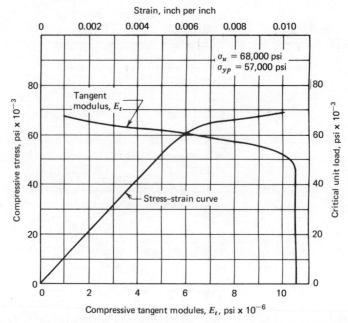

FIGURE 16.8. Stress-strain and compressive tangent modulus curves for 7075-T7351 aluminum alloy. (See ref. 6, page 3–232)

a column in the critical buckling range. Thus, we define the safety factor as

$$n_b = \frac{P_{cr}}{P_{\text{applied}}} \tag{16-28}$$

If we use the Euler equation (16-23), we calculate the critical buckling load to be

$$P_{cr} = \frac{\pi^2 EI}{L_e^2} = \frac{\pi^2 (10.5 \times 10^6)(0.31)}{(36)^2} \tag{16-29}$$

or

$$P_{cr} = 24{,}750 \text{ lb} \tag{16-30}$$

and the estimated factor of safety on buckling is therefore

$$(n_b)_{\text{Euler}} = \frac{24{,}750}{18{,}000} = 1.37 \tag{16-31}$$

If the Euler-Engesser relationship of (16-26) is used to estimate the safety factor, one proceeds in much the same way except that the proper tangent modulus must be used. From the stress-strain curve of Figure 16.8, one may graphically determine the tangent modulus at many points over the range of stress and plot a curve of E_t as a function of critical unit load, as illustrated also in Figure 16.8. It is important to recognize that the combination of E_t and P_{cr}/A, used in (16-26), must simultaneously lie on the E_t curve of Figure 16.8 to obtain a solution. Thus, an iterative process is indicated. The result of (16-30) may be used as a starting point for the iteration. Using (16-26) then gives

$$\frac{P_{cr}}{A} = \frac{\pi^2 E_t}{(L_e/k)^2} = \frac{\pi^2}{(36/1.05)^2} E_t \tag{16-32}$$

or

$$\frac{P_{cr}}{0.28} = 0.00935\, E_t \tag{16-33}$$

which must be solved simultaneously with the curve of Figure 16.8 through an iterative procedure. This yields the solution

$$\left.\begin{array}{l} E_t = 6.28 \times 10^6 \text{ psi} \\ P_{cr} = 16{,}500 \text{ lb} \end{array}\right\} \tag{16-34}$$

It may be observed that this value of critical load is substantially lower than the Euler prediction of (16-30). In fact, the safety factor is

$$(n_b)_{\text{Engesser}} = \frac{16{,}500}{18.000} = 0.92 \tag{16-35}$$

and, since this safety factor is less than unity, failure is predicted to occur. Since we note that inelastic behavior is expected for the applied load and given material, it is reasonable to assess the Engesser prediction as more accurate than the simple Euler prediction, and, indeed, such is the case.

Now, using the secant formula with very small eccentricity, say 0.001 inch, from (16-27) we may write

$$\frac{P_{cr}}{0.28} = \frac{57{,}000}{1 + \dfrac{(0.001)(1.5)}{(1.05)^2} \sec\left[\dfrac{36}{2(1.05)}\sqrt{\dfrac{P_{cr}}{0.28 E_t}}\right]} \tag{16-36}$$

If E_t is taken equal to 8.0×10^6 psi, which must be verified finally from Figure 16.8, (16-36) becomes

$$P_{cr}\left[1 + 0.00136 \sec 1.15\sqrt{P_{cr} \times 10^{-4}}\right] = 15{,}950 \tag{16-37}$$

Solving this equation by a trial and error process, one obtains

$$P_{cr} = 15{,}750 \text{ lb} \tag{16-38}$$

and checking Figure 16.8, the assumed value of E_t is reasonably close to the final value read from the E_t curve at a critical unit load value of $(P_{cr}/A) = 56{,}300$ psi. Thus, the estimated safety factor is

$$(n_b)_{\substack{\text{secant} \\ \text{small ecc.}}} = \frac{15{,}750}{18.000} = 0.87 \tag{16-39}$$

and buckling would be predicted at about 87 percent of the desired applied load.

Finally, if an eccentricity of 0.15 inch exists, the secant formula becomes

$$\frac{P_{cr}}{0.28} = \frac{57{,}000}{1 + \dfrac{0.15(1.5)}{(1.05)^2} \sec\left[\dfrac{36}{2(1.05)}\sqrt{\dfrac{P_{cr}}{0.28E_t}}\right]} \tag{16-40}$$

and if E_t now is assumed to be 10.5×10^6 psi from Figure 16.8, the buckling equation is

$$P_{cr}\left[1 + 0.204 \sec \sqrt{P_{cr} \times 10^{-4}}\right] = 15{,}950 \tag{16-41}$$

Again, using a trial and error technique, the solution to (16-41) is

$$P_{cr} = 11{,}250 \text{ lb} \tag{16-42}$$

The safety factor for this eccentrically loaded column then is

$$(n_b)_{\substack{\text{secant} \\ \text{0.15 ecc.}}} = \frac{11{,}250}{18.000} = 0.63 \tag{16-43}$$

and buckling would be predicted at about 63 percent of the desired applied load.

16.7 LATERAL BUCKLING OF DEEP, NARROW BEAMS SUBJECTED TO BENDING

Numerous cases of buckling are observed in practice that are not column buckling, as just described, but that may be analyzed in much the same way. For example, if a thin, deep beam is subjected to a pure bending moment, as shown in Figure 16.9, it will buckle laterally in a twisting action when the moment reaches a critical value. This may be visualized by recognizing that the fibers along the compressive edge of the deep beam may be thought of as a compressively loaded column that buckles laterally when the critical load is reached.

To develop an expression for the critical buckling moment, it is necessary to find the smallest value of applied moment that will cause the beam to remain buckled after a small lateral deflection has been induced.

By St. Venant's principle, it may be assumed that the applied moment vector $\mathbf{M}_0$ will follow the deflected beam as it buckles, and at an arbitrary section, such as $A\text{-}A$ in Figure 16.9, $\mathbf{M}_0$ may be resolved into components $\mathbf{M}_x$, $\mathbf{M}_y$, and $\mathbf{M}_z$, where

$$M_x = M_0 \cos\varphi_1 \cos\theta \tag{16-44}$$

$$M_y = M_0 \cos\varphi_2 \sin\beta \tag{16-45}$$

$$M_z = M_0 \cos\varphi_1 \sin\theta \tag{16-46}$$

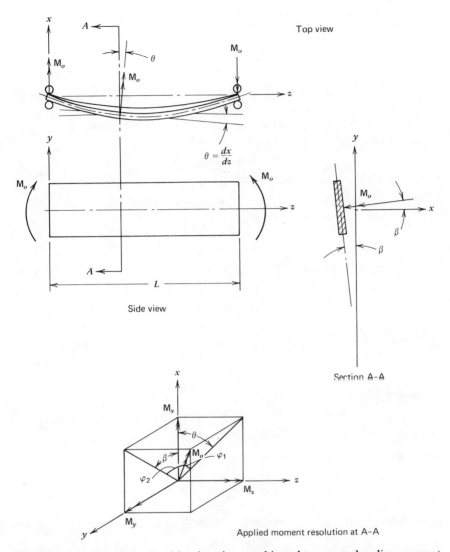

FIGURE 16.9.   Buckling of a thin, deep beam subjected to a pure bending moment.

Since all angles are assumed to be small, we may write

$$\cos \beta = \cos \theta = \cos \varphi_1 = \cos\varphi_2 = 1 \qquad (16\text{-}47)$$

and

$$\sin \beta = \beta \qquad (16\text{-}48)$$

$$\sin \theta = \frac{dx}{dz} \qquad (16\text{-}49)$$

Thus, the moment components become

$$M_x = M_0 \equiv M_{bx} \qquad (16\text{-}50)$$

$$M_y = M_0\beta \equiv M_{by} \qquad (16\text{-}51)$$

$$M_z = M_0\frac{dx}{dz} \equiv M_{tz} \qquad (16\text{-}52)$$

where $\mathbf{M}_x$ and $\mathbf{M}_y$ produce bending, and $\mathbf{M}_z$ produces torsion.
For torsion, then, we may write

$$\frac{d\beta}{dz} = \frac{M_{tz}}{GJ_e} \qquad (16\text{-}53)$$

and for bending about the y axis, which is the buckling direction,

$$\frac{d^2x}{dz^2} = -\frac{M_{by}}{EI_y} \qquad (16\text{-}54)$$

Combining (16-52) with (16-53) then yields

$$\frac{d\beta}{dz} = \frac{M_0}{GJ_e}\frac{dx}{dz} \qquad (16\text{-}55)$$

and combining (16-51) with (16-54) gives

$$\frac{d^2x}{dz^2} = -\frac{M_0}{EI_y}\beta \qquad (16\text{-}56)$$

Differentiating (16-55) with respect to z results in

$$\frac{GJ_e}{M_0}\frac{d^2\beta}{dz^2} = \frac{d^2x}{dz^2} \qquad (16\text{-}57)$$

which may be equated to (16-56) to give

$$\frac{d^2\beta}{dz^2} + \frac{M_0^2}{GJ_e EI_y}\beta = 0 \qquad (16\text{-}58)$$

If we define

$$k_1^2 = \frac{M_0^2}{GJ_e EI_y} \qquad (16\text{-}59)$$

then (16-58) may be written as

$$\frac{d^2\beta}{dz^2} + k_1^2\beta = 0 \qquad (16\text{-}60)$$

which is formally the same as (16-11) for column buckling with the general
solution

$$\beta = A\cos kx + B\sin kx \qquad (16\text{-}61)$$

The boundary conditions are

$$\beta = 0 \quad \text{at} \quad z = 0$$
$$\beta = 0 \quad \text{at} \quad z = L$$

(16-62)

which are also formally the same as (16-13) for column buckling.

Therefore, the solution for buckling of the thin, deep beam is formally the same as (16-18) for column buckling, whence

$$kL = \pi$$

(16-63)

and from from (16-59) at the point of instability where $M_0 = M_{cr}$

$$\sqrt{\frac{M_{cr}^2}{GJ_e EI_y}} \, L = \pi$$

(16-64)

and solving for the critical buckling moment M_{cr}

$$M_{cr} = \frac{\pi \sqrt{GJ_e EI_y}}{L}$$

(16-65)

where M_{cr}=critical bending moment about x axis (constant over the whole
           beam)
E=Young's molulus of elasticity
I_y=area moment of inertia about y axis
G=shear modulus of rigidity
J_e=property of the cross section relating torque to angle of twist in
$\theta = M_t L / GJ_e$. (for a thin rectangle $J_e = dt^3/3$)
L=length of beam
d=depth of beam
t=thickness of beam

Note that this solution is based on ends that are constrained to remain vertical but free to rotate about the vertical y axis. If the ends are fixed to rotation about the y axis also, the effective length of the beam is $L/2$, and the buckling critical bending moment is doubled.

Buckling critical loads for lateral buckling of thin, deep beams subjected to variable bending moments have also been calculated. A generalized equation for such conditions may be written as*

$$P_{cr} = \frac{K \sqrt{GJ_e EI_y}}{L^2}$$

(16-66)

where the value of k for several cases is given in Table 16.2.

*See p. 626 of ref 1.

Table 16.2.   Buckling Constant K for Various Beam Loading Conditions

Type of Beam	Type of Loading	K
Cantilever	P (lb) concentrated at free end	4.013
Cantilever	q (lb/in.) distributed over length L	12.85
Simply supported	P (lb) concentrated at center	16.93
Simply supported	q (lb/in.) distributed over length L	28.3

16.8 LATERAL BUCKLING OF THIN, CIRCULAR SHAFT SUBJECTED TO TORSION

Another interesting example of the buckling phenomenon is observed if a thin, circular shaft is subjected to a pure torsional moment, as illustrated in Figure 16.10. If a critical moment is exceeded, the thin shaft will buckle into a helical configuration, as shown.

To develop an expression for the critical buckling moment, it is again necessary to find the smallest value of applied torsional moment that will cause the shaft to remain buckled. By St. Venant's principle, it may again be assumed that the applied moment vector $\mathbf{M}_{t0}$ will follow the deflected shaft; and at an arbitrary section, such as A in Figure 16.10, $\mathbf{M}_{t0}$ may be resolved into components $\mathbf{M}_x$, $\mathbf{M}_y$, and $\mathbf{M}_z$, where

$$M_x = M_{t0} \cos \varphi_1 \sin \theta_1 \qquad (16\text{-}67)$$

$$M_y = M_{t0} \cos \varphi_1 \cos \theta_1 \qquad (16\text{-}68)$$

$$M_z = M_{t0} \cos \varphi_2 \sin \theta_2 \qquad (16\text{-}69)$$

Since all angles are assumed to be small

$$\cos \varphi_1 = \cos \varphi_2 = \cos \theta_1 = 1 \qquad (16\text{-}70)$$

and

$$\sin \theta_1 = \frac{dx}{dy} \qquad (16\text{-}71)$$

$$\sin \theta_2 = \frac{dz}{dy} \qquad (16\text{-}72)$$

whence the moment components become

$$M_x = M_{t0} \frac{dx}{dy} \equiv M_{bx} \qquad (16\text{-}73)$$

$$M_y = M_{t0} \equiv M_{ty} \qquad (16\text{-}74)$$

$$M_z = M_{t0} \frac{dz}{dy} \equiv M_{bz} \qquad (16\text{-}75)$$

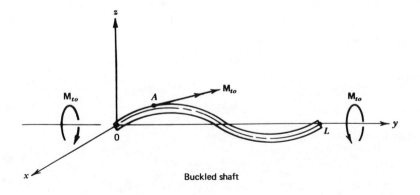

Buckled shaft

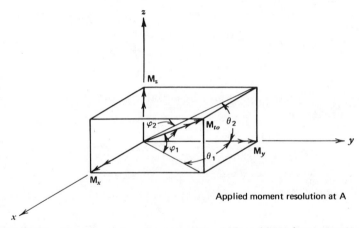

Applied moment resolution at A

FIGURE 16.10. Buckling of a thin, circular shaft subjected to a pure torsional moment.

Then, for bending about the x axis,

$$\frac{d^2z}{dy^2} = -\frac{M_{bx}}{EI} \tag{16-76}$$

and for bending about the z axis

$$\frac{d^2x}{dy^2} = \frac{M_{bz}}{EI} \tag{16-77}$$

which may be written as

$$\frac{d^2x}{dy^2} + k_2\frac{dz}{dy} = 0 \tag{16-78}$$

and

$$\cdot \frac{d^2 z}{dy^2} - k_2 \frac{dx}{dy} = 0 \qquad (16\text{-}79)$$

where

$$k_2 \equiv \frac{M_{t0}}{EI} \qquad (16\text{-}80)$$

The general solutions for x and z are

$$x = C_1 \sin ky + C_2 \cos ky + C_3 \qquad (16\text{-}81)$$

$$z = -C_1 \cos ky + C_2 \sin ky + C_4 \qquad (16\text{-}82)$$

as may be verified by substitution into (16-78) and (16-79), respectively.
The boundary conditions are

$$
\begin{array}{lll}
x = 0 & \text{at} & y = 0 \\
x = 0 & \text{at} & y = L \\
z = 0 & \text{at} & y = 0 \\
z = 0 & \text{at} & y = L
\end{array}
\qquad (16\text{-}83)
$$

from which (16-81) and (16-82) may be evaluated at the boundaries to obtain
for the nontrivial case

$$\frac{(1 - \cos k_2 L)^2}{\sin k_2 L} + \sin k_2 L = 0 \qquad (16\text{-}84)$$

which reduces to

$$\cos k_2 L = 1 \qquad (16\text{-}85)$$

The smallest nonzero value of k_2 that satisfies (16-85) is that which makes
the argument equal to 2π; that is,

$$k_2 L = 2\pi \qquad (16\text{-}86)$$

or, utilizing (16-80),

$$\frac{(M_{t0})_{cr} L}{EI} = 2\pi \qquad (16\text{-}87)$$

Solving for the critical torsional moment for lateral buckling of the thin
shaft then gives

$$(M_{t0})_{cr} = \frac{2\pi EI}{L} \qquad (16\text{-}88)$$

A similar development * has been made for the case of a thin, circular shaft
simultaneously subjected to an axial load P and a torsional moment M_{t0} to

*See p. 167ff. of ref. 7.

obtain the buckling equation

$$\frac{M_{to}^2}{4(EI)^2} + \frac{P}{EI} = \frac{\pi^2}{L^2} \qquad (16\text{-}89)$$

It may be noted that when P is zero, (16-89) degenerates to (16-88) for pure torsion only; and when M_{t0} is zero, it degenerates to the Euler equation (16-20) for pure axial load only. It is also of interest that for a shaft in tension the sign of P in (16-89) changes, and the torsional moment M_{t0} required for buckling instability is increased. Thus, placing the shaft in tension improves its resistance to buckling.

16.9   OTHER BUCKLING PHENOMENA

Numerous other cases of both primary buckling and local buckling may be encountered in specific engineering applications. Problems of buckling in curved beams, rings, arches, thin plates, panels, thin shells with and without internal pressurization, domes, thin tubes, and flanged beams, all under various loading conditions and for many different geometrical configurations, may be important. Although a detailed discussion of these buckling phenomena is beyond the scope of this text, excellent developments covering many of these cases are readily available in the literature.[*]

Another phenomenon of interest is *creep buckling*. Sometimes a structure may be properly designed to resist buckling only to find that under certain operating conditions creep behavior is induced, which leads to significant dimensional changes and buckling instability. This phenomenon is called *creep buckling*, and the time required for a loaded structure to reach the critical buckling configuration is called the *critical time*. Although solutions to a few creep buckling configurations have been developed,[†] the analyses are very complex, and general solutions are not readily available. A detailed study of creep buckling is beyond the scope of this text.

QUESTIONS

1.   Referring to the pinned mechanism with lateral spring at point B, shown in Figure 16.1, do the following:
(a)   Repeat the derivation leading to (16-5) using the concepts of upsetting moment and resisting moment to find an expression for critical load.
(b)   Use an energy method to again find an expression for critical load in the

[*]See, for example, refs. 1, 4, 5, and 7.
[†]See p. 860ff. of ref. 8.

mechanism of Figure 16.1 by equating change in potential energy of P_a to strain energy stored in the spring. (Hint: use the first two terms of the series expansion for cos α to approximate cos α.)

(c)   Compare results of parts (a) and (b) of this problem.

2.   Verify the value of effective length $L_e = 2L$ for a column fixed at one end and free at the other (Figure 16.4b) by writing and solving the proper differential equation for this case and comparing the result with the solution of (16-20) for a pinned-pinned column.

3.   A solid cylindrical steel bar is 2.0 inches in diameter and 12 feet long. If both ends are pinned, estimate the axial load required to cause the bar to buckle.

4.   If the same amount of material used in the steel bar of problem 3 had been formed into a hollow cylindrical bar of the same length and supported at the ends in the same way (i.e., same cross-sectional area), what would the critical buckling load be if the tube wall thickness were made (a) 1/4 inch, (b) 1/8 inch, and (c) 1/16 inch. What conclusions would you draw from these results?

5.   If the solid bar of problem 3 were fixed at the ends, to what value would the critical buckling load be changed?

6.   The compressive stress-strain curve for 7075-T7351 aluminum alloy is given in Figure 16.8. A hollow cylindrical bar of this material has a 4-inch outside diameter with a 1/8-inch wall thickness. If a 9-ft-long column is constructed with one end fixed and one end free (as in Figure 16.4b), calculate the value of critical buckling load according to

(a)   Euler's equation

(b)   the Euler-Engesser equation

(c)   the secant formula, assuming zero eccentricity

(d)   the secant formula, assuming a load eccentricity of 1/8 inch off the axial center line of the column

(e)   the secant formula, assuming a load eccentricity of 1 inch off the axial center line of the column.

7.   Repeat problem 6 with everything the same except the column length is 4 feet rather than 9 feet. Compare results with problem 6 results.

8.   A steel pipe 4 inches in outside diameter and 0.226-inch wall thickness is used to support a tank of water weighing 10,000 lbs when full. The pipe is set vertically in a heavy, rigid concrete base, as shown in Figure Q16.8. The pipe material is 1045 cold-drawn steel with $\sigma_u = 103,000$ psi and $\sigma_{yp} = 90,000$ psi. A safety factor of 2 on load is desired.

(a)   Derive a design equation for the maximum safe height H above ground level that could be used for this application. (Use the approximation $I \doteq \pi D^3 t/8$.)

(b)   Compute a numerical value for H.

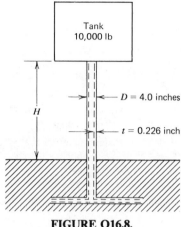

FIGURE Q16.8.

(c)   Would compressive yielding be a problem in this design? Justify your answer.

9.   Instead of using a steel pipe for supporting the tank of problem 8, it is being proposed to use an 8 I18.4 I-beam for the support and a plastic line to carry the water. This I-beam has a cross section of 8.0 inches depth, 4.0 inches flange width, area of 5.34 inches^2, $I_x = 56.9$ inches^4, and $I_y = 3.8$ inches^4; it weighs 18.4 lb/foot. (a) Compute the maximum safe height H above ground level for this I-beam, and compare the result with problem 8b. (b)   Based on a desire to achieve maximum height with a column of minimum weight, would you recommend the pipe or the I-beam?

10.   (a)   A steel wire of 0.1 inch diameter is subjected to torsion. If the material has a tensile yield strength of 100,000 psi and the wire is 10 feet long, find the torque at which it will fail and identify the mode of failure. (b)   If a tensile force of 10 lbs is applied to the wire, what change would you predict in torque at which failure occurs? Would the mode of failure remain the same?

11.   A sheet-steel cantilevered bracket of rectangular cross section 0.125 inch by 4.0 inches is fixed at one end with the 4.0-inch dimension vertical. The bracket, which is 14 inches long, must support a vertical load P at the free end. (a)   What is the maximum load that should be placed on the bracket if a safety factor of 2 is desired? The steel has a yield strength of 45,000 psi. (b)   Identify the governing failure mode.

12.   A hollow tube is to be subjected to torsion. Derive an equation that gives the length of this tube for which failure is equally likely by yielding or by elastic instability.

REFERENCES

1. Shanley, F.R., *Strength of Materials*, McGraw-Hill, New York 1957.
2. Popov, E.P., *Introduction to Mechanics of Solids*, Prentice-Hall, Englewood Cliffs, N.J., 1958.
3. Crandall, S.H., and Dahl, N.C., *An Introduction to the Mechanics of Solids*, McGraw-Hill, New York, 1957.
4. Timoshenko, S.P., and Gere, J.M., *Theory of Elastic Stability*, McGraw-Hill, New York, 1961.
5. Horton, W.H., Bailey, S.C., and McQuilkin, B.H., *An Introduction to Instability*, Stanford University Paper No. 219, ASTM Annual Meeting, June 1966.
6. *Military Standardization Handbook, Metallic Materials and Elements for Aerospace Vehicle Structures, MIL-HDBK-5B*, Superintendent of Documents, Washington, D.C., September 1971.
7. Timoshenko, S.P., *Theory of Elastic Stability*, McGraw-Hill, New York, 1936.
8. Faupel, J.H., *Engineering Design*, John Wiley & Sons, New York, 1964.
9. Shanley, F.R., *Mechanics of Materials*, McGraw-Hill, New York, 1967.

CHAPTER 17

Wear, Corrosion, and Other Important Failure Modes

17.1 INTRODUCTION

Most of the failure modes listed in Chapter 2 have now been examined in this text. Throughout these discussions the objective has consistently been to present a brief qualitative description of the failure phenomena, followed by quantitative design procedures useful to practicing design engineers. Wear, corrosion, and related failure modes have been left for discussions in this final chapter not because they are unimportant but because the state-of-the-art knowledge in these areas has not yet developed widely accepted or well defined analytical or empirical design procedures or life prediction techniques.

It is only in the past decade that substantial progress has been made in quantitative predictions of wear. This late development of wear prediction techniques does not imply that wear is unimportant. Indeed, wear and corrosion probably account for a majority of mechanical failures in the field, and an extensive research literature has built up within the areas of wear and corrosion. In spite of these facts, however, quantitative life prediction techniques have not yet been developed into good design tools.

17.2 WEAR

Wear may be defined as the undesired cumulative change in dimensions brought about by the gradual removal of discrete particles from contacting surfaces in motion, due predominantly to mechanical action. It should be further recognized that corrosion often interacts with the wear process to change the character of the surfaces of wear particles through reaction with the environment.

Wear is, in fact, not a single process but a number of different processes that may take place independently or in combination. It is generally accepted that there are at least five major subcategories of wear,* including adhesive

*See p. 120 of ref. 1; see also ref. 2.

wear, abrasive wear, corrosive wear, surface fatigue wear, and deformation wear. In addition, the categories of fretting wear and impact wear (3, 4, 5) have been recognized by wear specialists. Erosion and cavitation are sometimes considered to be categories of wear as well. Each of these types of wear proceeds by a distinctly different physical process and must be separately considered, although the various subcategories may combine their influence either by shifting from one mode to another during different eras in the operational lifetime of a machine or by simultaneous activity of two or more different wear modes.

The complexity of the wear process may be better appreciated by recognizing that many variables are involved, including the hardness, toughness, ductility, modulus of elasticity, yield strength, fatigue properties, and structure and compositon of the mating surfaces, as well as geometry, contact pressure, temperature, state of stress, stress distribution, coefficient of friction, sliding distance, relative velocity, surface finish, lubricants, contaminants, and ambient atmosphere at the wearing interface. Clearance versus contact-time history of the wearing surfaces may also be an important factor in some cases. Although the wear processes are complex, progress has been made in recent years toward development of quantitative empirical relationships for the various subcategories of wear under specified operating conditions. Much experimental work remains to be completed, however, before these relationships are widely accepted.

Adhesive Wear

Adhesive wear is often characterized as the most basic or fundamental subcategory of wear since it occurs to some degree whenever two solid surfaces are in rubbing contact and remains active even when all other modes of wear have been eliminated. The phenomenon of adhesive wear may be best understood by recalling that all real surfaces, no matter how carefully prepared and polished, exhibit a general waviness upon which is superposed a distribution of local protuberances or asperities. As two surfaces are brought into contact, therefore, only a relatively few asperities actually touch, and the *real* area of contact, A_r, is only a small fraction of the *apparent* contact area, A_a. It has been deduced* from electrical conductance experiments that for the usual range of engineering design loads the ratio of real to apparent contact area, A_r/A_a, is in the range from 10^{-2} to 10^{-5}. Thus, even under very small applied loads the local pressures at the contact sites become high enough to exceed the yield strength of one or both surfaces, and local plastic flow ensues. If the contacting surfaces are clean and uncorroded, the very intimate contact generated by this local plastic flow brings the atoms of the two contacting surfaces close enough together to call into play strong adhe-

---

*See Chap. 1 of ref. 6 and Chap. II of ref. 7.

sive forces. This process is sometimes called *cold welding*. Then if the surfaces are subjected to relative sliding motion, the cold-welded junctions must be broken. Whether they break at the original interface or elsewhere within the asperity depends upon surface conditions, temperature distribution, strain-hardening characteristics, local geometry, and stress distribution. If the junction is broken away from the original interface, a particle of one surface is transferred to the other surface, marking one event in the adhesive wear process, as shown in Figure 17.1. Later sliding interactions may dislodge the transferred particles as a loose wear particle, or it may remain attached. If this adhesive wear process becomes severe and large-scale metal transfer takes place, the phenomenon is called *galling*. If the galling becomes so severe that two surfaces adhere over a large region so that the actuating forces can no longer produce relative motion between them, the phenomenon is called *seizure*. If properly controlled, however, the adhesive wear rate may be low and self-limiting, often being exploited in the "wearing-in" process to improve

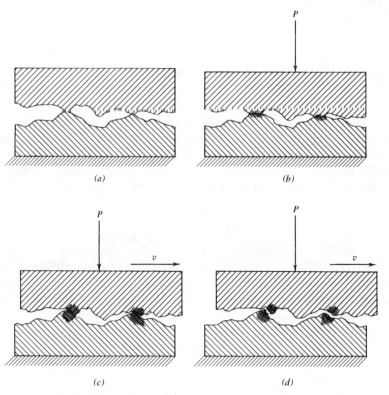

FIGURE 17.1. Contact between two solid bodies and the transfer of a particle by adhesive wear. (*a*) Unloaded surfaces in contact. (*b*) Applied load *P* causes plastic flow and cold welding. (*c*) Sliding motion and load produce strain hardening. (*d*) Particles transferred by rupture of asperity below weld junction.

mating surfaces such as bearings or cylinders so that full film lubrication may be effectively used.

One quantitative estimate of the amount of adhesive wear may be made as follows:* Assuming that the local compressive flow strength, which because of the multiaxial state of stress is approximately equal to three times the uniaxial yield strength $(3\sigma_{yp})$, is exceeded upon initial contact of the touching asperities, the material will flow locally until the compressive flow strength is nowhere exceeded. Thus, if the real contact area is A_r, the compressive flow strength is $3\sigma_{yp}$, and the normal force pressing the surfaces together is W, it may be written that

$$W = A_r(3\sigma_{yp}) \tag{17-1}$$

In accordance with the Archard development (9), it may be assumed that each time two asperities come into contact to form a junction, there is a constant probability that an adhesive wear particle will be formed. It is further assumed that each particle is hemispherical, of diameter d, equal to the junction diameter, and that all junctions are of the same size. If n junctions exist at any instant, the real contact area A_r is

$$A_r = n\left(\frac{\pi d^2}{4}\right) \tag{17-2}$$

Combining (17-1) and (17-2), then, we get

$$n = \frac{4A_r}{\pi d^2} = \frac{4W}{3\pi\sigma_{yp}d^2} \tag{17-3}$$

It is next assumed that each junction remains intact through a sliding distance equal to junction diameter d, after which it is broken and a new junction forms. Thus, each junction must be reformed $1/d$ times per unit sliding distance, and the total number of junctions N formed per unit sliding distance is

$$N = n\left(\frac{1}{d}\right) = \frac{4W}{3\pi\sigma_{yp}d^3} \tag{17-4}$$

If the probability of wear particle formation is k, the total number of junctions formed per unit sliding distance is N, and the formed particles are each hemispherical with volume $\pi d^3/12$, the wear particle volume ΔV formed per unit distance slid ΔL is given by

$$\frac{\Delta V}{\Delta L} = kN\left(\frac{\pi d^3}{12}\right) \tag{17-5}$$

Combining (17-4) with (17-5), we get

$$\frac{\Delta V}{\Delta L} = \frac{kW}{9\sigma_{yp}} \tag{17-6}$$

*See ref. 1 and Chaps. 2 and 6 of ref. 8.

Integrating (17-6) over the total sliding distance L_s then gives an expression for volumetric adhesive wear as

$$V_{adh} = \frac{kWL_s}{9\sigma_{yp}} \tag{17-7}$$

If d_{adh} is the average wear depth and A_a the apparent contact area, (17-7) may be rewritten as

$$d_{adh} = \frac{V_{adh}}{A_a} = \left(\frac{k}{9\sigma_{yp}}\right)\left(\frac{W}{A_a}\right)L_s \tag{17-8}$$

or

$$d_{adh} = k_{adh}\, p_m L_s \tag{17-9}$$

where $p_m = W/A_a$ is the mean nominal contact pressure between bearing surfaces and $k_{adh} = k/9\sigma_{yp}$ a wear coefficient that depends on the probability of formation of a transferred fragment and the yield strength (or hardness) of the softer material. Typical values of the wear constant k for several material pairs are shown in Table 17.1, and the influence of lubrication on the wear constant k is indicated in Table 17.2.

Table 17.1. Archard Adhesive Wear Constant k for Various Unlubricated Material Pairs in Sliding Contact*

Material Pair	Wear Constant k
Zinc on zinc	160×10^{-3}
Low-carbon steel on low-carbon steel	45×10^{-3}
Copper on copper	32×10^{-3}
Stainless steel on stainless steel	21×10^{-3}
Copper (on low-carbon steel)	1.5×10^{-3}
Low-carbon steel (on copper)	0.5×10^{-3}
Bakelite on bakelite	0.02×10^{-3}

*From Chap. 6 of ref. 8, with permission from John Wiley & Sons, Inc.

Table 17.2. Order of Magnitude Values for Adhesive Wear Constant k Under Various Conditions of Lubrication*

Lubrication Condition	Metal (on metal)		Nonmetal (on metal)
	Like	Unlike	
Unlubricated	5×10^{-3}	2×10^{-4}	5×10^{-6}
Poorly lubricated	2×10^{-4}	2×10^{-4}	5×10^{-6}
Average lubrication	2×10^{-5}	2×10^{-5}	5×10^{-6}
Excellent lubrication	2×10^{-6} to 10^{-7}	2×10^{-6} to 10^{-7}	2×10^{-6}

*From Chap. 6 of ref. 8, with permission from John Wiley & Sons, Inc.

Noting from (17-9) that

$$k_{adh} = \frac{d_{adh}}{p_m L_s} \qquad (17\text{-}10)$$

it may be observed that if the ratio $d_{adh}/(p_m L_s)$ is experimentally found to be constant, (17-9) should be valid. Experimental evidence has been accumulated* to confirm that for a given material pair this ratio is constant up to mean nominal contact pressures approximately equal to the uniaxial yield strength. Above this level the adhesive wear coefficient increases rapidly, with attendant severe galling and seizure.

Thus the average wear depth for adhesive wear conditions may be estimated as

$$\left. \begin{array}{l} d_{adh} = k_{adh}\, p_m L_s \text{ for } p_m < \sigma_{yp} \\ \text{unstable galling and seizure for } p_m \geq \sigma_{yp} \end{array} \right\} \qquad (17\text{-}11)$$

The problem in the practical application of this expression lies in finding the appropriate value of the adhesive wear constant k_{adh} to use for a given application. Values of k_{adh} range from approximately 10^{-5} to 10^{-13} in ^{2}/lb for various material combinations. Specific experimental data must often be developed for a given application, although useful data for approximating k_{adh} have been generated for many instances,[†] some of which are shown in Tables 17.1 and 17.2. The application of the adhesive wear equation (17-11) is further complicated by the fact that other types of wear often occur simultaneously, and under some circumstances they dominate the overall wear behavior.

In the selection of metal combinations to provide resistance to adhesive wear, it has been found that the sliding pair should be composed of mutually insoluble metals and that at least one of the metals should be from the B subgroup of the periodic table.[‡] The reasons for these observations are that the number of cold-weld junctions formed is a function of the mutual solubility, and the strength of the junction bonds is a function of the bonding characteristics of the metals involved. The metals in the B subgroup of the periodic table are characterized by weak, brittle covalent bonds. These criteria have been verified experimentally, as shown in Table 17-3, where 114 of 123 pairs tested substantiated the criteria.

Three major wear control methods have been defined, as follows:[§] *Principle of protective layers*, including protection by lubricant, surface film, paint, plating, phosphate, chemical, flame-sprayed, or other types of interfacial layers; *Principle of conversion*, in which wear is converted from destructive to

---

*See pp. 124–125 of ref. 1.
[†]See pp. 124–125 of ref. 1 and pp. 139–140 of ref. 8.
[‡]See p. 31 of ref. 10.
[§]See p. 36 of ref. 10.

Table 17.3. Adhesive Wear Behavior of Various Pairs*

Description of Metal Pair	Material Combination				Remarks
	Al Disk	Steel Disk	Cu Disk	Ag Disk	
Soluble pairs with poor adhesive wear resistance	Be	Be	Be	Be	These pairs substantiate the criteria of solubility and B subgroup metals
	Mg	—	Mg	Mg	
	Al	Al	Al	—	
	Si	Si	Si	Si	
	Ca	—	Ca	—	
	Ti	Ti	Ti	—	
	Cr	Cr	—	—	
	—	Mn	—	—	
	Fe	Fe	—	—	
	Co	Co	Co	—	
	Ni	Ni	Ni	—	
	Cu	—	Cu	—	
	—	Zn	Zn	—	
	Zr	Zr	Zr	Zr	
	Nb	Nb	Nb	—	
	Mo	Mo	Mo	—	
	Rh	Rh	Rh	—	
	—	Pd			
	Ag	—	Ag	—	
	—	—	Cd	Cd	
	—	—	In	In	
	Sn	—	Sn	—	
	Ce	Ce	Ce	—	
	Ta	Ta	Ta	—	
	W	W	W	—	
	—	Ir	—	—	
	Pt	Pt	Pt	—	
	Au	Au	Au	Au	
	Th	Th	Th	Th	
	U	U	U	U	
Soluble pairs with fair or good adhesive wear resistance. (F) = Fair	—	Cu(F)	—		These pairs do not substantiate the stated criteria
	Zn(F)	—	—		
	—	—	Sb(F)		
Insoluble pairs, neither from the B subgroup, with poor adhesive wear resistance		Li			These pairs substantiate the stated criteria
		Mg			
		Ca			
		Ba			

Table 17.3. (*Continued*)

Description of Metal Pair	Material Combination				Remarks
	Al Disk	Steel Disk	Cu Disk	Ag Disk	
Insoluble pairs, one from the B subgroup, with fair or good adhesive wear resistance. (F) = Fair	—	C(F)	—	—	These pairs substantiate the stated criteria
	—	—	—	Ti(F)	
	—	—	Cr(F)	Cr(F)	
	—	—	—	Fe(F)	
	—	—	—	Co(F)	
	—	—	Ge(F)	—	
	—	Se(F)	Se(F)	—	
	—	—	—	Nb(F)	
	—	Ag	—	—	
	Cd	Cd	—	—	
	In	In	—	—	
	—	Sn(F)	—	—	
	—	Sb(F)	Sb	—	
	Te(F)	Te(F)	Te(F)	—	
	Tl	Tl	Tl	—	
	Pb(F)	Pb	Pb	—	
	Bi(F)	Bi	Bi(F)	—	
Insoluble pairs, one from the B subgroup, with poor adhesive wear resistance	C	—	C	C	These pairs do not substantiate the stated criteria
	—	—	—	Ni	
	Se	—	—	—	
	—	—	—	Mo	

*See p. 34–35 of ref. 10.

permissible levels through better choice of metal pairs, hardness, surface finish, or contact pressure; and *Principle of diversion*, in which the wear is diverted to an economical replaceable wear element that is periodically discarded and replaced as "wear out" occurs. These general wear control methods pertain to not only adhesive wear but abrasive wear as well.

Abrasive Wear

In the case of abrasive wear, the wear particles are removed from the surface by the plowing and gouging action of the asperities of a harder mating surface or by hard particles entrapped between the rubbing surfaces. This type of wear is manifested by a system of surface grooves and scratches, often called *scoring*. The abrasive wear condition in which the hard asperities of one surface wear away the mating surface is commonly called *two-body wear*, and the condition in which hard abrasive particles between the two surfaces cause the wear is called *three-body wear*.

Abrasive wear may be subcategorized as high-stress gouging, low-stress scratching, or erosion. In the case of gouging, the high stress levels are often caused by impact loading, and the cutting action is accompanied by noticeable surface deformation. Low-stress scratching is usually caused by entrapped grit particles that crush to form tiny, sharp fragments that in turn scratch and abrade the wearing surfaces. Erosion is caused by a stream of particles moving generally parallel to a bounding surface under the driving force of a fluid, such as air or water, or is caused by a body force field, such as gravity.

A simple abrasive wear model has been developed (8) for the case in which the shape of the cutting asperities or particles is assumed to be conical, as shown in Figure 17.2. Considering first the action of a single conical asperity, we may deduce that the depth of penetration into the softer surface may be estimated by noting that penetration ceases when the portion of the load carried by a single asperity, W', divided by the projected contact area of the asperity in a horizontal plane, A_{ph}, becomes equal to the flow stress, or when

$$\frac{W'}{A_{ph}} = 3\sigma_{yp} \tag{17-12}$$

or

$$W' = (3\sigma_{yp})(\pi r^2) \tag{17-13}$$

where r is the radius of the hard penetrating cone at the elevation of the soft metal surface.

The cross-sectional area of the V groove produced by the cone plowing through the soft metal may now be expressed as

$$A_{pv} = rh = r^2 \tan\theta \tag{17-14}$$

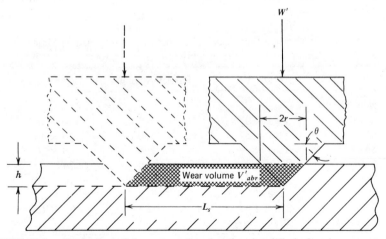

FIGURE 17.2. Conical particle assumption for the simplified abrasive wear model. (After ref. 8; adapted with permission from John Wiley & Sons, Inc.)

and the total wear volume produced for a sliding distance L_s is

$$V_{abr} = A_{pv} L_s = L_s r^2 \tan \theta \qquad (17\text{-}15)$$

Substituting for r^2 from (17-13) then yields for a single asperity

$$V'_{abr} = \frac{W' L_s \tan \theta}{3 \pi \sigma_{yp}} \qquad (17\text{-}16)$$

and for the aggregation of all asperities

$$V_{abr} = \frac{W L_s (\tan \theta)_m}{3 \pi \sigma_{yp}} \qquad (17\text{-}17)$$

where W is total applied load, $(\tan \theta)_m$ is a weighted mean value for all asperities, L_s is total distance of sliding, and σ_{yp} is the uniaxial yield point strength for the softer material.

Comparing (17-17) for abrasive wear volume with (17-7) for adhesive wear volume, we note that they are formally the same except the constant $k/3$ in the adhesive wear equation is replaced by $(\tan \theta)_m/\pi$ in the abrasive wear equation. Typical values of the wear constant $3(\tan \theta)_m/\pi$ for several materials are shown in Table 17.4.

An average abrasive wear depth d_{abr} may then be estimated as

$$d_{abr} = \frac{V_{abr}}{A_a} = \frac{(\tan \theta)_m}{3 \pi \sigma_{yp}} \left(\frac{W}{A_a} \right) L_s \qquad (17\text{-}18)$$

or

$$d_{abr} = k_{abr} P_m L_s \qquad (17\text{-}19)$$

where $p_m = W/A_a$ is mean nominal contact pressure between bearing surfaces, L_s is total sliding distance, and $k_{abr} = (\tan \theta)_m/3 \pi \sigma_{yp}$ is an abrasive wear coefficient that depends on the roughness characteristics of the surface and

Table 17.4. Abrasive Wear Constant $3(\tan \theta)_m/\pi$ for Various Materials in Sliding Contact as Reported by Different Investigators*

Materials	Wear type	Particle size, μ	$3(\tan \theta)_m/\pi$
Many	Two-body	–	180×10^{-3}
Many	Two-body	110	150×10^{-3}
Many	Two-body	40–150	120×10^{-3}
Steel	Two-body	260	80×10^{-3}
Many	Two-body	80	24×10^{-3}
Brass	Two-body	70	16×10^{-3}
Steel	Three-body	150	6×10^{-3}
Steel	Three-body	80	4.5×15^{-3}
Many	Three-body	40	2×10^{-3}

*See p. 169 of ref. 8. (Reprinted with permission from John Wiley & Sons, Inc.)

the yield strength (or hardness) of the softer material. Values of k_{abr} must be experimentally determined for each material combination and surface condition of interest, although useful data for approximating k_{abr} have been generated for several cases, some of which are shown in Table 17.4. As indicated in Table 17.4, experimental evidence shows that k_{abr} for three-body wear is typically about an order of magnitude smaller than for the two-body case, probably because the entrapped particles tend to roll much of the time and cut only a small part of the time.

As was the case for adhesive wear, the problem in practical application of (17-19) is finding the appropriate value of the abrasive wear constant. If proper surface finish is used, the two-body wear case rarely governs. Three-body wear generated by grit particles from external sources, such as dust and dirt in the atomsphere, is usually predominant; and since these particles are variable in composition, size, geometry, and quantity, the resulting abrasive wear is often widely variable. If particles from the external environment produce significant abrasive wear in a given application, steps must be taken to seal, filter, or otherwise exclude the offending grit particles.

In selecting materials for abrasive wear resistance, it has been established that both hardness and modulus of elasticity are key properties. Increasing

Table 17.5. Values of (Hardness/Modulus of Elasticity) for Various Materials (1)[†]

Material	Condition	BHN*/$(E \times 10^{-6})$ (in mixed units)
Alundum (Al_2O_3)	Bonded	143
Chrome plate	Bright	83
Gray iron	Hard	33
Tungsten carbide	9% Co	22
Steel	Hard	21
Titanium	Hard	17
Aluminum alloy	Hard	11
Gray iron	As cast	10
Structural steel	Soft	5
Malleable iron	Soft	5
Wrought iron	Soft	3.5
Chromium metal	As cast	3.5
Copper	Soft	2.5
Silver	Pure	2.3
Aluminum	Pure	2.0
Lead	Pure	2.0
Tin	Pure	0.7

*Brinell hardness number.
[†]Reprinted from copyrighted work with permission; courtsey of Elsevier Publishing Company.

wear resistance is associated with higher hardness and lower modulus of elasticity since both the amount of elastic deformation and the amount of elastic energy that can be stored at the surface are increased by higher hardness and lower modulus of elasticity.

Table 17.5 tabulates several materials in order of descending values of (hardness)/(modulus of elasticity). Well controlled experimental data are not yet available, but general experience would provide an ordering of materials for decreasing wear resistance compatible with the array of Table 17.5.

Corrosion Wear

When the conditions for adhesive or abrasive wear exist together with conditions that lead to corrosion, the two processes persist together and often interact synergistically. If the corrosion product is hard and abrasive, dislodged corrosion particles entrapped between contacting surfaces will accelerate the abrasive wear process. In turn, the wear process may remove the "protective" surface layer of corrosion product to bare new metal to the corrosive atmosphere, thereby accelerating the corrosion process. Thus, the corrosion wear process may be self-accelerating and may lead to high rates of wear.

On the other hand, some corrosion products, for example metallic phosphates, sulfides, and chlorides, form as soft lubricative films that actually improve the wear rate markedly, especially if adhesive wear is the dominant phenomenon.

Although some attempts have been made to quantitatively predict depth of wear under corrosive wear conditions,* good predictive models are not yet available.

Surface Fatigue Wear

When two surfaces operate in rolling contact, the wear phenomenon is quite different from the wear of sliding surfaces just described, although recent studies of sliding wear have resulted in a theory of sliding wear, called the "delamination" theory (13), which is very similar to the mechanism of wear between rolling surfaces in contact as described here. Rolling surfaces in contact result in Hertz contact stresses that produce maximum values of shear stress slightly below the surface.† As the rolling contact zone moves past a given location on the surface, the subsurface peak shear stress cycles from zero to a maximum value and back to zero, thus producing a cyclic stress field. From Chapters 7, 8, and 9, it is clear that such conditions may lead to fatigue failure by the initiation of a subsurface crack that propagates under repeated cyclic loading and that may ultimately propagate to the surface to

*See p. 190 of ref. 8.
†See, for example, p. 389 of ref. 14.

spall out a macroscopic surface particle to form a wear pit. This action, called *surface fatigue wear*, is a common failure mode in antifriction bearings, gears, cams, and all machine parts that involve rolling surfaces in contact. Empirical tests by bearing manufacturers have shown the life N in cycles to be approximately

$$N = \left(\frac{C}{P}\right)^3 \qquad (17\text{-}20)$$

where P is the bearing load and C is a constant for a given bearing. It may be noted that the life is inversely proportional to approximately the cube of the load. The constant C has been defined by the Antifriction Bearing Manufac-turer's Association (AFBMA) as the *basic load rating*[*], which is the radial load C that a group of identical bearings can sustain for a rating life of 1 million revolutions of the inner ring with a 90 percent reliability. It must be noted that since surface fatigue wear is basically a fatigue phenomenon, all the influencing factors of Chapter 7, 8, and 9 must be taken into account.

Surface durability and determination of gear-tooth wear load criteria must also consider the surface fatigue wear phenomenon. In some types of gearing, such as worm gears and hypoid gears, a combination of rolling and sliding exists; and adhesive wear, abrasive wear, corrosive wear, and surface fatigue wear are all potential failure modes. Only through proper design, good manufacturing practice, and use of proper lubricant can the desired design life be attained, and experimental life-testing is essential in a complex application such as this if proper field operation is to be expected. A more detailed discussion of surface fatigue wear of gear teeth may be found in any good machine design textbook.[†]

Deformation Wear, Fretting Wear, and Impact Wear

Deformation wear arises as a result of repeated plastic deformations at the wearing surfaces that may induce a matrix of cracks that grow and coalesce to form wear particles, or that may produce cumulative permanent plastic deformations that finally grow into an unacceptable surface indentation or wear scar. Deformation wear is generally caused by conditions that lead to impact loading between the two wearing surfaces. Although some progress has been made in deformation wear analysis,[‡] the techniques are specialized and beyond the scope of this discussion. Fretting wear, which has received renewed attention in the recent literature,[§] has already been discussed in Chapter 14. Impact wear is a term reserved for impact-induced repeated elastic deformations at the wearing surfaces that produce a matrix of cracks

[*]See, for example, p. 389 of ref. 14.
[†]See, for example, p. 510 ff. of ref. 14 or p. 444 ff. of ref. 15.
[‡]See p. 120 of ref. 5.
[§]See p. 55 of ref. 16 and p. 75 of ref. 8.

that grow in accordance with surface fatigue phenomena. Under some circumstances impact wear may be generated by purely normal impacts, and under other circumstances the impact may contain elements of rolling and/or sliding as well. The severity of the impact is generally measured or expressed in terms of the kinetic energy of the striking mass. The geometry of the impacting surfaces and the materials properties of the two contacting surfaces play a major role in the determination of severity of impact wear damage. The objective of a designer faced with impact wear as a potential failure mode is to predict the size of the wear scar, or its depth, as a function of the number of repetitive load cycles. Although empirical procedures have been established for making such predictions in some cases,* the details of such procedures are beyond the scope of this text.

17.3   EMPIRICAL MODEL FOR ZERO WEAR

In the discussions of various types of sliding wear given in Section 17.2, it was indicated that engineering predictive models for sliding wear were not well developed. However, an *empirical* approach to the prediction of sliding wear has been developed (17), and the pertinent empirical constants have been evaluated for a wide variety of materials and lubricant combinations for various operating conditions. This empirical development permits the designer to specify a design configuration to assure "zero wear" during the specified design lifetime. *Zero wear* is defined to be wear of such small magnitude that the surface finish is not significantly altered by the wear process. That is, the wear depth for zero wear is of the order of one-half the peak-to-peak surface finish dimension.

If a *pass* is defined to be a distance of sliding, W, equal to the dimension of the contact area in the direction of sliding, N is the number of passes, τ_{max} is the maximum shearing stress in the vicinity of the surface, τ_{yp} is the shear yield point of the specified material, and γ_r is a constant for the particular combination of materials and lubricant, then the empirical model asserts that there will be "zero wear" for N passes if

$$\tau_{max} \leq \left[\frac{2 \times 10^3}{N} \right]^{1/9} \gamma_r \tau_{yp} \qquad (17\text{-}21)$$

or, to interpret it differently, the number of passes that can be accommodated without exceeding the zero wear level is given by

$$N = 2 \times 10^3 \left[\frac{\gamma_r \tau_{yp}}{\tau_{max}} \right]^9 \qquad (17\text{-}22)$$

It may be noted that the constant γ_r is referred to 2000 passes and must be

*See ref. 5 and p. 401 of ref. 16.

experimentally determined. For quasi-hydrodynamic lubrication, γ_r ranges between 0.54 and unity. For dry or boundary lubrication, γ_r is 0.54 for materials with low susceptability to adhesive wear and 0.20 for materials with high susceptability to adhesive wear. In Table 17.6 values of γ_r are given for several combinations of materials and lubricants.*

Calculation of the maximum shear stress τ_{max} in the vicinity of the contacting surface must include both the normal force and the friction force. Thus, for conforming geometries, such as a flat surface on a flat surface or a shaft

Table 17.6. Values of Constant γ_r and Coefficient of Friction μ for Various Combinations of Materials and Lubricants (17)

Material Combination	Lubrication (see Table 17.7 for key)	γ_r	μ
52100 steel on 302 stainless	Dry	0.2	1.00
	A	0.2	0.19
	B	0.2	0.16
	C	0.2	0.21
52100 steel on 1045 steel	Dry	0.20	0.67
	A	0.54	0.15
	B	0.20	0.17
	C	0.20	0.28
	D	0.54	0.08
52100 steel on 52100 steel	Dry	0.20	0.60
	A	0.20	0.21
	B	0.54	0.16
	C	0.20	0.21
	D	0.54	0.10
52100 steel on 356 aluminum	Dry	0.20	1.40
	A	0.54	0.22
	B	0.54	0.17
	C	0.54	0.23
	D	0.54	0.10
52100 steel on sintered bronze	Dry	0.20	0.26
	A	0.20	0.23
	B	0.20	0.11
	C	0.20	0.18
52100 steel on chrome plating on 1018 steel substrate	Dry	0.20	0.51
52100 steel on anodized aluminum On 2024 substrate	Dry	0.54	0.16

*Abridged table from data of ref. 17.

Table 17.6. (*Continued*)

Material Combination	Lubrication (see Table 17.7 for key)	γ_r	μ
302 stainless on 302 stainless	Dry	0.20	1.02
	A	0.20	0.16
	B	0.20	0.15
	C	0.20	0.17
302 stainless on 1045 steel	Dry	0.20	0.17
	A	0.20	0.16
	B	0.54	0.14
	C	0.54	0.15
	D	0.54	0.11
302 stainless on 356 aluminum	Dry	0.20	1.78
	A	0.54	0.18
	B	0.54	0.21
	C	0.54	0.18
	D	0.54	0.10
302 stainless on Teflon	Dry	0.54	0.09
	A	0.54	0.15
	B	0.54	0.11
	C	0.54	0.12

in a journal bearing, a critical point at the contacting interface may be analyzed by the maximum shear stress theory* to determine τ_{max}. Since only a normal component and a friction-generated shear component of stress exist, the state of stress is approximately biaxial and

$$\tau_{max} = K_e\sqrt{\left(\frac{\sigma_n}{2}\right)^2 + \tau_f^2} \qquad (17\text{-}23)$$

where the normal stress σ_n is equal to the normal pressure p_o, the friction generated shear stress τ_f is equal to the coefficient of friction μ times normal pressure p_0, and K_e is a stress concentration factor to account for the shape of the corners and edges, to be experimentally determined. Typical values of K_e range from 2 or 3 for well rounded edges to as large as 1000 for sharp corners. Thus, (17-23) may be written as

$$\tau_{max} \doteq K_e\sqrt{\left(\frac{p_0}{2}\right)^2 + (\mu p_o)^2} \qquad (17\text{-}24)$$

*See Chap. 6.

Table 17.7. Characteristics of Lubricants A, B, C, and D used in Table 17.6*

	OIL A		OIL B	
	(Socony Vaccuum Gargoyle PE 797)		(Esso Standard Oil Millcot K-50)	
Type of stock	-Paraffin	Type of stock	-Napthenic	
Flash point (open cup)	-405°F	Flash point (open cup)	-435°F	
Pour point	-20°F	Pour point	-15°F	
Gravity	-33.0 API	Gravity	-23.1 API	
Viscosity index	-105	Viscosity index	-77	
Neutralization no.	-0.05	Neutralization no.	-0.03	
Type of additive	-Oxidation and corrosion	Type of additive	-Oxidation and tackiness	

	OIL C		OIL D	
	(Texaco MIL-0-5606)		Oil A doped with 0.2% by weight of Stearic acid	
Type of stock	-Paraffin			
Flash point (open cup)	-200°F			
Pour point	- -75°F			
Gravity	-1.15-1.18 (specific gravity)			
Viscosity index	-188			
Neutralization no.	-0.20			
Type of additive	-V. I. improver-anti-wear			

*From ref. 17.

or

$$\tau_{\max} \doteq K_e p_0 \sqrt{\left(\frac{1}{2}\right)^2 + \mu^2} \tag{17-25}$$

For flat sliders with nominal contact area A_0 under normal load P

$$p_0 = \frac{P}{A_0} \tag{17-26}$$

and for a shaft in a journal bearing of the same nominal diameter d and bearing length l

$$p_0 = \frac{P}{ld} \tag{17-27}$$

Maximum shearing stress expressions for other geometries may be found in the literature.*

The number of passes will usually require expression as a function of the number of cycles, strokes, oscillations, or hours of operation in the design lifetime. In most sliding pairs one element will typically remain fully loaded during operation, whereas the second element will be periodically unloaded as the load passes by, or the second element may be only partially unloaded if the motion is oscillatory. For the fully loaded element the number of passes n per unit operation is

$$n_{fl} = \frac{D_s}{W} \tag{17-28}$$

where D_s is the sliding distance per unit operation and W is the width of the contact area in the direction of sliding. The number of passes per unit operation for the fully unloaded element is

$$n_{fu} = \text{number of loadings per unit operation} \tag{17-29}$$

For the oscillatory motion case of partially unloading the second element, the number of passes per single complete oscillation for the partially unloaded member is

$$n_{pu} = \frac{D_s}{W} \tag{17-30}$$

and for the loaded element

$$n_l = 2 \tag{17-31}$$

Utilizing these definitions and a proper stress analysis at the wear interface allows one to design for "zero wear" by use of (17-21) or (17-22), together with empirical constants of the type shown in Table 17.6. Good success has been obtained using this technique (17).

*See, for example, pp. 4–8 of ref. 17.

17.4   USING THE IDEAS

An experiment was performed using a cylindrical 1045 steel slider heat treated to a hardness of Rockwell C-45 (σ_{yp} = 128,000 psi) pressed endwise against a 52100 steel disk with no lubricant. It was found that for a relative sliding velocity of 0.67 feet per second the 0.031-inch-diameter slider loaded by a 40-lb axial force produced a slider wear volume of 5.8×10^{-8} cubic inches during a test of 40 minutes duration.

1. If the same material combination is to be used in a slider bearing application at a sliding velocity of 3.0 feet per second under a bearing load of P = 100 lb, and if the slider is to be square, what side dimension s should it have to assure a lifetime of 1000 hours if a maximum wear depth of 0.050 inch can be tolerated?

2. If it is desired to design the slider bearing for zero wear, of what dimensions should the square slider be made?

To determine the slider dimensions to meet the criteria of the first problem, either (17-11) or (17-19) may be rearranged to give for the slider bearing application.

$$p_m = \frac{P}{A_a} = \frac{P}{s^2} = \frac{d_w}{k_w L_s} \qquad (17\text{-}32)$$

or

$$s^2 = \frac{P k_w L_s}{d_w} \qquad (17\text{-}33)$$

The load P = 100 lb and wear depth d_w = 0.050 inch are known design requirements, and sliding distance L_s may be calculated as

$$L_s = (1000 \text{ hours})\left(3\frac{\text{feet}}{\text{second}}\right)\left(3600\frac{\text{seconds}}{\text{hour}}\right)\left(12\frac{\text{inches}}{\text{foot}}\right) \qquad (17\text{-}34)$$

or

$$L_s = 1.296 \times 10^8 \text{ inches} \qquad (17\text{-}35)$$

Then

$$s = \sqrt{\frac{(100)(1.296)10^8 k_w}{0.050}} = 5.09 \times 10^5 \sqrt{k_w} \qquad (17\text{-}36)$$

The value of the wear constant k_w may be determined from an equation of the form of (17-10) and the experimental data as

$$k_w = \frac{d_w}{p_m L_s} \qquad (17\text{-}37)$$

where

$$d_w = \frac{V}{A_a} = \frac{5.8 \times 10^{-8}}{\left[\pi \dfrac{(0.031)^2}{4}\right]} = 7.68 \times 10^{-5} \text{ inches} \tag{17-38}$$

$$p_m = \frac{P}{A_a} = \frac{40}{\left[\pi \dfrac{(0.031)^2}{4}\right]} = 5.3 \times 10^4 \text{ psi} \tag{17-39}$$

$$L_s = \left(0.67 \frac{\text{feet}}{\text{second}}\right)\left(60 \frac{\text{seconds}}{\text{minute}}\right)(40 \text{ minutes})\left(12 \frac{\text{inches}}{\text{foot}}\right)$$

$$L_s = 1.93 \times 10^4 \text{ inches} \tag{17-40}$$

whence

$$k_w = \frac{7.68 \times 10^{-5}}{(5.3 \times 10^4)(1.93 \times 10^4)} = 0.75 \times 10^{-13} \frac{\text{inches}^2}{\text{pound}} \tag{17-41}$$

and from (17-36), then,

$$s = 5.09 \times 10^5 \sqrt{0.75 \times 10^{-13}} = 0.14 \text{ inch} \tag{17-42}$$

so the slider would be tentatively made 9/64 inch on a side.

It still remains to verify the limiting conditions in the use of (17-32) that for the equation to be valid it must be true that

$$p_m \le \sigma_{yp} \tag{17-43}$$

or

$$\frac{P}{A_a} = \frac{100}{(0.14)^2} = 5100 \le 128,000 \tag{17-44}$$

Since (17-43) is satisfied, the design is valid and the 9/64-inch square slider is adopted as the final design.

Proceeding to the zero wear design, we may use (17-21). The value of $\gamma_r = 0.20$ may be read from Table 17.6 for 52100 steel on 1045 steel under dry operating conditions. Thus, (17-21) may be written, assuming that the maximum shearing stress theory is sufficiently accurate, as

$$\tau_{\max} = \left[\frac{2 \times 10^3}{N}\right]^{1/9} (0.20)\left(\frac{128,000}{2}\right) \tag{17-45}$$

or

$$\tau_{\max} = 2.98 \times 10^4 \left[\frac{1}{N}\right]^{1/9} \tag{17-46}$$

Since the slider is fully loaded, the number of passes may be calculated from (17-28) as

$$N = \frac{L_s}{s} \frac{1.296 \times 10^8}{s} \qquad (17\text{-}47)$$

noting that the sliding distance per unit operation is the total sliding distance calculated in (17-35), assuming the unit operation to be the entire design life, and for the same reason the number of passes per unit operation is just the total number of passes N.

Thus, (17-46) becomes

$$\tau_{max} = 2.98 \times 10^4 \left[\frac{s}{1.296 \times 10^8} \right]^{1/9} \qquad (17\text{-}48)$$

or

$$\tau_{max} = 3.74 \times 10^3 (s)^{1/9} \qquad (17\text{-}49)$$

Then, from (17-25) and (17-26), the expression of (17-49) may be rewritten as

$$\frac{K_e P}{A_0} \sqrt{\left(\frac{1}{2}\right)^2 + \mu^2} = 3.74 \times 10^3 (s)^{1/9} \qquad (17\text{-}50)$$

The edge stress concentration factor K_e may be assumed to be 3 for well rounded edges; the coefficient of friction $\mu = 0.67$ may be read from Table 17.6; $P = 100$ lb is a design requirement, and, since the slider is square, $A_o = s^2$. Thus, (17-50) becomes

$$\frac{3(100)}{s^2} \sqrt{\left(\frac{1}{2}\right)^2 + (0.67)^2} = 3.74 \times 10^3 (s)^{1/9} \qquad (17\text{-}51)$$

to meet the condition of zero wear.

Solving for the dimension s,

$$s = \left[\frac{3(100)}{3.74 \times 10^3} \sqrt{\left(\frac{1}{2}\right)^2 + (0.67)^2} \right]^{1/2.111}$$

or

$$s = 0.278 \text{ inch} \qquad (17\text{-}52)$$

and the slider would be made 9/32 inch square with well rounded edges to meed the zero wear criterion.

17.5   CORROSION

Corrosion may be defined as the undesired deterioration of a material through chemical or electrochemical interaction with the environment, or destruction of materials by means other than purely mechanical action.

Failure by corrosion occurs when the corrosive action renders the corroded device incapable of performing its design function. Corrosion often interacts synergistically with another failure mode, such as wear or fatigue, to produce the even more serious combined failure modes, such as corrosion wear or corrosion fatigue. Failure by corrosion and protection against failure by corrosion has been estimated to cost in excess of eight billion dollars annually in the United States alone.* Although much progress has been made in recent years in understanding and controlling this important failure mode, much remains to be learned. It is important for the mechanical engineering designer to become acquainted with the various types of corrosion so that corrosion-related failures can be avoided.

The complexity of the corrosion process may be better appreciated by recognizing that many variables are involved, including environmental, electrochemical, and metallurgical aspects. For example, anodic reactions and rate of oxidation, cathodic reactions and rate of reduction, corrosion inhibition, polarization or retardation, passivity phenomena, effect of oxidizers, effect of velocity, temperature, corrosive concentration, galvanic coupling, and metallurgical structure all influence the type and rate of the corrosion process.

Corrosion processes have been categorized in many different ways. One convenient classification divides corrosion phenomena into the following types:† direct chemical attack, galvanic corrosion, crevice corrosion, pitting corrosion, intergranular corrosion, selective leaching, erosion corrosion, cavitation corrosion, hydrogen damage, biological corrosion, and stress corrosion cracking. Depending upon the types of environment, loading, and mechanical function of the machine parts involved, any of the types of corrosion may combine their influence with other failure modes to produce premature failures. Of particular concern are interactions that lead to failure by corrosion wear, corrosion fatigue, fretting wear, fretting fatigue, and corrosion-induced brittle fracture.

Direct Chemical Attack

Direct chemical attack is probably the most common type of corrosion. Under this type of corrosive attack the surface of the machine part exposed to the corrosive media is attacked more or less uniformly over its entire surface, resulting in a progressive deterioration and dimensional reduction of sound load-carrying net cross section. The rate of corrosion due to direct attack can usually be estimated from relatively simple laboratory tests in which small specimens of the selected material are exposed to a well simulated actual environment, with frequent weight change and dimensional measurements carefully taken. The corrosion rate is usually expressed in mils per year (mpy)

*See p. 1 of ref. 10.
†See p. 28 of ref. 19 and p. 85 of ref. 20.

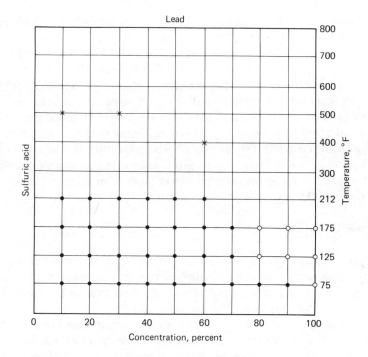

FIGURE 17.3. Nelson's method for summarizing corrosion rate data for lead in sulfuric acid environment as a function of concentration and temperature. (See ref. 21; reprinted with permission of McGraw Hill Book Company.)

and may be calculated as*

$$R = \frac{534W}{\gamma A t} \qquad (17\text{-}53)$$

where R is rate of corrosion penetration in mils[†] per year (mpy), W is weight loss in milligrams, A is exposed area of the specimen in inches^2, γ is density of the specimen in grams per cubic centimeter, and t is exposure time in hours. Use of this corrosion rate expression in predicting corrosion penetration in actual service is usually successful if the environment has been properly simulated in the laboratory. Corrosion rate data for many different

*See p. 133 of ref. 19.
†One mil is 0.001 inch.

combinations of materials and environments are available in the literature.*
Figure 17.3 illustrates one presentation of such data.

Direct chemical attack may be reduced in severity or prevented by any one
or a combination of several means, including selecting proper materials to
suit the environment; using plating, flame spraying, cladding, hot dipping,
vapor deposition, conversion coatings, and organic coatings or paint to
protect the base material; changing the environment by using lower temperature or lower velocity, removing oxygen, changing corrosive concentration, or
adding corrosion inhibitors; using cathodic protection in which electrons are
supplied to the metal surface to be protected either by galvanic coupling to a
sacrificial anode or by an external power supply; or adopting other suitable
design modifications.

Galvanic Corrosion

Galvanic corrosion is an accelerated electrochemical corrosion that occurs
when two dissimilar metals in electrical contact are made part of a circuit
completed by a connecting pool or film of electrolyte or corrosive medium.
Under these circumstances, the potential difference between the dissimilar
metals produces a current flow through the connecting electrolyte, which
leads to corrosion, concentrated primarily in the more anodic or less noble
metal of the pair. This type of action is completely analogous to a simple
battery cell. Current must flow to produce galvanic corrosion, and, in general,
more current flow means more serious corrosion. The relative tendencies of
various metals to form galvanic cells, and the probable direction of the
galvanic action, are illustrated for several commercial metals and alloys in
seawater in Table 17.8.† Ideally, tests in the actual service environment should
be conducted; but, if such data are unavailable, the data of Table 17.8 should
give a good indication of possible galvanic action. The farther apart the two
dissimilar metals are in the galvanic series, the more serious the galvanic
corrosion problem may be. Material pairs within any bracketed group exhibit
little or no galvanic action. It should be noted, however, that there are
sometimes exceptions to the galvanic series of Table 17.8, so wherever
possible corrosion tests should be performed with actual materials in the
actual service environment.

The accelerated galvanic corrosion is usually most severe near the junction
between the two metals, decreasing in severity at locations farther from the
junction. The ratio of cathodic area to anodic area exposed to the electrolyte
has a significant effect on corrosion rate. It is *desirable* to have a *small ratio* of
cathode area to anode area. For this reason, if only *one* of two dissimilar
metals in electrical contact is to be coated for corrosion protection, the *more*
noble or more corrosion-resistant metal should be coated. Although this at

*See, for example, refs. 21, 22, and 23.
†See p. 32 of ref. 19 or p. 86 of ref. 20.

Table 17.8. Galvanic Series of Several Commercial Metals and Alloys in Seawater*

↑	Platinum
	Gold
Noble or	Graphite
cathodic	Titanium
(protected	Silver
end)	⎡ Chlorimet 3 (62 Ni, 18 Cr, 18 Mo) ⎤
	⎣ Hastelloy C (62 Ni, 17 Cr, 15 Mo) ⎦
	⎡ 18-8 Mo stainless steel (passive) ⎤
	⎢ 18-8 stainless steel (passive) ⎥
	⎣ Chromium stainless steel 11-30% Cr (passive) ⎦
	⎡ Inconel (passive)(80 Ni, 13 Cr, 7 Fe) ⎤
	⎣ Nickel (passive) ⎦
	Silver solder
	⎡ Monel (70 Ni, 30 Cu) ⎤
	⎢ Cupronickels (60-90 Cu, 40-10 Ni) ⎥
	⎢ Bronzes (Cu-Sn) ⎥
	⎢ Copper ⎥
	⎣ Brasses (Cu-Zn) ⎦
	⎡ Chlorimet 2 (66 Ni, 32 Mo, 1 Fe) ⎤
	⎣ Hastelloy B (60 Ni, 30 Mo, 6 Fe, 1 Mn) ⎦
	⎡ Inconel (active) ⎤
	⎣ Nickel (active) ⎦
	Tin
	Lead
	Lead-tin solders
	⎡ 18-8 Mo stainless steel (active) ⎤
	⎣ 18-8 stainless steel (active) ⎦
	Ni-Resist (high Ni cast iron)
	Chromium stainless steel, 13% Cr (active)
	⎡ Cast iron ⎤
	⎣ Steel or iron ⎦
Active or	2024 aluminum (4.5 Cu, 1.5 Mg, 0.6 Mn)
anodic	Cadmium
(corroded	Commercially pure aluminum (1100)
end)	Zinc
↓	Magnesium and magnesium alloys

*See p. 32 of ref. 19. (Reprinted with permission of McGraw-Hill Book Company.)

first may seem the wrong metal to coat, the area effect, which produces anodic corrosion rates of $10^2 - 10^3$ times cathodic corrosion rates for equal areas, provides the logic for this assertion.

Galvanic corrosion may be reduced in severity or prevented by one or a combination of several steps, including the selection of material pairs as close together as possible in the galvanic series, preferably in the same bracketed group; electrical insulation of one dissimilar metal from the other as completely as possible; maintaining as small a ratio of cathode area to anode area as possible; proper use and maintenance of coatings; the use of inhibitors to decrease the aggressiveness of the corroding medium; and the use of cathodic protection in which a third metal element anodic to both members of the operating pair is used as a sacrificial anode that may require periodic replacement.

Crevice Corrosion

Crevice corrosion is an accelerated corrosion process highly localized within crevices, cracks, and other small-volume regions of stagnant solution in contact with the corroding metal. For example, crevice corrosion may be expected in gasketed joints; clamped interfaces; lap joints; rolled joints; under bolt and rivet heads; and under foreign deposits of dirt, sand, scale, or corrosion product. Until recently, crevice corrosion was thought to result from differences in either oxygen concentration or metal ion concentration in the crevice compared to its surroundings. More recent studies seem to indicate, however, that the local oxidation and reduction reactions result in oxygen depletion in the stagnant crevice region, which leads to an excess positive charge in the crevice due to increased metal ion concentration. This, in turn, leads to a flow of chloride and hydrogen ions into the crevice, both of which accelerate the corrosion rate within the crevice. Such accelerated crevice corrosion is highly localized and often requires a lengthy incubation period of perhaps many months before it gets under way. Once started, the rate of corrosion accelerates to become a serious problem. To be susceptible to crevice corrosion attack, the stagnant region must be wide enough to allow the liquid to enter but narrow enough to maintain stagnation. This usually implies cracks and crevices of a few thousandths to a few hundredths of an inch in width.

To reduce the severity of crevice corrosion, or prevent it, it is necessary to eliminate the cracks and crevices. This may involve caulking or seal welding existing lap joints; redesign to replace riveted or bolted joints by sound, welded joints; filtering foreign material from the working fluid; inspection and removal of corrosion deposits, or using nonabsorbent gasket materials.

Pitting Corrosion

Pitting corrosion is a very localized attack that leads to the development of an array of holes or pits that penetrate the metal. The pits, which typically are about as deep as they are across, may be widely scattered or so heavily concentrated that they simply appear as a rough surface. The mechanism of pit growth is virtually identical to that of crevice corrosion described, except that an existing crevice is not required to initiate pitting corrosion. The pit is probably initiated by a momentary attack due to a random variation in fluid concentration or a tiny surface scratch or defect. Some pits may become inactive because of a stray convective current, whereas others may grow large enough to provide a stagnant region of stable size, which then continues to grow over a long period of time at an accelerating rate. Pits usually grow in the direction of the gravity force field since the dense concentrated solution in a pit is required for it to grow actively. Most pits, therefore, grow downward from horizontal surfaces to ultimately perforate the wall. Fewer pits are formed on vertical walls, and very few pits grow upward from the bottom surface.

Measurement and assessment of pitting corrosion damage is difficult because of its highly local nature. Pit depth varies widely and, as in the case of fatigue damage, a statistical approach must be taken in which the probability of a pit of specified depth may be established in laboratory testing. Unfortunately, a significant size effect influences depth of pitting, and this must be taken into account when predicting service life of a machine part based on laboratory pitting corrosion data.

The control or prevention of pitting corrosion consists primarily of the wise selection of material to resist pitting or, since pitting is usually the result of stagnant conditions, imparting velocity to the fluid. Increasing its velocity may also decrease pitting corrosion attack.

Intergranular Corrosion

Because of the atomic mismatch at the grain boundaries of polycrystalline metals, the stored strain energy is higher in the grain boundary regions than in the grains themselves. These high-energy grain boundaries are more chemically reactive than the grains. Under certain conditions depletion or enrichment of an alloying element or impurity concentration at the grain boundaries may locally change the composition of a corrosion-resistant metal, making it susceptible to corrosive attack. Localized attack of this vulnerable region near the grain boundaries is called intergranular corrosion. In particular, the austenitic stainless steels are vulnerable to intergranular corrosion if *sensitized* by heating into the temperature range from 950°F to 1450°F, which causes depletion of the chromium near the grain boundaries as chromium carbide is precipitated at the boundaries. The chromium-poor regions then corrode because of local galvanic cell action, and the grains literally fall out of the matrix. A special case of intergranular corrosion, called "weld decay," is generated in the portion of the weld-affected zone, which is heated into the sensitizing temperature range.

To minimize the susceptibility of austenitic stainless steels to intergranular corrosion, the carbon content may be lowered to below 0.03 percent, stabilizers may be added to prevent depletion of the chromium near the grain boundaries, or a high-temperature solution heat treatment, called quench-annealing, may be employed to produce a more homogeneous alloy.

Other alloys susceptible to intergranular corrosion include certain aluminum alloys, magnesium alloys, copper-based alloys, and die-cast zinc alloys in unfavorable environments.

Selective Leaching

The corrosion phenomenon in which one element of a solid alloy is removed is termed selective leaching. Although the selective leaching process may occur in any of several alloy systems, the more common examples are *dezincification* of brass alloys and *graphitization* of gray cast iron.

Dezincification may occur as either a highly local "plug-type" or a broadly distributed layer-type attack. In either case, the dezincified region is porous, brittle, and weak. Dezincification may be minimized by adding inhibitors such as arsenic, antimony, or phosphorus to the alloy; by lowering oxygen in the environment; or by using cathodic protection.

In the case of graphitization of gray cast iron, the environment selectively leaches the iron matrix to leave the graphite network intact to form an active galvanic cell. Corrosion then proceeds to destroy the machine part. Use of other alloys, such as nodular or malleable cast iron, mitigates the problem because there is no graphite network in these alloys to support the corrosion residue. Other alloy systems in adverse environments that may experience selective leaching include aluminum bronzes, silicon bronzes, and cobalt alloys.

Erosion Corrosion

Erosion corrosion is an accelerated, direct chemical attack of a metal surface due to the action of a moving corrosive medium. Because of the abrasive wear action of the moving fluid, the formation of a protective layer of corrosion product is inhibited or prevented, and the corroding medium has direct access to bare, unprotected metal. Erosion corrosion is usually characterized by a pattern of grooves or peaks and valleys generated by the flow pattern of the corrosive medium. Most alloys are susceptible to erosion corrosion, and many different types of corrosive medium may induce erosion corrosion, including flowing gases, liquids, and solid aggregates. Erosion corrosion may become a problem in such machine parts as valves, pumps, blowers, turbine blades and nozzles, conveyors, and piping and ducting systems, especially in the regions of bends and elbows.

Erosion corrosion is influenced by the velocity of the flowing corrosive medium, turbulence of the flow, impingement characteristics, concentration of abrasive solids, and characteristics of the metal alloy surface exposed to the flow. Methods of minimizing or preventing erosion corrosion include reducing the velocity, eliminating or reducing turbulence, avoiding sudden changes in the direction of flow, eliminating direct impingement where possible, filtering out abrasive particles, using harder and more corrosion-resistant alloys, reducing the temperature, using appropriate surface coatings, and using cathodic protection techniques.

Cavitation Corrosion

Cavitation often occurs in hydraulic systems, such as turbines, pumps, and piping, when pressure changes in a flowing liquid give rise to the formation and collapse of vapor bubbles at or near the containing metal surface. The impact associated with vapor bubble collapse may produce high-pressure shock waves that may plastically deform the metal locally or destroy any protective surface film of corrosion product and locally accelerate the corrosion process. Further, the tiny depressions so formed act as a nucleus for subsequent vapor bubbles, which continue to form and collapse at the same site to produce deep pits and pock marks by the combined action of mechanical deformation and accelerated chemical corrosion. This phenomenon is called cavitation corrosion. Cavitation corrosion may be reduced or prevented by eliminating the cavitation through appropriate design changes. Smoothing the surfaces, coating the walls, using corrosion-resistant materials, minimizing pressure differences in the cycle, and using cathodic protection are design changes that may be effective.

Hydrogen Damage

Hydrogen damage, although not considered to be a form of direct corrosion, is often induced by corrosion. Any damage caused in a metal by the presence of hydrogen or the interaction with hydrogen is called hydrogen damage. Hydrogen damage includes hydrogen blistering, hydrogen embrittlement, hydrogen attack, and decarburization.

Hydrogen blistering is caused by the diffusion of hydrogen atoms into a void within a metallic structure where they combine to form molecular hydrogen. The hydrogen pressure builds to a high level that, in some cases, causes blistering, yielding, and rupture. Hydrogen blistering may be minimized by using materials without voids, by using corrosion inhibitors, or by using hydrogen-impervious coatings.

Hydrogen embrittlement is also caused by the penetration of hydrogen into the metallic structure to form brittle hydrides and pin dislocation movement to reduce slip, but the exact mechanism is not yet fully understood. Hydrogen embrittlement is more serious at the higher strength levels of susceptible alloys, which include most of the high-strength steels. Reduction and prevention of hydrogen embrittlement may be accomplished by "baking out" the hydrogen at relatively low temperatures for several hours, use of corrosion inhibitors, or use of less susceptible alloys.

Decarburization and hydrogen attack are both high-temperature phenomena. At high temperatures hydrogen removes carbon from an alloy, often reducing its tensile strength and increasing its creep rate. This carbon-removing process is called *decarburization*. It is also possible that the hydrogen may lead to the formation of methane in the metal voids, which may expand to form cracks, another form of hydrogen attack. Proper selection of alloys and coatings is helpful in prevention of these corrosion-related problems.

Biological Corrosion

Biological corrosion is a corrosion process or processes that result from the activity of living organisms. These organisms may be microorganisms, such as aerobic or anaerobic bacteria, or they may be macroorganisms, such as fungi, mold, algae, or barnacles. The organisms may influence or produce corrosion by virtue of their processes of food ingestion and waste elimination. There are, for example, sulfate-reducing anaerobic bacteria, which produce iron sulfide when in contact with buried steel structures, and aerobic sulfur-oxidizing bacteria, which produce localized concentrations of sulfuric acid and serious corrosive attack on buried steel and concrete pipe lines. There are also iron bacteria, which ingest ferrous iron and precipitate ferrous hydroxide to produce local crevice corrosion attack. Other bacteria oxidize ammonia to nitric acid, which attacks most metals, and most bacteria produce carbon dioxide, which may form the corrosive agent carbonic acid. Fungi and mold assimilate organic matter and produce organic acids. Simply by their presence, fungi may provide the site for crevice corrosion attacks, as does the presence of attached barnacles and algae. Prevention or minimization of biological corrosion may be accomplished by altering the environment or by using proper coatings, corrosion inhibitors, bactericides or fungicides, or cathodic protection.

17.6 STRESS CORROSION CRACKING

Stress corrosion cracking is an extremely important failure mode because it occurs in a wide variety of different alloys. This type of failure results from a field of cracks produced in a metal alloy under the combined influence of tensile stress and a corrosive environment. The metal alloy is not attacked over most of its surface, but a system of intergranular or transgranular cracks propagate through the matrix over a period of time.

Stress levels that produce stress corrosion cracking are well below the yield strength of the material, and residual stresses as well as applied stresses may produce failure. The lower the stress level, the longer is the time required to produce cracking, and there appears to be a threshold stress level below which stress corrosion cracking does not occur.*

The chemical compositions of the environments that lead to stress corrosion cracking are highly specific and peculiar to the alloy system, and no general patterns have been observed. For example, austenitic stainless steels are susceptible to stress corrosion cracking in chloride environments but not in ammonia environments, whereas brasses are susceptible to stress corrosion cracking in ammonia environments but not in chloride environments. Thus, the "season cracking" of brass cartridge cases in the crimped zones was found to be stress corrosion cracking due to the ammonia resulting from decomposition of organic matter. Likewise, "caustic embrittlement" of steel boilers, which resulted in many explosive failures, was found to be stress corrosion cracking due to sodium hydroxide in the boiler water.

Stress corrosion cracking is influenced by stress level, alloy composition, type of environment, and temperature. Crack propagation seems to be intermittent, and the crack grows to a critical size, after which a sudden and catastrophic failure ensues in accordance with the laws of fracture mechanics. Stress corrosion crack growth in a statically loaded machine part takes place through the interaction of mechanical strains and chemical corrosion processes at the crack tip. The largest value of plane strain stress intensity factor for which crack growth does not take place in a corrosive environment is designated K_{Iscc}. In many cases, corrosion fatigue behavior is also related to the magnitude of K_{Iscc} (25).

Prevention of stress corrosion cracking may be attempted by lowering the stress below the critical threshold level, choice of a better alloy for the environment, changing the environment to eliminate the critical corrosive element, use of corrosion inhibitors, or use of cathodic protection. Before cathodic protection is implemented care must be taken to ensure that the phenomenon is indeed stress corrosion cracking because hydrogen embrittlement is accelerated by cathodic protection techniques.

Much remains to be learned about stress corrosion cracking and its prevention. Major research efforts are in progress throughout the world to better control this important failure mode.

17.7 CLOSURE

The seventeen chapters of this text have been composed to provide insight to the important failure modes often encountered by the mechanical designer. Where possible, quantitative prediction techniques have been presented as tools by which mechanical failure may be predicted and prevented. These

*See p. 96 of ref. 19.

techniques and tools, sometimes precise, sometimes empirical, sometimes oversimplified, are constantly being questioned and improved by the profession. Even as the reader completes the reading of this text, new techniques have already been developed; and the designer must again search the literature, and continue to do so, for new and more accurate tools needed for the responsible practice of the profession.

> *The woods are lovely, dark and deep,*
> *But I have promises to keep,*
> *And miles to go before I sleep,*
> *And miles to go before I sleep.**
> —Robert Frost

QUESTIONS

1.   Give a definition of *wear failure* and list the major subcategories of wear.

2.   A slider block of 1020 steel, 2 inches wide, 4 inches long, and 0.5 inch thick, is placed in contact with a large plate of hardened 4340 steel and loaded with a 1000-lb force normal to the contacting interface. If the yield strength of the 1020 steel is 48,000 psi, estimate the ratio of real contact area to apparent contact area. Does this ratio fall into the expected range?

3.   One part of the mechanism in a new metering device for a seed packaging machine is shown in Figure Q17.3. Both the slider shoe and the rotating wheel are to be made of stainless steel, with a yield strength of 40,000 psi. The contact area of the shoe is 1 inch long by 0.5 inch wide. The rotating wheel is 10 inches in diameter and rotates at 30 rpm. The spring is set to exert a constant normal force at the wearing interface of 15 lb.

(a)   If no more than 0.050-inch wear of the shoe can be tolerated, and no lubricant may be used, estimate the maintenance interval in operating hours between shoe replacements. (Assume that adhesive wear predominates.)

(b)   Would this be an acceptable maintenance interval?

(c)   If it were possible to use a lubrication system that would provide "excellent" lubrication to the contact interface, estimate the potential improvement in maintenance interval and comment on its acceptability.

(d)   Suggest other ways to improve the design from the standpoint of reducing wear rate.

4.   It has been determined that the reciprocating sliding wear between two steel parts is primarily abrasive in nature and may be regarded as three-body

*From "Stopping by Woods on a Snowy Evening" from *The Poetry of Robert Frost* edited by Edward Connery Lathem. Copyright 1923, (©) 1969 by Holt, Rinehart and Winston. Copyright 1951 by Robert Frost. Reprinted by permission of Holt, Rinehart and Winston, Publishers.

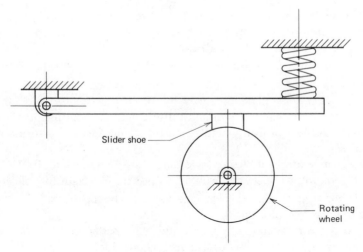

FIGURE Q17.3.

wear due to the debris produced. Further, analysis of the wear debris indicates an average particle size of about 100 μ. The contacting interface between the parts is 1 inch wide by 2 inches long (in the direction of the motion) and the normal force between the parts is 30 lb. The peak-to-peak magnitude of the reciprocating motion is 4 inches. If the device operates at a speed of 300 cycles per minute, estimate the time required to cause the shoe to wear away 0.25 inch. The steel used has a yield strength of 120,000 psi.

5. In a cinder block manufacturing plant the blocks are transported from the casting machine on rail carts supported by ball-bearing equipped wheels. The carts are currently being stacked full, six blocks high, and bearings must be replaced on a 1-year maintenance schedule because of ball-bearing failures. To increase production, a second casting machine is to be installed, but it is desired to use the same rail cart transport system with the same number of carts by merely stacking the blocks 12 high. What bearing replacement interval would you predict might be necessary under this new procedure?

6. A 1-inch diameter steel shaft of 52100 steel (σ_{yp} = 150,000 psi) reciprocates back and forth continuously in a sintered bronze bushing (σ_{yp} = 20,000 psi) with well rounded edges and a flow of lubricant (oil C of Table 17.7) is provided. The bronze bushing is 1.5 inches long, and the normal load carried by the shaft as it moves against the bushing is 800 lbs. The peak-to-peak motion of the reciprocating shaft is 1.0 inch, and the operating speed is 1000 cycles per minute. If the allowable wear may not significantly alter the surface finish, about how long would you expect the device to operate before trouble develops because of wear?

7. The shaft and bushing design of problem 6 is used in a rotating mode instead of a reciprocating mode, with all loads and dimensions remaining the

same. The shaft, however, rotates at 1200 rpm with no reciprocation. Using the same failure criterion, how long would you expect the device to operate before significant wear is observed?

8. Give a definition of *corrosion failure* and list the major subcategories of corrosion.

9. A rectagular lead plate 6 inches by 10 inches by 0.20 inch thick is submersed in 80 percent concentrated sulfuric acid at room temperature.
(a) What weight loss of lead would you estimate because of corrosion over a period of 90 days?
(b) If the temperature of the sulfuric acid were increased to 100°F above room temperature, what change would you predict in weight loss of lead?

10. It is planned to thread a bronze valve body into a cast iron pump housing to provide a bleed port.
(a) From the corrosion standpoint would it be better to make the bronze valve body as large as possible or as small as possible?
(b) Would it be more effective to put an anticorrosion coating on the bronze valve or on the cast iron housing?
(c) What other steps might be taken to minimize corrosion of the unit?

REFERENCES

1. Burwell, J.T., Jr., "Survey of Possible Wear Mechanisms," *Wear*, 1 (1957): 119–141.

2. Peterson, M.B., Gabel, M.K., and Derine, M.J., "Understanding Wear"; Ludema, K.C., "A Perspective on Wear Models"; Rabinowicz, E., "The Physics and Chemistry of Surfaces"; McGrew, J., "Design for Wear of Sliding Bearings"; Bayer, R. G., "Design for Wear of Lightly Loaded Surfaces," *ASTM Standardization News*, 2, No. 9 (September 1974): 9–32.

3. Waterhouse, R.B., *Fretting Corrosion*, Pergamon Press, New York, 1972.

4. Engel, P.A., "Predicting Impact Wear," *Machine Design*, Penton/IPC, Cleveland, May 26, 1977, pp. 100–105.

5. Engel, P.A., *Impact Wear of Materials*, Elsevier, New York, 1976.

6. Bowden, F.P., and Tabor, D., *Friction and Lubrication of Solids*, Oxford University Press, London, 1950.

7. Bowden, F.P., and Tabor, D., *Friction and Lubrication*, Methuen & Co., Ltd., London, 1967.

8. Rabinowicz, E., *Friction and Wear of Materials*, John Wiley & Sons, New York, 1966.

9. Archard, J.F., "Contact and Rubbing of Flat Surfaces," Journal of Applied Physics, 24, (1953): 981–988.

10. Lipson, C., *Wear Considerations in Design*, Prentice-Hall, Englewood Cliffs, N.J., 1967.

11. Davies, R., "Compatability of Metal Pairs," *Engineering Approach to Surface Damage*, University of Michigan Press, Ann Arbor, 1958.

12. Bayer, R.G., "Understanding the Fundamentals of Wear," *Machine Design*, 44, No. 31 (Dec. 28, 1952): 73–76.

13. Suh, N.P., "The Delamination Theory of Wear," *Wear*, 25, (1973): 111–124.

14. Shigley, J.E., *Mechanical Engineering Design*, 2nd ed., McGraw-Hill, New York, 1972.

15. Spotts, M.F., *Design of Machine Elements*, Prentice-Hall, Englewood Cliffs, N.J., 1978.

16.   Glaeser, W.A., Ludema, K.C., and Rhee, J.K. (eds.), *Wear of Materials*, American Society of Mechanical Engineers, New York, April 25–28, 1977.

17.   MacGregor, C.W. (ed.), *Handbook of Analytical Design for Wear*, Plenum Press, New York, 1964.

18.   Suh, N.P., et al., "The Delamination Theory of Wear—II," *Annual Report Nr* 229-011, Office of Naval Research, Arlington, Va., 1975.

19.   Fontana, M.G., and Greene, N.D., *Corrosion Engineering*, McGraw-Hill, New York, 1967.

20.   Seabright, L.S., and Fabian, R.J., "The Many Faces of Corrosion," *Materials in Design Engineering*, **57**, No. 1 (January 1963).

21.   Nelson, G., "Corrosion Data Survey," National Association of Corrosion Engineers, Houston, Texas 1972.

22.   Uhlig, H.H. (ed.), *Corrosion Handbook*, John Wiley & Sons, New York, 1948.

23.   Rabald, E., *Corrosion Guide*, Elsevier, New York, 1951.

24.   Rhodin, T.N., *Physical Metallurgy of Stress Corrosion Fracture*, Interscience Publishers, New York, 1959.

25.   Rolfe, S.T., and Barsom, J.M., *Fracture and Fatigue Control in Structures*, Prentice-Hall, Englewood Cliffs, N.J., 1977.

Index

Abrasive wear, 8, 11, 494, 584, 590-594, 601, 602, 610

Adhesive wear, 8, 11, 494, 583, 584-590, 592, 601, 602

Asperites, surface, 479, 481-484, 491, 584, 590, 591

Basic load rating, bearing, 595

Beach marks, 168

Bearings, basic load rating, 595

Beltrami theory, see Total strain energy failure theory

Bond, metallic, 15

Brinnelling, 8, 9

Brittle behavior, 144, 149, 229, 230, 421

Brittle fracture, 8, 9, 33, 36, 60, 69, 70, 149, 604

Buckling, 8, 13, 557-579
  column, 559-572
  critical load, 557, 560, 562, 567, 569
  critical unit load, 567
  effective column length, 563-565
  end support influence, 563-565
  Euler-Engesser equation, 566, 567, 569-571
  Euler's critical load, 562
  Euler's equation, 562, 565, 569, 570
  inelastic, 565-568
  lateral, beams in bending, 572-576
    shafts in torsion, 576-579
  local, 568, 579
  primary, 568, 579
  resisting moment, 557, 560
  secant formula, 567, 569, 571
  slenderness ratio, 567
  upsetting moment, 557, 560

Burgers circuit, 42

Burgers vector, 40, 43, 45

Cathodic protection, 608

Caustic embrittlement, 613

Cavitation, 8, 10, 604
  corrosion, 8, 10, 604

Central limit theorem, 327, 340

Characteristic life parameter, Weibull 332, 348, 350

Chi-squared distribution, 324, 327, 328

Chi-squared statistic, 321

Cladding, effects on fatigue, 196

Coating, effects on fatigue, 196, 197

Cold-rolling, effects on fatigue, 169, 200, 201, 203, 486, 488, 489

Cold welding, 585

Column buckling, 559-572

Combined creep and fatigue, 8, 14, 437, 451-470
  frequency modified stress and strain range method, 460, 461, 470
  interaction damage rule, 468
  interspersed fatigue with creep method, 461, 463-467, 470
  strain-range partitioning method, 461, 464-466, 468-470
  total time to fracture vs. number of cycles method, 460, 463, 464
  total time to fracture vs. time of one cycle method, 460, 462, 463

Condon-Morse curves, 16, 17, 19, 20

Confidence intervals, 342-344

Confidence limits, 340-343, 372, 374

Contact:
  apparent area, 584
  real area, 584

Corrosion, 8, 10, 583, 603-613
  biological, 8, 10, 604, 612
  cathodic protection, 608
  caustic embrittlement, 613

cavitation, 8, 10, 604, 611
crevice, 8, 10, 604, 608
decarburization, 612
dezincification, 610
direct chemical attack, 8, 10, 604-606
effects on fatigue, 205, 206
erosion, 8, 10, 604, 610, 611
galvanic, 8, 10, 604, 606-608
graphitization, 610
hydrogen damage, 8, 10, 604, 611, 612
intergranular, 8, 10, 604, 609,
    610
pitting, 8, 10, 604, 608, 609
rate, 604, 605
season-cracking, 613
selective leaching, 8, 10, 604, 610
stress corrosion, 8, 10, 13, 48, 604,
    612, 613
Corrosion-fatigue, 8, 14, 604
Corrosion-wear, 8, 11, 13, 14, 584, 594,
    604
Corten-Dolan theory, 254-263, 269, 271,
    273
Crack:
  detection, 288, 296, 299, 300
  displacement modes, 52, 53
  growth, *see* Crack, propagation; Fatigue
    cracks, propagation
  growth laws, *see* Fatigue cracks, propaga-
    tion
  growth rate, *see* Fatigue cracks, propa-
    gation
  length, critical, 51, 288, 296, 300, 302
    detectable, 288, 296, 299, 300
    equation for, 290
  opening displacement (COD), 67, 68
  propagation, 33, 36, 48, 50, 56, 69,
    275. *See also* Fatigue cracks
    propagation
    slow, 33, 50
    spontaneous (rapid), 36, 50, 56, 69,
    275
  size, critical, 51, 288, 296, 300, 302
  tip, plasticity, 51, 56, 57, 67, 293-295
Cracks:
  central through-the-thickness, 60, 61
  critical size, 51, 288, 296, 300, 302
  double edge through-the-thickness,
    60, 63
  emanating from hole, 62, 66
  Griffith, 35
  self-propagating, 35
  single edge through-the-thickness, 60,
    62, 64, 65, 68, 302

surface flaw shape parameter, 62, 67,
    70
  surface type, 62, 67, 70
Creep, 8, 12, 48, 435-470
  abridged test, 438
  combined with fatigue, *see* Combined
    creep and fatigue
  cumulative, 437, 449-451, 456
  diffusional, 23, 31
  dynamic, 437
  general expression for, 446
  Larson-Miller parameter, 440, 441
  life fraction rule, 450-452
  logarithmic, 444
  log-log stress-time law, 445
  Manson-Haferd parameter, 440-442
  mechanical acceleration test, 438, 439
  multiaxial stress, 437, 447-449
  parabolic, 444
  prediction, 442, 437, 448, 449
  steady-state (secondary), 12
  strain-hardening rule, 449, 451
  tertiary, 12
  testing, 438-440, 443-445, 453-457
  thermal acceleration test, 439, 440
  time-hardening rule, 449, 450
  transient (primary), 12
  uniaxial stress, 442-447
Creep buckling, 8, 13, 437, 579
  critical time for, 579
Creep deformation, 435, 437, 448, 464
Creep-limited static stress, 458
Creep rate, constant, 442
  equations, 447
Creep rupture, 435, 436
Creep strain, 448
Critical region, statistical, 338-340, 354,
    372
Cumulative creep damage, 437, 449-451,
    456
Cumulative damage, in fatigue, 164, 170,
    213, 214, 240-276, 288, 380, 395,
    397
Cumulative distribution function, 322,
    492, 493
  chi-squared, 328, 329
  F (Snedcor), 332-335
  normal (Gaussian), 324, 326
  t (Student), 330, 331
  Weibull, 336, 337
Cycle counting methods, 282-286, 395,
    396
Cycle-dependent stress relaxation,
    276, 281

Cycle ratio, 241-246, 262
Cyclic hardening, 276, 281, 381, 382
Cyclic softening, 276, 281, 381, 382

Damage, cumulative, 164, 170, 213, 214,
    240-276, 288, 380, 390, 395, 397
Damage fraction, 241-244, 246, 262
Damage lines, constant, 263
Damage tolerance, 297, 300, 301
Damage tolerant structure, 297
Damping, 518, 520-524
Decarburization, 10, 612
Deformation:
    elastic, 23, 34
    plastic, 23, 33, 35, 37, 47, 100, 111, 112,
        120, 479, 514
Degrees of freedom, statistical, 328, 330,
    355, 356
Delamination theory (wear), 594
Design:
    for column buckling, 568-572
    combined stress theories, 149
    for creep, 445, 446
    damage tolerant, 297, 300, 301
    engineering, 2
    fail-safe, 164, 296, 298
    for fatigue, 214, 222-224, 230-234, 268-
        275, 301, 393-397, 426-430
    for fracture, 60
    for impact, 507, 547, 549
    leak-before-break, 298
    objectives of, 3, 4
    "perfection," 3
    safe life, 298
    for wear, 601-603
    for yielding, 60
Design of experiment, statistical,
    320
Design stress, 149, 158
Detection of cracks, 288, 296, 299,
    300
Dilatation energy, 137
Direction cosines, 79, 80, 83, 85, 91-94,
    107
Discontinuities, see Stress concentration
Dislocations:
    Burgers vector, 40, 43, 45
    climb, 44, 48
    cross slip, 44, 48
    density, 46, 47
    edge, 37, 38, 41, 43, 45
    Frank-Read source, 47
    generation, 45, 47, 48
    geometry, 37-40

hybrid (mixed), 37, 40, 41, 43
interaction, 45, 48
line, 38-40, 43, 45
line tension, 45
loops, 45, 48, 168
mobility, 37, 42-44, 167
pinning, 45, 46
screw, 37, 39, 41, 43, 45
tangled forest, 48
theory of, 22, 36, 37, 49
types, 37
Distortion energy, 137, 140
    failure theory, 114, 128, 137, 140, 142-
        144, 149, 153, 156
    fatigue failure theory, 227, 231, 232,
        429, 430
Distributions:
    of populations, 322-326, 361
    of samples, 327-336, 361
Ductile behavior, 144, 149, 229, 230,
    421
Ductile rupture, 8, 9, 33, 149

Elastica, 562
Elastic-plastic fracture mechanics, 51, 67
Elliptic equation, fatigue, 218,
    219
Endurance limit, see Fatigue limit
Energy absorbing capacity, under impact,
    547, 549
Equivalent completely reversed stress,
    230, 286, 287
Equivalent stress, 391, 447
Equivalent total strain range, 391
Erosion-corrosion, 8, 10, 591, 604, 610,
    611
Error, Type I statistical, 338
    Type II, statistical, 338
Estimation, of population parameters,
    340-342, 352, 354, 374
Euler-Engesser equation, 566, 567, 569-
    571
Euler's critical load, 562
Euler's equation, 562, 565, 569, 570

Factor of safety, see Safety factor
Fail-safe design, 164, 296, 298
Failure, combined stress theories, 126-
        158. See also Failure theory
    evaluation
    inducing agents, 6, 7
    locations, 6, 7
    manifestations of, 6
    mechanical, 1, 6

Failure modes, 1, 6, 60, 70, 71, 76
  glossary, 9-14
  observed in practice, 7-14
  synergistic, 8
Failure modulus, 60
Failure prediction, 60, 71, 613, 614
  brittle behavior, 35, 36, 56, 60, 128,
    144, 155, 225, 226, 229
  combined creep and fatigue, 458
  ductile behavior, 60, 130-134, 137, 140-
    144, 146, 156, 226, 228, 229, 231,
    429, 430
  fatigue, 221, 222, 225, 226, 228, 229,
    231, 241, 268, 429, 430
  multiaxial stress, 126-128, 130-134, 137,
    140-143, 146, 155, 156, 225, 226,
    228, 229, 231, 429, 430
Failure strength, 50, 60, 150, 386, 387
Failure theories, see Failure, combined
    stress theories; Fatigue, failure
    theories
Failure theory evaluation, 144, 149, 229,
    230
Fatigue, 8, 9, 164-234, 240-304, 319-
    357, 360-376, 379-397, 403, 404,
    416-430, 451-470, 479-491
  amplitude, ratio, 170
  beach marks, 168
  brittle materials, 229, 230
  combined with creep, See Combined
    creep and fatigue
  corrosion, 8, 14, 604
  damage, see Cumulative damage, in
    fatigue
  data distribution, 181, 319-357, 360-
    376
  ductile materials, 229, 230
  endurance limit, see Fatigue limit
  failure theories, 225-234, 429, 430
  fretting-, 8, 12, 207, 208, 296, 479-491,
    604
  full scale testing, 164, 295, 296
  high cycle, 8, 9, 164-234, 379
  impact-, 8, 11, 500
  loading spectra, 170, 240, 245, 295,
    394, 395
  low cycle, 8, 9, 165, 379-397
  nonzero mean strain, 387-390, 395,
    397
  nonzero mean stress, 169, 171,
    212, 214-234, 282, 286, 287,
    380, 387, 397, 424, 427,
    428

  scatter, 181, 202, 203, 319, 357,
    360, 417, 418
  S-N curves, see S-N curves; S-N-P
    curves
  statistical analysis, 319-357, 360-376
  stress concentration factor, 277, 404,
    420, 422, 427
    for brittle materials, 421
    for ductile materials, 421, 424
    for finite life, 420, 421, 425
  stress ratio, 170, 292
  surface-, 8, 9, 584, 594, 595. see also
    Wear, surface-fatigue
  testing machines, 174-180
  testing methods, 181, 360-376
  thermal, 8, 9, 391-393
  transition life, 385, 387
Fatigue cracks:
  acceleration, 293-295
  initation, 167, 168, 254, 258, 266, 268,
    274, 275-288, 296, 490
  interaction effects, 293-295
  nuclei, 167, 168, 254, 258, 296, 490,
    491
  propagation, 33, 36, 48, 167, 254, 266,
    268, 275, 288-296, 300, 303, 490,
    491
  retardation, 293-295
  rogue, 299
  stage I growth, 168
  stage II growth, 168
Fatigue ductility, 384, 386, 387,
    390
  coefficient, 384, 390
  exponent, 384
Fatigue limit, 183, 184, 246-249, 367,
    369, 372, 422
Fatigue strength, 184, 214, 372, 487
  modification factors, 214, 487, 490
Fatigue stress:
  alternating, 169, 170
  completely reversed, 169
  equivalent completely reversed, 230,
    286, 287
  maximum, 170, 231, 232
  mean, 170, 231, 232, 281, 282
  minimum, 170, 231, 232
  multiaxial, 169, 224-234, 380, 391,
    427-430
  nonzero mean, 169, 171, 212-234, 282,
    286, 287, 380, 387, 397, 424,
    427, 428
  range of, 170

released, 171, 172
spectra, 170, 172, 173, 245. *See also*
    Fatigue, loading spectra
Fatigue testing, 174-181, 360-376
    all or nothing method, 363-365
    constant stress level method, 361-363,
        365
    extreme value method, 374, 375
    least-of-n method, 374, 375
    mortality method, 363-365
    probit method, 363, 365
    Prot method, 367-369
    quantal response method, 363-365
    response method, 363-365
    staircase method, 369-374
    standard method, 360, 361
    step-test method, 365-367
    survival method, 363-365
    up-and-down method, 369-374
F-distribution, 324, 327, 351, 354, 355
Flaw size, initial, 51, 300, 490,
    491
Force flow, 400, 461
Force-induced elastic deformation, 8,
    9
Forces, interatomic, 15
Forman equation, 291
Fractile, 321
Fracture:
    brittle, 8, 9, 33, 36, 60, 69, 70, 149,
        604
    critical normal stress for, 33, 545
    cup and cone, 34
Fracture control, 49, 240, 297, 299,
    301
Fracture ductility, 386, 387
Fracture mechanics, 36, 49-71, 275, 288-
    295, 490, 491
    elastic-plastic, 51, 67
Fracture stress (strength), 50, 60, 150,
    386, 387
Fracture toughness, 50, 51, 56, 60, 291,
    292, 301, 540
    dynamic, 540, 541
    plane strain, 58, 59, 67-70, 301, 302,
        540, 541
    plane stress, 58, 68, 301
Frank-Read source, 47
Freedom, degrees of, statistical, 328, 330,
    355, 356
Frequency function, *see* Probability
    density function
Fretting, 8, 11, 296, 479-497

damage, 481, 485, 487, 491-493,
    496, 497
preventing or minimizing, 496, 497
residual stress effects, 485-490
use of fracture mechanics, 490, 491
variables, 480
Fretting action, 11, 479
Fretting corrosion, 8, 12, 479, 480, 494-
    496
Fretting fatigue, 8, 12, 207, 208, 296,
    479, 480, 481-491, 604
Fretting mechanisms:
    asperity-contact microcrack
        initiation, 481-484, 491
    friction-generated cyclic stress, 481-484
    pit-digging, 481, 482, 484, 491
    surface delamination, 481, 483, 484,
        491
Fretting wear, 8, 12, 479, 484, 491-
    494, 584, 595, 604
    depth of, 492, 493
    weight loss, 491
F-statistic, 321
F-test, 351, 354, 355
Full scale testing, fatigue, 164, 295, 296
Fundamental assertion of multiaxial
    stress fatigue, 234

Galling, 8, 13, 585, 588
Gatts theory, 249-253, 269, 271, 272
Gaussian distribution, 323-327, 342
Geometrical discontinuities, *see*
    Stress concentration
Gerber equation, 214, 218
Glide plane, 43, 44
Goodman diagram, 219, 456
Goodman equations, 214, 218, 230, 286,
    287
Grain direction, effects on fatigue, 185,
    187, 188
Grain size, effects on fatigue, 185, 186
Griffith-Irwin-Orowan criterion, 36
Griffith theory, 34, 35

Hartman-Schijve equation, 292
Heat affected zone, 188
Heat treatment, effects on fatigue, 188-
    190
Henry theory, 246-249, 269, 271, 272
High cycle fatigue, 8, 9, 164-234, 379
Histogram, 181, 182
Hooke's law, 104, 107, 113, 132, 136,
    137, 139, 501

Huber-Von Mises-Hencky theory, *see*
    Distortion energy, failure theory
Hydrogen:
  attack, 611, 612
  blistering, 611
  damage, 8, 10, 604, 611, 612
  embrittlement, 611
Hypothesis:
  alternative, 336
  null, 336, 354, 355, 356
  statistical, 336
  testing, 338, 340
Hysteresis loops, 280, 281, 380-382
Hysteretic damping, 518, 520-524

Impact, 8, 11, 500-552
  critical velocity, 542
  duration, 529-531
  effect on material properties, 536-543
  energy methods, 501-507
  fracture, 8, 11, 505
  one-dimensional wave equation, 509
  particle velocity, 510-515, 518, 527
  rise time, 500
  with stress and strain concentration,
    546-549
  stress wave:
    attenuation, 520-524
    propagation, elastic, 507-536, 543-
     546
    propagation, plastic, 535, 536, 542
  travelling waves, 511, 524-535
  velocity, critical, 542
    particle, 510-515, 518, 527
    wave propagation, 510-515, 518, 535,
     536, 542-546
Impact deformation, 8, 11, 503, 504
Impact factor, 503
Impact fatigue, 8, 11, 500
Impact fretting, 8, 11, 494, 501
Impact loading:
  rate effects, 536-543
  suddenly applied, 503, 504, 518-520
  types, 500
Impact stress, 503, 504, 514, 516-535
  damping effects, 518, 520-524
Impact wear, 8, 11, 500, 584, 595, 596
Impulse, 500, 513
Incremental strain theory, 112
Initial flaw size, 51, 300, 490, 491
Inspection:
  intervals, 296, 300
  procedures, 299

Instability:
  elastic, 557-579
  plastic, 98, 100, 103, 104, 119, 120
Interaction effects, 293-295
Interatomic spacing, 19, 20
Internal friction, 518, 520-524

J-integral, 67, 68

Kurtosis, 321

Larson-Miller parameter, 440, 441
Lattice:
  classification, Bravais, 17, 18
  crystal, 16-18, 23, 37
  defects, 37
Leak-before-break design, 298
Life prediction, 240, 275, 302-304, 380,
    422-426, 466, 468
Limit load, 296
Linear damage rule:
  creep, 449
  fatigue, 241-244, 265, 268-271, 288, 390
Linear elastic fracture mechanics, *see*
    Fracture mechanics
Loading:
  rate effects, 536-543
  spectra, 170, 240, 245, 295, 394, 395
Local stress-strain approach, 275-288
Log-normal distribution, 353, 361, 363
Low cycle fatigue, 8, 9, 165, 379-397
Lubrication, effects on wear, 587

Manson-Coffin relationship, 287, 384, 385,
    461, 466
Manson double linear rule, 266-269, 274,
    275
Manson-Haferd parameter, 440-442
Marco-Starkey theory, 243
Marin theory, 263-266, 269
Master diagram, 215-217
Material composition, effects on fatigue,
    184, 185
Material properties:
  cyclic, 277-279
  impact, 536-543
  monotonic (static), 276-279
Maximum normal strain, failure theory,
    128, 132-134, 143
Maximum normal stress:
  failure theory, 128-130, 142-144,
    149, 153
  fatigue failure theory, 225

Maximum shearing stress:
  failure theory, 128, 130-132, 142-
    144, 149, 153, 156
  fatigue failure theory, 226
Mean:
  arithmetic, 321
  geometric, 321
  statistical, 181, 321, 323, 341, 348, 350,
    372
Mean deviation, 321
Means, comparison of, 330, 351
Median, 321, 350
Miner's rule, *see* Palmgren-Miner hypothesis
Minimum Life parameter, Weibull, 332,
  348-350
Mode, statistical, 321
Modes of failure, 1, 6, 60, 70, 71, 76
  glossary, 9-14
  observed in practice, 7-14
  synergistic, 8
Modulus:
  of elasticity, 105
  mechanical, 76, 126, 224
  of plasticity, 105
Mohr's circle, 145-147, 156, 157
Mohr's failure theory, 128, 143, 146-149,
  153, 156, 157
Momentum, 513
Morrow relationship, 384, 385
Multiaxial stress, *see* Fatigue stress, multi-
  axial; stress, multiaxial

Necking, 34
Neuber's rule, 277, 390
Non-zero mean strain, 387-390, 395, 397
Non-zero mean stress, 169, 171, 212, 214-
  234, 282, 286, 287, 380, 387, 397,
  424, 427, 428
Normal distribution, 323-327, 342
  standard, 324-326
Normal varible, standard, 324
Notch effects, *see* Stress concentration
Notch sensitivity index, 416-420, 422, 424

Octahedral shear strain, 114, 115
Octahedral shear stress, 114, 115
  failure theory, 114, 120, 141. *See also*
    Distortion energy, failure strain
Optimization, design, 2
Overloads, in fatigue, 281, 282, 293- 295

Palmgren-Miner hypothesis, 241-244, 265,
  269-271, 288, 390, 395, 397

Parameter:
  Larson-Miller, 440, 441
  Manson-Haferd, 440-442
  of population, 319, 323
Paris-Erdogan equation, 289
Plane strain, minumum thickness for, 58
Plastic zone:
  adjustment factor, 56, 57
  at crack tip, 51, 56, 57, 67
Plating, effects on fatigue, 196, 198
Poisson's ratio, 106, 113
Population:
  mean, 181, 319, 323, 341, 342, 350,
    352, 353
  parameter, 319, 323
  statistical, 319
  variance, 181, 354
Power of test, statistical, 338
Prediction of failure, *see* Failure
    prediction
Preloading, 213
Presetting, effects on fatigue, 200,
    202
Pressure vessel, thin-walled, 116-118
Principal planes, 81, 83, 93, 94, 98
  shearing stresses, 86, 89, 91, 94, 130
  shear planes, 86, 89-91, 94
  strain, 98, 110
  stresses, normal, 81, 85, 93, 128, 131,
    133, 136, 154, 155, 231
Probability density function, 322, 357
  chi-squared, 328
  continuous, 322-324
  discrete, 322, 323
  F, 332
  Gaussian, 323-325
  normal, 323-325
  t, 330
  Weibull, 332, 336
Probability of failure, 183
Probability paper, 344-351, 361, 362,
    364, 366
  log-normal, 344, 346, 361, 362
  normal, 344, 345, 348, 364, 366
  plotting procedure, 344, 347-349
  Weibull, 344, 348-351
Proportional deformation theory, 112

Quantile, 321
Quasi-static loading, 500, 501

Radiation damage, 8, 13
Randomness, 320

Rain-flow cycle counting, 282, 284-286, 395, 396

Range, statistical, 321

Range-pair cycle counting, 282, 284, 286

Rankine's theory, *see* Maximum normal stress, failure theory

Recovery, 28, 29, 30, 48

R-curve, 67, 68

Relaxation:
   cycle dependent, 276
   thermal, 8, 12, 437

Reliability, 183, 319

Residual stresses, 34
   effects on fatigue, 169, 199-203, 282, 485-490, 497
   effects on fretting-fatigue, 485-490, 497

Resilience, 98, 99
   modulus of, 99

Reversals, fatigue, 384

Richart-Newmark theory, 243

Robinson's hypothesis (creep), 449

Rogue cracks, 299

Rupture, ductile, 8, 9, 33, 149

Safe-life design, 298

Safety factor, 69, 150, 151, 156, 158, 223, 233, 234, 394, 395, 397, 425, 426, 430, 445, 446, 570-572

Sample:
   mean, 181, 320, 341, 348, 353
   size, 338, 342-344, 357
   statistical, 319
   variance, 181

Sampling distributions, 327-336, 340

Sampling procedures, 320

Scabbing, 8, 13, 543-546

Scatter:
   in fatigue, 181, 202, 203, 319, 357, 360, 417, 418
   reduction of, 202, 203, 486

Scatter factor, 299

Schmid's Law, 27

Scoring (wear), 590

Season-cracking, 613

Secant formula, buckling, 567, 569, 571

Seizure, 8, 13, 585, 588

Sequential loading, 276. *See also* Fatigue, loading spectra

Service loading, fatigue, 295, 296, 491

Shape parameter, Weibull, 332, 348, 350

Shear modulus of elasticity, 110

Shock, *see* Impact

Shot-peening, effects on fatigue, 169, 199, 200, 203, 486, 488, 489

Significance, statistical, 338-340, 342, 354, 356, 372

Simulation of service loading spectra, 295, 296, 491

Size, effects on fatigue, 170, 198, 199

Skewness, 321

Slenderness ratio, 567

Sliding, grain boundary, 23, 31

Slip:
   bands:
      coarse, 23, 24, 167
      extrusion, 167
      fine, 167, 168
   critical shear stress for, 21, 25, 27, 34
   direction, 23, 26
   lines, 23, 24, 48
   plane, 26, 43, 48
   step, 43, 46, 48
   system, 23

Slope, Weibull, 332, 348, 350

S-N curves, 166, 180-214, 360-362, 364, 380, 420, 421
   approximation of, 251-253, 422, 423
   factors that affect, 184-214, 420, 421

Snedcor's F-distribution, 324, 327, 351, 354, 355

S-N-P curves, 180-214, 357, 360, 363, 365, 375, 420, 421
   factors that affect, 184-214, 420-421

Soderberg equations, 214, 218, 219

Sohnck's Law, 33

Spacing, interatomic, 19, 20

Spalling, 8, 13, 543-546

Spectrum loading, 170, 240, 245, 295, 394, 395

Speed, effects on fatigue, 207, 209, 210

Staircase testing, 369-374

Standard deviation, 321, 323, 341, 348, 353, 367, 372. *See also* Variance

Standard normal variable, 324

Statistics:
   descriptive, 320, 321
   of sample, 319
   use in fatigue analysis, 181, 319-357, 360-376

Strain:
   effect of rate, 32, 34, 536-543
   elastic, 17, 19, 20, 60
   engineering, 19, 98-100
   multiaxial, 114
   normal, 105, 106

plastic, 21, 46, 113, 116, 120, 435, 437, 464
principal, 98, 110
shear, 107, 108
state of, 98
true, 98, 100, 102, 104, 113, 116, 118
Strain amplitude:
    cyclic, 380, 382
    elastic, 385, 387
    plastic, 382
    total, 382
Strain concentration, under impact, 546-549
Strain concentration factor, 277, 413, 414
Strain control, 382
Strain cubic equation, 111
Strain cycling, 380-384
Strain energy:
    per unit volume, 60, 134, 135
    release rate, 58
Strain hardening, 29, 30, 48, 276, 281, 381, 382
    cyclic, 276, 281, 381, 382
    as function of temperature, 29
    as function of time, 29, 30
    linear, 28
    parabolic, 28
Strain hardening exponent:
    cyclic, 279, 281, 385, 395
    static, 101, 102, 110, 279
Strain-life relationships, 287, 383, 384-390
Strain range:
    plastic, 382
    total, 382, 391
Strain range partitioning, 461, 464-466, 468-470
Strain softening, cyclic, 276, 281, 381, 382
Strength coefficient:
    cyclic, 279, 281, 385
    static, 101, 102, 110, 279
Strength reduction factor, 420, 422
Stress:
    attenuation factor, 523, 524
    biaxial, 126, 142-144
    design, 149, 158
    engineering, 98-101
    equivalent, 391, 447
    hydrostatic, 130, 131, 133
    multiaxial, 114, 126, 128, 131-134, 137 144-146, 149, 153. See also Fatigue stress, multiaxial

normal, 77, 78, 105, 106
principal normal, 81, 85, 93, 94
principal shear, 86, 89, 91, 94
residual, see Residual stresses
shear, 77, 78, 107
state of, 76, 77, 481, 485-490, 501
triaxial, 76, 126, 128, 131, 141
true, 98, 100-102
uniaxial, 114, 126
Stress concentration:
    effects on fatigue, 169, 193-195, 277, 400-430
    elastic range, 404-413
    highly local, 402
    plastic range, 413, 414
    under impact, 546-549
    widely distributed, 402
Stress concentration factor, 277, 403, 404-413, 422, 598, 603
    brittle materials, fatigue, 421
    ductile materials, fatigue, 421, 424
    fatigue, 277, 404, 420, 422, 427
    finite life, fatigue, 420, 421, 425
    graphs for, 405-411
    multiaxial state of stress, 426-428
    multiple notches, 415, 416
Stress corrosion, 8, 10, 13, 48, 604, 612, 613
Stress cubic equation, 83, 91, 107, 154, 428
Stress intensity factor, 54, 60
    critical, 54-57, 288, 302
    range, 289, 290
    threshold, 291, 292
Stress invariants, 83
Stress raisers, see Stress concentration
Stress rupture, 8, 12, 435, 436, 438, 440-442, 449-453, 456
    Larson-Miller parameter, 440, 441
    life fraction rule, 450-452
    Manson-Haferd parameter, 440-442
    strain hardening rule, 449, 450
    time hardening rule, 449, 450
Stress-strain curve:
    cyclic, 278, 280, 382
    static, 278, 280, 382
Stress-time pattern, effects on fatigue, 209-212
Stress wave:
    attenuation, 520-524
    propagation, elastic, 507-536, 543-546
    propagation, plastic, 535, 536, 542
    reflection at boundaries, 515-519, 525, 526, 532, 533, 536, 543-546

Structure:
  arrangements, 298
  classifications, 298
  damage tolerant, 297
Student's t-distribution, 324, 327, 330, 351, 354
St. Venant's theory, *see* Maximum normal strain, failure theory
Suddenly applied load, 503, 504, 518-520, 524
Suh delamination theory (wear), 594
Surface cracks, 62, 67, 70
Surface fatigue, 8, 9, 584, 595. *See also* Fatigue, surface; Wear, surface-fatigue
Surface finish, effects on fatigue, 170, 195-197

Tangent modulus equation, buckling, 566
t-distribution, 324, 327, 330, 351, 354
Temperature, effects on fatigue, 170, 203, 204
Temperature-induced elastic deformation, 8, 9
Theories of failure, *see* Failure, combined stress theories
Thermal expansion, 20
Thermal fatigue, 8, 9, 391-393
Thermal relaxation, 8, 12, 437
Thermal shock, 8, 12
Total strain energy failure theory, 128, 136, 137, 143
Toughness, 98
  modulus of, 99
Toughness merit number, 100
Transition, ductile to brittle, 34
Transition life, fatigue, 385, 387
Tresca-Guest theory, *see* Maximum shearing stress, failure theory
Truncation, of service loading spectra, 295, 296
t-statistic, 321, 330, 355, 356
t-test, 351, 354
Twinning, mechanical, 23, 30, 31

Ultimate strength, 103
Universe, statistical, 319. *See also* Population, statistical
Up-and-down testing procedure, 369-374

Variance:
  analysis of, 332
  reduction in fatigue, 202, 203
  statistical, 181, 321, 354
Velocity:
  critical impact, 542
  particle, 510-515, 518, 527
  wave propagation, 510-515, 518, 535, 536, 542-546
Von Mises theory, *see* Distortion energy, failure theory

Wave equation, one-dimensional, 509
Wave front, 512, 518
  propagation velocity, 510-515, 518, 535, 536, 542-546
Waves travelling, 511, 524-535
Wear, 8, 10, 11, 12, 479, 480, 583-603, 610
  abrasive, 8, 11, 584, 590-594, 601, 602, 610
  adhesive, 8, 11, 494, 583, 584-590, 592-601, 602
  apparent contact area, 584
  Archard equation, 586, 587
  constants:
    for abrasive, 592
    for adhesive lubricated, 587
    for adhesive unlubricated, 587
  control methods, 588, 590
  corrosive, 8, 11, 13, 14, 584, 594, 604
  deformation, 8, 11, 584, 595
  delamination theory, 594
  depth, 588, 592, 601, 602
  fretting-, 8, 12, 479, 480, 491-494, 584, 595, 604
  impact-, 8, 11, 500, 584, 595, 596
  particle formation, 586
  prediction, 583
  principle of conversion, 588, 590
  principle of diversion, 590
  principle of protective layers, 588
  real contact area, 584
  resistance, 588-590, 593, 594
  scoring, 590
  surface-fatigue, 8, 11, 584, 594, 595
  three-body, 590, 593
  two-body, 590, 593
  variables, 584
  zero wear, 596
    model, 596-600, 602, 603
Weibull distribution, 324, 332, 336

Welding, effects on fatigue, 188, 190-192
Wheeler model, 294
Wöhler diagram, 166. *See also* S-N curves

Yielding, 8, 9, 48, 69, 146, 149, 156, 549
  large scale, 51, 57

  small scale, 51, 56, 57
Yield strength, 50, 146
Youngs modulus, of elasticity, 105

Zero wear, 596
  model, 596-600, 602, 603